Lipids

Lipids: Biochemistry, Biotechnology and Health

SIXTH EDITION

(formerly Lipid Biochemistry: An Introduction, Editions 1–5)

BY

Michael I. Gurr

John L. Harwood

Keith N. Frayn

Denis J. Murphy

Robert H. Michell

WILEY Blackwell

Registered office: John Wiley & Sons, Ltd, The Atrium, Southern Gate, Chichester, West Sussex, PO19 8SQ, UK

Editorial offices: 9600 Garsington Road, Oxford, OX4 2DQ, UK
The Atrium, Southern Gate, Chichester, West Sussex, PO19 8SQ, UK
111 River Street, Hoboken, NJ 07030-5774, USA

For details of our global editorial offices, for customer services and for information about how to apply for permission to reuse the copyright material in this book please see our website at www.wiley.com/wiley-blackwell.

The right of the author to be identified as the author of this work has been asserted in accordance with the UK Copyright, Designs and Patents Act 1988.

Designations used by companies to distinguish their products are often claimed as trademarks. All brand names and product names used in this book are trade names, service marks, trademarks or registered trademarks of their respective owners. The publisher is not associated with any product or vendor mentioned in this book.

Limit of Liability/Disclaimer of Warranty: While the publisher and author(s) have used their best efforts in preparing this book, they make no representations or warranties with respect to the accuracy or completeness of the contents of this book and specifically disclaim any implied warranties of merchantability or fitness for a particular purpose. It is sold on the understanding that the publisher is not engaged in rendering professional services and neither the publisher nor the author shall be liable for damages arising herefrom. If professional advice or other expert assistance is required, the services of a competent professional should be sought.

Library of Congress Cataloging-in-Publication Data

Names: Gurr, M. I. (Michael Ian), author. | Harwood, John L., author. | Frayn
   K. N. (Keith N.), author. | Murphy, Denis J., author | Michell, R. H., author.
   Lipid biochemistry. Preceded by (work):
Title: Lipids:Biochemistry, Biotechnology and Health / by Michael I. Gurr , John
   L. Harwood, Keith N. Frayn, Denis J. Murphy, and Robert H. Michell.
Description: 6th edition. | Chichester, West Sussex ; Hoboken, NJ : John
   Wiley & Sons Inc., 2016. | Preceded by Lipid biochemistry / by Michael I.
   Gurr, John L. Harwood, and Keith N. Frayn. 5th ed. 2002. | Includes
   bibliographical references and index.
Identifiers: LCCN 2016000533 (print) | LCCN 2016002203 (ebook) | ISBN
   9781118501139 (pbk.) | ISBN 9781118501085 (Adobe PDF) | ISBN
   9781118501108 (ePub)
Subjects: | MESH: Lipids
Classification: LCC QP751 (print) | LCC QP751 (ebook) | NLM QU 85 | DDC
   572/.57–dc23
LC record available at http://lccn.loc.gov/2016000533

A catalogue record for this book is available from the British Library.

Wiley also publishes its books in a variety of electronic formats. Some content that appears in print may not be available in electronic books.

Cover images: Left panel: An artery partially occluded by an atherosclerotic plaque (Section 10.5.1). The red stain is for macrophages that are present in the plaque and become foam cells. The green stain is for smooth muscle cells in the arterial wall and capping the plaque. Photo courtesy of Thomas S. Davies and Susan Chazi, Cardiff University, UK from work funded by the British Heart Foundation.
Middle panel: An enterocyte from human jejunum displaying multiple lipid droplets a few hours after consuming a fatty meal (Section 7.1.3). The figure also shows mitochondria (dark) and the microvilli (brush border). Electron micrograph courtesy of Dr M Denise Robertson, University of Surrey, UK from work funded by the Biotechnology and Biological Sciences Research Council (BBSRC). Reproduced, with permission from BMJ Publishing Group Ltd, from MD Robertson, M Parkes, BF Warren et al. (2003) Mobilization of enterocyte fat stores by oral glucose in man. Gut 6: 833–8.
Right panel: Distribution of different molecular species of phosphatidylcholine within developing oilseed rape embryos as revealed by MALDI-MS imaging (Section 9.3.1). Red shows high concentrations and green low. Photo courtesy of Helen Woodfield and Drew Sturtevent from work funded by the BBSRC in Prof. Kent Chapman's laboratory at the University of North Texas, USA.
Background: Gettyimages/manuela schewe-behnisch / eyeem

Set in 8.5/12pt, MeridienLTStd-Roman by Thomson Digital, Noida, India

1   2016

Contents

Preface

Our main aims in writing this book have been, as ever, to aid students and other researchers in learning about lipids, to help staff in teaching the subject and to encourage research in this field. Since the publication of the Fifth Edition in 2002, there have been huge advances in our knowledge of the many aspects of lipids, especially in molecular biology. Far more is now known about the genes coding for proteins involved in lipid metabolism and already techniques of biotechnology are making use of this knowledge to produce specialized lipids on an industrial scale. The new knowledge has also had a far-reaching influence on medicine by revealing the role of lipids in disease processes to a much greater extent than hitherto and allowing for advances in diagnosis and disease prevention or treatment. We have endeavoured to reflect as many of these advances as possible in this new edition. Although modern textbooks of general biochemistry or biology now cover lipids to a greater extent than when our first edition was published in 1971, a book devoted entirely to lipids is able to go into far more detail on all these diverse aspects of the subject and to discuss exciting new developments with greater authority. It should be emphasized here that we have referred to a wide range of organisms – including archaea, bacteria, fungi, algae, 'higher' plants and many types of animals and not restricted ourselves to mammalian lipids.

Because of this research activity, we have rewritten large parts of the book and have given it a new title that reflects the fact that it is increasingly difficult to identify old boundaries between subjects such as biochemistry, physiology and medicine. This runs in parallel with changes in university structure: away from narrow 'departments' of 'biochemistry', 'zoology', 'botany' and the like, towards integrated 'schools' of biological sciences or similar structures. The increasing diversity of the subject requires greater specialist expertise than is possible with one or two authors. Accordingly, we have brought two new colleagues on board and one of the original authors has been given the role of coordinating editor to assure, as far as possible, consistency of style, so that we could avoid identifying authors with chapters. The authors have consulted widely among colleagues working in lipids and related fields to ensure that each chapter is as authoritative as possible. We are grateful for their help, which is recorded in the acknowledgements section. As a result, advances in such topics as enzymes of lipid metabolism, lipids in cell signalling, lipids in health and disease, molecular genetics and biotechnology have been strengthened.

The need to include new material has had to be balanced against the need to keep the book to a moderate size, with a price within most students' budgets. Some things had to go! As in the Fifth Edition, we decided to restrict some material of historical interest. Nevertheless, we thought that the inclusion of many short references to historical developments should remain, to add interest and to put certain aspects of lipidology in context. We have also removed some of the material that dealt with analytical procedures so that we could focus more on metabolic, physiological, clinical and biotechnological aspects. Chapter 1 now summarizes lipid analytical methods, with ample references to more specialist literature but has a section on lipidomics to highlight modern approaches to lipid profiling in biological fluids and tissues. This introductory chapter also contains a guide to finding your way around the book, which we hope students will find useful. We shall appreciate comments and suggestions so that future editions can be further improved.

MI Gurr
JL Harwood
KN Frayn
DJ Murphy
RH Michell

Acknowledgements

Over the years, we have received invaluable assistance from many colleagues in the compilation of this book and our thanks have been recorded in the previous five editions. Their contributions are still significant in this new edition and we are also grateful to the following for helping us with new material.

In Chapter 1, Jules Griffin provided valuable assistance with the lipidomics section. The substantial section on fatty acid biosynthesis has been brought up to date with help from Stuart Smith and his colleague Marc Leibundgut, whose huge expertise has been much appreciated. Many other aspects of Chapter 3 have benefited from the help of John Cronan Jr., Michael Schweizer, Marc Leibundgut and Ivo Fuessner. Bill Christie's wide knowledge of lipid chemistry, nomenclature and analysis has been invaluable throughout the book. Deficiencies in our knowledge of fat-soluble vitamins have been rectified by David Bender (Chapter 6); recent advances in comparative aspects of lipid metabolism by Caroline Pond (Chapter 7); lipids in immunity by Parveen Yaqoob and Philip Calder (Chapter 10); lung surfactant by Fred Possmayer (Chapters 4 & 10) and lipoproteins in human metabolism and clinical practice by Fredrik Karpe and Sophie Bridges (Chapters 7 & 10). Gary Brown and Patrick Schrauwen helped with information on inborn errors of lipid metabolism; Jenny Collins with cancer and lipid metabolism; and Sara Suliman with understanding lipodystrophies (Chapter 10).

Our thanks are due to the Wiley-Blackwell team for guiding us through the intricacies of the publication process. Particular mention should be made of Nigel Balmforth, who has been associated with *Lipid Biochemistry* from its early days with Chapman & Hall, then Blackwell and finally Wiley. Finally, after the enormous amount of work that goes into writing a book of this complexity, the authors conclude that all 'i's and 't's must have been dotted and crossed. It takes an expert, conscientious and helpful copy-editor to put a stop to this complacency and create a much better product. Martin Noble has done just that. Thank you all.

About the authors

Michael I. Gurr was Visiting Professor in Human Nutrition at Reading and Oxford Brookes Universities, UK.

John L. Harwood is Professor of Biochemistry in the School of Biosciences, Cardiff University, UK.

Keith N. Frayn is Emeritus Professor of Human Metabolism at the University of Oxford, UK.

Denis J. Murphy is Professor of Biotechnology in the School of Applied Sciences, University of South Wales, UK.

Robert H. Michell is Emeritus Professor of Biochemistry in the School of Biosciences, University of Birmingham, UK.

About the companion website

www.wiley.com/go/gurr/lipids

The website includes:

- Powerpoint slides of all the figures from the book, to download
- Pdfs of all tables and boxes from the book, to download
- Updates to Further Reading and additional figures to download

CHAPTER 1

Lipids: definitions, naming, methods and a guide to the contents of this book

1.1 Introduction

Lipids occur throughout the living world in microorganisms, fungi, higher plants and animals. They occur in all cell types and contribute to cellular structure, provide energy stores and participate in many biological processes, ranging from transcription of genes to regulation of vital metabolic pathways and physiological responses. In this book, they will be described mainly in terms of their functions, although on occasion it will be convenient, even necessary, to deal with lipid classes based on their chemical structures and properties. In the concluding section of this chapter, we provide a 'roadmap' to help students find their way around the book, so as to make best use of it.

1.2 Definitions

Lipids are defined on the basis of their solubility properties, not primarily their chemical structure.

The word 'lipid' is used by chemists to denote a chemically heterogeneous group of substances having in common the property of insolubility in water, but solubility in nonaqueous solvents such as chloroform, hydrocarbons or alcohols. The class of natural substances called 'lipids' thus contrasts with proteins, carbohydrates and nucleic acids, which are chemically well defined.

The terms 'fat' and 'lipid' are often used interchangeably. The term fat is more familiar to the layman for substances that are clearly fatty in nature, greasy in texture and immiscible with water. Familiar examples are butter and the fatty parts of meats. Fats are generally solid in texture, as distinct from oils which are liquid at ambient temperatures. Natural fats and oils are composed predominantly of esters of the three-carbon alcohol *glycerol* with *fatty acids*, often referred to as 'acyl lipids' (or more generally, 'complex lipids'). These are called triacylglycerols (TAG, see Section 2.2: often called 'triglycerides' in older literature) and are chemically quite distinct from the oils used in the petroleum industry, which are generally hydrocarbons. Alternatively, in many glycerol-based lipids, one of the glycerol hydroxyl groups is esterified with phosphorus and other groups (phospholipids, see Sections 2.3.2.1 & 2.3.2.2) or sugars (glycolipids, see Section 2.3.2.3). Yet other lipids are based on sphingosine (an 18-carbon amino-alcohol with an unsaturated carbon chain, or its derivatives) rather than glycerol, many of which also contain sugars (see Section 2.3.3), while others (isoprenoids, steroids and hopanoids, see Section 2.3.4) are based on the five-carbon hydrocarbon isoprene.

Chapter 2 deals mainly with lipid structures, Chapters 3 and 4 with biochemistry and Chapter 5 with lipids in cellular membranes. Aspects of the biology and health implications of these lipids are discussed in parts of Chapters 6–10 and their biotechnology in Chapter 11. The term 'lipid' to the chemist thus embraces a huge and chemically diverse range of fatty substances, which are described in this book.

1.3 Structural chemistry and nomenclature

1.3.1 Nomenclature, general

Naming systems are complex and have to be learned. The naming of lipids often poses problems. When the subject was in its infancy, research workers gave names to substances that they had newly discovered. Often, these

Lipids: Biochemistry, Biotechnology and Health, Sixth Edition. Michael I. Gurr, John L. Harwood, Keith N. Frayn, Denis J. Murphy and Robert H. Michell.
© 2016 John Wiley & Sons, Ltd. Published 2016 by John Wiley & Sons, Ltd.

substances would turn out to be impure mixtures but as the chemical structures of individual lipids became established, rather more systematic naming systems came into being and are still evolving. Later, these were further formalized under naming conventions laid down by the International Union of Pure and Applied Chemistry (IUPAC) and the International Union of Biochemistry (IUB). Thus, the term 'triacylglycerols' (TAGs – see Index – the main constituents of most fats and oils) is now preferred to 'triglyceride' but, as the latter is still frequently used especially by nutritionists and clinicians, you will need to learn both. Likewise, outdated names for phospholipids (major components of many biomembranes), for example 'lecithin', for phosphatidylcholine (PtdCho) and 'cephalin', for an ill-defined mixture of phosphatidylethanolamine (PtdEtn) and phosphatidylserine (PtdSer) will be mostly avoided in this book, but you should be aware of their existence in older literature. Further reference to lipid naming and structures will be given in appropriate chapters. A routine system for abbreviation of these cumbersome phospholipid names is given below.

1.3.2 Nomenclature, fatty acids

The very complex naming of the fatty acids (FAs) is discussed in more detail in Chapter 2, where their structures are described. Giving the full names and numbering of FAs (and complex lipids) at each mention can be extremely cumbersome. Therefore a 'shorthand' system has been devised and used extensively in this book and will be described fully in Section 2.1, Box 2.1. This describes the official system for naming and numbering FAs according to the IUPAC/IUB, which we shall

use routinely. An old system used Greek letters to identify carbon atoms in relation to the carboxyl carbon as C1. Thus, C2 was the α-carbon, C3 the β-carbon and so on, ending with the ω-carbon as the last in the chain, furthest from the carboxyl carbon. Remnants of this system still survive and will be noted as they arise. Thus, we shall use '3-hydroxybutyrate', *not* 'β-hydroxybutyrate' etc.

While on the subject of chain length, it is common to classify FAs into groups according to their range of chain lengths. There is no standard definition of these groups but we shall use the following definitions in this book: short-chain fatty acids, 2C–10C; medium-chain, 12C–14C; long-chain, 16C–18C; very long-chain >18C. Alternative definitions may be used by other authors.

1.3.3 Isomerism in unsaturated fatty acids

An important aspect of unsaturated fatty acids (UFA) is the opportunity for isomerism, which may be either positional or geometric. Positional isomers occur when double bonds are located at different positions in the carbon chain. Thus, for example, a 16C monounsaturated (sometimes called monoenoic, see below) fatty acid (MUFA) may have positional isomeric forms with double bonds at C7-8 or C9-10, sometimes written $\Delta7$ or $\Delta9$ (see Box 2.1). (The position of unsaturation is numbered with reference to the first of the pair of carbon atoms between which the double bond occurs, counting from the carboxyl carbon.) Two positional isomers of an 18C diunsaturated acid are illustrated in Fig. 1.1(c,d).

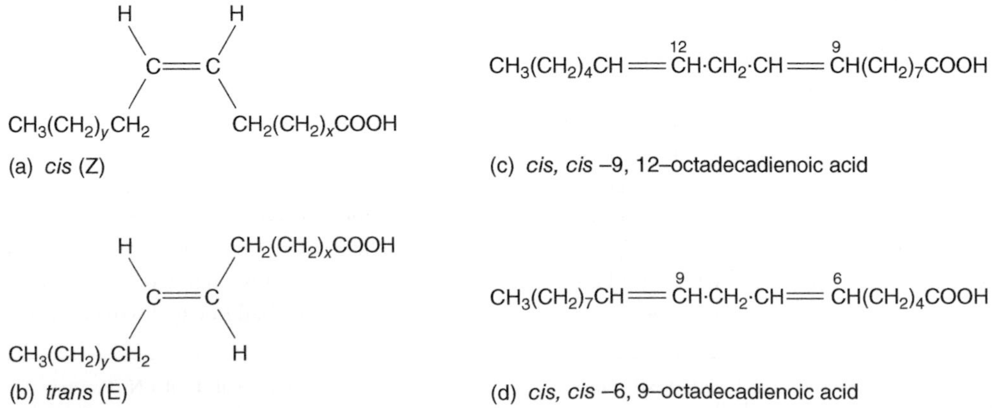

(a) *cis* (Z)

(b) *trans* (E)

(c) *cis, cis* –9, 12–octadecadienoic acid

(d) *cis, cis* –6, 9–octadecadienoic acid

Fig. 1.1 Isomerism in fatty acids. (a) *cis*-double bond; (b) a *trans*-double bond; (c) *c,c*-9,12-18:2; (d) *c,c*-6,9-18:2.

Geometric isomerism refers to the possibility that the configuration at the double bond can be *cis* or *trans*. (Although the convention *Z/E* is now preferred by chemists instead of *cis/trans*, we shall use the more traditional and more common *cis/trans* nomenclature throughout this book.) In the *cis* form, the two hydrogen substituents are on the same side of the molecule, while in the *trans* form they are on opposite sides (Fig. 1.1a,b). *Cis* and *trans* will be routinely abbreviated to *c,t* (see Box 2.1).

1.3.4 Alternative names

Students also need to be aware that the term 'ene' indicates the presence of a double bond in a FA. Consequently, mono-, di-, tri-, poly- (etc.) unsaturated FAs may also be referred to as mono-, di-, tri- or poly- (etc.) enoic FAs (or sometimes mono-, di-, tri- or poly-enes). Although we have normally used 'unsaturated' in this book, we may not have been entirely consistent and '-enoic' may sometimes be encountered! Furthermore it is important to note that some terms are used in the popular literature that might be regarded as too unspecific in the research literature. Thus shorthand terms such as 'saturates', 'monounsaturates', 'polyunsaturates' etc. will be avoided in much of this text but, because some chapters deal with matters of more interest to the general public, such as health (Chapter 10) and food science or biotechnology (Chapter 11), we have introduced them where appropriate, for example when discussing such issues as food labelling.

1.3.5 Stereochemistry

Another important feature of biological molecules is their stereochemistry. In lipids based on glycerol, for example, there is an inherent asymmetry at the central carbon atom of glycerol. Thus, chemical synthesis of phosphoglycerides yields an equal mixture of two stereoisomeric forms, whereas almost all naturally occurring phosphoglycerides have a single stereochemical configuration, much in the same way as most natural amino acids are of the L (or S) series. Students interested in the details of the stereochemistry of glycerol derivatives should consult previous editions of this book (see Gurr *et al.* (1971, 1975, 1980, 1991, 2002) and other references in **Further reading)**. The IUPAC/IUB convention has now abolished the DL (or even the more recent RS) terminology and has provided rules for the unambiguous numbering of the glycerol carbon atoms. Under this system, the phosphoglyceride, phosphatidylcholine, becomes 1,2-diacyl-*sn*-glycero-3-

phosphorylcholine or, more shortly, 3-*sn*-phosphatidylcholine (PtdCho; Fig. 1.2). The letters *sn* denote '*stereochemical numbering*' and indicate that this system is being used. The stereochemical numbering system is too cumbersome to use routinely in a book of this type and, therefore, we shall normally use the terms 'phosphatidylcholine' etc. or their relevant abbreviations, but introduce the more precise name when necessary.

1.3.6 Abbreviation of complex lipid names and other biochemical terms

Students will appreciate that the official names of complex lipids (and many other biochemicals) are cumbersome and research workers have evolved different systems for abbreviating them. In this latest edition we have incorporated all abbreviations into the index. At the first mention of each term in the text, we shall give the full authorized name followed by the abbreviation in parentheses. This will be repeated *at the first mention in each subsequent chapter*. Students should be aware that, unlike the IUB/IUPAC naming system, which is now generally accepted and expected to be used, the abbreviation system is still very much a matter of personal choice. Therefore students may expect to find alternative phospholipid abbreviations in some publications, for example PC, PE, PS and PI for

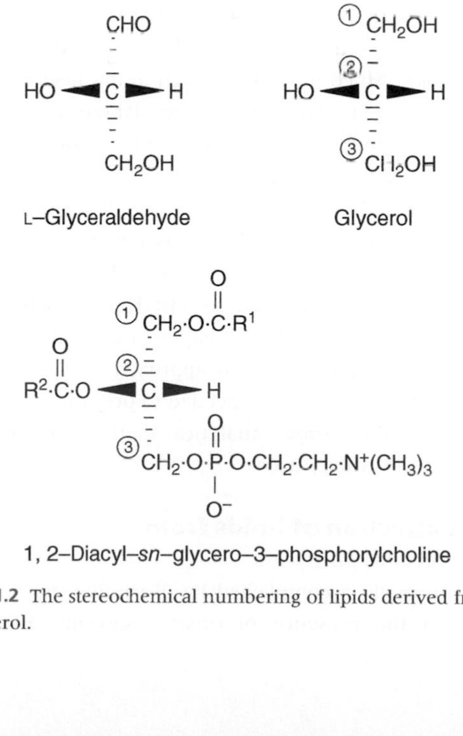

1, 2–Diacyl–*sn*–glycero–3–phosphorylcholine

Fig. 1.2 The stereochemical numbering of lipids derived from glycerol.

phosphatidylcholine, -ethanolamine, -serine and –inositol, instead of the PtdCho, PtdEtn, PtdSer and PtdIns used here. With very few exceptions we have not defined abbreviations for well-known substances in the general biochemical literature, such as ATP, ADP, NAD(H), NADP(H), FMN, FAD etc.

Another field in which nomenclature has grown up haphazardly is that of the enzymes of lipid metabolism. This has now been formalized to some extent under the Enzyme Commission (EC) nomenclature. The system is incomplete and not all lipid enzymes have EC names and numbers. Moreover, the system is very cumbersome for routine use and we have decided not to use it here. You will find a reference to this nomenclature in **Further reading** should you wish to learn about it.

Since the last edition was published in 2002, there have been huge advances in molecular biology and, in particular, in identifying the genes for an ever-increasing number of proteins. Where appropriate, we have referred to a protein involved in human lipid metabolism, of which the gene has been identified and have placed the gene name in parentheses after it (protein name in Roman, gene name in Italic script).

1.4 Lipidomics

1.4.1 Introduction

Since the last edition of this book in 2002, there have been very considerable advances in analysing and identifying natural lipids. Much modern research in this field is concerned with the profiling of lipid molecular species in cells, tissues and biofluids. This has come to be known as 'lipidomics', similar to the terms 'genomics' for profiling the gene complement of a cell or 'proteomics' for its proteins.

Some older methods of lipid analysis, presented in previous editions, will be described only briefly here and the student is referred to **Further reading** for books, reviews and original papers for more detail. Before describing the modern approach to lipidomics, we describe briefly the steps needed to prepare lipids for analysis and the various analytical methods, many of which are still widely used.

1.4.2 Extraction of lipids from natural samples

This is normally accomplished by disrupting the tissue sample in the presence of organic solvents. Binary mixtures are frequently used, for example chloroform and methanol. One component should have some water miscibility and hydrogen-bonding ability in order to split lipid-protein complexes in the sample, such as those encountered in membranes (Chapter 5). Precautions are needed to avoid oxidation of, for example, UFAs. Control of temperature is important, as well as steps to inhibit breakdown of lipids by lipases (see Sections 4.2 & 4.6). The extract is finally 'cleaned up' by removing water and associated water-soluble substances (see **Further reading**).

1.4.3 Chromatographic methods for separating lipids

Once a sample has been prepared for analysis, chromatography can be used to separate its many lipid constituents. A chromatograph comprises two immiscible phases: one is kept stationary by being held on a microporous support; the other (moving phase) percolates continuously through the stationary phase. The stationary phase may be located in a long narrow bore column of metal, glass or plastic (column chromatography), coated onto a glass plate or plastic strip (thin layer chromatography, TLC, see Fig. 1.3) or it may simply be a sheet of absorbent paper (paper chromatography).

The principle of chromatography is that when a lipid sample (often comprising a very large number of molecular species) is applied to a particular location on the stationary phase (the origin) and the moving phase percolates through, the different components of the mixture partition differently between the two phases according to their differing chemical and physical properties. Some will tend to be retained more by the stationary phase, while others tend to move more with the moving phase. Thus, the components will move apart as the moving phase washes through the system (see Christie, 1997; Christie & Han 2010; and Hammond 1993 in **Further reading** for more details of the theory of chromatography).

Many types of adsorbent solid can be used as the stationary phase (e.g. silica, alumina). The moving phase may be a liquid (liquid chromatography, LC) or a gas (gas chromatography, GC – the original term gas-liquid chromatography, GLC, is now less used). Particularly good separations may now be achieved by GC (see Fig. 1.4) with very long thin columns packed with an inert support for the stationary phase or in which the stationary phase is coated on the wall of the column. This is useful for volatile compounds or those that can be converted

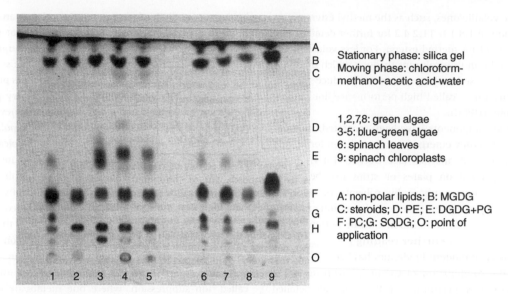

Fig. 1.3 Separation of lipid classes by thin-layer chromatography (TLC).

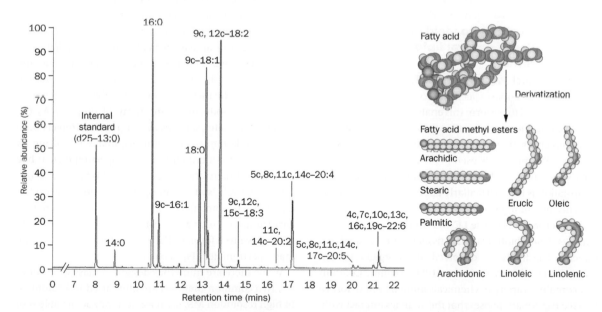

Fig. 1.4 Separation of fatty acid methyl esters by gas chromatography (GC). The figure shows the FA composition of a lipid extract of heart tissue as measured by GC on a capillary column. To the right of the chromatogram is depicted the conversion of a complex lipid into FA methyl esters in preparation for chromatography. The peaks on the chromatogram are labelled with shorthand abbreviations for FAs (see Box 2.1 for details). Detection is by a flame ionization detector. From JL Griffin, H Atherton, J Shockcor & L Atzori (2011) Metabolomics as a tool for cardiac research. *Na Rev Cardiol* **8**: 630–43; p. 634, Fig. 3a. Reproduced with permission of Nature Publishing Group.

into more volatile ones, such as the methyl esters of FAs (see Sections 2.1.8.1 & 11.2.4.2 for further details of the preparation of FA methyl esters). For less volatile complex lipids, LC in thin columns through which the moving phase is passed under pressure can produce superior separations: this is called high performance liquid chromatography (HPLC).

Once the components have been separated, they can be collected as they emerge from the column for further identification and analysis (see Section 1.4.4). Compounds separated on plates or strips can be eluted from the stationary phase by solvents or analysed in situ by various means. (Further information on methods of detection can be found in Christie & Han (2010) and Kates (2010) in **Further reading**.)

The power of modern lipidomics has been made possible by the combination of GC or LC with improved methods of mass spectrometry (MS) to provide detailed and sophisticated analyses of complex natural lipid mixtures and this is the subject of the next section.

1.4.4 Modern lipidomics employs a combination of liquid chromatography or gas chromatography with mass spectrometry to yield detailed profiles of natural lipids – the 'lipidome'

While individual FAs can be readily measured by gas chromatography-mass spectrometry (GC-MS), the commonest method to perform this analysis relies on cleaving FAs from the head groups that they are associated with and converting them into methyl esters by transesterification. This process is used to make the FAs volatile at the temperature used by GC-MS, but during this process information is lost, particularly about which lipid species are enriched in a given FA.

An alternative is to use LC-MS. In this approach, lipid extracts from biofluids and tissues can be analysed directly. The lipids are dissolved in an organic solvent and injected directly onto the HPLC column. Columns can contain a variety of chemicals immobilized to form a surface (stationary phase) that the analytes interact with. For the analysis of lipids, columns containing long chains of alkyl groups are most commonly used, in particular 8C and 18C columns, which have side-chain lengths of 8 and 18 carbons, respectively. The most commonly used HPLC method is referred to as 'reverse phase', whereby

lipids are initially loaded onto a HPLC column and then the HPLC solvent is varied from something that is predominantly aqueous to a solvent that is predominantly organic, across what is termed a gradient. The solvents are referred to as the mobile phases. During this process, lipids are initially adsorbed on to the stationary phase, until their solubility increases to the point that they begin to dissolve in the mobile phase. In this manner, polar and nonpolar lipids can readily be separated and typically, in a lipid extract, lipid molecular species would elute in the order of nonesterified fatty acids (NEFAs), phospholipids, cholesteryl esters and TAGs. The chromatography serves two important purposes. Firstly, it reduces the complexity of the subsequent mass spectra generated by the mass spectrometer, making metabolite identification more convenient. Secondly, some metabolites can ionize more readily than others and this can produce an effect called 'ion suppression' where one metabolite ionizes more easily and reduces the energy available for the ionization of other species. As a result, the mass spectrometer may detect only the metabolite that ionizes readily and miss the other metabolites that do not readily form ions.

LC-MS is most commonly used with 'electrospray ionization' where the analytes are introduced to the mass spectrometer in the form of a spray of solvent. They are accelerated over an electric field across the capillary that introduces them into the mass spectrometer and the nebulization of the spray is often assisted by the flow of an inert gas. The inert gas causes the solvent to evaporate (desolvate), producing a fine spray of droplets. As the solvent evaporates, charges build up in the droplets until they explode into smaller droplets, finally producing an ion that is introduced into the mass spectrometer. While this may sound relatively destructive, this form of ionization is relatively 'soft', ensuring that the molecule itself or an adduct (a combination of the molecule and another charged species such as H^+, Na^+, K^+ or other ions present in the solvent) is formed. The ions are then detected by the mass spectrometer (Fig. 1.5).

While there are numerous designs of mass spectrometer, two common methods are often used in lipidomics. In high resolution MS, the mass accuracy achievable is so great that chemical formulae can be determined with reasonable precision. This is because only carbon-12 has a mass of exactly 12 atomic mass units, while other nuclides all have masses that slightly differ from a whole number. These mass deficits can be used to predict what

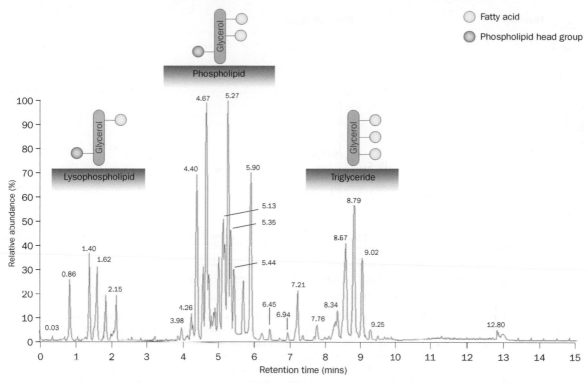

Fig. 1.5 Separation and identification of heart lipidome by liquid chromatography-mass spectrometry (GC-MS). Intact lipids from an extract of heart tissue have been separated, detected and identified by GC-MS. Chromatography separates the intact lipids according to their polarity and high resolution MS identifies individual lipid molecular species. From JL Griffin, H Atherton, J Shockcor & L Atzori (2011) Metabolomics as a tool for cardiac research. *Nat Rev Cardiol* **8**: 630–43; p. 634, Fig. 3b. Reproduced with permission of Nature Publishing Group.

nuclides are present and estimate a small number of chemical formulas that may be responsible for the ion. The accuracy of modern high resolution mass spectrometers is so high, often less than 3–5 parts per million, that it is possible in lipidomics to determine what species are being detected by their exact mass and references to databases such as LIPID MAPS (http://www.lipidmaps.org/). However, even in cases where only one formula is identified this could still belong to a range of potential lipid species. For example, if we take the PtdCho (36:2 – i.e. total FA chains of 36 carbon atoms with a total of 2 double bonds), this could be due to a PtdCho containing two C18:1 FAs, one C18:0 and one C18:2 or a variety of other isomers. To further define the chemical structure, fragmentation can be performed. In this process the ion is accelerated through a low pressure of inert gas, producing collisions and fragmentation of the parent ion. The daughter fragments can then be used to work out the parent structure, with head groups and FAs commonly being lost in the process (Fig. 1.6).

In the other form of commonly used lipidomics, a triple quadrupole mass spectrometer is used. In this instrument, the mass spectrometer consists of three electromagnet gates called quadrupoles. The first is used to select for one ion, which is usually the parent ion of the lipid species being detected. The second quadrupole acts as a fragmentation cell where the ion is fragmented. The third quadrupole then selects a particular fragment ion. While many lipid species may have the same parent mass, it is very unlikely that they will fragment in the same manner and thus this method is highly selective. In addition, these instruments can be made to be quantitative and are particularly appropriate for targeted analyses where a limited number of species is to be assayed. Furthermore, in an approach termed 'shotgun lipidomics', the assay can be set up to scan for particular lipid

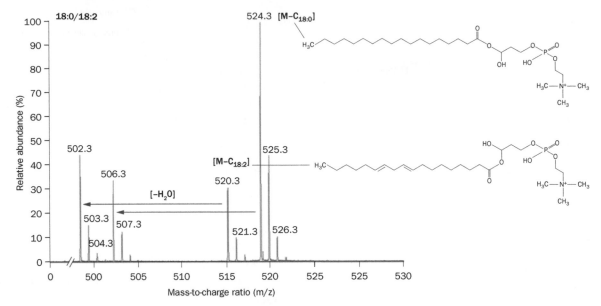

Fig. 1.6 Fragmentation of two phosphocholines derived from phosphatidylcholines (PtdChos). This figure demonstrates the further characterization of the lipidome by the technique of 'tandem MS'. One of the main challenges of LC-MS is lipid identification because of the large numbers of isomers present. In this technique, chromatography is dispensed with altogether and the sample is directly infused into a high resolution MS instrument. Figure 1.6 illustrates the identification of two phosphocholine isomers produced by fragmentation of PtdChos that would have been esterified with 18:0/18:2 and 18:1/18:1 respectively. From JL Griffin, H Atherton, J Shockcor & L Atzori (2011) Metabolomics as a tool for cardiac research. *Nat Rev Cardiol* **8**: 630–43; p. 636, Fig. 4a. Reproduced with permission of Nature Publishing Group.

species either by virtue of the head group present (e.g. scanning for PtdCho species) or particular FAs (e.g. identifying lipid species that contain a particular FA such as arachidonic acid, all-c5,8,11,14-20:4, n-6).

More detailed accounts of these methods can be found in **Further reading.**

1.5 A guide to the contents of this book

The purpose of this section is to provide a 'roadmap' to enable students to find their way around and make best use of the information provided in this book.

Continuing the scheme adopted in this chapter, each chapter is divided into numbered sections; the first number of the section will indicate the chapter number. There will be extensive cross-referencing between sections within chapters and between chapters. Although there are several ways we could have arranged the succession of chapters, we hope that the one we have chosen will be a logical one.

At the end of each chapter there is a '**Key points**' section that provides a concise summary of the

principal information in the chapter. This is followed by a section on '**Further reading**', which provides a selection of useful reviews and also some original research publications to give students a flavour of important and exciting recent advances. Although items in **Further reading** will be referenced throughout each chapter, there are limited references to specific pieces of literature in the main text. The number of references in **Further reading** could not be unlimited. We have attempted to cite those most useful that were available at the time of writing but additional references and/or diagrams are available on the companion website. Information in the text will be supplemented with figures and tables, and 'boxes' will be used to provide more detail on specific topics where inclusion in the text might interfere with the flow.

Chapter 2 introduces the chemical structures of the different types of lipids in three sections. These deal with (1) FAs, (2) lipids mainly involved in energy storage and (3) those predominantly associated with cellular membranes and also involved in physiological processes such as cell signalling. Of course there will be overlap between these functions: it is impossible (and undesirable) fully to compartmentalize lipid forms and functions. These sections discuss how the chemical structures of lipids relate to their physical and physiological properties and point the way to aspects of their metabolism, function and utilization in subsequent chapters. The FA section contains Box 2.1, which provides useful information on the complex topic of FA nomenclature.

Chapter 3 covers the metabolism of the FAs. This starts with their biosynthesis and discusses up-to-date knowledge of biosynthetic pathways, the enzymes involved in their biosynthesis and the genes coding for them. The degradation of FAs by oxidative pathways is then discussed in detail with particular reference to the generation of metabolic energy. A key section in this and other chapters concerns the all-important matter of how these metabolic pathways are controlled and integrated.

In the discussions of the biosynthesis of the polyunsaturated fatty acids (PUFAs) and their subsequent oxidation to form physiologically active products such as the 'eicosanoids', reference will be made to later chapters that describe the role of such molecules in cell signalling (Chapter 8) and as mediators in such physiological processes as immunity and the implications for health and disease (Chapter 10). Some PUFAs described in Chapter 3 are essential components of the diet ('essential fatty acids', EFAs) and their roles will be discussed further in Chapter 6.

Chapter 4. Just as Chapter 3 discusses the metabolism of the FAs, Chapter 4 deals with the metabolism of the complex lipids. Many of these (TAGs, phosphoglycerides, glycosylglycerides etc.) are FA esters of glycerol but the chapter also covers the sphingolipids (derivatives of the base sphingosine rather than glycerol, many of which incorporate sugars in the molecule) and the isoprenoids (also called terpenoids), in which the sterols, such as cholesterol and the plant steroids, are included.

The formation of TAGs is related to their role in energy storage in adipose tissue. This has ramifications for influence of dietary fats on the fat stores (Chapter 6) and on the relationships between the energy stores and health problems such as obesity, insulin resistance, diabetes, immune function, cancer and cardiovascular diseases, all of which are discussed in more detail in Chapter 10. Numerous seed oils of commercial importance store TAGs as an energy source. This too, has implications for the type and amounts of lipids in the diet (Chapter 6), their implications for health (Chapter 10) and their

biotechnological modification to provide useful products (Chapter 11).

An important section in this chapter discusses the many lipases (see Sections 4.2 & 4.6) that degrade lipids. Some are involved in the digestion of dietary lipids (Chapter 7), many others are involved in modifying the FA composition of lipids to suit the needs of particular cell types and cell structures (Chapters 4, 5, 7, 9 & 10), others are utilized in biotechnological processes (Chapter 11) and yet others are involved in the release of components of lipid molecules that are destined to become cell-signalling molecules (Chapters 8 & 10). Failure to degrade certain glycolipids, mainly owing to gene defects, can result in several lethal diseases of the nervous system that are addressed in Chapter 10.

Failure in the regulation of the metabolism of cholesterol in human beings, as a result of gene defects (Chapters 4 & 7) or dietary imbalance (Chapters 7 & 10) has implications for cardiovascular diseases that are also explored in Chapter 10.

Chapter 4 also mentions the biosynthesis of lipids that have specific functions and points the way to more detailed discussion in later chapters – for example the platelet activating factor and the lung surfactant lipid in Chapter 10.

Chapter 5 discusses the various ways in which different lipids can associate with each other and with proteins as a result of their chemical and physical properties. Such lipid assemblies are crucial to the structure and function of cells and cell organelles and in this chapter, we explore what is currently known about how lipids have shaped the evolution of living cells. Light is cast on the way in which, for example, the evolution of the bacteria and the archaea depended on the development of lipids with quite different chemical structures. Of particular importance is the development of different types of membranes whose lipid composition is crucial to their functions. Membranes are important for the topics discussed in each of the chapters of this book because of their role in cell structure, function and integrity, as a location for many metabolic pathways, their involvement in inter- and intracellular signalling processes, in the trafficking of biochemical substances within and between cells and because the development of many disease processes results from defects in the integrity of many membranes. As well as their presence in membranes, lipids accumulate as droplets (LDs) in cells (see

Section 5.5) where they act as energy stores or sources of molecules involved in the mediation of metabolic processes. Some lipid assemblies are involved in processes outside cells, for example in the formation of surface layers with barrier properties (see Section 2.2.4) or, as lipoproteins, in the transport of lipids in the bloodstream (see Sections 7.2 & 10.5).

Chapter 6 discusses the types of lipids in food and the diet and their biological roles. (Chapter 7, which follows, explains how these dietary lipids are digested, absorbed and the digestion products transported in the blood to the tissues of the body.) These two chapters are devoted mainly to human diets but there is also discussion of other simple-stomached animals, such as rats, mice and pigs, which are often used as so-called 'animal models'. This is because, in animal studies, procedures can be more easily controlled and the experimental design can be more rigorous. The disadvantage is that the biochemistry and physiology may sometimes differ between species, leaving open some doubt as to their relevance to Man.

Much of the food we now eat is processed in some way – industrially and domestically. There is some discussion here of how such processes may affect dietary lipids but reference is made to Chapter 11, which provides more detail on food processing and bio-technological developments. Dietary fats provide metabolic energy and although the subject is introduced here, readers will find more detailed information in Chapter 9. Dietary fats also supply many essential nutrients. This chapter picks up on the EFAs – PUFAs that are essential for health but cannot be made in the body – that were first introduced in Chapters 2 and 3. Also essential for good health are the fat-soluble vitamins, which are required in only milligram or microgram quantities as distinct from the gram or almost gram quantities of the EFAs. While knowledge of them developed in the late 19th and early 20th centuries, it is only in the last few decades that the full extent of their physiological roles as, for example, hormones and signalling molecules and regulators of metabolism has been realized. The molecular biology revolution has indicated the key involvement of some of them in the regulation of gene expression. The chapter ends with a thorough discussion of the role of lipids in foetal and postnatal development.

Chapter 7 describes in detail the processes by which lipid components of the diet are digested and the digestion products absorbed from different parts of the intestinal tract. Once within the intestinal absorbing cells (enterocytes) they are 'remodelled' and combined with proteins ('lipoproteins') for transport around the body in the bloodstream. The proteins not only help to solubilize the lipids but also direct them to sites of further metabolism. The different types of lipoproteins are described and also the elaborate system for the control of their metabolism and their movement to appropriate tissues. Such a complex system is vulnerable to defects either from gene mutations or from 'dietary overload' and the reader is pointed to Chapter 10, which describes the involvement of various lipids in health and disease.

Chapter 8 is concerned entirely with molecules that send signals to different cells of the body. The emphasis in this chapter is mainly on two types: the phosphoinositides and the sphingolipids. Before the mid-1960s, lipids were thought of as having three main biological functions: as structural components of membranes, as energy stores and as a barrier against the environment or providers of insulation. Phosphatidylinositol (PtdIns) was already known as a widespread membrane component but everything changed when it was discovered that inositol phospholipids with additional phosphate groups esterified in different positions on the inositol ring could, when cells were stimulated by agonist molecules such as hormones, be catabolized to yield compounds that sent signals across the membrane that then resulted in a variety of metabolic changes. Even a molecule such as diacylglycerol (DAG), it was then discovered, could act as a 'messenger'. Similar roles were discovered also for a variety of sphingolipids. Several other lipid molecules with signalling functions are described in other chapters, for example: platelet activating factor (PAF, an 'ether' phospholipid) in Chapters 4 and 10; endocannabinoids in Chapters 4 and 8.

Chapter 9 is devoted entirely to the role of lipids as energy stores in animals and plants. The first part goes into detail in the animal storage organs – white and brown adipose tissue. The white form is the main storage tissue for TAGs; it is widely dispersed around the body rather than being a discrete organ like the liver or brain. It contains smaller amounts of other lipid molecules and as well as a storage organ it is now known to have endocrine properties, producing hormones. The uptake of TAGs into the fat cells and their mobilization for energy supply is discussed in relation to the biochemistry already described in Chapters 3 and 4.

The cells of the brown form of adipose tissue contain many small LDs (in contrast to white adipose tissue's unilocular droplet) and these are surrounded by mitochondria that accept FAs released from the fat droplets and oxidize them by the process of β-oxidation, which is described in detail in Chapter 3.

Lipid storage by some plants is important for supplying the metabolic energy for seed development and germination. The different storage locations – fruits, seeds and pollen grains – and the types of lipids involved, are described. Plant storage fats are important in diets (Chapter 6) and require industrial processing (Chapter 11). New methods of introducing genes for the biosynthesis of specific lipids that may not be native to a particular plant are now becoming available (Chapter 11).

Chapter 10 addresses the subject of lipids in health and disease. It opens with a discussion of various inborn errors of metabolism, describing the genetic background and the implications for dietary lipids. There are relevant pointers to other chapters in which the biochemical basics are discussed (Chapters 3, 4, 7 & 9). A section on cancer examines the influence of dietary lipids (both in development and treatment), the roles of specific lipids in physiological functions associated with cancer development and the involvement of the immune system. A whole section is devoted to the ways in which lipids may be involved in aspects of immune function, including their modification of gene expression. Once again there is comprehensive referencing to the biochemistry of lipids in Chapters 3, 4, 5, 7 and 9. The conditions of obesity and diabetes (see also Chapter 9) and disorders of lipoprotein metabolism (in their association with cardiovascular diseases, Chapter 7) are similarly related to preceding biochemical background (Chapters 3 & 4).

Chapter 11 discusses the industrial processing of lipids and lipid-containing foods as well as how biotechnology is being applied in the development of new products with very specific properties. Nonfood aspects include the properties and production of soaps, detergents, biofuels and oleochemicals. Many of these topics are introduced for the first time but reference is made back to Chapter 7 when discussing the detergent properties of the bile salts. The functional properties of lipid-based foods such as spreads are discussed in terms of their enhancement of palatability and their role as carriers for fat-soluble vitamins, with reference back to Chapters 6 and 10. Foods that supply different types of FAs and their relevance to health (e.g the *n*-3 PUFAs and plant sterols) and disease (e.g. the *trans*-FAs) are discussed with reference back to Chapters 6 and 10. Finally, recent advances in the use of genetic modification to produce crops and livestock with novel lipid profiles are described.

KEY POINTS

- In contrast to carbohydrates, proteins and nucleic acids, lipids are defined on the basis of their physical properties (insolubility in water) rather than on the basis of consistent chemical features. For this reason, the student will need to learn and remember a wide range of different chemical types and their rather complex nomenclature.

- Lipids can usually be extracted easily from tissues by making use of their hydrophobic characteristics. However, such extractions yield a complex mixture of different lipid classes which have to be purified further for quantitative analysis. Moreover, the crude lipid extract may be contaminated by other hydrophobic molecules, e.g. by intrinsic membrane proteins, and need to be 'cleaned up'.

- Of the various types of separation, thin layer and column chromatography are most useful for intact lipids. A powerful tool for quantitation of volatile lipids or derivatives is GC but HPLC has become increasingly used.

- Current research is increasingly concerned with identifying complete profiles of the extremely complex lipid constituents of living tissues – the so-called 'lipidome'. Modern 'lipidomics' utilizes a combination of either GC-MS or LC-MS to define the lipidome.

- With this background to what lipids are and how they are studied, the 'roadmap' then guides the student through the remaining ten chapters.

Further reading

Ceve G, ed., (1993) *Phospholipids Handbook*, Marcel Dekker, Basel.

Christie WW (1989) *Gas Chromatography and Lipids*. The Oily Press, Ayr, UK.

Christie WW, ed., (1997) *Advances in Lipid Methodology*, **4** vols. The Oily Press, Ayr, UK.

Christie WW & Han X (2010) *Lipid Analysis*, 4th edn. The Oily Press, Bridgwater, UK.

Fahy E, Subramaniam S, Murphy RC, *et al.* (2005) A comprehensive classification system for lipids. *J Lipid Res*, **46**:839–61.

Griffin JL, Atherton H, Shockcor J & Atzori L (2011) Metabolomics as a tool for cardiac research. *Na Rev Cardiol* **8**:630–43.

Gross RW & Han X (2009) Shotgun lipidomics of neutral lipids as an enabling technology for elucidation of lipid-related diseases, *Am J Physiol Endoc-M* **297** E297–303.

Gunstone FD, Harwood JL& Dijkstra AJ, eds. (2007) *The Lipid Handbook*, 3rd edn. CRC Press, Boca Raton, USA.

Gurr MI, James AT, Harwood JL, *et al.* (1971, 1975, 1980, 1991, 2002) *Lipid Biochemistry: An Introduction*, Editions 1–4, Chapman & Hall, London; Edition 5, Blackwell Science, Oxford, UK.

Hamilton RJ& Hamilton S, eds. (1992) *Lipid Analysis: A Practical Approach*, IRL Press, Oxford, UK.

Hammond EW (1993) *Chromatography for the Separation of Lipids*. CRC Press, Boca Raton, USA.

International Union of Biochemistry and Molecular Biology (1992) *Biochemical Nomenclature and Related Documents*, 2nd edn. Portland Press, London, UK.

IUPAC–IUB Commission on Biochemical Nomenclature (1989) *Eur J Biochem* **186**:429–58.

Kates M (1986) *Techniques of Lipidology*, 2nd edn. Elsevier Science, Amsterdam, The Netherlands. (This classic book lacks details about recent advances (e.g. HPLC) but still contains a wealth of basic information.)

Kates M (2010) *Techniques of Lipidology*, 3rd edn. Newport Somerville Innovation Ltd, Ottawa, Canada.

Leray C (2013) *Introduction to Lipidomics: From Bacteria to Man*. CRC Press, Boca Raton, USA.

Lipid Library (http://lipidlibrary.aocs.org)

Nomenclature Committee of the International Union of Biochemistry (1984) *Enzyme Nomenclature*, Academic Press, London, UK. (The most up-to-date information on enzyme nomenclature can be found by accessing: http://www.chem.qmw.ac.uk/iubmb/enzyme/ (last accessed 4 December 2015).)

Nygren H, Seppänen-Laakso T, Castillo S, Hyötyläinen T & Orešič M (2011) Liquid chromatography-mass spectrometry (LC-MS)-based lipidomics for studies of body fluids and tissues. *Method Mol Biol* **708**:247–57.

Roberts LD, Koulman A & Griffin JL (2014) Methods for performing lipidomics in white adipose tissue, *Method Enzymol* **538**:211–31.

CHAPTER 2

Important biological lipids and their structures

2.1 Structure and properties of fatty acids

In lipid biochemistry – as in most other fields of science – various trivial names and shorthand nomenclatures have come into common usage. Thus, the poor student is faced with a seemingly endless series of seemingly illogical names and symbols for the various new compounds to be learnt. Fatty acid (FA) names are no exception and there are trivial names for most commonly occurring FAs. A shorthand system to avoid repeating cumbersome chemical names for FAs is explained in Box 2.1. This derives from the use of gas chromatography (GC; see Section 1.4.4) where separations can be achieved owing to both carbon chain length and degree of unsaturation. This shorthand system has the merit of not attributing a more precise identification than the gas chromatograph alone can give.

2.1.1 Saturated fatty acids

Most saturated FAs (SFAs) are straight chain structures with an even number of carbon atoms. Acids from 2C to longer than 30C have been reported but the most common lie in the range 12–22C. Some of the more important naturally occurring straight chain SFAs are shown in Table 2.1, together with further information. In general, FAs do not exist as free carboxylic acids because of their marked affinity for many proteins. (One result of this is an inhibitory action on many enzymes.) Where non-esterified fatty acids (NEFAs) are reported as major tissue constituents, they are usually artefacts owing to cell damage, which allows lipases (see Section 4.2) to break down the endogenous acyl lipids (see Sections 2.2 & 2.3). Exceptions to this statement are the albumin-bound FAs of mammalian blood. Free acids used to be referred to as FFA (free fatty acids) but are now preferably termed NEFAs.

The configuration of a typical saturated chain is shown in Fig. 2.1. Because of continuous thermal motion in living systems and the free rotation about the carbon–carbon bonds, the FAs are capable of adopting a huge number of possible configurations, but with a mean resembling an extended straight chain. Steric hindrance and interactions with other molecules in Nature will, of course, restrict the motion of NEFAs and the acyl chains of complex lipids. The physical properties of acyl lipids are affected by their individual FAs – a most obvious one being the melting point. As a general rule, membranes are incapable of operating with lipids whose acyl chains are solid (in the gel phase). Thus, for mammals, this means the acyl chains must be fluid at about 37 °C and for poikilotherms (organisms unable to regulate their own temperature) at temperatures between about −10 °C and over 100 °C, depending on the environment.

Although it is possible for the membranes of, for example, a mammal to contain lipids with acyl chains whose melting points are slightly above 37 °C, the presence of other lipid types ensures that the mixture is in fact semiliquid. While most natural FAs are even-numbered as a result of their mode of biosynthesis, odd-numbered acids do occur. The formation of both types is discussed in Section 3.1. For further discussion of lipids in membranes see Chapter 5.

2.1.2 Branched-chain fatty acids

Although branched-chain FAs are usually also saturated, they are discussed separately here. Two distinct series, which are often found in bacteria, are the iso-series where the terminal group is:

$$\overset{\displaystyle CH_3}{\underset{\displaystyle CH_3 - CH-}{|}}$$

and the anteiso-series where the terminal group is:

$$\overset{\displaystyle CH_3}{\underset{\displaystyle CH_3 - CH_2 - CH-}{|}}$$

Lipids: Biochemistry, Biotechnology and Health, Sixth Edition. Michael I. Gurr, John L. Harwood, Keith N. Frayn, Denis J. Murphy and Robert H. Michell.
© 2016 John Wiley & Sons, Ltd. Published 2016 by John Wiley & Sons, Ltd.

Box 2.1 The naming and representation of fatty acids

Most FAs have a 'common name' reflecting their origin: e.g. palmitic acid is a major constituent of palm oil, oleic of olive oil (coming from the Latin *oleum*, oil). These common names are listed in Tables 2.1–2.3. However, common names tell us nothing about their chemistry, so that a more systematic nomenclature is required.

FAs are characterized by (1) chain length (number of carbon atoms), (2) degree of unsaturation (number of double bonds), (3) position and geometry (*cis* or *trans*) of double bonds and (4) presence of other substituents, such as branched chains, ring systems and oxygen groups.

For chain length, we shall use the simple notation of C to designate a FA carbon chain and *n* for the number of carbon atoms. Thus, we shall refer to a '10C *fatty acid*' etc. To refer to a *specific carbon* atom in the chain, we shall write (e.g.) 'the substituent at C10'. The *number* of double bonds will be indicated simply as n:0 for a SFA (e.g. palmitic would be 16:0); n:1 for MUFA (e.g. 18:1 for oleic); n:2 for dienoic acids (e.g. 18:2 for linoleic) etc. This gives no information on the *positions* of the double bonds, which is indicated relative to one end of the chain of carbon atoms. There are two systems. One numbers from the carboxyl carbon as C1 and is the standard system adopted by the IUB/IUPAC (see **Further reading**). Thus linoleic acid could be written as *cis* (Δ)9, *cis* (Δ)12-18:2, or (*cis*, *cis*) 9, 12-octadecadienoic acid, or all-*cis*-9,12-octadecadienoic acid, to indicate that it is an 18-carbon FA with *cis* double bonds 9 and 12 carbons from the carboxyl end. In this book, we usually abbreviate '*cis*' and '*trans*' to '*c*' and '*t*'. Whereas most naturally occurring UFAs have their double bonds in the *cis* configuration, *trans* unsaturation does also occur. Thus, oleic acid will be written as c9-18:1 and vaccenic acid, with a *trans* double bond (produced by fermentation in the rumen) will be written *t*11-18:1.

The use of Δ to indicate unsaturation is an old system whose origin is obscure and, because the IUB/IUPAC numbering system is unambiguous in defining the number of each carbon in relation to the carboxyl (see Section 1.3.5), we shall omit the 'Δ' in this edition, although readers will come across it in other publications. It is retained in many publications in connection with the desaturase enzymes that introduce double bonds: thus, a Δ9-desaturase (in this book, 9-desaturase) introduces a double bond between carbons 9 and 10 (counting from the carboxyl end).

The other system counts from the noncarboxyl (methyl) end (sometimes called the ω-carbon, ω being the last letter of the Greek alphabet). Thus an ω-3 FA has its first double bond, counting from the methyl end, between carbons 3 and 4. This is useful because many FAs fall into families when classified in this way and their metabolic connections can be more easily appreciated (see Section 3.1.8 & Fig. 3.54). The ω- terminology is often found now in connection with dietary fats. However, standardized terminology uses '*n*' rather than ω; thus, we have families of *n*-3, *n*-6 PUFAs.

Finally, *substitution* in the chain can be denoted thus: br-16:0 for a branched chain 16C acid or OH-16:0 for a hydroxy 16C acid. These more complicated structures are spelt out in more detail in Sections 2.1.2–2.1.5.

However, branch points can also be found in other positions. The presence of the side-chain has a similar effect on fluidity as the presence of a *cis* double bond (i.e. lowers the melting temperature). Branched-chain acids occur widely, but mainly at low concentrations in animal fats and some marine oils. They are rarely found in plant lipids. Butter fats, bacterial and skin lipids contain significant amounts. Branched-chain FAs are major components of the lipids of Gram-positive bacteria and more complex structures with several branches may be found in the waxy outer coats of mycobacteria (see Section 5.6.1).

Polymethyl FAs include those of isoprenoid origin such as phytanic and pristanic acids (Fig. 2.2), which are formed from dietary chlorophyll (see Section 3.2.2). A different pattern is seen in FAs from bird uropygial (preen) glands where methyl groups are found on alternate carbons.

2.1.3 Unsaturated fatty acids

2.1.3.1 Monounsaturated (monoenoic) fatty acids

Over one hundred naturally occurring monounsaturated fatty acids (MUFA), alternatively named *monoenoic* acids, have been identified but most of these are extremely rare. In general, the more common compounds have an even number of carbon atoms, a chain length of 16C–22C and a double bond with the *cis* configuration. Often the *cis* bond is at the 9-position. *Trans* isomers are less common but do exist, one of the most interesting being *trans*-3-hexadecenoic acid (*t*3-16:1), a major fatty acid esterified to phosphatidylglycerol (PtdGro) in the photosynthetic membranes of higher plants and algae.

The presence of a double bond causes a restriction in the motion of the acyl chain at that point. Furthermore, the *cis* configuration introduces a kink into the average molecular shape (Fig. 2.1b) while the *trans* double bond ensures that the FA has an extended conformation and

Table 2.1 Some naturally occurring straight chain saturated acids.

No. of carbon atoms	Systematic name	Common name	Melting point (°C)	Occurrence
2	n-Ethanoic	Acetic	−16	As alcohol acetates in many plants, and in some plant triacylglycerols. At low levels widespread as salt or thiolester. At higher levels in the ruminant as salt.
3	n-Propanoic	Propionic	−21	At high levels in the rumen.
4	n-Butanoic	Butyric	−8	At high levels in the rumen, also in milk fat of ruminants.
6	n-Hexanoic	Caproic	−3	In milk fat.
8	n-Octanoic	Caprylic	16	Very minor component of most animal and plant fats. Major component of many milks and some seed triacylglycerols. Flavouring uses.
10	n-Decanoic	Capric	31	Widespread as a minor component. Major component of many milk and some seed triacylglycerols. Uses in industry.
12	n-Dodecanoic	Lauric	44	Widely distributed, a major component of some seed fats (e.g. palm kernel or coconut oil). Industrial uses.
14	n-Tetradecanoic	Myristic	54	Widespread; occasionally found as a major component (e.g. in nutmeg).
16	n-Hexadecanoic	Palmitic	63	The most common saturated fatty acid in animals, plants and microorganisms. Enriched in palm oil.
18	n-Octadecanoic	Stearic	70	Major component in animals and some fungi, minor constituent in plants (but major in a few, e.g. cocoa butter).
20	n-Eicosanoic	Arachidic	77	Widespread minor component, occasionally a major component (e.g. ground nut).
22	n-Docosanoic	Behenic	81	Fairly widespread as minor component in seed triacylglycerols and plant waxes.
24	n-Tetracosanoic	Lignoceric	88	Fairly widespread as minor component in seed triacylglycerols and plant waxes.
26	n-Hexacosanoic	Cerotic	89	Widespread as component of plant and insect waxes. Found in some bacterial lipids.
28	n-Octacosanoic	Montanic	90	Major component of some plant waxes.

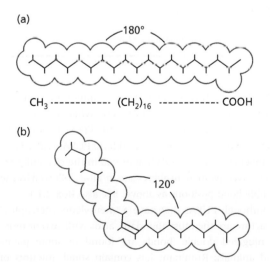

(a)

180°

CH₃ ------------ (CH₂)₁₆ ------------ COOH

(b)

120°

Fig. 2.1 Preferred conformation of (a) a saturated (stearic) and (b) a monosaturated (oleic) acid.

properties nearer to that of an equivalent chain length SFA (Tables 2.1 & 2.2). Because the *cis* forms are less stable thermodynamically than the *trans* forms, they have lower melting points than the latter or their saturated counterparts.

In addition to the normal ethylenic double bonds, some FAs possess acetylenic (triple) bonds. These occur in a number of rare seed oils and a few mosses and in marine organisms such as sponges. Many of these acids contain additional double or triple bonds and would thus be categorized as polyunsaturated (polyenoic) FAs (below).

2.1.3.2 Polyunsaturated (polyenoic) fatty acids

FAs with more than one double bond are called polyunsaturated (PUFAs) or 'polyenoic'. These may be subdivided into dienoic, trienoic, tetraenoic, pentaenoic, hexaenoic etc. with 2, 3, 4, 5, 6 double bonds respectively.

All dienoic acids are derived from monoenoic acids, the position of the second bond being a function of the

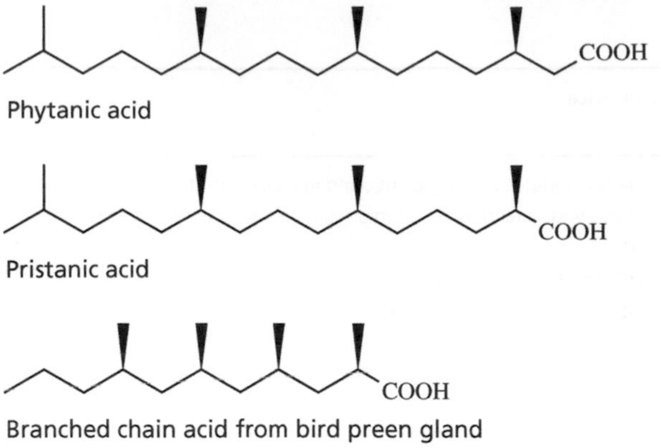

Phytanic acid

Pristanic acid

Branched chain acid from bird preen gland

Fig. 2.2 Examples of branched fatty acids. Phytanic acid (derived from the phytol sidechain of chlorophyll) and pristanic acid are minor components of fish oils. Fatty acids produced by the preen (uropygial) gland of birds have a different pattern with methyl groups found on alternative (usually) even carbons.

Table 2.2 Some naturally occurring monoenoic fatty acids.

No. of carbon atoms	Systematic name	Common name	Melting point (°C)	Occurrence
16	*trans*-3-hexadecenoic		53	Plant leaves;eukaryotic algae;specifically as component of phosphatidylglycerol in chloroplasts.
16	*cis*-5-hexadecenoic			Ice plant, Bacilli, sponges.
16	*cis*-7-hexadecenoic			Algae; higher plants, bacteria.
16	*cis*-9-hexadecenoic	Palmitoleic	1	Widespread: animals, plants, microorganisms. Major component in some seed oils and many marine organisms. Has increasing nonfood uses.
18	*cis*-6-octadecenoic	Petroselenic	33	Found in umbelliferous seed oils.
18	*cis*-9-octadecenoic	Oleic	16	Most common monoenoic fatty acid in plants and animals. Also found in most microorganisms.
18	*trans*-9-octadecenoic	Elaidic	44	Ruminant fats, hydrogenated margarines.
18	*trans*-11-octadecenoic	*trans*-Vaccenic	44	Found in rumen fats via biohydrogenation of polyunsaturated fatty acids.
18	*cis*-11-octadecenoic	Vaccenic	15	*E. coli* and other bacteria.
20	*cis*-11-eicosenoic	Gondoic	24	Seed oil of rape; fish oils.
22	*cis*-13-docosenoic	Erucic	34	Seed oil of *Cruciferae* (rape, mustard, etc.).

biochemical system. Thus, mammals have desaturases that are capable of removing hydrogens only from carbon atoms between an existing double bond and the carboxyl group. Because of this, further desaturations may need to be preceded by chain elongation. Higher plants, by contrast, can carry out desaturation mainly between the existing double bond and the terminal methyl group. In either case, the double bonds are almost invariably separated from each other by a methylene grouping, as for example in linoleic acid (*c*9,*c*12-18:2,n-6) (Table 2.3). In animals, the very long-chain PUFAs such as arachidonic (all-*c*5,8,11,14-20:4,*n*-6), eicosapentaenoic (EPA; all-*c*5,8,11,14,17-20:5,*n*-3) and docosahexaenoic (DHA;

all-*c*4,7,10,13,16,19-22:6.*n*-3) have important roles as precursors of signalling molecules, while α-linolenate (all-*c*-9,12,15-18:3,*n*-3) is used in the lipoxygenase pathway of plants (see Section 3.4). (n.b.: in PUFAs with a large number of double bonds that are **all** in the *cis* configuration, it is convenient to abbreviate to 'all-*c*', followed by the double bond positions as above. See also Box 2.1.)

Although PUFAs that have a 'methylene-interrupted' structure are the most abundant, acids with two or more 'conjugated double' bonds are found in some plants and animals. Ruminant fats contain small amounts of conjugated linoleic acids. These are formed by biohydrogenation of linoleic or α-linolenic acids in the rumen (see

Table 2.3 Some naturally occurring polyunsaturated fatty acids.

No. of carbon atoms	Systematic name	Common name (or abbreviation)	Melting point (°C)	Occurrence
Dienoic acids				
18	*cis, cis*-6,9-octadecadienoic	Petroslinoleic	−11	Minor component in animals.
	cis, cis-9,12-octadecadienoic	Linoleic	−5	Major component in plant lipids. In mammals it is derived only from dietary vegetables, and plant and marine oils.
Trienoic acids (methylene interrupted)				
16	All-*cis*-7,10,13-hexadecatrienoic			Algae and many higher plants
18	All-*cis*-6,9,12-hexadecatrienoic	γ-Linolenic		Minor component in animals and some algae. Important constituent of some plants, e.g. evening primrose oil.
	All-*cis*-9,12,15-hexadecatrienoic	α-Linolenic	−11	Higher plants and algae, especially as component of galactosyl diacylglycerols.
Trienoic acids (conjugated)				
18	*cis*-9, *trans*-11, *trans*-13-octadecatrienoic	Eleostearic	49	Some seed oils, especially Tung oil.
Tetraenoic acids				
16	All-*cis*-4,7,10,13-hexadecatetraenoic			Green algae and *Euglena gracilis*
20	All-*cis*-5,8,11,14-eicosatetraenoic	Arachidonic	50	A major component of animal phospholipids. Major component of many marine algae and some terrestrial species, such as mosses.
Pentaenoic acids				
20	All-*cis*-5,8,11,14,17-eicosapentaenoic	EPA		Major component of some marine algae, fish oils.
22	All-*cis*-7,10,13,16,19-docosapentaenoic	Clupanodonic		Animals, especially as phospholipid component. Widespread in fish as minor component
Hexaenoic acids				
22	All-*cis*-4,7,10,13,16,19-docosahexaenoic	DHA		Animals, especially as phospholipid component. Abundant in fish and some marine algae.

Section 3.1.9). The main component is 9*c*,11*t*-18:2 (rumenic acid). Some of the naturally occurring conjugated plant FAs, such as α-eleostearic acid (9*c*,11*t*,13*t*-18:3), have uses in industry.

There are also many rare FAs found in diverse organisms containing allenic (−HC=C=CH−) or cumulenic (−HC=C=C=CH−) structures. These core structures have double bonds in the *cis* configuration.

2.1.4 Cyclic fatty acids
These acids are rather uncommon but, nevertheless, examples provide important metabolic inhibitors. The ring structures are usually either cyclopropyl or cyclopentyl. Cyclopropane and cyclopropene FAs are produced

by many bacteria and are also found in some plants and fungi (Table 2.4). Some organisms from extreme environments, such as hot springs, produce FAs with a terminal cyclohexyl group (up to 90% of their total acids).

2.1.5 Oxy fatty acids
A great range of keto, hydroxy and epoxy acids has been identified in recent years. The most widely occurring epoxy acid is vernolic acid. Hydroxy acids do not occur very extensively, although they are found in some sphingolipids. However, they are major components of surface waxes, cutin and suberin of plants (Table 2.4). Many of the oxygen-containing FAs that occur in seed oils have important uses in industry.

Table 2.4 Examples of some substituted fatty acids.

Structure	Name	Notes
Cyclic fatty acids		
$CH_3(CH_2)_5\overset{\overset{\displaystyle CH_2}{\triangle}}{CH\text{-}CH}(CH_2)_9COOH$	Lactobacillic	Produced by Lactobacilli, *Agrobacterium tumefaciens*. Found in many bacteria, some protozoa and in invertebrates from their diet.
$CH_3(CH_2)_7\overset{\overset{\displaystyle CH_2}{\triangle}}{C=C}(CH_2)_7COOH$	Sterculic	Found in plants of the Malvales family. Inhibits stearate conversion to oleate.
$CH_2(CH_2)_{11}COOH$ (cyclopentene ring)	Chaulmoogric	From seeds of the tree *Hydnocarpus wightiana*. *Used for leprosy treatment.*
Epoxy-fatty acids		
$CH_3(CH_2)_4\overset{\overset{\displaystyle O}{\triangle}}{CH\text{-}CH}CH_2CH=CH(CH_2)_7COOH$	Vernolic	Epoxy derivative of oleate used as an industrial biofeedstock. Found in seeds of *Vernonia* and *Euphorbia* plants.
Hydroxy-fatty acids		
$CH_3(CH_2)_{13}\overset{\overset{\displaystyle OH}{\mid}}{CH}COOH$	α-Hydroxy palmitate	Found in galactosyl-cerebrosides
$HOCH_2(CH_2)_{14}COOH$	ω-Hydroxy palmitate	Constituent of suberin coverings of plants
$CH_3(CH_2)_5\overset{\overset{\displaystyle OH}{\mid}}{CH}CH_2CH=CH(CH_2)_7COOH$	Ricinoleic	Represents >90% of the total acids of castor bean oil. Used in paint, varnish and other industrial uses.

2.1.6 Fatty aldehydes and alcohols

Many tissues contain appreciable amounts of fatty alcohols or aldehydes whose chain length and double bond patterns reflect those of the FAs from which they can be derived. Sometimes the alcohols form esters with FAs and these 'wax esters' are important in marine waxes such as sperm whale oil or its plant equivalent, jojoba wax.

2.1.7 Some properties of fatty acids

The short-chain FAs (i.e. of chain lengths up to 10C) are poorly water-soluble. In solution they easily associate and thus tend not to exist as single molecules. Indeed, the actual solubility (particularly of longer-chain acids) is often very difficult to determine because it is influenced considerably by the pH and also because the tendency for FAs to associate leads to monolayer or micelle formation.

Micelle formation is characteristic of many lipids. McBain and Salmon many years ago introduced the concept of micelles. The most striking evidence for the formation of micelles in aqueous solutions of lipids lies in extremely rapid changes in physical properties over a limited range of concentrations, the point of change being known as the critical micellar concentration (CMC) and this exemplifies the tendency of lipids to self-associate rather than remain as single molecules. The CMC is not a fixed value but a small range of concentrations and is markedly affected by the presence of other ions, neutral molecules, etc. The value of the CMC can be conveniently measured by following the absorbance of a lipophilic dye such as rhodamine in the presence of increasing 'concentrations' of the lipid. The tendency of lipids to form micelles or other structures in aqueous solution often means that

Table 2.5 The melting points of a series of saturated fatty acids.

Fatty acid	Chain length	Melting point (°C)
Butanoic	C4	−8
Pentanoic	C5	−35
Hexanoic	C6	−3
Heptanoic	C7	−9
Octanoic	C8	16
Nonanoic	C9	15
Decanoic	C10	31
Undecanoic	C11	28
Dodecanoic	C12	44
Tridecanoic	C13	45
Tetradecanoic	C14	54
Pentadecanoic	C15	53
Hexadecanoic	C16	63

study of enzyme kinetics is difficult. Thus, for example, enzymes metabolizing lipids often do not display Michaelis-Menten kinetics and phrases such as 'apparent' K_m must be used. In some cases the enzymes prefer to work at interfaces rather than with free solutions (see Section 4.6.1).

FAs are easily extracted from solution or suspension by lowering the pH to form the uncharged carboxyl group and extracting with a nonpolar solvent such as light petroleum. In contrast, raising the pH increases solubility because of the formation of alkali metal salts, which are the familiar soaps. Soaps have important properties as association colloids (i.e. the dispersed phase consists of molecules that have hydrophobic and hydrophilic parts) and are surface-active agents (see also Section 11.2.13).

The influence of FA structure on its melting point has already been mentioned with branched chains and *cis* double bonds lowering the melting points of equivalent saturated chains. Interestingly, the melting point of FAs depends on whether the chain is even or odd numbered (Table 2.5).

SFAs are very stable, but unsaturated fatty acids (UFAs) are susceptible to oxidation; the more double bonds the greater the susceptibility. UFAs, therefore, have to be handled under an atmosphere of inert gas (e.g. nitrogen) and kept away from (photo) oxidants or substances giving rise to free radicals. Antioxidant compounds have to be used frequently in the biochemical laboratory just as organisms and cells have to utilize similar compounds to prevent potentially harmful attacks on acyl chains in vivo (see Section 3.3).

2.1.8 Quantitative and qualitative fatty acid analysis

2.1.8.1 General principles

Chapter 1, Section 1.4.4 referred to general methods for lipid analysis. For the separation and identification of FAs by GC (an example of a separation is shown in Fig. 2.3), sample preparation is important. FA methyl esters can be routinely prepared by refluxing the FA sample with methanolic-HCl. However, for labile acids, such as cyclic acids, base-catalysed methanolysis is needed. It is advisable to analyse the same sample using different stationary phases to overcome problems such as the overlapping of certain FAs on some columns. Quantitation can be accomplished by adding a known amount of an 'internal standard' to the sample, for example 17:0 or 21:0, which are rarely found in natural mixtures. (For more details see Christie & Han (2010), Gunstone *et al.* (2007), Kates (2010) or The Lipid Library in **Further reading**.)

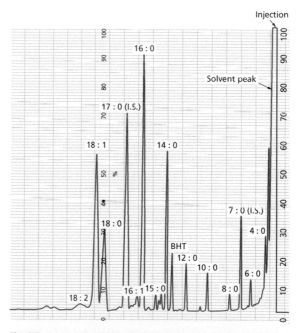

Fig. 2.3 Separation of fatty acid methyl esters by gas-liquid chromatography. The oven temperature was programmed from 65 °C to 185 °C at 8 °C min⁻¹. The sample consists of fatty acids derived from a milk diet for calves. Heptanoic and heptadecanoic acids were added as 'internal standards' (IS). BHT (butylatedhydroxytoluene) is an antioxidant routinely added to prevent oxidation of unsaturated fatty acids. Reproduced with kind permission of Mr JD Edwards-Webb.

2.1.8.2 Determination of the structure of an unknown acid

Where GC-MS (see Section 1.4.4) is not readily available, co-chromatography with authentic standards on different types of columns will serve to provide a provisional identification. Chain-lengths of unknown components can be estimated from comparisons with known components, since there is a simple linear relationship between the log of retention time and the carbon number for each particular type of acid on a given column. Double bond positions in UFAs can be determined by oxidizing that acid with permanganate or ozone. The chain splits at each double bond, yielding fragments. So, for example, $c9$-18:1 gives 9:0 and a 9C dicarboxylic acid. Infrared spectrometry can distinguish between *cis* and *trans* double bonds. The addition of silver nitrate to silica as the stationary phase in either GC or TLC is useful for the separation of UFAs, which are separated in bands according to the total number of double bonds, monoenes emerging first.

GC-MS provides the most comprehensive information, however, as discussed in Section 1.4.4.

2.2 Storage lipids – triacylglycerols and wax esters

2.2.1 Introduction

Many plants and animals need to store energy for use at a later time. Lipid fuel stores are mainly triacylglycerols (TAGs) or wax esters.

Energy needs are of two main types. One is a requirement to maintain, throughout life, on a minute-to-minute basis, the organism's essential biochemistry (see Section 9.1). Either carbohydrates or FAs can fulfil this function, although tissues do vary in their ability to use different fuels: the mammalian brain, for example, is normally reliant on glucose as a fuel source, since it cannot use FAs directly. (However, during prolonged starvation, the brain can make use of metabolites of FAs called 'ketone bodies'.) The second type of energy requirement is for a more intermittent and specialized store of fuel that can be mobilized when required (see Sections 9.1–9.3). Thus, animals may need to use such reserves after a period of starvation or hibernation or for the energy-demanding processes of pregnancy and lactation, while plants need energy for seed germination. Animals and plants each make use of different types of fuels that may be stored

in different tissues. Thus, practically all animals use lipids as a long-term form of stored energy and almost all use TAGs as the preferred lipid. Some marine animals, however, use wax esters as their energy store. Most animals store their energy in a specialized tissue, the adipose tissue (see Section 9.1), but some fish use their flesh or the liver as a lipid store. In general, plants store fuel required for germination in their seeds either as lipid or polysaccharide. Oil-bearing seeds usually contain TAGs, but wax esters are used as fuel by some desert plants, such as jojoba. Some fruits and pollen grains also store TAGs as an energy supply (see Section 9.3.1).

2.2.2 The naming and structure of the acylglycerols (glycerides)

Acylglycerols are esters of the trihydric alcohol glycerol, in which one, two or three hydroxyls may be esterified with FAs.

2.2.2.1 Introduction

TAGs are the major components of natural fats and oils; partial acylglycerols are usually intermediates in the breakdown or biosynthesis of TAGs.

The preferred term for esters of glycerol (a trihydric alcohol) with FAs is now 'acylglycerols', although you will certainly come across the synonym 'glycerides' in other literature. The terms TAG, diacylglycerol (DAG) and monoacylglycerol (MAG) (Fig. 2.4) will be used here for the specific compounds in which three, two or one of the glycerol hydroxyl groups are esterified, rather than the older names triglyceride, diglyceride and monoglyceride.

TAGs are the chief constituents of natural fats (solids) and oils (liquids). Fats and oils are sometimes used as terms that are equivalent to TAGs but it is important to remember that natural fats and oils also contain minor proportions of other lipids, for example sterols and carotenoids. The most abundant FAs in natural acylglycerols are palmitic (16:0), stearic (18:0), oleic ($c9$-18:1) and linoleic ($9c,12c$-18:2); plant acylglycerols usually have a relatively higher proportion of UFAs. Coconut and palm kernel oils are exceptions in having a predominance of SFA with chain lengths in the range 8C–12C (see Section 6.1 and Table 2.6). Milk fats have a much higher proportion, depending on species, of short-chain (2C–10C) FAs than other animal fats (see Section 6.1 & Table 2.7). Odd-chain or branched-chain FAs are only minor constituents of acylglycerols. Some seed

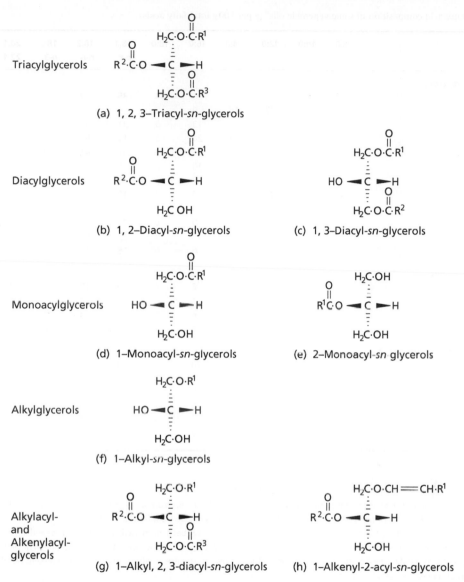

Fig. 2.4 Structures of acylglycerols and alkylglycerols. To emphasize the stereochemistry of the central carbon atom, imagine that this carbon is in the plane of the paper. The groups linked by dotted lines are to be thought of as behind, and those linked by solid lines are to be thought of as in front of the plane of the paper. R represents a long hydrocarbon chain. Examples (a)–(e) illustrate structure of acylglycerols. In (f)–(h), one of the groups has an alkyl (or 'ether') linkage. In example (f), when R^1 is derived from long-chain alcohols corresponding to 16:0, 18:0 and 18:1, the glycerol monoethers are named chimyl, batyl and selachyl alcohols, respectively. In example (h) the group containing a *cis* double bond, $-O-CH=CHR^1$ is sometimes called a 'vinyl ether' and is also found in the class of phospholipids called plasmalogens (see Chapter 4).

oils contain a variety of unusual FAs such as those with oxygen-containing groups and ring systems (see Section 2.1.5).

Alkylglycerols contain hydrocarbon chains linked to glycerol by an ether rather than an ester linkage. Chimyl,

batyl and selachyl alcohols are glycerol monoethers derived from long-chain alcohols corresponding to 16:0, 18:0 and 18:1, respectively (Fig. 2.4f). These are the major naturally occurring glycerol ethers which may be esterified, in addition, with one or two FAs (Figs. 2.4g,h).

Table 2.6 The fatty acid composition of some vegetable oils* (g per 100 g total fatty acids).

	8:0	10:0	12:0	14:0	16:0	18:0	18:1 n-9	18:2 n-6	18:3 n-3	20:1+ 22:1	Others
A. Major edible oil crops											
Cocoa butter					26	34	35	3			2
Coconut (3.0)	7	7	47	18	9	3	6	2			1
Corn (2.1)					13	3	31	52	1		–
Cottonseed (5.0)				1	24	3	19	53			–
Groundnut[a] (4.5)					13	3	37	41	2		4
Olive fruit (2.7)					10	2	78	7	1		2
Palm fruit (32.8)				1	43	4	41	10			1
Palm kernel (seed) (3.8)	4	4	45	18	9	3	15	2			–
Rape (Canola) (16.1)					4	2	56	26	10	2	–
Sesame (0.8)					9	6	38	45	1	1	–
Soybean (32.6)					11	4	22	53	8	1	1
Sunflower (9.1)					6	6	18	69			1
B. Major industrial oils											
Camelina			–	tr	6	3	19	17	32	15	8
Castor[b] (0.5)					1		3	4			92
Linseed (0.6)					6	3	17	14	60		–
Rape (high erucic)					3	1	16	14	10	55	1
C. Other oils and fats of interest											
Avocado					20	1	60	18			1
Cuphea viscosissima	9	76	3	1	3		2	5			1
Evening primrose[c]					7	2	9	72			10
Rice bran					16	2	42	37	1		2
Safflower (high oleic)					6	2	74	16			2
Safflower (high linoleic)					7	3	14	75			1

[a] Also called peanut.

[b] Castor oil contains 90% of ricinoleic acid (see Table 2.4).

[c] Evening primrose oil contains about 10% of γ-linolenic acid (see Table 2.3).

* The vegetable oils listed according to whether their main commercial use is in (A) foods, (B) chemicals. Some crops mainly used in foods also contribute to the chemical industry (e.g. use of palm oil in making 'biodiesel', a vehicle fuel made from fatty acid esters).

Category (C) includes the fat-rich fruit, avocado, two varieties of safflower, which have been much used in nutrition studies and three crops for which there is increasing specialist interest: evening primrose for its high γ-linolenic acid content, *Cuphea* species for their high content of medium-chain fatty acids and rice bran for its good balance of oleic and linoleic acids. The numbers in parentheses after each crop name represent world production in 2010 in millions of tonnes.

Usually both alkyl and alkenyl ethers occur together in varying proportions. They are abundant in the liver oils of marine animals such as sharks, in the central nervous system of mammals and are characteristically elevated in tumours (see Section 10.2). Only very small quantities of alkyl glycerols are found in plants.

MAGs and DAGs can be thought of as metabolic intermediates usually present only in small amounts. However, both have other functions such as in cell signalling where DAG has a role in regulating protein kinases (see Section 8.1.3) and 2-arachidonyl-glycerol is an endocannabinoid (see also Sections 4.6.6 & 8.2).

2.2.2.2 All natural oils are complex mixtures of molecular species

Natural acylglycerols contain a vast range of different FAs. Even if the number of FAs was restricted to three, it would be possible to have 18 different arrangements of these FAs within the TAG molecule. There are, however, many hundreds of FAs and, therefore, the potential number of possible molecular species of TAGs is enormous.

The pioneering work that provided the solid basis for a knowledge of natural TAG structure was that of Hilditch and his associates beginning in 1927. They

Table 2.7 The fatty acid composition of some animal storage fats* (g per 100 g total fatty acids).

	4:0-12:0	14:0	16:0	16:1	18:0	18:1	18:2 n-6	20:1+ 22:1	20:5 n-3	22:6 n-3	total
Adipose tissue											
Cow		3	26	9	8	45	2				93
Human (1)		2	19	7	3	48	13				92
Human (2)		2	20	4	5	39	24				94
Lamb		3	21	4	20	41	5				94
Pig (1)		1	29	3	14	43	9				99
Pig (2)		1	21	3	12	46	16				99
Poultry		1	27	9	7	45	11				100
Milk											
Cow	10	12	26	3	11	29	2				93
Goat	20	11	26	3	10	26	2				98
Egg yolk											
Hen			29	4	9	43	11				96
Fish oils											
Cod (liver)		6	8	10	3	17	3	25	10	11	93
Mackerel (flesh)		8	16	9	2	13	1	26	8	8	91
Herring (flesh)		9	15	8	1	17	1	39	3	3	96
Sardine (flesh)		8	18	10	1	13	1	7	18	9	85

*The figures are rounded to the nearest whole number; therefore the content of an individual fatty acid of less than 0.5% is not recorded. The figures across a row rarely sum to 100% because each sample contains a large number of components, each contributing less than 1%. Thus, samples for ruminants (cow, goat, lamb) contain a wide variety of odd- and branched-chain fatty acids. Hen's egg yolk contains about 1% 20:4n-6, which is not normally found in other storage fats and is not listed individually in this table. Data on human milk composition are presented in Table 6.2. Human adipose tissue (1) is a sample from British adults, whereas human adipose tissue (2) is a sample from Israeli adults, whose diet contains more linoleic acid than the British diet. Pig adipose tissue (1) is a sample taken from pigs given a normal cereal-based diet, whereas pig adipose tissue (2) is a sample from pigs given diets to which have been added oilseeds rich in linoleic acid. Contrast the linoleic acid content from storage fats of simple-stomached animals (pig, poultry, human) with those from ruminants in which much of the polyunsaturates in the animals' diets have been reduced by biohydrogenation in the rumen. Note also that whereas fish oils are mainly of interest because of their content of n-3 PUFAs, cod, herring and mackerel are particularly rich in long-chain monoenes.

had to rely on the painstaking use of fractional crystallization, low pressure fractional distillation and counter-current distribution. At that time they assumed that FAs were randomly distributed among the three positions of TAG. Modern procedures of lipase hydrolysis and separations by different types of chromatography (see Section 1.4.3 & Box 2.2) have revealed that the FAs are not normally distributed in a random fashion. In most TAG, there is a stereospecific arrangement of FAs at the three positions. This is because of the acyl selectivity of the enzymes involved in their biosynthesis. Box 2.2 illustrates how such a stereospecific structure may be determined.

Examples of common fats with important stereospecific FA distributions are:

1 Milk fats, in which the characteristic short-chain FAs accumulate at position sn-3 (for example, butyrate (4:0) in cow's milk fat). Long-chain SFAs tend to occur at position sn-1 and UFAs at position sn-2. A notable exception is human milk fat in which palmitic acid occurs mainly at position 2.

2 Animal depot (adipose tissue, see Sections 6.1.1 & 9.1) fats usually have SFAs at position 1 with short-chain and UFAs at position 2. A notable exception is lard (pig adipose tissue) in which palmitic acid is mainly located at position 2. Position 3 seems to have a more random population, although PUFA tend to concentrate at position 3 in mammals, but at position 2 in fish and invertebrates.

3 Certain seed oils contain acetate residues that occur only at position 3. This inherent asymmetry may lead to optical activity, which although extremely small, is measurable, especially in the extreme cases of acylglycerols containing acetate.

Box 2.2 Stereospecific structural analysis of triacylglycerols

By combining chromatographic separations with specific enzymic hydrolyses of the separated triacylglycerol components, it is possible to identify the fatty acids on each of the three positions, *sn*-1, -2 and -3, which are stereochemically distinct. An initial separation by argentation TLC (see diagram below and Chapter 1) separates triacylglycerol species according to the total degree of unsaturation of the molecule. A more recent development is argentation HPLC, which is capable of some extremely good resolutions and lends itself more readily to quantitative analysis than TLC. In the example shown, palm (A), olive (B), groundnut (C) and cottonseed (D) oils are separated on silica gel impregnated with silver nitrate and developed in isopropanol:chloroform, 1.5:98.5 (v/v). Spots in marker lanes are located by spraying with 50% sulphuric acid and charring. The numbers along the right-hand side indicate the total number of bonds per molecule in each separated species. Areas of gel corresponding to each molecular species are removed from lanes *that are not sprayed with sulphuric acid* and the triacylglycerols eluted with diethyl ether. They are then treated with lipases to liberate fatty acids from specific positions as shown in the diagram. In this example, the fatty acid composition at position 3 is determined by difference. A slight disadvantage of methods involving lipase hydrolysis is that they may not always yield a random sample of fatty acids from any one position. Thus very long chain polyunsaturated fatty acids characteristic of fish oils (e.g. 20:5, 22:6) are more slowly hydrolysed, whereas short- and medium-chain fatty acids (4–10C) are more rapidly hydrolysed than normal chain length fatty acids (12–18C).

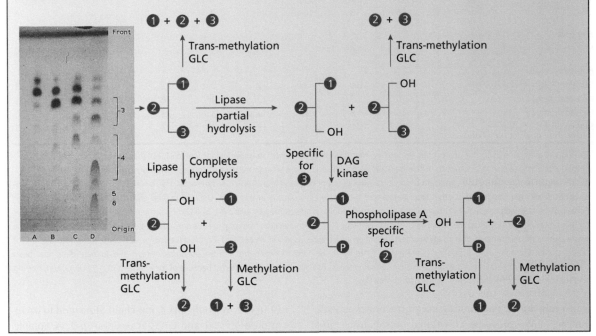

4 Older varieties of rapeseed oil, which contain the unusual FA, erucic acid (22:1*n*-9), have this FA exclusively at positions 1 and 3, but none at position 2.

There are many other examples of stereospecific distribution arising because of the selectivity of the various acyltransferase enzymes responsible for the esterification of FAs at each of the glycerol hydroxyl groups as discussed in Section 4.1.1.

2.2.2.3 General comments about storage triacylglycerols in animals and plants

The TAGs of animals are mainly stored in adipose tissue but can accumulate in other tissues either normally or under pathological conditions (see Sections 5.5, 9.1 & 10.4.1.3). Their FA composition will reflect both the animal's diet but also the contribution of metabolism. For example, the very long-chain (>18C) PUFAs which provide the health benefits of fish oils (see Sections 10.3.2, 10.5.2.5, 10.5.3, 10.5.4.3 & 11.4.3) are almost entirely produced by algae on which the fish feed. Some typical FA compositions of animal storage fats are shown in Table 2.7.

Although plants can accumulate TAGs in various tissues (such as leaves) under stress conditions, the only major storage site is the seed (or fruit) of oleaginous species (see also Section 9.3). The seed TAG is a source of

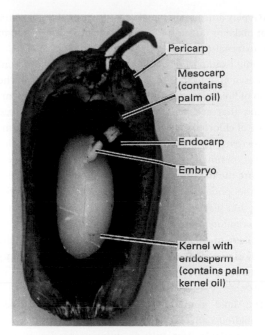

Fig. 2.5 A section through an oil palm fruit. Note the location of two quite distinct types of oil, palm oil and palm kernel oil. Reproduced with kind permission of Dr LH Jones.

energy during germination while fruit oils can act as an attractant for animals, and, hence, aid dispersal. Although most (85%) plant fats and oils are used for human or animal food, there is an increasing interest in plants as renewable sources of chemicals to replace the dwindling supplies of petroleum. This has even included biofuel – where hasty legislation by some governments to include biofuels in the overall fuel supply has led to competition for agricultural land between biofuel and food production. Nevertheless, production of biofuel on marginal land using specialized crops (e.g. camelina (*Camelina sativa* or 'false flax') or in nonagricultural areas (e.g. algae) is being explored and may be used in the future (see also Section 11.2.3).

Examples of plant oils are shown in Table 2.6 and a section through an oil palm fruit (the world's major source of edible oil) is shown in Fig. 2.5. Further details of the use of storage lipids are given in Chapters 9 and 11.

2.2.3 Wax esters

These esters of long-chain FAs and long-chain fatty alcohols provide energy sources, as an alternative to TAGs, in some animals, plants and microorganisms. Wax esters have the general formula R^1COOR^2 and are esters of long-chain FAs (R^1COOH) with long-chain fatty alcohols (R^2OH). They occur in some species of bacteria, notably the mycobacteria and corynebacteria. However, they are mainly important for providing a form of energy reserve that is an alternative to TAGs in certain seed oils (e.g. jojoba), the oil of the sperm whale, the flesh oils of several deep-sea fish (e.g. orange roughy) and in zooplankton. Thus, a major part of jojoba oil consists of wax esters, 70% of sperm whale oil and as much as 95% of orange roughy oil. In global terms, however, the zooplankton are most important – they may synthesize and store massive amounts of wax esters, as much as 70% of dry weight.

The component FAs and alcohols of wax esters have an even number of between 10 and 30 carbon atoms. Normally, they are straight-chain saturated or mono-unsaturated chains of 16C–24C. Branched and odd-numbered chains are rare in both constituents except for the bacterial waxes.

Wax esters from cold-water organisms exist as oils down to about 0 °C and their physicochemical properties differ from those of the TAGs in that they are more hydrophobic and less dense, which may help to provide greater buoyancy for marine animals.

2.2.4 Surface lipids include not only wax esters but a wide variety of lipid molecules

Although true waxes are esters of long-chain FAs and fatty alcohols, the term wax is often used to refer to the whole mixture of lipids found on the surfaces of leaves, or on the skin or the fur of animals. These surface lipids comprise a complex mixture including wax esters, long-chain hydrocarbons, NEFAs, alcohols and sterols. The surface lipids are responsible for the water-repellent character of the surface and are important in conserving the organism's water balance as well as providing a barrier against the environment. Some components, notably the long-chain NEFAs, have antimicrobial properties and this also contributes to their protective role in a physiological as well as a physical way.

The hydrocarbon components of the surface lipids (see Section 5.6.1) have very long chain lengths (around 30C) and are formed from long-chain FAs (see Section 4.4.4) by two sorts of mechanisms. Early experiments suggested that FAs could be decarboxylated to yield hydrocarbons. Now it is thought that odd-chain alkanes are derived from aldehyde intermediates by a novel decarbonylation reaction (CO released).

2.3 Membrane lipids

2.3.1 General introduction

One of the most important general functions for lipids is their role as constituents of cellular membranes. These membranes not only separate cells from the external environment, but also compartmentalize cells and provide a special milieu for many important biochemical processes.

Lipids form the basis of the bilayer structure of natural membranes. For eukaryotic cells there are different membranes associated with organelles as well as the bounding cytosolic (plasma) membrane. Each of these membranes has a characteristic lipid composition which is maintained within rather tight limits. In some cases, the characteristic lipid composition has been linked to specific functions in that membrane and presumably, that is why each membrane is different. Even in prokaryotes, internal membranes (when present) have a distinct lipid composition compared to the cytosolic membrane (see Sections 5.3 & 5.4).

Almost all membrane lipids are amphipathic molecules with hydrophobic and hydrophilic regions on the same molecule. This allows amphipathic lipids spontaneously to form bilayers under many conditions. Phospholipids and glycolipids predominate and nonpolar lipids such as

TAGs play little part in membrane structure. In organisms that make or utilize sterols, these tend to be enriched in the external (cytoplasmic) membrane.

Some examples of the lipid compositions of different tissues or membranes are given in Tables 2.8 and 2.9. The ratio of lipid to protein can differ considerably. In membranes that are particularly metabolically active (e.g. those of chloroplasts or mitochondria) proteins predominate (over 70%), while myelin, which has a major insulation function, contains 75% lipids. Phosphoglycerides are the major membrane lipids in most membranes but the photosynthetic membranes of cyanobacteria, algae and plants, which contain glycosylglycerides, are an exception. Because such membranes are the most abundant in the world, glycosylglycerides are actually the most prevalent membrane lipids – even though many mammalian-focused scientists have never heard of them!

The lipid composition of microbial membranes, although usually dominated by phosphoglycerides, is highly variable. Not only do different species of, say, Gram negative bacteria show a huge range of compositions (Table 2.9) but the lipids of microbes are changed by growth conditions. Factors that have a major influence are temperature, oxygen availability, nutrient supply and stages of development.

Table 2.8 The lipid composition of different eukaryotic tissues and their membranes.

	Animals				Plants		Yeast	Fungus
	Rabbit brain	Bovine liver	Rat liver mitochondria	Rat liver plasma membrane	Chloroplast thylakoid	Barley root ER	(S. cerevisiae)	(Neurospora crassa)
PtdCho	31	54	41	46	5	58	39	45
Plasmal.[b]	1	2						
PtdEtn	13	9	34	25	–	21	30	28
Plasmal.[b]	23	4						
PtdIns	3	8	7	7	1	10	16	10
PtdSer	16	4	1	4	–	1	5	7
DPG	2	4	15	1	–	0	5	9
PtdGro	–				10	6	1	–
Sphingo.	12	6	2	17	–	–	–	–
Others[a]					85		4	

In some membranes (e.g. plasma membranes) sterols may be significant constituents.

[a] In chloroplast thylakoids or the photosynthetic membranes of cyanobacteria, glycosylglycerides are the major components rather than phospholipids (see Tables 2.11 and 2.13). Pigments are also present.

[b] The PtdCho, PtdEtn (and PtdSer) fractions may contain plasmalogen derivative in animals but these are not usually separated during analysis except where they may be major components, as in the examples shown.

PtdCho, phosphatidylcholine; Plasmal., plasmalogen derivative; PtdEtn, phosphatidylethanolamine; PtdIns, phosphatidylinositol; PtdSer, phosphatidylserine; DPG, diphosphatidylglycerol (cardiolipin); PtdGro, phosphatidylglycerol; Sphingo., sphingomyelin.

Table 2.9 The phosphoglyceride composition of different bacterial cytosolic membranes.

	% total phosphoglycerides Gram positive			
	Microccus roseus	*Staphylcoccus aureus*	*Streptococcus agalactia*[a]	*Bacillus cereus*
PtdEtn	0	0	0	53
PtdGro	74	57	23	31
DPG	26	5	45	8
LysylPtdGro	–	38	14	–
	Gram negative			
	Escherichia coli	*Salmonella typhimureum*	*Pseudomonas cepacia*	*Psychrobacter oleovorans* [b]
PtdEtn	82	60	87	36
PtdGro	6	33	13	19
DPG	12	7	–	45

Only major components are listed. PtdEtn, phosphatidylethanolamine; PtdGro, phosphatidylglycerol; DPG, diphosphatidylglycerol (cardiolipin); LysylPtdGro, *sn*-3 lysyl derivative of PtdGro (i.e. the amino acid lysine linked to the *sn*-3 position of the head group glycerol).
[a] Phosphoglycolipids (16%) major components
[b] Used to be called *Micrococcus cryophilus*

In the following section we will describe the most common types of membrane lipids but it should be borne in mind that there are many lipids with a limited or specialized distribution. Indeed, new molecules are continually being discovered!

2.3.2 Glycerolipids

Three important classes of membrane lipid are widely distributed – glycerolipids, sphingolipids and sterols. Of these, glycerolipids are quantitatively by far the most important group. They can be conveniently divided into two main subgroups – those containing phosphorus (phosphoglycerides) and those without phosphorus but containing a sugar constituent (glycosylglycerides). Phosphoglycerides dominate in higher animals whereas glycosylglycerides are important in the thylakoids of chloroplasts, algae and cyanobacteria.

2.3.2.1 Phosphoglycerides are the major lipid components of most biological membranes

The stereochemistry of phosphoglycerides was discussed in Section 1.3.5. The phosphoglycerides comprise a very widespread and diverse group of structures. In most membranes they are the main lipid components and, indeed, the only general exceptions to this statement are the photosynthetic membranes of plants, algae and cyanobacteria and membranes of the archaea.

Usually, phosphoglycerides contain FAs esterified at positions *sn*-1 and -2 of glycerol. They are, thus, diacylphosphoglycerides. These lipids are named after the moiety which is attached to the phosphate esterified at

position *sn*-3 of glycerol. Thus, the compounds can be thought of as derivatives of DAG in which the hydroxyl on carbon atom 3 is esterified with phosphoric acid, which in turn is esterified with a range of molecules – organic bases, amino acids, alcohols.

The simplest phosphoglyceride contains only phosphoric acid attached to DAG and is called phosphatidic acid (PtdOH; Table 2.10). Where additional 'X' groups are esterified to the phosphate moiety, the lipids are called phosphatidyl-X. Major types of such diacylphosphoglycerides are shown in Table 2.10 where relevant comments about their distribution and properties are made also.

2.3.2.2 Phosphonolipids constitute a rare class of lipids found in a few organisms

In 1953, Rouser and his associates first identified a phosphonolipid in biological extracts from a sea anemone. Subsequent work has shown that such lipids, which contain a C—P bond, are significant constituents of lower animals such as molluscs, coelenterates and protozoa. Phosphonolipids can also be detected in bacteria and mammals, but only in very low quantities Two types of glycerophosphonolipids have been found (Fig. 2.6) and there are also other structures that contain a sphingosine (and, sometimes, a galactosyl) residue instead of glycerol. Usually, organisms tend to accumulate phosphonolipids based on either glycerol or on sphingosine rather than both types. In protozoa such as *Tetrahymena*, the glycerophosphonolipids are concentrated in the ciliary membranes. This is perhaps because the phosphonolipids are particularly resistant to chemical as well as enzymic attack.

Table 2.10 Structural variety of different diacylglycerophospholipids.

X	Name of phospholipid	Source	Remarks
H	Phosphatidic acid (PtdOH)	Animals, higher plants, microorganisms	Only small amounts normally found. Main importance as a biosynthetic intermediate. Also a signalling molecule.
$OH \cdot CH_2CH_2N^+(CH_3)_3$ Choline	Phosphatidylcholine (lecithin) (PtdCho)	Animals; first isolated from egg yolks; higher plants; rare in microorganisms.	Most abundant animal phospholipid and most prevalent in plant membranes except chloroplast thylakoids
$OH \cdot CH_2CH_2NH_3^+$ Ethanolamine	Phosphatidylethanolamine (PtdEtn)	Animals; higher plants; microorganisms.	Widely distributed. N-acyl derivatives found in animals and plants can be metabolized to the bioactive N-acylethanolamines (endocannabinoids).
$OH \cdot CH_2 \cdot CH \cdot NH_3^+$ $\quad\quad\quad\;\; \vert$ $\quad\quad\quad COO^-$ Serine	Phosphatidylserine (PtdSer)	Animals; higher plants; microorganisms.	Widely distributed but in small amounts. Serine as L isomer. Lipid usually in salt form with K^+, Na^+, Ca^{2+}.
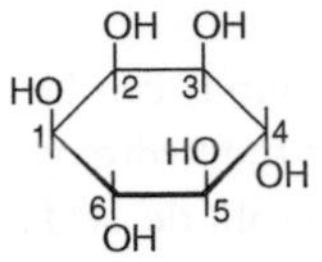 *Myo*-inositol (structure)	Phosphatidylinositol (PtdIns)	Animals, higher plants, microorganisms	The natural lipid is found as a derivative of *myo*-inositol-1-phosphate only. The major inositol-containing phosphoglyceride (about 80% of total inositol lipids in animals).
Inositol-4-phosphate	Phosphatidylinositol 4-phosphate (PtdIns 4P)	Animals; small amounts in yeast; plants.	Mainly nervous tissue, but also plasma membranes of other cells.
Inositol-4, 5-*bis*phosphate	Phosphatidylinositol 4.5- *bis*phosphate (PtdIns(4, 5) P_2)	Animals; small amounts in yeast; plants.	Distribution as above. Both compounds have very high rates of turnover. Note: many other inositol phosphoglycerides present in small amounts, e.g. 3-phosphate derivatives.
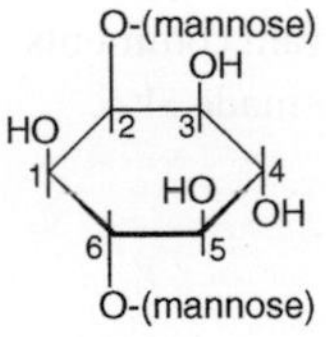 Inositol mannoside (structure)	Phosphatidylinositol mannoside X = 0, monomannoside X = 1, dimannoside etc	Microorganisms (*M. phlei, M. tuberculosis*).	
$CH_2OH \cdot CHOH \cdot CH_2OH$ Glycerol	Phosphatidylglycerol (PtdGro)	Mainly higher plants and micro-organisms	Free' glycerol has opposite stereochemical configuration to the acylated glycerol, i.e. 1,2-diacyl-1-*sn*-glycerol. Only significant phosphoglyceride in thylakoids of cyanobacteria and chloroplasts. Probably most abundant phospholipid because of its prevalence in photosynthetic membranes. The glycerol head group can have an amino acid attached to its terminal hydroxyl. Such derivatives are found in seeds and lysyl-PG is a major lipid of some bacteria.
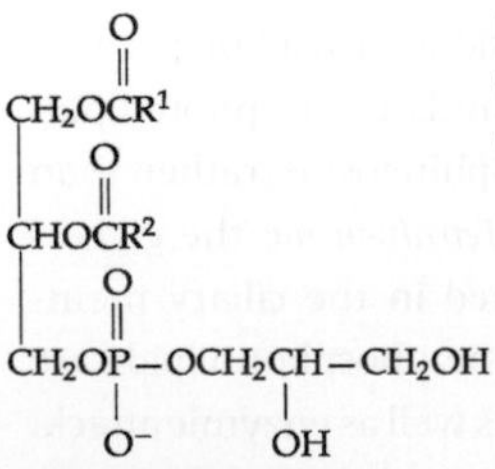 Phosphatidylglycerol (structure)	Diphosphatidylglycerol (*bis*phosphatidylglycerol, cardiolipin)	Animal; higher plants; microorganisms.	Major component of many bacteria, localized in inner mitochondrial membrane of eukaryotes.

Myo-inositol structure:

OH OH (positions 2, 3); HO at 1; HO, OH at 4; OH at 5, 6

Inositol mannoside structure:

O-(mannose), OH (positions 2, 3); HO at 1; HO, OH at 4; O-(mannose) at 5, 6

Phosphatidylglycerol structure:

$$CH_2OCR^1$$
$$\vert$$
$$CHOCR^2$$
$$\vert$$
$$CH_2OP-OCH_2CH-CH_2OH$$

with O and OH substituents

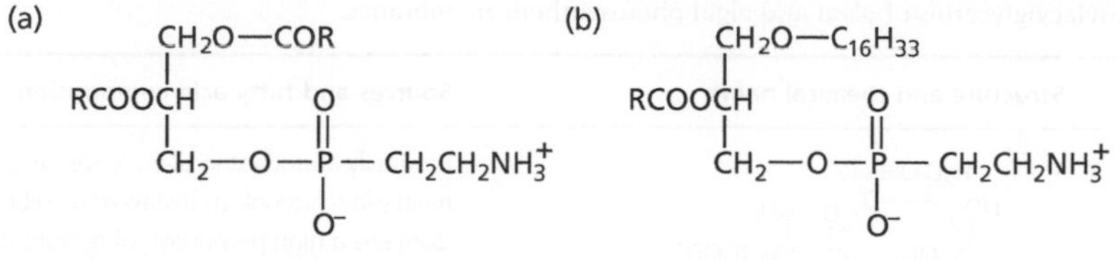

(a)   (b)

$$CH_2O\!-\!COR \qquad\qquad\qquad CH_2O\!-\!C_{16}H_{33}$$

$$RCOOCH \qquad\qquad\qquad\qquad RCOOCH$$

$$CH_2\!-\!O\!-\!\underset{\underset{O^-}{|}}{\overset{\overset{O}{\|}}{P}}\!-\!CH_2CH_2NH_3^+ \qquad CH_2\!-\!O\!-\!\underset{\underset{O^-}{|}}{\overset{\overset{O}{\|}}{P}}\!-\!CH_2CH_2NH_3^+$$

Fig. 2.6 Glycerophosphonolipids: (a) phosphatidylethylamine; (b) an analogue, which is a derivative of chimyl alcohol rather than a diacylphosphonolipid. It is named 1-hexadecyl-2-acyl-*sn*-glycero(3)-2-phosphonoethylamine.

2.3.2.3 Glycosylglycerides are particularly important components of photosynthetic membranes

Glycolipids that are based on glycerol have been found in a wide variety of organisms. However, whereas in animals they are found only in very small quantities, they are major constituents of some microorganisms and are the main lipid components of the photosynthetic membranes of algae (including blue-green algae or cyanobacteria) and plants (Table 2.11). Their structure is analogous to that of glycerophospholipids with the sugar (s) attached glycosidically to position *sn*-3 of glycerol and FAs esterified at the other two positions.

Since the membranes of higher plant chloroplasts are the most prevalent on land, and the photosynthetic membranes of marine algae are the most common in the seas and oceans, it follows that these glycosylglycerides are the most abundant membrane lipids in Nature, in spite of the sparse attention paid to them in most standard biochemistry textbooks! The two galactose-containing lipids – monogalactosyldiacylglycerol (MGDG) and digalactosyldiacylglycerol (DGDG) – represent about 40% of the dry weight of photosynthetic membranes of higher plants. Position 1 of the galactose ring has a β-link to glycerol whereas, in DGDG, there is an α, 1-6 bond between the sugars. Galactose is almost the only sugar found in the glycosylglycerides (specifically the glycosyldiacylglycerols of higher plants) but other sugars, such as glucose, may be found in algae, particular marine species. In bacteria several combinations of residues may be found in glycosyldiacylglyerols (Table 2.12). The most common combinations are two glucose, two galactose or two mannose residues linked α,1-2 or β,1-6. Such glycosylglycerides do not form a large proportion of the total lipids in bacteria, but are found more frequently in the Gram-positives or photosynthetic Gram-negatives. In addition, bacteria may contain higher homologues with up to seven sugar residues.

Apart from the galactose-containing lipids, a third glycosylglyceride is found in chloroplasts (Table 2.11). This is the plant sulpholipid. It is more correctly called sulphoquinovosyldiacylglycerol (SQDG) and contains a sulphonate constituent on carbon 6 of a deoxyglucose residue. This sulphonic acid group is very stable and also highly acidic so that the plant sulpholipid is a negatively charged molecule in Nature. Although this sulpholipid occurs in small amounts in photosynthetic bacteria and some fungi, it is really characteristic of the photosynthetic membranes of chloroplasts and cyanobacteria.

All three chloroplast glycolipids usually contain large amounts of α-linolenic acid (18:3 – Table 2.13) in plants. In fact, MGDG may have up to 97% of its total acyl groups as this one component in some plants! The reason for this exceptional enrichment is not known although speculations have been made. In algal chloroplasts, MGDG is highly enriched with a variety of PUFA, but with little α-linolenate. Moreover, a unifying theory connecting α-linoleate with photosynthesis is not possible since many cyanobacteria do not make this acid and marine algae contain little. In the plant sulpholipid, the most usual combination of acyl groups is palmitate/α-linolenate. Interestingly, unlike animal lipids (which tend to have SFAs/UFAs at the *sn*-1/*sn*-2 positions) most palmitate is esterified at position 2 and most of the α-linolenate at position 1. This distribution of FAs stems from the special features of plant FA metabolism (see Section 3.1.8.2) and because SQDG is assembled via the so-called 'prokaryotic pathway' of metabolism. Since the position at which SFAs and UFAs are attached to the glycerol has a marked effect on their melting properties, FA distribution may have functional significance. It has also been speculated that the different distribution of acyl moieties between plants and animals may be related to specific interactions with membrane proteins.

Bacteria contain a number of phosphoglycolipids. These compounds are confined to certain types of Gram-positive

Table 2.11 Major glycosylacylglycerols of plant and algal photosynthetic membranes.

Common name	Structure and chemical name	Sources and fatty acid composition
Monogalactosyl-diacylglycerol (MGDG)	 1,2-diacyl-[β-D-galactopyranosyl-(1'→3)-sn-glycerol	Especially abundant in plant leaves, algae and cyanobacteria; mainly in (chloroplast) thylakoid membranes. Contains a high proportion of polyunsaturated fatty acids. In most plants MGDG contains about 90% 18:3 but in some species (e.g. spinach) it contains about 25% 16:3 and 70% 18:3 Also found in the central nervous systems of several animals in small quantities.
Digalactosyldiacylglycerol (DGDG)	 1,2-diacyl-[α-D-galactopyranosyl-(1'→6')-β-D-galactopyranosyl-(1'→3)]-sn-glycerol	Usually found together with MGDG in chloroplasts of higher plants and algae and thylakoids of cyanobacteria. Not quite so abundant as MGDG. Also has high proportion of polyunsaturated fatty acids, especially 18:3 in plants. In both lipids the glycerol has the same configuration as in the phospholipids.
Plant sulpholipid (sulphoquinovosyl-diacylglycerol; SQDG)	 1,2-diacyl-[6-sulpho-α-D-quinovopyranosyl-(1'→3)-sn-glycerol. [D-quinovose is 6-deoxy-D-glucose. Note the carbon–sulphur bond]	Usually referred to as a 'sulpholipid' as distinct from a 'sulphatide', which is reserved for cerebroside sulphates. Typical lipid of chloroplast membranes but present elsewhere in some marine algae. Also found in cyanobacterial photosynthetic membranes and, to a lesser extent, in purple photocynthetic bacteria. Contains more saturated fatty acids (mainly palmitic) than the galactolipids, e.g. spinach leaf sulpholipid has 39% 16:0, 7% 18:2, 53% 18:3.

Table 2.12 Some glycosylglycerides found in bacteria.

Glyceride	Structure of glycoside moiety	Occurrence
Monoglucosyldiacylglycerol	α-D-Glucopyranoside	*Pneumonoccus, Mycoplasma*
Diglucosyldiacylglycerol	β-D-Glucopyranosyl-(1→6)-O-β-D-glucopyranoside	*Staphylococcus*
Diglucosyldiacylglycerol	α-D-Glucopyranosyl-(1→2)-O-α-D-glucopyranoside	*Mycoplasma, Streptococcus*
Dimannosyldiacylglycerol	α-D-Mannopyranosyl-(1→3)-O-α-D-mannopyranoside	*Micrococcus lysodeikticus* (now *Psychrobacter oleovorans*)
Galactofuranosyldiacylglycerol	β-D-Galactofuranoside	*Mycoplasma, Bacteroides*
Galactosylglucosyldiacylglycerol	α-D-Galactopyranosyl-(1→2)-O-α-D-glucopyranoside	*Lactobacillus*
Glucosylgalactosylglucosyldiacylglycerol	α-D-Glucopyranosyl-(1→6)-O-α-D-galactopyranosyl-(1→2)-O-α-D-glucopyranoside	*Lactobacillus*

Table 2.13 Fatty acid compositions of glycosylglycerides in two plants.

Plant leaf		16:0	16:3	18:1	18:2	18:3
		(% total fatty acids)				
Spinach	MGDG	trace	25	1	2	72
('16:3-plant')	DGDG	3	5	2	2	87
	SQDG	39	–	1	7	53
Pea	MGDG	4	–	1	3	90
('18:3-plant')	DGDG	9	–	3	7	78
	SQDG	32	–	2	5	58

'16:3-plants' contain hexadecatrienoate in their monogalactosyl-diacylglycerol while '18:3-plants' contain α-linolenate instead. The reason for this is provided by the differences in fatty acid metabolism between these two types of plants (see Section 3.1.8.2). For abbreviations see Table 2.11.
16:3 is all *cis* 7, 10, 13-hexadecatrienoic acid, 18:1 is oleic acid, 18:2 is linoleic acid and 18:3 is α-linolenic acid (see Tables 2.2 & 2.3).

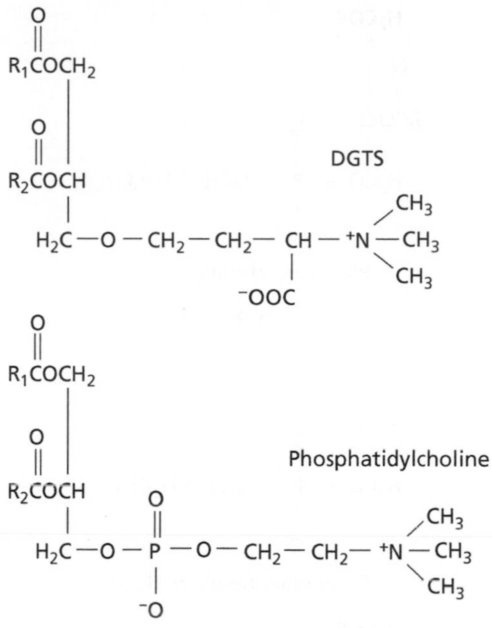

Fig. 2.7 Comparison of the structure of the betaine lipid, DGTS with phosphatidylcholine. Abbreviation: DGTS, 1,2-diacylglycerol-3-O-4'(N,N,N-trimethyl)-homoserine.

organisms such as the streptococci or mycoplasmas. In addition, different species of algae, bacteria and archaea contain various sulphoglycolipids – usually with the sulphur present in a sulphate ester attached to the carbohydrate moiety. Lipids containing taurine seem to be significant components of some protozoa.

2.3.2.4 Betaine lipids are important in some organisms

Betaine ether-linked glycerolipids are naturally occurring lipids which are important in many lower plants (e.g. bryophytes), algae, fungi and protozoa. Because of their structure (Fig. 2.7) and properties, they are often envisaged as substituting for phosphatidylcholine (PtdCho – Table 2.10) in membranes. Several structures have been identified but diacylglyceryltrimethylhomoserine (DGTS) is the most common. Typically, in those species containing it, DGTS represents 10–20% of all membrane lipids but, in some mosses, it may be around 40% of the total.

2.3.2.5 Ether-linked lipids and their bioactive species

Apart from the betaine ether-linked glycerolipids (see Section 2.3.2.4), there are many lipids that have ether links to carbon position *sn*-1 (and sometimes position *sn*-2) of the glycerol backbone. In mammals, the ether linkages are almost exclusively found in the choline or ethanolamine glycerolipid classes. If the ether link is unsaturated (Fig. 2.8),

the lipid is termed a plasmalogen. Indeed, of the ether-linked phosphoglycerides, ethanolamine plasmalogen is the most widespread. It is especially abundant in nervous tissues and in white blood cells and platelets. The ether lipids are not confined to mammals but are also found in lower animals, some plants and many bacteria.

As with diacylphosphoglycerides, the ether-linked phospholipids contain different combinations of aliphatic species at the *sn*-1 position and acyl groups at the *sn*-2 position. However, plasmalogens are noted in mammals for containing large amounts of arachidonate or DHA at the *sn*-2 position. They form important sources of these acids for signalling purposes (see Section 3.5).

Particular interest has recently been paid to the biologically active platelet activating factor (PAF) (Fig. 2.8) and its derivatives (see Sections 4.5.11 & 10.3.7.2). PAF has a similar structure to the ether derivative of PtdCho (plasmanylcholine) except that the FA at the *sn*-2 position is replaced by an acetyl group.

Although they are neither phospho- nor glycosylglycerides, the nonpolar phytanyl ethers are major membrane lipids in the archaea (see Section 2.3.5).

H₂COR

O
‖
RCOCH   O
            ‖
H₂CO —— P —— OCH₂CH₂N⁺(CH₃)₃
            |
            O⁻

Plasmanyl choline

H₂COCH ══ CHR

O
‖
RCOCH   O
            ‖
H₂CO —— P —— OCH₂CH₂N⁺H₃
            |
            O⁻

Ethanolamine plasmalogen

H₂COR

O
‖
CH₃COCH   O
            ‖
H₂CO —— P —— OCH₂CH₂N⁺(CH₃)₃
            |
            O⁻

Platelet activating factor (PAF)

Fig. 2.8 Some important ether lipids.

2.3.3 Sphingolipids

The lipids described in Section 2.3.2 were based on a glycerol backbone. However another important group of acyl lipids have sphingosine-based structures (Fig. 2.9).

Both glycolipids and phospholipids are found, with some compounds capable of dual classification, e.g. phospholipids containing sugar residues. More than 300 different types of complex sphingolipids have been reported and this does not take account of variations in the ceramide backbone! Sphingolipids are also found in essentially all animals, plants and fungi but are infrequent in bacteria and archaea. Generally, they tend to be concentrated in the outer leaflet of the cytosolic membrane where they are often associated with membrane rafts (see Section 5.3.3).

Many types of sphingolipid bases that are found in Nature and some important ones are illustrated in Fig. 2.10. They are all long-chain amino alcohols but differ from each other in their chain lengths (12C–22C), the number of hydroxyl groups, the number and position of double bonds and whether they contain branched (methyl) groups. Both the amino and the alcohol moieties can be substituted to produce the various sphingolipids (Fig. 2.9).

Figure 2.10 includes some sphingosine bases common in organisms other than mammals. Thus, 14C-sphingoid bases are common in insects, those with extra double bonds in plants and methyl branches in fungi. Three structures constitute the main bases in animals with sphinganine and sphingosine being present in the sphingomyelin of most tissues, while a trihydroxylated derivative of sphinganine, called phytosphingosine (abbreviated t18:0 in Fig. 2.10) is found in kidney. This base is also prominent in 'phytoglycolipid', which is a sphingolipid exclusive to plants that contains inositol as well as several other monosaccharide residues. In the yeast *Saccharomyces cerevisiae* (a useful experimental model organism), the initial ceramide compounds can accept inositol and mannose residues to form several prevalent sphingolipids.

Attachment of an acyl moiety to the amino group yields a ceramide (Fig. 2.9). This acyl link is resistant

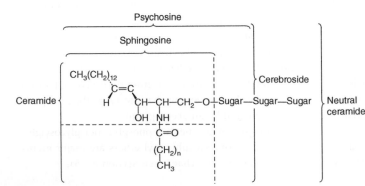

Fig. 2.9 Basic structure of sphingolipids.

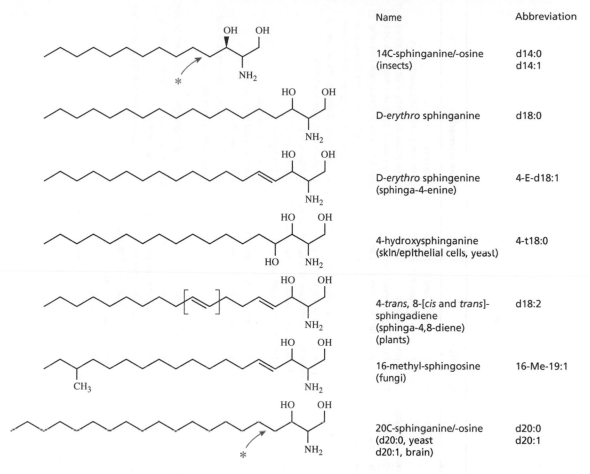

Name	Abbreviation
14C-sphinganine/-osine (insects)	d14:0 d14:1
D-*erythro* sphinganine	d18:0
D-*erythro* sphingenine (sphinga-4-enine)	4-E-d18:1
4-hydroxysphinganine (skin/epithelial cells, yeast)	4-t18:0
4-*trans*, 8-[*cis* and *trans*]-sphingadiene (sphinga-4,8-diene) (plants)	d18:2
16-methyl-sphingosine (fungi)	16-Me-19:1
20C-sphinganine/-osine (d20:0, yeast d20:1, brain)	d20:0 d20:1

Fig. 2.10 Structures of major animal sphingolipid bases and examples from other eukaryotes. *saturated or unsaturated bond.

to alkaline hydrolysis and, therefore, can be easily distinguished from the O-esters found in glycerol-based acyl lipids. The simplest glycosphingolipids are the monoglycosylceramides (also called cerebrosides). In animals the galactosylcerebroside (GalCer) is the most common. Further attachment of hexosides to the glucosylcerebroside (GlcCer) yields more complex glycosylsphingolipids (often referred to as 'neutral' ceramides) (Table 2.14). These are usually written with a shorthand nomenclature, e.g. GlcGalGlcCer would be glucosyl-(1-4)galactosyl-(1-4)glucosyl-ceramide. The deacylated product of galactosylcerebroside, O-sphingosylgalactoside is called psychosine.

The mammalian glycosphingolipids usually have a glucose linked to the ceramide. Although, in theory, there is a huge number of possible structures, Nature

has been kind to biochemists and certain combinations are common! These are called the root structures. A good example is shown in Fig. 2.11 where all the gangliosides (sphingolipids that contain sialic acid groups, illustrated below) have the same glycosyl residues in order. Were it not for this, the nine common monosaccharides could give rise to nearly 20 million different tetra-hexosides!

Some glycosphingolipids contain one or more molecules of sialic acid linked to one or more of the sugar residues of a ceramide oligosaccharide. These lipids are called gangliosides (Table 2.15). Sialic acid is N-acetyl neuraminic acid (NANA or NeuAc; Fig. 2.12). Glycolipids with one molecule of sialic acid are called monosialogangliosides; those with two sialic residues are disialogangliosides and so on. (Instead of an N-acetyl group, other modifications of the sialic acid are possible.)

Table 2.14 The structure of some glycosyl ceramides.

Accepted nomenclature	Old trivial name	Structure	Description	General remarks
MONOGLYCOSYLCERAMIDES	Cerebroside	Cer-gal Cer-glc In all glycosyl ceramides and gangliosides there is an *O*-glycosidic linkage between the primary hydroxyl of sphingosine and C-1 of the sugar.	'Cerebroside' originally used for galactosyl ceramide of brain but now widely used for monoglycosyl ceramides. Sugar composition depends largely on tissue. Brain cerebroside has mainly galactoside while serum has mainly glucose. In animals the highest concentration is in brain. Monogalactosyl ceramide is largest single component of myelin sheath of nerve. Intermediate concentrations in lung and kidney. Also found in liver, spleen, serum with trace amounts in almost all tissues examined.	*Fatty acids:* cerebrosides containing galactose are characterized by having large concentrations of (a) hydroxy acids, (b) very long-chain odd and even fatty acids, in comparison with other lipids. The hydroxy acids include α-hydroxy acids formed by α-oxidation. This oxidation mechanism is probably also responsible for odd-chain acids. Typical acids are: 22:0 (behenic); 24:0 (lignoceric); 24:1 (nervonic); α-OH-24:0 (cerebronic)
Sulphatide	Cerebroside sulphate	Cer-gal-3-sulphate	Very generally distributed like cerebrosides. Fatty acid and base composition similar.	Galactose usual monosaccharide. High in brain, kidney. Involved in neuronal cell differentiation and leukocyte adhesion to selectins.
DIGLYCOSYL CERAMIDES (a) Lactosyl ceramide	Cytosides Cytolipin H	Cer-glc-(4 ← 1)-gal	Major diglycosyl ceramide. Widely distributed. Key substance in glycosyl ceramide metabolism. It may accumulate, but as it is a precursor for both ceramide oligosaccharides or gangliosides it may be present in trace quantities only.	Base: d18:1 (sphingosine) is dominant. Smaller amounts of d18:0 (sphinganine) (especially leaves, wheat flour). Hydroxylated bases (e.g. t18:0) also occur (see Fig. 2.10).
(b) Digalactosyl ceramide		Cer-gal-(4 ← 1)-gal	Minor diglycosyl ceramide. Found especially in kidney (human and mouse).	

TRIGLYCOSYL CERAMIDES

| Digalactosyl glucosyl ceramide | Cer-glc(4←1)-gal-(4←1)-gal | Source: kidney, lung, spleen, liver. Most analyses have been on human tissue. |

TETRAGLYCOSYL CERAMIDES

(a) Aminoglycolipid globoside	Cer-glc(4←1)-gal-(4←1)-gal-(3→1)-β-N-acetylgalactosamine	Most abundant lipid in human erythrocyte stroma. acetyl galactosamine So-called 'Forssman Antigen'.
(b)	Cer-glc(4←1)-gal-(4←1)-gal-(3→1)-α-N-acetyl galactosamine	
(c) Asialoganglioside	Cer-glc(4←1)-gal-(4←1)-galNAc-(3←?)-gal	Basic ganglioside structure without N-acetyl neuramic acid (sialic acid) residues. Intermediate in ganglioside biosynthesis (rat, frog, brain).

In general, each *organ* has a dominant type of glycolipid but its nature may depend also on the *species*, e.g. monoglycosyl ceramides are dominant in brain; trihexosides and aminoglycolipids in red cells. A ceramide with a 30C ω-hydroxy fatty acid is a major skin component, where it is largely responsible for water impermeability. Modern analytical techniques are revealing that most types are widespread among tissues.

Cerebrosides are widespread, but minor components of higher plants, the best characterized being those of bean leaves and wheat flour where glucose is probably the only sugar.

Sphingolipids are rare in microorganisms.

Cer, ceramide; glc, glucose; gal, galactose; GalNAc, *N*-acetyl galactosamine

Nomenclature of sphingoid bases is not straightforward because the usual major mammalian base, sphingosine, is an unsaturated version of a simpler compound sphinganine (Fig. 2.10). Thus, sphingosine is sphing-4-enine. Shorthand nomenclatures use the letter 'd' to indicate two 'di-) hydroxyls and 't' (tri) for three hydroxyls. The chain lengths and numbers of double bonds are indicated as a:b (e.g. 18:0) in a similar fashion to fatty acids (Chapter 1). See Fig. 2.10 for some examples with abbreviations.

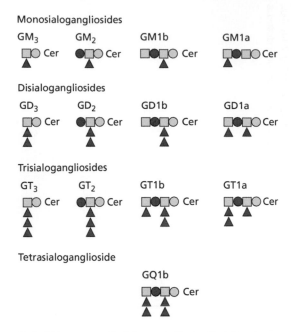

Monosialogangliosides

GM₃ GM₂ GM1b GM1a

Disialogangliosides

GD₃ GD₂ GD1b GD1a

Trisialogangliosides

GT₃ GT₂ GT1b GT1a

Tetrasialoganglioside

GQ1b

Not all known gangliosides are indicated but the above structures include the major compounds found in animals

Symbols: ○, glucose; □, galactose;
●, N-acetylgalactosamine; ▲, sialic acid.

Fig. 2.11 Diagrammatic depiction of ganglioside structures.

Disialogangliosides may have each sialic acid moiety linked to a separate sugar residue or both sialic acids may be linked to each other and one of them linked to a central sugar residue. Because of their complex structure and, hence, cumbersome chemical names, many shorthand notations have been employed. One of the most commonly used is that introduced by the Swedish scientist, Svennerholm. The parent molecule is denoted as GMI and other derived structures are as shown in Table 2.15. The subscripts M, D, T and Q refer to monosialo-, disialo-, trisialo- and quatra (tetra)sialo-gangliosides, respectively. A more recent nomenclature system, due to the German, Wiegandt, is also included in Table 2.15. The latter method has the advantage that, once the symbols have been learnt, the structure can be worked out from the shorthand notation and it can be applied to nonsialic acid-containing glycolipids as well. However, it has yet to supersede the Svennerholm notation, which will be used routinely in this book.

Instead of glycosylation of the alcohol moiety of the sphingosine base, esterification with phosphocholine can

take place. The phospholipid produced thus is called sphingomyelin (Fig. 2.13).

Physical techniques have revolutionized the analysis and identification of sphingolipids. In many cases a combination of GC, mass spectrometry (MS) and nuclear magnetic resonance spectrometry (NMR) provides sufficient information to elucidate a complete structure. The position of attachment of the sugar residues can be worked out from the use of glycosidase enzymes or by permethylation analysis. On the other hand, the molecular weight and the sequence of some quite complex glycosphingolipids can often be established in one step by the use of the gentler techniques of MS, such as fast-atom bombardment.

As a generalization, sphingolipids have a rather specific distribution in cells where they are concentrated in the outer (external) leaflet of the plasma membrane. Their important functions on the cell surface are discussed in Sections 8.3 & 10.2.2).

2.3.4 Sterols and hopanoids

Sterols are common in the membranes of eukaryotic organisms, where they are important in providing stability, but are rare in prokaryotes. In the membranes of higher animals, cholesterol is the almost exclusive sterol constituent whereas in plants, other sterols, such as β-sitosterol, predominate.

Sterols belong to a group of steroids which have one or more hydroxyl groups and a hydrocarbon side-chain. Steroids themselves have a ring system based on 1, 2-cyclopentanoperhydrophenanthrene.

2.3.4.1 Major sterols

Cholesterol and the functionally related sterols of fungi and plants are significant components of many organisms, especially of their external plasma membranes. However, they are not needed by all types of organisms nor, indeed, are they components (or, therefore, necessary) for all classes of membranes, even in organisms with an appreciable sterol content. Sterols are common in eukaryotes, but rare in prokaryotes (see also Section 5.2.2). Whereas vertebrates synthesize cholesterol, invertebrates rely mainly on their food for an external sterol supply.

Yeast and fungi, in the main, have side-chain alkylated compounds (in most yeasts, ergosterol is dominant), while plants and algae contain β-sitosterol and stigmasterol as their most abundant sterols (Fig. 2.14). However, it should be emphasized that, as with most situations in biology, it is unwise to make too many generalizations.

Table 2.15 Structure of some gangliosides.

Class	Structure	Wiegandt shorthand	Svennerholm shorthand	Composition and Occurrence
MONOSIALOGANGLIOSIDE	(a) Cer-glc-(4←1)-gal-(4←1)-galNAc-(3←1)-gal (β, β, β) with 3 / (↑)2 / NANA	$G_{GNT}1$	G_{M1}	*Major bases*: C_{18} and C_{20} sphingosines. Minor amounts of dihydroanalogues. *Fatty acids*: large amounts of f 18:0 (86–95% in brain). *Occurrence*: mainly in grey matter of brain but also in spleen, erythocytes, liver kidney. Modern analytical techniques have shown them to be present in a much wider range of tissues than previously realized. Main gangliosides of human brain are $G_{GNT}1$, 2a, 2b, 3a. Gangliosides appear to be confined to the animal kingdom. In man, cattle, horse, main ganglioside outside brain is $G_{LACT}1$. *N-glycolyl-neuraminic acid is chief sialic acid in erythrocyte and spleen gangliosides of horse and cattle.* Physical properties: *insoluble in nonpolar solvents; form micelles in aqueous solution.*
Tay-Sachs Ganglioside	(b) Cer-glc-(4←1)-gal-(4←1)-galNAc (β, β) with 3 / (↑)2 / NANA	$G_{GNTrII}1$	G_{M2}	
Haematoside	(c) Cer-glc-(4←1)-gal-(3←2)-NANA (β)	$G_{LACT}1$	G_{M3}	
DISIALOGANGLIOSIDE	(a) Cer-glc-(4←1)-gal-(4←1)-galNAc-(3←1)-gal (β, β, β) with 3 / (↑)2 / NANA and 3 / (↑)2 / NANA	$G_{GNT}2a$	G_{D1a}	
	(b) Cer-glc-(4←1)-gal-(4←1)-galNAc-(3←1)-gal (β, β, β) with 3 / (↑)2 / NANA-(8←2)-NANA			
TRISIALOGANGLIOSIDE	Cer-glc-(4←1)-gal-(4←1)-galNAc-(3←1)-gal (β, β, β) with 3 / (↑)2 / NANA and NANA-(8←2)-NANA	$G_{GNT}3a$	G_{T1}	
		$G_{GNT}2b$	G_{D1b}	

Abbreviations in column 2 are the same as in Table 2.14, NANA, N-acetylneuraminic acid (sialic acid). Wiegandt abbreviations: G, ganglioside. Subscript denotes sialic-free oligosaccharide.

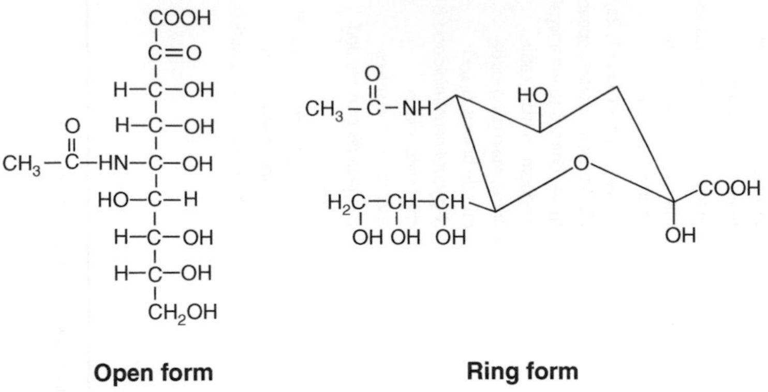

Open form   **Ring form**

Fig. 2.12 *N*-acetylneuraminic acid (NeuAc) or sialic acid.

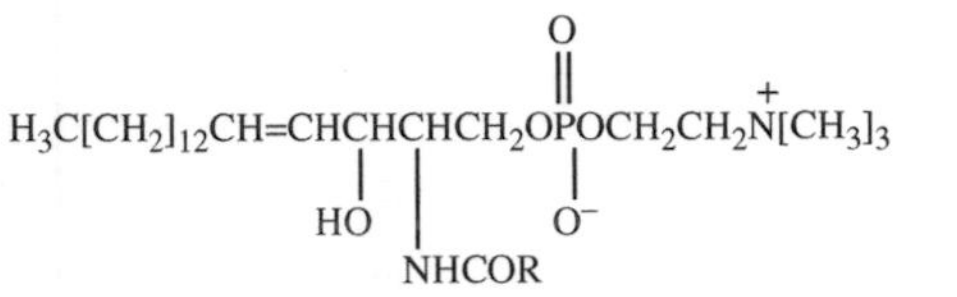

$$H_3C[CH_2]_{12}CH=CHCHCHCH_2OPOCH_2CH_2N^+[CH_3]_3$$

Fig. 2.13 Sphingomyelin (containing sphingosine as the sphingosyl moiety).

For example, cholesterol, which in many animal tissues comprises over 95% of the sterol fraction, cannot be regarded as exclusive to animals. Furthermore, marine invertebrates, whether or not they are sterol auxotrophs, discriminate much less than the mammalian intestine in the sterols that they absorb and, therefore, have compositions dependent on the general availability of sterols. Moreover, whereas plants typically have C24-ethyl sterols such as stigmasterol, some species of red algae contain only cholesterol.

It is generally believed that the main function for membrane sterols is in the modulation of fluidity (see Sections 5.1–5.4). The latter is mediated by the interaction of sterol with the glycerolipid components. For that purpose, cholesterol seems to be optimally adapted. Any changes in the structure appear to reduce its effect on membranes (Fig. 2.15). Points to note are:

1 All *trans*-fusion of the rings gives a planar molecule capable of interacting on both faces.
2 The 3-hydroxy function permits orientation of cholesterol in the bilayer.
3 In naturally occurring sterols, the side chain is in the (17β, 20R) configuration, which is thermodynamically preferred and permits maximal interaction with phosphoglycerolipids.
4 Methyl groups at C10 and C18 are retained and provide bridgeheads.

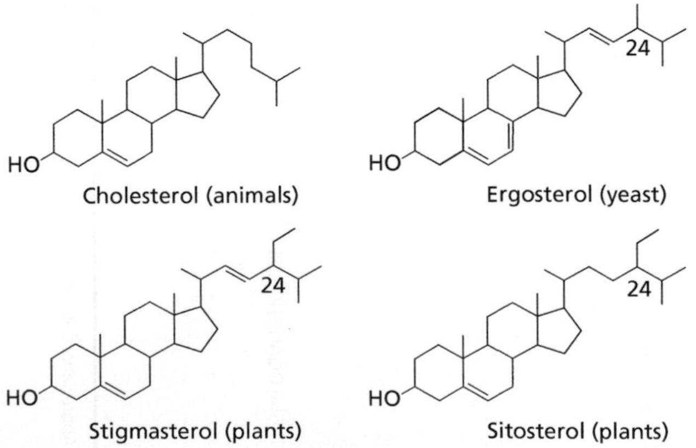

Cholesterol (animals)

Ergosterol (yeast)

Stigmasterol (plants)

Sitosterol (plants)

Fig. 2.14 Major types of membrane sterols.

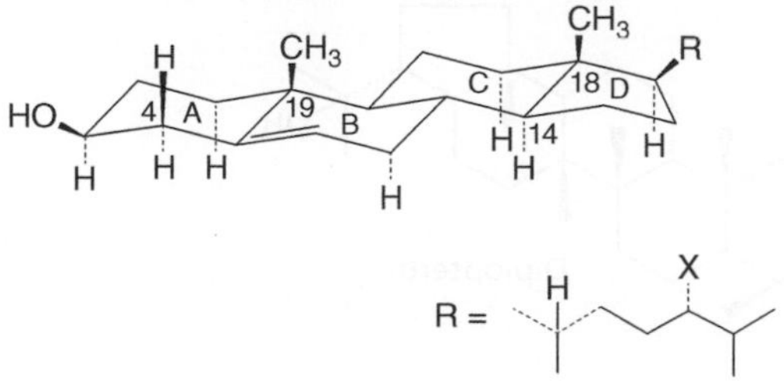

Fig. 2.15 The fused-ring structure of sterols showing important features. Reproduced with kind permission of K. Bloch from (1983) *Critical Reviews in Biochemistry*, **17**, 47–92.

5 An unmodified isooctyl side chain renders the core of the bilayer relatively fluid (cf. substitutions at C24 in ergosterol or stigmasterol).

6 The tetracyclic ring is uniquely compact and rigid. Other molecules of comparable hydrophobicity are much less restrained conformationally. Because of these properties such sterol molecules are able to separate or laterally displace both the acyl chains and polar head groups of membrane phospholipids.

As mentioned above, cholesterol is, by far, the most important sterol in mammalian tissues. So far as plants and algae are concerned, the major structures are β-sitosterol (~70%), stigmasterol (~20%), campesterol (~5%) and cholesterol (~5%). Yeasts can accumulate large amounts of sterols (up to 10% of the dry weight) and phycomycetes contain almost exclusively ergosterol. This compound is the major sterol of other yeasts and mushrooms except the rust fungi in which it is absent and replaced by various 29C sterols. Sterols are absent from archaea and from all except three bacterial lineages (see also Section 5.2.2)

The 3-hydroxyl group on ring A can be esterified with a FA. Sterol esters are found in plants as well as animals. Although sterols are present in most mammalian body tissues, the proportion of sterol ester to free sterol varies markedly. For example, blood plasma, especially that of humans, is rich in sterols and, like most plasma lipids, they are almost entirely found as components of the lipoproteins (see Section 7.2). About 60–80% of this sterol is esterified. In the adrenals, too, where cholesterol is an important precursor of the steroid hormones, over 80% of the sterol is esterified. By contrast, in brain and other nervous tissues, where cholesterol is an important component of myelin, virtually no cholesteryl esters are present.

The 3-hydroxyl group can also form a glycosidic link with the 1-position of a hexose sugar (usually glucose).

Such sterol glycosides are widespread in plants and algae and the 6-position of the hexose can be esterified with a FA to produce an acylated sterol glycoside.

2.3.4.2 Other sterols and steroids

A full description of the various types of steroids, their metabolism and function, is beyond the scope of this book. Cholesterol is the essential precursor for bile acids, corticoids, sex hormones and vitamin D-derived hormones, as has been well established for all vertebrates. Their biosynthesis is outlined briefly in Section 4.9.6 and Box 7.1.

2.3.4.3 Hopanoids and related lipids

Bacteria and other prokaryotes (e.g. cyanobacteria) do not, in general, contain sterols. Instead, they contain related molecules containing five fused rings called 'hopanoids'. These are named after the plant genus *Hopea* from whose resin they were originally isolated. However, only a few higher or lower plants contain hopanoids and these are usually of a more complex structure.

The hopanoids have many different structures. The simplest 30C hopanoid is diploptene usually found with diplopterol or hopan-22-ol. The most abundant compound in living organisms is tetrahydroxybacteriohopane (Fig. 2.16). Hopanoids are most common in aerobic bacteria including cyanobacteria, heterotrophs and methanotrophs. They are found in some anaerobic bacteria but not in the achaea (see Section 2.3.5).

The 30C hopanoids are believed to have functions in membranes similar to those of sterols. That is, they modulate 'fluidity' by interacting with other lipids to increase order. They may also influence membrane permeability to gases such as oxygen and help adaptation to extreme conditions. They have been located in both the thylakoids and cytoplasmic membranes of cyanobacteria and in the outer and cytoplasmic membranes of bacteria. Like sterols, they have been found at high concentrations in bacterial detergent-resistant domains (equivalent to membrane rafts, see Section 5.3.3). In some species they appear essential to growth. Because the pentacyclic ring structure of hopanoids is very stable, they are abundant in geological deposits (such as shales) and serve as biological markers for ancient life on Earth (i.e. paleobiology).

Several types of compounds related to hopanoids, such as those derived from grammacerane (where the fifth ring is 6-membered), are found in Nature but their description is beyond the scope of this book.

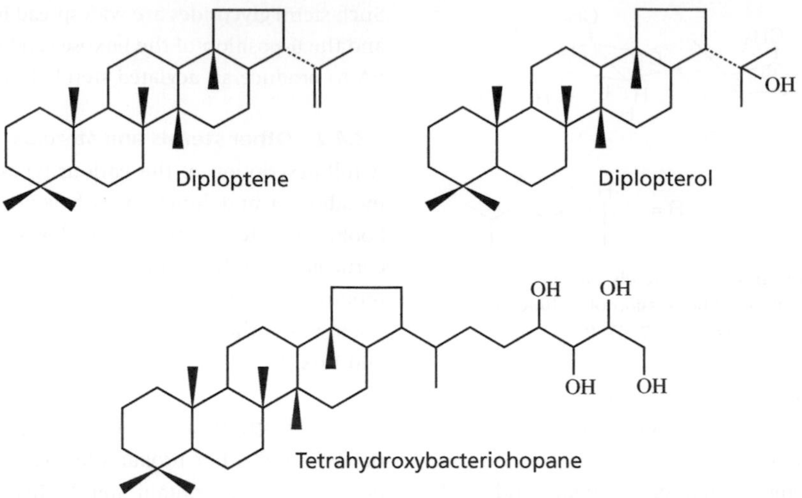

Fig. 2.16 Structure of the simplest hopanoids, diploptene and diplopterol and the most abundant, tetrahydroxybacteriohopane.

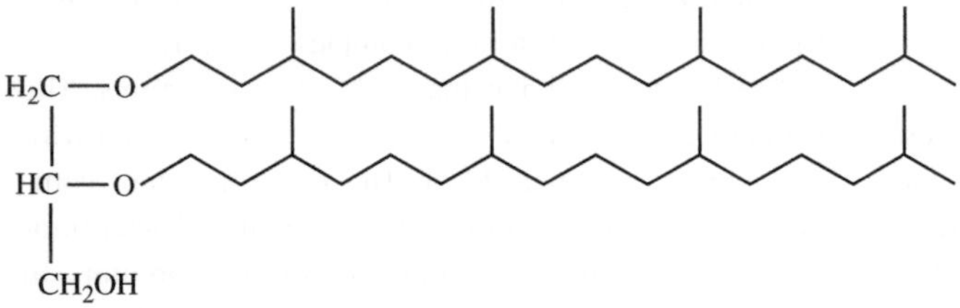

Fig. 2.17 Structure of the diphytanylglycerol ether, archaeol. Variants in the structure found in some archaea include 25C chains (instead of 20C) and hydroxyl substituents on C3 of one or other of the chains. (See also Section 5.2.3 and Fig. 5.9.)

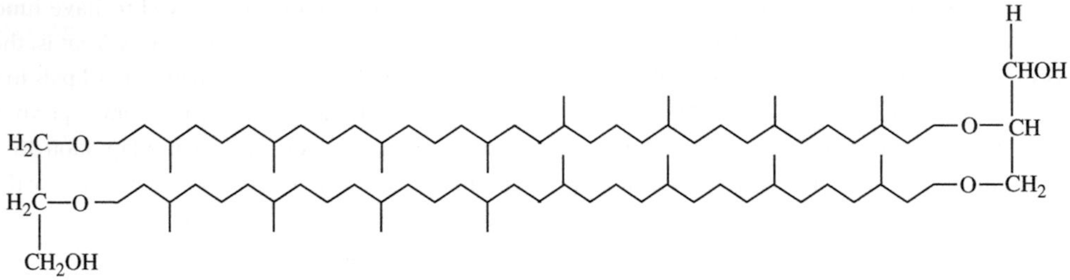

Fig. 2.18 The dibiphytanyldiglyceroltetraether, caldarchaeol typical of many archaea. In such lipids two diether lipids are linked head to head and the molecule can completely span the membrane (see Section 5.2.3 and Fig. 5.9).

2.3.5 Membrane lipids of the archaea

The archaea (which used to be called archeabacteria) have now been recognized as a distinct kingdom. Their 16S ribosomal RNA sequences, cell wall structures and membrane lipid patterns distinguish them from other organisms. These factors also help to classify the three most studied groups of archaea – extreme halophiles, methanogens and thermoacidophiles. Whereas eubacteria and eukaryotes (animals, plants, etc.) contain mostly derivatives of DAGs as membrane lipids, the archea contain variants of a 20C,20C. isoprenyl glycerol diether and its dimer.

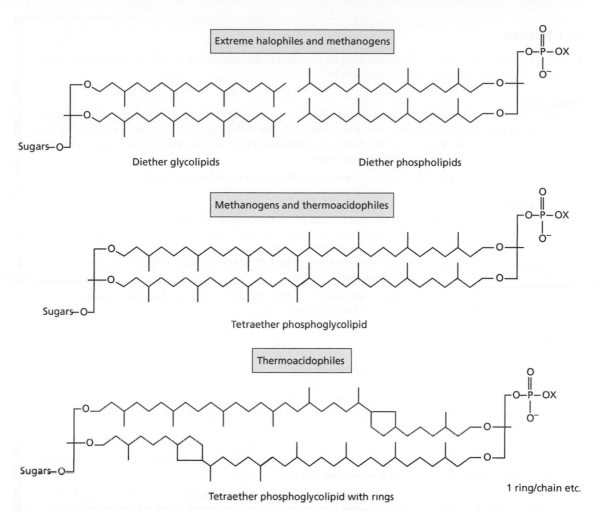

Fig. 2.19 Distribution of lipids in different types of archaea. The extreme halophiles only contain archaeol-based lipids while the methanogens also contain caldarchaeol-based compounds. Extreme thermophiles contain mostly lipids based on caldarchaeol and nonitolcardarchaeol, with or without rings, and with phosphoinositol polar headgroups as well as sugars. In the diagram X can be H, glycerol, inositol, ethanolamine, serine or glycerol-phosphate-O-methyl.

The core lipids (archaeol, caldarchaeol) are shown in Figs. 2.17 and 2.18, respectively. Variants include different chain lengths as well as those with cyclopentane rings. A huge number of phospho-, glyco-, sulpho- and other substitutions are also found in the different groups of archea. Where present, the glycerophosphate group has the reversed stereochemical configuration to that in the phosphoglycerides of higher organisms (see Section 5.2.3). Some examples of these variations in relation to the three groups of archaeal organisms are shown in Fig. 2.19. As a generalization, extreme halophiles contain only archaeol-derived lipids. Methanogens have both archaeol- and caldarchaeol-derived lipids (in ratios of between 2:1 and 1:1). Thermoacidophiles contain mainly caldarchaeol-derived lipids (and variants) with less than 10% archaeol-derived compounds. Further discussion of archaeal lipids in relation to membrane structure can be found in Sections 5.2.2 and 5.2.3. See also Lombard *et al.* (2012) in **Further reading**.

KEY POINTS

- Because lipids do not have a chemical definition, there are a number of major types as well as a host of different minor classes. However, most lipids contain FAs of various chain lengths.

Fatty acids

- The most common FAs are straight chains with an even carbon number. Major types are saturated and unsaturated. Although over 1000 FAs are known, the main ones in Nature are limited to 20 or so. Palmitic acid is the main SFA, while oleic is the main MUFA. In PUFAs, the double bonds are usually separated by a methylene group. Two main groups are the n-3 (e.g. α-linolenic, EPA, DHA) and the n-6 (e.g. linoleic, arachidonic acids) series.

- Other types of FAs include odd-chain length, branched-chain, cyclic and oxygen-substituted molecules.

Storage lipids

- For most organisms that store lipids, TAGs are the compounds used. Partial acylglycerols – MAG, DAG – are also found in small amounts. Because TAGs contain three esterified FAs, a huge range of molecular species is found naturally. Some are solid at room temperature (fats) and some are liquid (oils). In general, most TAGs contain the common FAs (16C or 18C) but TAGs from individual tissues or species can contain large amounts of unusual FAs such as short or medium chains (coconut, mammalian milk fat) or substituted FAs (e.g. ricinoleic acid in castor bean TAG).

- Ether lipids are sometimes found. These may be alkyl glycerols or, more commonly, wax ethers. The latter are particularly important in marine zooplankton where they are important in food chains.

Membrane lipids

- Most membrane lipids are amphipathic molecules. In quantitative terms the phosphoglycerides and glycosylglycerides are the most important. Other major groups are the sphingolipids and sterols.

- Phosphoglycerides are the main lipids in most membranes. The most widespread are PtdCho, PtdEtn, PtdIns and PtdSer. Cardiolipin is important in many bacteria and in mitochondria while PtdGro is virtually the only phosphoglyceride in chloroplast thylakoids. Individual membranes contain a characteristic pattern of phosphoglycerides which may include other derivatives such as plasmalogens.

- In chloroplasts and the thylakoids of cyanobacteria the main lipids are glycosylglycerides – MGDG, DGDG and SQDG. These are only found in very small amounts elsewhere. However, because of the prevalence of photosynthetic tissues on land and in the oceans, the glycosylglycerides are the most abundant lipids in Nature.

- Sphingolipids contain a sphingosine base instead of glycerol. A fatty acid is usually attached to the amino group at C2. There are a large number of different bases found in Nature, but most are 12C–22C. The alcohol moiety at C1 of the sphingosine base can be linked to carbohydrate moieties (glucose, galactose). In turn, further groups can be added to generate sulphatides, complex neutral sphingolipids and gangliosides (containing scalic acid).

- Sterols are found in the membranes of eukaryotic organisms but are rare in prokaryotes. Depending on the organism, different sterols are prevalent (e.g. cholesterol in animals, ergosterol in yeast, sitosterol or stigmasterol in plants). Like the sphingolipids, sterols are concentrated in plasma membranes and their lipid rafts. In prokaryotes, hopanoids are found which serve a similar function to sterols in eukaryotes.

- The archaea contain ether lipids in their membranes. These are variants of a 20C,20C-isoprenyl glycerol diether and its dimer. The latter span the membrane to form a monomolecular structure rather than the usual bilayer.

Further reading

Ansell GB & Hawthorne JN, eds. (1982) *Phospholipids*, Elsevier, Amsterdam.

Ceve G, ed. (1993) *Phospholipids Handbook*, Marcel Dekker, Basel.

Christie WW & Han X (2010) *Lipid Analysis*, 4th edn. The Oily Press, Bridgewater.

Dembitsky VM (1996) Betaine ether-linked glycerolipids: chemistry and biology. *Prog Lipid Res* **35**:1–52.

Gunstone FD, Harwood JL & Dijkstra AJ, eds. (2007) *The Lipid Handbook*, 3rd edn. CRC Press, Boca Raton.

Hirabashi Y, Igarashi Y, Merrill AH Jr. *et al.*, eds. (2006) *Sphingolipid Biology*. Springer, Tokyo.

Kates M, ed. (1990) *Glycolipids, Phosphoglycolipids and Sulphoglycolipids*. Plenum Press, New York.

Kates M (1993) Membrane lipids of archaea. In M Kates, ed. *The Biochemistry of the Archaea (Archaebacteria)* Elsevier, Amsterdam, pp. 261–95.

Kates M (2010) *Techniques of Lipidology*, 3rd edn. Newport Somerville Innovation Ltd, Ottawa.

Leray C (2013) *Introduction to Lipidomics: From Bacteria to Man.* CRC Press, Boca Raton.

Lipid Library (http://lipidlibrary.aocs.org).

Lombard J, López-García P & Moreira D (2012) Phylogenomic investigation of phospholipid synthesis in archaea. Archaea doi 10.1155/2012/630910: 1–13.

Murphy DJ, ed. (2005) *Plant Lipids: Biology, Utilisation and Manipulation*, Blackwell Science, Oxford.

Ratledge CK & Wilkinson SG, eds. (1989) *Microbial Lipids, 2 vols*, Academic Press, London.

Vance DE & Vance JE, eds. (2008) *Biochemistry of Lipids, Lipoproteins and Membranes*, 5th edn. Elsevier, Amsterdam.

Supplementary reading

Bloch K (1983) Sterol structure and membrane function. *Crit Rev Biochem* **17**:47–92.

Fahy E, Subramanian S, Brown HA *et al.* (2005) A comprehensive classification system for lipids. *J. Lipid Res.* **46**:839–61.

Kaya K (1992) Chemistry and biochemistry of taurolipids. *Prog Lipid Res* **31**:87–108.

IUPAC-IUB Commission on Biochemical Nomenclature (1978) *Biochem J* **171**:21–35.

IUPAC-IUB Commission on Biochemical Nomenclature (1987) *Eur J Biochem* **167**:181–4.

IUPAC-IUB Commission on Biochemical Nomenclature (1989) *Eur J Biochem* **186**:429–58.

Merrill AH Jr, Wang MD, Park M & Sullards MC (2007) (Glyco) sphingolipidology: an amazing challenge and opportunity for systems biology. *Trends Biochem Sci* **32**:457–68.

Nomenclature Committee of the International Union of Biochemistry (1984) *Enzyme Nomenclature*. Academic Press, London.

Rezanka T & Sigler K (2009) Odd-numbered very-long chain fatty acids from the microbial, animal and plant kingdoms. *Prog Lipid Res* **48**:206–38.

CHAPTER 3
Fatty acid metabolism

In the next two chapters we will discuss the metabolism of lipids. To begin with, because most lipids contain fatty acids (FAs, i.e. acyl lipids – Chapter 2), we will start with the biosynthesis of FAs, their metabolism and how they can give rise to important signalling molecules.

3.1 The biosynthesis of fatty acids

Naturally occurring FAs normally, but not exclusively, have straight even-numbered hydrocarbon chains. This is explained by their principal mode of biosynthesis from acetate (2C) units.

3.1.1 Conversion of fatty acids into metabolically active thioesters is often a prerequisite for their metabolism

For most of the metabolic reactions in which FAs take part, whether they be anabolic (synthetic) or catabolic (degradative), thermodynamic considerations dictate that the acids be 'activated'. For these reactions thioesters are generally utilized. The active form is usually the thioester of the FA with the complex nucleotide, coenzyme A (CoA) or the small protein known as acyl carrier protein (ACP – see Fig. 3.1). These molecules contain a thioester and, at the same time, render the acyl chains water-soluble.

For many tissues, a preliminary step involves the uptake of FAs. In animals, the latter are transported between organs either as nonesterified fatty acids (NEFAs) complexed to albumin or in the form of complex lipids associated with lipoproteins. NEFAs are released from the triacylglycerol (TAG) component of the complex lipids by lipoprotein lipase (LPL; see Sections 4.2.2, 7.2.5 & 7.2.6). The mechanism by which NEFAs enter cells is undefined. Both protein-mediated and diffusion processes have been implicated. While the former predominates at low (nanomolar) concentrations of FAs, passive diffusion is effective at higher concentrations. In artificial membrane systems, un-ionized FAs can move across bilayers rapidly and it has been theorized that desorption from the membrane may be helped by fatty acid binding proteins (FABPs).

In contrast to simple diffusion, there is increasing evidence for protein-mediated transport. Of the various putative transport proteins, FABPs, fatty acid translocase (FAT) and fatty acid transport protein (FATP) have been identified in a variety of cell types. All three proteins are regulated to some degree by members of the PPAR (peroxisome proliferator-activated receptor) family of transcription factors. (The PPAR transcription factors are part of the steroid hormone receptor superfamily and affect many genes of lipid metabolism – see Sections 3.1.11 & 7.3.2 for further details.) Several isoforms of FABPs and FATPs are also regulated by insulin. Moreover, expression, regulation and activity of the various transport proteins can often (but not always) be correlated with increased FA movements in animal models of fat metabolism such as in Ob (obese) mice mutants. Null mutants, e.g. for FAT, have also been created and these show abnormal FA metabolism. (For further discussion of the physiological roles of the various proteins in FA transport across membranes, see Chapter 7.)

Once inside cells, the FAs can diffuse or are transported into different organelles such as mitochondria. The discovery of small (14–15 kDa) proteins, termed FABPs in animals led to the suggestion that they are important as carriers and for intracellular storage, in addition to transport (above). Similar proteins have been identified in other organisms such as plants. Some of the FABPs will bind other ligands, in particular acyl-CoAs. The physiological significance of this is unknown because separate acyl-CoA binding proteins have also been reported.

Lipids: Biochemistry, Biotechnology and Health, Sixth Edition. Michael I. Gurr, John L. Harwood, Keith N. Frayn, Denis J. Murphy and Robert H. Michell.
© 2016 John Wiley & Sons, Ltd. Published 2016 by John Wiley & Sons, Ltd.

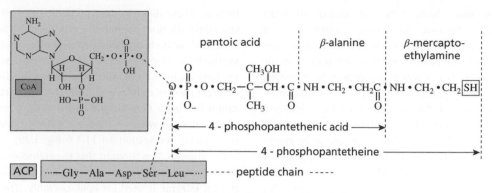

Fig. 3.1 The structures of coenzyme A and the acyl carrier protein from *E. coli*.

Nevertheless, knockout mice lacking heart FABP have difficulty utilizing FAs, show exercise intolerance and develop cardiac hypertrophy. Similarly, mice lacking liver FABP have reduced FA oxidation and ketogenesis so there is no doubt that FABPs are important for normal FA metabolism (see also Section 7.1.3).

In animals, FABPs are divided into four subfamilies. Apart from Subfamily I, which has a preference for retinol and retinoic acids, all the other FABPs bind long- and very-long-chain FAs but with differences in ligand selectivity, binding affinity and mechanism. They all contain a characteristic β-sheet. The binding pocket is within the barrel. Apart from their role in FA transport, the FABPs allow targeting of FAs to specific metabolic pathways and participate in the regulation of gene expression and cell growth.

Small acyl-CoA-binding proteins (ACBPs) of around 10 kDa molecular mass function in acyl-CoA transport within cells as well as in protecting cytosolic enzymes in mammals, plants and yeast. They have important roles in maintenance of the acyl-CoA pool and for glycerolipid biosynthesis. There are also large ACBPs (up to 75 kDa) in mammals and plants. Recent investigations of the large ACBPs have shown that, in addition to their function in lipid metabolism, they also participate in a range of biological activities including development (e.g. embryogenesis, skin and hair development, senescence), stress responses (e.g. cold or heavy metal tolerance), vesicular trafficking and signalling pathways. In plants, the ACBPs are also involved in oxidative stress and pathogen resistance. The roles of FABPs and ACBPs in lipid transport in humans are discussed further in Section 7.1.3.

3.1.1.1 Acyl-CoA thioesters were the first types of activated fatty acids to be discovered

The formation of acyl-CoA is catalysed by several acyl-CoA synthetases (ACSs). These belong to a large protein family found in a wide range of, if not all, organisms, which itself is part of the superfamily of AMP-binding proteins (due to their AMP-forming reaction).

The ACS reaction occurs in two steps:

$$E + R - COOH + ATP \xrightarrow{Mg^{2+}} (E : R - CO - AMP) + PPi$$
$$(E : R - CO - AMP) + CoASH \rightarrow E + R - CO - SCoA + AMP$$

Both reactions are freely reversible but the reaction is driven to the right because pyrophosphate is rapidly hydrolysed. Evidence for the overall reaction has come mainly from work with heart acetyl-CoA and butyryl-CoA synthetases, but it is presumed that all acyl-CoA synthetases work via a similar mechanism.

The ACSs differ from each other with respect to their subcellular locations and their specificities for FAs of different chain lengths. Their overlapping chain length specificities and their tissue distributions mean that most saturated (SFA) or unsaturated (UFA) fatty acids in the range 2C–22C can be activated in animal tissues, although at different rates.

Now that molecular biology can be used to identify cDNAs for ACS, we are becoming aware of just how many enzyme isoforms are likely to exist. Originally it was thought that short-, medium- and long-chain isoforms were often present. However, in oilseed rape, 12 cDNAs have already been reported, while in *Mycobacterium tuberculosis* (the causative agent of tuberculosis) 36 putative ACS genes have been identified!

The first short-chain ACS was studied in heart mitochondria. Termed acetyl-CoA synthetase, it is most active with acetate and is found in many mitochondria. A cytosolic acetyl-CoA synthetase is present in the cytosol of some tissues where it functions by activating acetate for FA biosynthesis (in contrast to the mitochondrial isoform where FAs are activated for oxidation). Indeed, the presence of an acetyl-CoA synthetase in chloroplasts has also enabled [^{14}C] acetate to be extensively used in biochemical experiments with plants.

Acetyl-CoA synthetase (which also activates propionate at slower rates), along with other short-chain ACSs, is important in ruminants. The rumen microorganisms in these animals generate huge amounts of short-chain acids including acetate. Other sources of acetate in animals include the oxidation of ethanol by the combined action of alcohol and aldehyde dehydrogenases and from β-oxidation. Under fasting conditions or in diabetes, the ketotic conditions cause a rise in the production of both ketone bodies and acetate (via the action of acetyl-CoA hydrolase). The absence of acetyl-CoA synthetase in liver mitochondria allows acetate to be transported from this tissue and to be activated and oxidized in mitochondria of other tissues.

A propionyl-CoA synthetase may be especially important in ruminants where propionate is formed by rumen microorganisms and is an important substrate for gluconeogenesis.

Medium-chain ACSs have been demonstrated in the mitochondria from several mammalian tissues. In heart mitochondria the synthetase works with FAs of 3C–7C but is most active with butyrate. The liver enzyme works with 4–14C substrates, of which octanoate is the best. A second liver isoform has been found which activates medium-chain substrates including branched or hydroxy FAs. This isoform also activates aromatic carboxylic acids such as benzoate and salicylate. The latter function is needed for the formation of glycine conjugates of aromatic carboxylic acids.

The long-chain ACSs in animals are firmly membrane-bound and can only be solubilized by the use of detergents. Within the cell, activity has been detected in endoplasmic reticulum (ER) and the outer mitochondrial membrane with small amounts in peroxisomes (when the latter are present). There is some dispute as to whether the activity present in mitochondrial and ER (separated as a microsomal fraction) membranes is due to the same enzyme but, in any case, five isozymes are present. These differ in their properties as well as their subcellular distribution. Because long-chain FA activation is needed for both catabolism (β-oxidation) and for biosynthesis (acylation of complex lipids) it would be logical if the long-chain ACSs of mitochondria and the ER formed different pools of cellular acyl-CoA. This compartmentalization has been demonstrated with yeast (*Candida lipolytica*) mutants where it plays a regulatory role in lipid metabolism (see Section 3.1.11.3 & Fig. 3.29).

For animals, it is thought that the mitochondrial and peroxisomal enzymes are used for FA oxidation while the ER enzyme is used for lipid assembly. The original long-chain ACSs that were studied in detail were those from rat liver mitochondria and ER. These were found to be very similar (if not identical) as judged by several molecular and catalytic properties. They worked efficiently for both SFAs and UFAs in the range 10C–20C.

Molecular studies have revealed the presence of five long-chain ACSs in mammals. These differ from each other in their tissue distribution, intracellular location, substrate selectivity, kinetics and regulation. In addition, in some tissues, notably liver, there is a very-long-chain ACS that functions with FAs of 22C or longer and is associated with the ER and peroxisomes.

In *Eshcherichia coli*, ACS activity is inducible when the bacterium is grown in a medium with long-chain FAs as the carbon source. Two synthetase isoforms have been identified. While the first is cytosolic, the second (Fab K) is induced anaerobically and is loosely bound to the inner membrane. The first ACS, termed FadD in *E. coli*, works cooperatively with an outer membrane FA transporter, FadL, to convert transported long-chain FAs into substrates for β-oxidation. Medium-chain FAs do not need FadL to enter cells and may cross the outer membrane by simple diffusion. FA transport and activation are directly coupled to transcriptional control of the structural genes of enzymes induced in FA metabolic pathways. This occurs through the fatty acyl-CoA responsive transcription factor, FadR (Fig. 3.2). In the absence of exogenous FAs, FadR is active in DNA binding and the *fad* (FA degrading) genes are expressed at high levels. The FadR conformation is changed by fatty acyl-CoAs that loosen the binding to DNA. This results in de-repression of the *fad* genes and reduces expression of some of the *fab* (FA biosynthesis) genes. This is a very nice example of gene regulation that fits exactly with the organism's metabolic requirements.

(a) In absence of exogenous free fatty acids

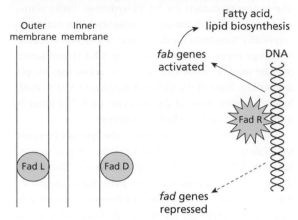

(b) High level of free fatty acids in growth medium

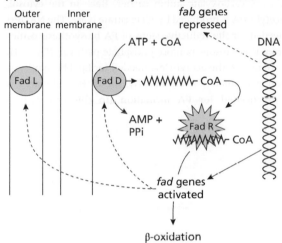

Fig. 3.2 How exogenous fatty acids control fatty acid metabolism in *E. coli*. *fab* genes are those involved in fatty acid biosynthesis while *fad* genes code proteins in fatty acid degradation. The respective proteins are called Fab or Fad, e.g. FadD, FadL. (a) In absence of exogenous free fatty acids. (b) High level of free fatty acids in growth medium. Adapted with kind permission of C.C. DiRusso and Elsevier Science from *Progress in Lipid Research*, **38** (1999), Fig. 8, p. 162.

The *E. coli* FadD ACS has been studied in some detail. Site-directed mutagenesis has been used to identify seven amino acids in a highly conserved region, which are essential for catalytic activity. The motif itself functions in part to promote FA chain selectivity. Moreover, the tertiary structure of *E. coli* ACS has been predicted by

comparing it to known structures for other members of the AMP-binding superfamily of proteins, e.g. luciferase.

Although most ACSs are ATP-dependent, some have been found that use GTP. The best known of these is one isomer of succinyl-CoA synthetase which functions in the tricarboxylic acid cycle. The physiological significance of the other GTP-dependent ACSs is uncertain.

3.1.1.2 Acyl-acyl carrier proteins can be found as distinct metabolic intermediates in some organisms

The other form of 'activated' FA (Fig. 3.1) is that of acyl-ACP. In these compounds the FA is also attached to a 4'-phosphopantetheine moiety but this in turn is connected via a conserved serine residue of a peptide chain (rather than to a nucleotide as in coenzyme A). The 4'-phosphopantetheine is transferred from CoA. ACPs can be isolated from plants (including algae), cyanobacteria and most bacteria and mitochondria which contain Type II fatty acid synthases (FASs; see Section 3.1.3.2). They are small, highly stable, acidic proteins of 9–11 kDa in size. An enzyme that converts FAs into acyl-ACP (see Section 3.1.3.2) esters has been isolated from *E. coli*. It is presumed that a similar activity exists in plants because FAs up to 12C can be efficiently used by plant FASs, which need acyl-ACPs (see Section 3.1.3.2). The *E. coli* enzyme uses ATP for the activation step.

3.1.2 The biosynthesis of fatty acids can be divided into *de novo* synthesis and modification reactions

Most naturally occurring FAs have even numbers of carbon atoms. Therefore, it was natural for biochemists to postulate that they were formed by condensation from 2C units. This suggestion was confirmed in 1944 when Rittenberg and Bloch isolated FAs from tissues of rats that had been fed acetic acid labelled with ^{13}C in the carboxyl group and ^{2}H in the methyl group. The two kinds of atoms were located at alternate positions along the chain, showing that the complete chain could be derived from acetic acid.

This stimulated interest in the mechanism of chain lengthening and, when the main details of the β-oxidation pathway (the means by which FAs are broken down two carbons at a time; see Section 3.2.1) were worked out in the early 1950s, it was natural for many biochemists to ask the question, 'Can β-oxidation be reversed in certain circumstances to synthesize FAs instead of breaking them down?'

The study of FA biosynthesis began in the laboratories of Gurin in the USA (who studied liver) and Popjak in London (using mammary gland). Several discoveries soon indicated that the major biosynthetic route to long-chain FAs was distinct from β-oxidation. In the first place, a pyridine nucleotide was involved; but it was NADPH, not NADH as would be expected by reversal of β-oxidation (see below and Section 3.2.1). Second, a requirement for bicarbonate or carbon dioxide was noticed by Wakil and his colleagues, who were studying FA biosynthesis in pigeon liver and by Brady and Klein studying rat liver and yeast, respectively. This observation led to the discovery that malonyl-CoA, formed by the carboxylation of acetyl-CoA, is a unique intermediate in the biosynthesis pathway.

FA biosynthesis can be conveniently divided into *de novo* synthesis, where a small precursor molecule (usually the 2C acetyl group) is gradually lengthened by 2C units to give rise, in most systems, to 16C and 18C products and various modifications that can take place once the long-chain FAs have been made (Fig. 3.3). Of the modifications, elongation and desaturation are the most important.

3.1.3 *De novo* biosynthesis

The source of carbon for FA biosynthesis varies somewhat, depending on the type of organism. For mammals (especially humans!) fat deposition is a typical response to energy excess (see Sections 9.1 & 10.4.1) and carbohydrate (and, to some extent, amino acids) may supply the carbon. Most of the carbon for *de novo* FA (and lipid) formation goes through the pyruvate pool. The latter is, of course, the end-product of glycolysis.

Although pyruvate is produced in the cytosol (the major site of *de novo* FA biosynthesis in animals), it is converted into acetyl-CoA mainly in the mitochondria. Under conditions favouring FA biosynthesis, pyruvate is transported into mitochondria where pyruvate dehydrogenase is activated. The acetyl-CoA product is combined with oxaloacetate to produce citrate, which leaves the mitochondria via a tricarboxylate anion carrier. Back in the cytosol, acetyl-CoA is produced by ATP: citrate lyase. The NADPH needed for the reductive steps of FA biosynthesis comes from the cytosolic pentose phosphate pathway (Fig. 3.4).

In plants, the acetyl-CoA needed for lipid biosynthesis comes ultimately from photosynthesis. How the acetyl-CoA needed for FA formation (in this case within

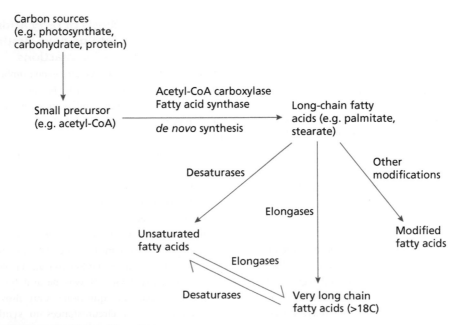

Fig. 3.3 A simplified picture of fatty acid synthesis.

plastids) is actually generated is not clear. In some plants, all the enzymes of glycolysis (as well as pyruvate dehydrogenase) appear to be present in the plastid. In other plants, such as spinach, acetyl-CoA appears to arrive via a circuitous route from the mitochondria. Thus, acetyl-CoA generated by mitochondrial pyruvate dehydrogenase is hydrolysed to free acetate. In its un-ionized form acetic acid can easily cross membranes and thus this 2C substrate moves to the plastid, where it is activated by acetyl-CoA synthetase (see Section 3.1.1.1 & Fig. 3.4). It may be that some plants can use both methods.

3.1.3.1 Acetyl-CoA carboxylase

The process of *de novo* biosynthesis to produce long-chain FAs involves the participation of two enzyme systems, acetyl-CoA carboxylase (ACC) and fatty acid synthase (FAS). Both of these are complex and catalyse multiple reactions. In early experiments on FA biosynthesis in pigeon liver, it was found that the soluble enzymes necessary could be split into two fractions by ammonium sulphate fractionation. The cofactor biotin was bound to one of the protein fractions. At that time it was known that biotin was involved in carboxylation reactions and, when Wakil and Brady independently

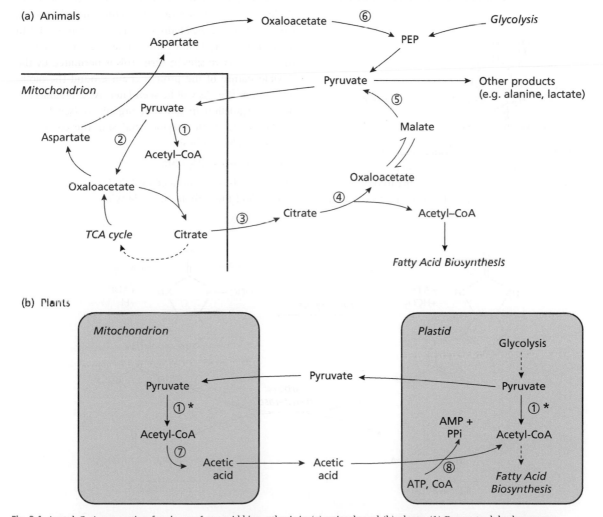

Fig. 3.4 Acetyl-CoA generation for *de novo* fatty acid biosynthesis in (a) animals and (b) plants. (1) Pyruvate dehydrogenase; (2) pyruvate carboxylase; (3) tricarboxylate anion carrier; (4) ATP:citrate lyase; (5) malic enzyme; (6) PEP carboxykinase; (7) acetyl-CoA thioesterase (or spontaneous cleavage); (8) acetyl-CoA synthase. *Note: in plants; if plastid pyruvate dehydrogenase is not present then the mitochondrial enzyme must be utilized.

discovered that malonate was an intermediate in FA biosynthesis, the essential features of the pathway began to emerge. The acetate must be in its 'activated form' as its CoA thioester (see Section 3.1.1 & Fig. 3.1). The reaction is catalysed by ACC, which could be identified in the pigeon liver fraction, where the biotin was bound. Confirmation that biotin was involved came from the inhibition of carboxylation by avidin. This protein was already known as a potent inhibitor of biotin, for it is the component of raw egg white that can bind extremely strongly to biotin thus causing a vitamin deficiency known as 'egg white injury'. The active carbon is attached to one of the ureido nitrogens of the biotin ring.

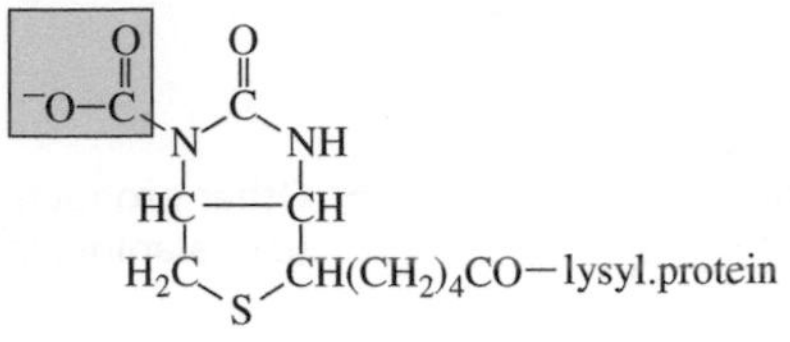

carboxylbiotinyl enzyme

ACC is a Type I biotin-containing enzyme. Such enzymes catalyse carboxylation reactions in two main steps (Fig. 3.5). In the first step, a biotin carboxylase uses ATP to facilitate the transfer of CO_2 to the N-1 nitrogen

atom in the ureido ring of biotin. The sequences of the biotin carboxylases which form part of the various Type I enzymes (pyruvate carboxylase, acetyl-CoA carboxylase etc.) are very similar and the reaction mechanism appears identical. It is the carboxyltransferase partial reaction that provides specificity. The main reactions appear to take place in steps and these are described in Knowles' 1989 review; see **Further reading**). The attachment of the carboxylate group to nitrogen on the biotin moiety, aids in its transfer to acetyl-CoA.

ACCs can have a multi-subunit structure or exist as multifunctional proteins. Different examples from the microbial, plant and animal kingdoms are shown in Table 3.1). In the case of the multifunctional proteins it is important for the biotin prosthetic group to be able to interact with the biotin carboxylase and the carboxyltransferase active sites in turn. This is permitted by the flexible nature of the protein chain around the biotin attachment site. As will be seen later, ACC is subject to precise regulation in different organisms (see Section 3.1.11) as befits an enzyme catalysing the committed step for FA biosynthesis.

3.1.3.2 Fatty acid synthase
The malonyl-CoA generated by ACC forms the source of nearly all of the carbons of the fatty acyl chain. In most

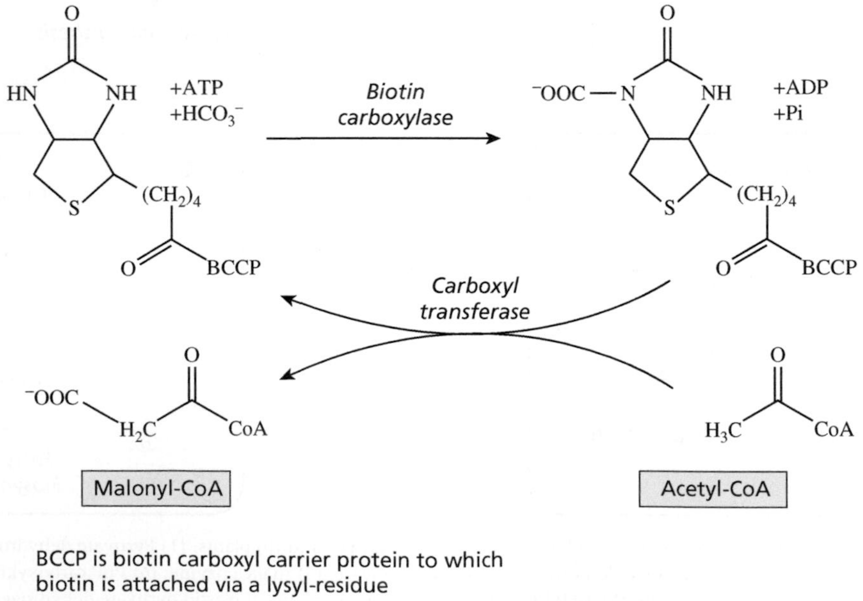

BCCP is biotin carboxyl carrier protein to which biotin is attached via a lysyl-residue

Fig. 3.5 The reactions of acetyl-CoA carboxylase. BCCP is biotin carboxyl carrier protein to which biotin is attached via a lysyl-residue.

Table 3.1 Examples of different acetyl-CoA carboxylases.

Species	Protein structure	Details
E. coli	Multiprotein complex	Four proteins: biotin carboxylase, BCCP and carboxyltransferase (a heterodimer). Transcription of all four *acc* genes is under growth rate control.
Yeast	Multifunctional protein	190-230 kDa. Activated but not polymerized by citrate.
Dicotyledon plants	Multiprotein complex	In chloroplasts. Similar properties to *E. coli* enzyme, although the four individual enzymes are bigger. Three of the subunits coded by nucleus, one by chloroplast.
	Multifunctional protein	Presumed to be cytosolic. Concentrated in epithelial cells in pea. Molecular mass 200-240 kDa. Functions as dimer. Graminicide insensitive.
Grasses (*Poaceae*)	Multifunctional proteins	Two isoforms, both 200-240 kDa, which function as dimers. Chloroplast isoform is graminicide sensitive but cytosolic form (concentrated in epithelial cells) is insensitive. Both are nuclear encoded.
Animals	Multifunctional proteins	Cytosolic. About 250 kDa but functions as polymer of up to 10^7 kDa. Aggregation increased by citrate. Also regulated by phosphorylation in response to hormones.

Adapted from JL Harwood (1999) Lipid biosynthesis. In FD Gunstone, ed., *Lipid Synthesis and Manufacture*, pp. 422–66, Sheffield Academic Press, Sheffield, with permission from Sheffield Academic Press.

cases, only the first two carbons arise from a different source – acetyl-CoA – and the latter is known as the 'primer' molecule.

$$\underbrace{CH_3CH_2}_{\text{from primer}}\ \underbrace{CH_2CH_2(CH_2)_{10}CH_2\ COOH}_{\text{from malonyl-CoA}}\ \text{(palmitic acid)}$$

The FA chain then grows in a series of reactions illustrated in Fig. 3.6. The individual enzymes for *E. coli* are listed in Table 3.2. The basic chemistry is similar in all organisms although the organization of the enzymes is not. Acetyl-CoA is not necessarily the only primer; for example, in mammalian liver and mammary gland, butyryl-CoA is more active, at least in vitro. Moreover, in goat mammary gland there are enzymes that can catalyse the conversion of acetyl-CoA into crotonyl-CoA by what is, essentially, the reverse of β-oxidation (see 3.2.1). Finally the use of propionyl-CoA or branched-chain primers permits the formation of odd chain length or branched-chain FAs, respectively (see Section 3.1.6).

Branched-chain FAs are particularly important for Gram-positive bacteria, which use them rather than unsaturated moieties to modulate membrane fluidity (see also Chapter 5). Branched-chain precursors of the amino acids, valine and isoleucine (i.e. isobutyryl-CoA and 2-methylvaleryl-CoA) are used. As in straight-chain FA formation, the primer is condensed with malonyl-ACP through the activity of condensing enzyme III (FabH).

The overall reaction of FA (palmitate) biosynthesis can be summarized:

$$CH_3CO-CoA + 7\ ^-OOCCH_2CO-CoA + 14NADPH$$
$$+ 14H^+ \longrightarrow CH_3(CH_2)_{14}COOH + 7CO_2 + 8CoASH$$
$$+ 14NADP^+ + 6H_2O$$

The steps have been elucidated mainly from studies of *E. coli*, yeast, the tissues of various animals and some higher plants. These groups of enzymes are known collectively as fatty acid synthases (FASs).

FASs can be divided mainly into Type I and Type II enzymes (Table 3.3). Type I synthases are multifunctional proteins in which the proteins catalysing the individual partial reactions are discrete domains on large polypeptides. This type includes the animal synthases and those from yeast and mycolic acid-producing bacteria. Type II synthases contain discrete proteins that can be separated, purified and studied individually. The Type II FAS occurs in lower bacteria, plant plastids and eukaryotic mitochondria and has been studied most extensively in *E. coli*. In addition, Type III synthases – occurring in different organisms – catalyse the addition of C2 units to preformed acyl chains and are better known as FA elongases (see Section 3.1.5). Although, historically, the reactions of the yeast synthases were deciphered first, we shall start by describing the separate reactions catalysed by the enzymes of *E. coli*.

The Americans, Goldman and Vagelos, discovered that the product of the first condensation reaction of FA

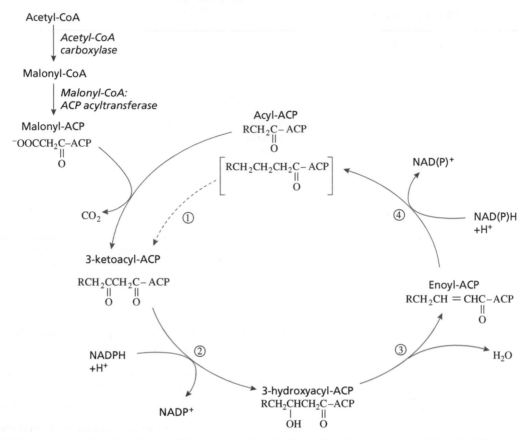

Fig. 3.6 The repeat cycle of reactions for the addition of two carbons by fatty acid synthase. Reactions of the cycle: (1) condensation (3-ketoacyl-ACP synthase); (2) reduction (3-ketoacyl-ACP reductase); (3) dehydration (3-hydroxyacyl-ACP dehydratase); (4) reduction (enoyl-ACP reductase).

biosynthesis in *E. coli* was bound to a protein through a thioester linkage: acetoacetyl-S-protein. When they incubated this product with the coenzyme NADPH and a crude protein fraction containing the synthase, they obtained butyryl-S-protein and when malonyl-CoA was included in the reaction mixture, long-chain FAs were produced. Subsequently, intermediates at all steps of the reaction were shown to be thioester-linked to this protein, which was therefore called acyl carrier protein (ACP). It is a small molecular mass protein (about 8.8 kDa), has one SH group per mole of protein and is very stable to heat and acid pH. Vagelos and his colleagues worked out the structure of the prosthetic group to which the acyl moieties were attached in the following way. They made acyl-CoA derivatives in which the acyl group was labelled with radioactive carbon atoms (such as [2-^{14}C]-malonyl-ACP) and then hydrolysed the acyl

protein with proteolytic enzymes. This yielded small radioactive peptides with structures that were fairly easy to determine. The prosthetic group structure, finally elucidated by Wakil, turned out to be remarkably similar to coenzyme A: the acyl groups were bound covalently to the thiol group of 4'-phosphopantetheine, which in turn was bound through its phosphate group to a serine hydroxyl of the protein (Fig. 3.1). To study the biosynthesis of ACP, Vagelos made use of a strain of *E. coli* that had to be supplied with pantothenate in order to grow. When the bacteria were fed with radioactive pantothenate, the labelled substance was incorporated into the cell's ACP and CoA. However, whereas the ACP concentration remained constant under all conditions, that of CoA was very much dependent on the concentration of pantothenate. When the cells were grown in a medium of high pantothenate concentration, the

Table 3.2 Reactions of fatty acid synthesis in *E. coli*.

(1) Malonyl transacylase	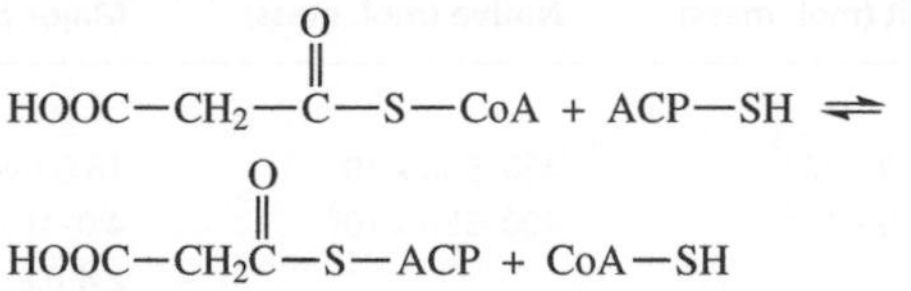 $HOOC-CH_2-\overset{\overset{O}{\|}}{C}-S-CoA + ACP-SH \rightleftharpoons$ $HOOC-CH_2\overset{\overset{O}{\|}}{C}-S-ACP + CoA-SH$	Specific for malonate: not a saturated acyl-CoA. Malonyl-S-pantetheins is also a substrate. Both enzymes (1) and (2) can be characterized by (i) the amount of ^{14}C or malonate transferred to ACP or (ii) chromatography of malonyl hydroxamate. Intermediate is acyl-S-enzyme.
(2) 3-ketoacyl-ACP synthase (*Fab B* or *Fab F*)	$CH_3-\overset{\overset{O}{\|}}{C}-S-ACP + HOOC-CH_2-\overset{\overset{O}{\|}}{C}-S-ACP \rightleftharpoons$ $CH_3-\overset{\overset{O}{\|}}{C}-CH_2-\overset{\overset{O}{\|}}{C}-S-ACP + CO_2 + ACP-SH$	Assayed by coupling with the next reaction and following NADPH oxidation spectrophotometrically. SH-enzyme with a key cysteine at the active site inhibited by iodoacetamide. Two isoforms in *E. coli* with different substrate selectivities. Both inhibited by cerulenin.
(3) 3-ketoacyl-ACP reductase (*Fab G*)	$CH_3-\overset{\overset{O}{\|}}{C}-CH_2\overset{\overset{O}{\|}}{C}-S-ACP + NADPH + H^+ \rightleftharpoons$ $D(-)CH_3-\overset{\overset{OH}{\|}}{CH}-CH_2-\overset{\overset{O}{\|}}{C}-S-ACP + NADP^+$	Also reacts with acyl-CoA or pantetheine esters, but much more slowly. Specifically produces the D(—)isomer.
(4) 3-hydroxyacyl-ACP dehydratase (*Fab A, Fab Z*)	$D(-)CH_3-\overset{\overset{OH}{\|}}{CH}-CH_2-\overset{\overset{O}{\|}}{C}-S-ACP \rightleftharpoons$ $\overset{H_3C}{}\diagdown C=C \diagup^{H}_{}$ $H\diagup\diagdown\underset{\|}{C}-S-ACP + H_2O$ $\overset{\|}{O}$	Measured by hydration of crotonyl-ACP accompanied by decrease in absorption at 263 nm. Does not metabolize model compounde. Stereo-specific for D(—)isomer. Fab Z is active on all chain lengths of saturated and unsaturated intermediates. Fab A is a dehydratase/isomerase needed for anaerobic unsaturated fatty acid synthesis.
(5) Enoyl-ACP reductase (*Fab I*)	$\overset{H_3C}{}\diagdown C=C\diagup^{H}_{}$ $H\diagup\diagdown\underset{\|}{C}-S-ACP + NADPH + H^+$ $\overset{\|}{O}$ $\rightleftharpoons CH_3-CH_2-CH_2\overset{\overset{O}{\|}}{C}-S-ACP + NADP^+$	Two enzymes occur: (i) NADPH-specific, short-chain acid preferred; inhibited by iodoacetamide, NEM, pCMB. Specific for ACP esters. (ii) NADH-specific, long-chain acids preferred. Uses CoA or ACP esters. NEM stimulates.
(6) Acetoacetyl-ACP synthase (*Fab H*)	$CH_3-\overset{\overset{O}{\|}}{C}-SCoA + HOOC-CH_2\overset{\overset{O}{\|}}{C}-S-ACP$ $\rightleftharpoons CH_3-\overset{\overset{O}{\|}}{C}-CH_2-\overset{\overset{O}{\|}}{C}-S-ACP + CO_2 + CoA$	Allows acetyl-CoA to be used directly (without acetyl transfer to ACP). Unlike 3-ketoacyl-ACP synthase it is not inhibited by cerulenin.

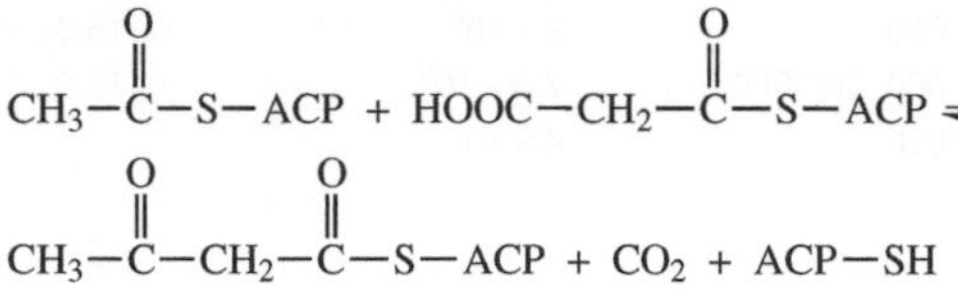

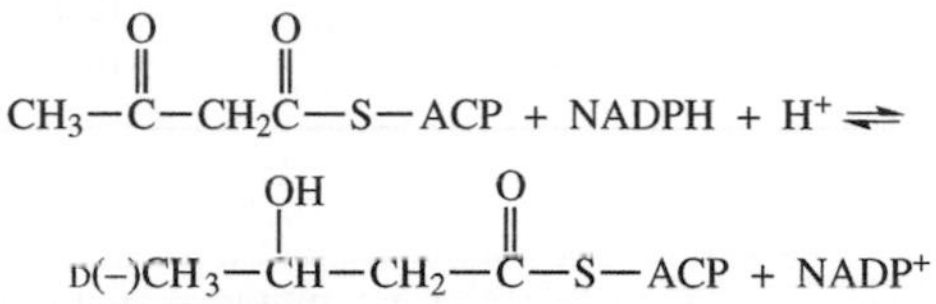

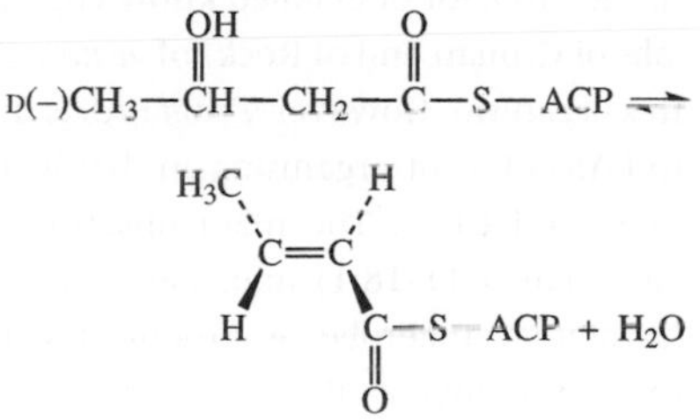

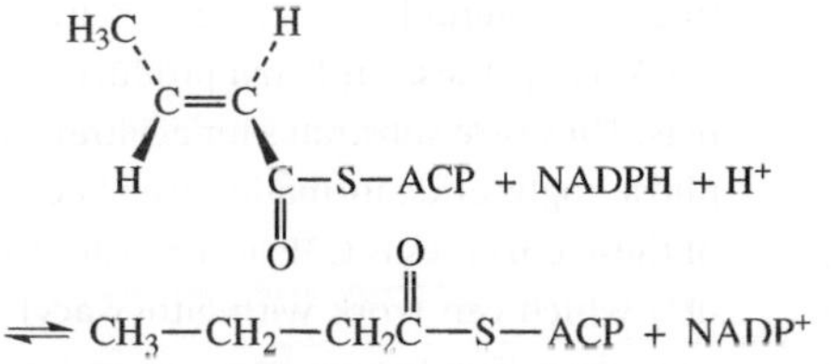

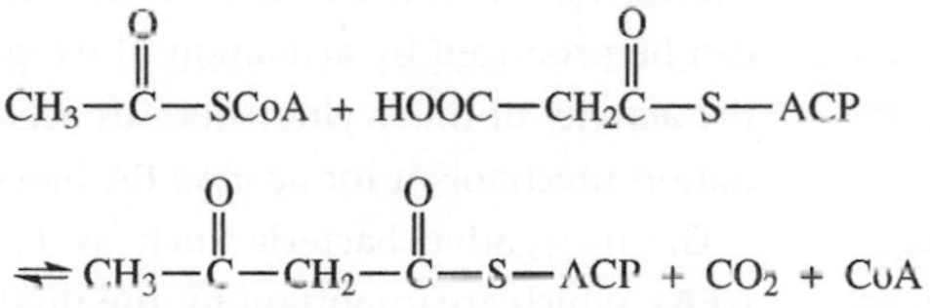

Abbreviations: NEM = *N*-ethylmaleimide; pCMB = *p*-chloromercuribenzoate.

concentration of CoA produced was also high; but if these cells were washed free of pantothenate and transferred to a medium with no pantothenate, then the CoA concentration dropped rapidly while that of ACP remained constant. In other words, ACP was synthesized at the expense of CoA. In *E. coli* the concentration of CoA is about eight times that of ACP. Even so ACP is an abundant protein, representing about 0.25% of the total soluble proteins. It seems that ACP biosynthesis is under tight control and this may be yet another factor involved

in the overall control of FA biosynthesis. The ACP pool must be severely depleted in order to affect the rate of FA biosynthesis in *E. coli*. Indeed, ACP is one of the most interactive proteins in bacterial physiology and plays many roles. Since ACPs from one bacterium can usually substitute for that in other species, *E. coli* ACP is conveniently used in assays of plant Type II FASs.

Two enzymes are involved in the turnover of the prosthetic group of ACP. Both enzymes are highly specific and even small modifications of the ACP structure

Table 3.3 Types of fatty acid synthases in different organisms.

Source	Subunit types	Subunit (mol. mass)	Native (mol. mass)	Major products
Type I: Multicatalytic polypeptides				
Mammalian, avian liver	α	$220–270 \times 10^3$	$450–550 \times 10^3$	16:0 free acid
Mammalian mammary gland	α	$200–270 \times 10^3$	$400–550 \times 10^3$	4:0–16:0 free acids
Goose uropygial gland	α			2,4,6,8-tetramethyl-10:0
M. smegmatis	α	290 000	2×10^6	16:0-, 24:0-CoA
S. cerevisiae	α, β	185 000, 180 000	2.3×10^6	16:0-, 18:0-CoA
Dinoflagellates	α	180 000	4×10^5	
Type II: Freely dissociable enzymes				
Higher plant chloroplasts	Separate enzymes	–	–	16:0-, 18:0-ACP
E. gracilis chloroplast	Separate enzymes			12:0-, 14:0-, 16:0-, 18:0-ACP
E. coli	Separate enzymes	–	–	16:0-, 18:1-ACP

prevents activity. The first enzyme catalyses the transfer of 4′-phosphopantetheine from CoA:

$$\text{Apo-ACP} + \text{CoA} \underset{holo-ACP\ synthase}{\overset{Mg^{++}}{\rightleftharpoons}} \text{Holo-ACP} + 3'5'\text{-ADP}$$

The second causes degradation and, hence, turnover of ACP.

$$\text{Holo-ACP} + H_2O \underset{ACP\ phosphodiesterase}{\rightleftharpoons} \text{Apo-ACP} + 4'\text{-phosphopantetheine}$$

In vivo, both enzymes have been demonstrated in *E. coli* and turnover of the 4′- phosphopantetheine group has been well established and studied in prokaryotes. In plants, also, the holo-ACP synthase has been detected. (Note that an apoprotein refers to the protein part of a holoenzyme that has had a second moiety (in this case, the prosthetic group) added.)

In *E. coli* the concentration of ACP is carefully regulated. Moreover, overproduction of ACP encoded by an inducible plasmid vector is lethal to *E. coli*. This is probably because most of the protein expressed in inducible systems is apo-ACP and the latter is a potent inhibitor of glycerol 3-phosphate (G3P) acyltransferase, which is a key enzyme for lipid assembly (see Sections 4.1.1 and 4.5.4). In fact, virtually all the ACP is normally maintained in its active holo-form in vivo.

Following the discovery of ACP, the different steps in the reaction sequence were quickly elucidated, first in the yeast *Saccharomyces cerevisiae* and later confirmed with purified enzymes from *E. coli*. The availability of mutants and the ability to manipulate gene expression in *E. coli*

has led to a lot of detailed knowledge (elucidated in the labs of Cronan and of Rock) of *de novo* FA biosynthesis in this organism. However, *E. coli* FAS is unusual, compared to FASs of most organisms, in that its products are both SFAs and UFAs. The main unsaturated product is *cis*-vaccenate (*c*11-18:1) and, perhaps because *E. coli* is a facultative anaerobe, is produced without a need for oxygen – whereas the normal mechanism for FA desaturation is aerobic (see Section 3.1.8).

FA biosynthesis in *E. coli* produces 16C and 18C products. These are substrates for acyltransferases involved in phosphoglyceride formation (see Section 4.5.8). The first of these enzymes is G3P acyltransferase (*plsB* gene product) which can work with either acyl-ACP or acyl-CoA substrates. The former come from FAS, while the latter can be produced by activation of exogenous FAs. Thus, the activity of the acyltransferases acts as a chain termination mechanism for *de novo* FA biosynthesis.

Gram-negative bacteria such as *E. coli* also contain UFAs, which are important for the fluidity characteristics of their membranes. Unlike most organisms (see Section 3.1.8), such bacteria can produce their unsaturated products anaerobically using FAS (Fig. 3.7). The *fabA*-encoded dehydratase is capable of producing both the normal *trans*-2 product but also the *cis*-3 UFA at the 10C stage. While *trans*-2-decenoyl-ACP is reduced in the normal way to a SFA (10:0), the *cis*-double bond alters the shape of the substrate molecule which, together with its changed position, prevents reduction. However, condensation can still take place so that a series of *cis*-UFAs is produced. Both FabB and FabF condensing enzymes are capable of working with either SFA or UFA substrates but

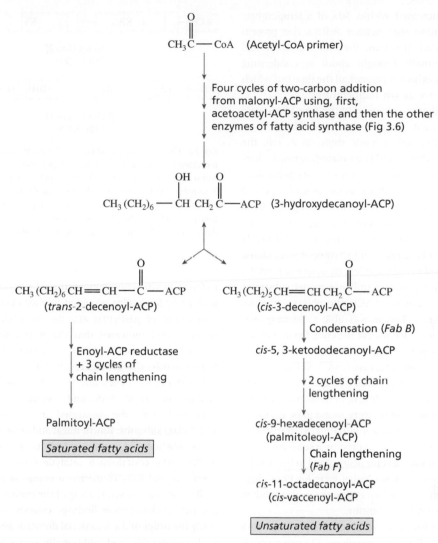

Fig. 3.7 Production of unsaturated fatty acids in *E. coli.*

FabF is much better than FabB in using palmitoleoyl-ACP. On the other hand, FabB is required for the elongation of the *cis*-3-decenoyl-ACP. Indeed, mutants lacking FabB require supplementation with UFAs for growth. Hence, the production of *cis*-vaccenate (*c*11-18:1), the main UFA in *E. coli*, is very much dependent on FabF activity.

Control of product quality is one of the most important aspects of microbial physiology. A good example is in maintenance of membrane fluidity at different temperatures. *E. coli* is a poikilotherm (cannot adjust its own

temperature) and, therefore, at lower growth temperatures it has to make more UFA because these have a lower transition temperature (see Tables 2.1 and 2.2). Thus, the activity of FabF (condensing enzyme II) in producing the main unsaturated product, *c*11-18:1, is vital for temperature adaption. Indeed, the observation that strains lacking condensing enzyme II also lacked thermal regulation originally demonstrated its importance. Recently, mutation of its gene (*fabF*) was shown to affect both enzyme activity and temperature regulation to prove the connection. Remarkably, *c*11-18:1

biosynthesis is increased within 30 s of a temperature downshift indicating that neither mRNA nor protein synthesis is needed. It seems, therefore, that thermal regulation is normally brought about by condensing enzyme II (FabF), which is present all the time but which is much more active at low temperatures.

The discovery that the enzymes of the Type I FAS were apparently in a tight complex presented difficulties in studying the individual enzymic steps. In *E. coli*, the individual intermediates could be isolated, purified, char-acterized and used as substrates to study the enzymology of each reaction. In yeast this was not possible because the intermediates remained bound to the enzymes all the time. Pioneering work in Feodor Lynen's laboratory established that the component enzymes were intimately associated and all of the reaction intermediates remained covalently attached to the complex. By mapping peptides labelled with substrates and utilizing model substrates that could mimic the natural ACP-phosphopantetheine-linked intermediates, Lynen was able to identify and assay the component enzymes of the complex and estab-lish the reaction sequence for the synthesis of the prod-ucts, palmitoyl- and stearoyl-CoA. Although the underlying chemistry was similar to that of the *E. coli* pathway (Table 3.2), the loading of malonyl moieties and the unloading of the product were found to be catalysed by the same enzyme, a malonyl/palmitoyl transferase and the acetyl primer was delivered by a separate acetyl transferase. Lynen's achievements resulted in his sharing with Konrad Bloch the 1964 Nobel Prize for Physiology or Medicine for their contributions to the understanding of FA and cholesterol metabolism.

Further insight into the workings of the fungal FAS were elucidated by Eckhart Schweitzer, a Lynen protégé, who identified and mapped mutants deficient in various components of the yeast FAS. These studies established that the FAS components were encoded by two different genes: acetyl transferase (*AT*), enoyl reductase (*ER*), dehydratase (*DH*) and malonyl/palmitoyl transferase (*MPT*) by *fas*-1, ketoacyl synthase (*KS*), ketoreductase (*KR*) and ACP by *fas*-2 (Fig. 3.8). (NB: gene names are always in italic print and are not listed in the Index to this book. The gene name *ER* should not be confused with the abbreviation for endoplasmic reticulum ER – Roman type – which may be found in the Index.) It was pro-posed, therefore, that yeast FAS consisted of two dis-similar multifunctional peptides α (encoded by *fas*-2) and

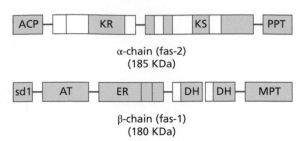

Fig. 3.8 The yeast fatty acid synthase. Functional domains are indicated in shaded areas; structural elements are in white. Abbreviations: On the α-chain ACP, acyl carrier protein; KR, 3-Ketoacyl-ACP reductase; KS, 3-ketoacyl-ACP synthase; PPT, phosphopantetheinyl transferase. On the β-chain sdI, structural domain I; AT, acetyltransferase; ER, enoyl reductase; DH, dehydratase; MPT, malonyl/palmitoyltransferase.

β (*fas*-1). In vitro complementation of appropriate pairs of mutant FAS proteins provided further evidence for the assignment of particular activities to each of the two subunits and indicated that the minimum functional unit contained two copies of each subunit (i.e. α2β2). Subsequently, the genetic results were confirmed inde-pendently in the Lynen and Wakil laboratories by disso-ciating the yeast FAS and assaying the activities associated with the separated α (~2.1 kDa) and β (~2.2 kDa) subunits. Biochemical and cross-linking stud-ies established that the fungal FAS has a molecular mass of 2.6 MDa, contained 6 catalytic centres for FA bio-synthesis and thus contained 6 copies of each subunit (α6β6). Ultimately, sequencing of the entire *fas*-1 and *fas*-2 genes validated these findings, established unambigu-ously the order of the functional domains associated with each polypeptide and additionally revealed that the α subunit also contained (at the C-terminus) a phospho-pantetheine transferase domain capable of post-transla-tional modification of the ACP domains (Fig. 3.8).

Early studies by Lynen's laboratory showed that the fungal FAS particles were large enough to be imaged by electron microscopy and revealed that these heterodo-decamers form an extremely large barrel-shaped struc-ture. Recently, as a result of a fortuitous accident in Tom Steitz's laboratory, while attempting to crystallize the yeast 40 S ribosomal subunit, crystals of FAS were obtained instead that co-sedimented at 40 S with the small ribosomal subunit. X-ray analysis of these crystals (by Steitz's and Nenad Ban's laboratories; see Fig. 3.9)

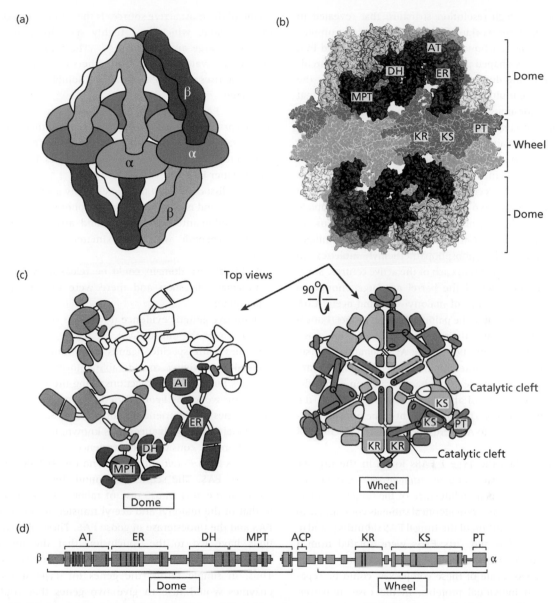

Fig. 3.9 Structural organization of the fungal-type fatty acid synthase from yeast. (a) Early electron microscopy studies revealed the protein as a 27 nm barrel-shaped particle consisting of a central wheel formed by the six alpha subunits flanked by domes formed by the six beta subunits, three on each side of the wheel. (b) Recent high-resolution crystal structure shown in surface representation. (c) Cartoon representations of the upper dome and the central wheel viewed from the top. Components of the beta chains comprising the domes are shown in dark grey, light grey and white, whereas components of the alpha chains comprising the wheel are shown in two different shades of blue depending on their spatial arrangement. The six alpha subunits in the central wheel form three alternating KS and KR dimers. In each dimer the active centres are orientated one towards the upper chamber, the other towards the lower chamber. For each alpha chain, the KR and KS active sites point into different reaction chambers. The six ACP domains, not shown for clarity, are orientated three towards each of the upper and lower chambers. Three pairs of phosphopantetheinyl transferase (PT) domains are located outside of the barrel, suggesting that activation of the ACP domains takes place prior to assembly of the structure. All the active centres of the beta

provided a high-resolution structure that revealed in detail the inner workings of this extraordinary complex in which 48 functional centres are accommodated in a single barrel-shaped particle. The highly complex architecture of the dodecameric particle is facilitated by the presence of multiple structural domains and linkers that account for almost half of the total protein. The barrel is divided into two chambers by a central wheel-like structure that contains the six ACP domains, three oriented towards each of the two chambers. The catalytic domains are embedded in the walls of the chamber with their active centres directed to the inside of the chamber so that each of the two chambers contains three complete centres for the assembly of a FA. The ACP domains are tethered to the central wheel region by flexible hinges that permit the phosphopantetheinyl moieties to deliver intermediates to each of the active centres. Small pores in the sides of the barrel are large enough to permit only the entry of malonyl-CoA and acetyl-CoA substrates and exit of the palmitoyl-CoA product. Fungal FASs also contain their own machinery for post-translational modification of the ACP domains. The six resident phosphopantetheinyl transferase domains are located on the outside of the barrel, three on each side, and out of range of the ACP domains caged in the internal chambers. Thus attachment of the phosphopantetheine moieties to ACP must occur prior to assembly of the complex.

Remarkably, the Type I FASs found in the mycolic acid-producing subgroup of *Actinomycetales* exhibit the same barrel-shaped architecture of the fungal counterpart but are hexamers of identical subunits comparable to a head-to- tail fusion of the fungal FAS subunits β und α.

In the 1960s and '70s FASs were isolated from a variety of animal species and tissues and, as with the fungal FASs, none of these complexess could be separated into individual proteins. In this case, however, the components were found to covalently linked on two identical polypeptides each of molecular mass ~270 kDa.

One of the most active sources is the goose uropygial (preen) gland, which is a highly specialized organ producing large amounts of FAs. (These are used to provide the waterproofing for feathers of water fowl) FAS comprises up to 30% of the total soluble protein in this gland.

Animal FAS complexes are homodimers of native molecular mass ~540 kDa. For some years there was considerable debate as to whether the animal enzymes were heterodimers, like yeast, or homodimers. Evidence for the latter is:

1 When dissociated into monomers, only a single polypeptide band is obtained on electrophoresis.

2 4′-Phosphopantetheine was found associated with the polypeptide with approximately 2 moles per dimer.

3 A thioesterase domain could be released by partial proteinase digestion and there were two domains per dimer.

Definitive genetic evidence now exists (i.e., a single gene encodes the entire FAS).

Studies with specific ligands and inhibitors have succeeded in giving considerable information about the sites for the partial reactions along the FAS molecule. The work has been carried out in a number of laboratories but Suriender Kumar and Salih Wakil made major initial advances. It is known, for example, that there is considerable sequence homology for the ACPs from *E. coli* and barley and the ACP domain of rabbit FAS. The sequence around the active site serine of the acyl transferase of rabbit FAS is similar to that of the malonyl and acetyl transferases of yeast FAS and the thioesterase of goose FAS. These observations have led to the conclusion that the multifunctional forms of FAS have arisen by gene fusion. Thus, in simple terms, the genes for Type II FAS enzymes would fuse to give two genes that could code for a yeast-type FAS. The latter would then fuse to give a single gene coding for the animal Type I FAS enzyme.

---

subunits are oriented towards the inside of either the upper or lower chamber of the barrel. (d) Domain map of the fungal FAS. Wide bars represent catalytic domains, narrower bars represent structural elements and the narrowest bars represent interdomain linkers. In yeast, the MPT domain is split and the catalytic domains are distributed on the two polypeptide chains. However, in many other fungi and in mycobacteria a single polypeptide chain harbors all domains.

However, the fusion of the gene for animal FAS must have occurred by independent events (rather than by the simple scheme described above) for several reasons.

1 The molecular mass of the two yeast FAS proteins combined is 50% greater than the mammalian FAS.

2 The chain termination mechanisms for yeast FAS (transfer to CoA) and mammalian FAS (liberation of NEFA) are different.

3 The second reductase of yeast FAS is unique in using FMN as a cofactor.

4 In mammalian FAS, a single active site transfers the acetyl and malonyl residues whereas in yeast FAS there are two active sites (see below).

5 Recent studies on the site of the partial reactions on mammalian FAS have shown how they are arranged on the individual polypeptides of the dimer (Fig. 3.10). This arrangement places the enoyl reductase between 3-ketoacyl synthase and ACP in mammalian FAS whereas the reductase is on a separate protein (β-unit) from the other two functions (α-unit) in yeast FAS (Fig. 3.8).

The FAS from *Mycobacterium smegmatis* may, however, have originated from fusion of two genes similar to those of yeast. It has a molecular mass of about 290 kDa (Table 3.3) and, like the yeast enzyme, uses FMN for its enoyl reductase and transfers the products to CoA.

The discovery that components of the animal and fungal type I FASs exhibited significant sequence similarity to their Type II counterparts led to the hypothesis that the multifunctional polypeptides of FAS have arisen by fusion of the genes encoding individual catalytic components. However, in eukaryotes, evolution of the Type I FASs appears to have proceeded along entirely different lines (see above), resulting in two distinct architectural forms, a 2600 kDa barrel-shaped structure in fungi and a 540 kDa X-shaped structure in animals. Tentative location of the catalytic domains on the animal FAS polypeptide was made initially using active-site labelling and peptide mapping and ultimately refined by sequencing and mutagenesis. These studies revealed that genes encoding the catalytic domains of the animal FAS have fused in a different order than in the fungal counterpart with minimal insertions of noncatalytic structural elements (Fig. 3.10). The two architectural forms of FAS also differ at the functional level in several ways. First, the loading of acetyl and malonyl substrates is catalysed by the same transferase in the animal FAS (MAT in Fig. 3.10) but by different transferases in the fungal form (AT and MPT, respectively: Fig. 3.8). Second, the chain-terminating reaction is catalysed by a thioesterase in the animal FAS (TE) but by an acyltransferase in the fungal form; this transferase activity is shared by the same domain responsible for loading of malonyl moieties in the fungal form (MPT). Thus the products are NEFAs in the animal FAS and acyl-CoA thioesters in the fungal form. Third, the enoylreductase reaction requires NADPH as a cofactor in the animal FAS, FMN in the fungal form. Finally, the fungal FAS is capable of auto phosphopantetheinylation using the resident PPTase (phosphopantetheine transferase) domain on the α-subunit, whereas the animal FAS is post-translationally modified by a freestanding PPTase that is responsible for servicing all proteins that require phosphopantetheinylation.

In the early 1980s, Wakil proposed that the two subunits of the animal FAS dimer were oriented in an antiparallel orientation and this 'head-to-tail' arrangement is commonly found today in textbooks of biochemistry. However, the development of technology for the engineering and expression of recombinant animal FAS mutants in Stuart Smith's laboratory enabled the application of mutant complementation and chemical cross-linking approaches to interrogate rigorously the head-to-tail model. These studies revealed that the substrate loading and condensation reactions are catalysed by cooperation of an ACP domain of one subunit with the AT or KS domains, respectively, of either subunit. The β-carbon-processing reactions, responsible for the complete reduction of the β-ketoacyl moiety following each condensation step, and the chain termination reaction, are catalysed by cooperation of an ACP domain with the KR, DH, ER and TE domains associated with the same subunit. Furthermore, the two N-terminal KS domains

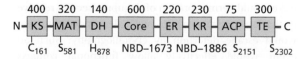

Fig. 3.10 Linear domain map for rat fatty acid synthase. The approximate number of residues in each domain is indicated above the map. Nucleophilic residues involved in covalent acyl-O-serine, acyl-S-cysteine and acyl-S-phosphopantethein intermediates and the dehydratase active-site histidine are shown below. The start of two glycine-rich motifs in the nucleotide-binding domains are shown as NBD. Data taken Joshi, Witkowski & Smith (1998) – see **Further reading**.

could be chemically cross-linked with spacers as short as 6 Å. These findings led to the proposal of an alternative model in which the subunits are coiled in a head-to-head orientation (Fig. 3.11).

More recently X-ray crystallographic studies by Nenad Ban and colleagues have provided a detailed insight into the architecture of the animal FAS at 3.2 Å resolution. They characterized the overall appearance of the FAS as X-shaped body in which the upper 'arms' contain all of the enzymes required for the β-carbon processing reactions (DH, ER and KR) and the lower 'legs' consist of the two enzymes required for substrate loading and chain-elongation (MAT and KS) (Fig. 3.12). This arrangement provides for two reaction chambers in the regions between the arm and leg sections, one on each side of the molecule. The crystal structure (see figure in **About the companion website**) revealed

(a)

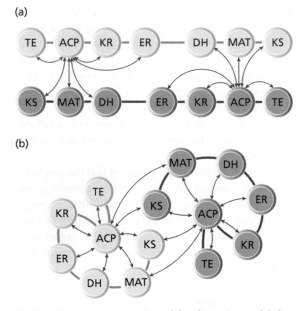

(b)

Fig. 3.11 Cartoon representations of the alternative models for the animal FAS. (a) In the original head-to-tail model the two subunits are fully extended in an antiparallel orientation. (b) In the new model the subunits are arranged in a head-to-head coiled configuration. Arrows indicate the functional interactions between the ACP domains and catalytic domains, both inter- and intrasubunit. Only in the new model are interactions possible between the ACP and the KS and MAT catalytic domains of both subunits. Abbreviations: KS, β-ketoacyl synthase; MAT, malonyl/acetyl transferase; DH, dehydratase; ER, enoylreductase; KR, β-ketoreductase; ACP, acyl carrier protein; TE, thioesterase.

that dimerization of the KS, ER and DH domains also contributes to stabilization of the overall dimeric structure of the FAS. The crystallographic findings corroborated the proposed head-to-head coiled arrangement of the FAS subunits and identified two nonenzymic domains within the previously unassigned structural core. One of these regions appears to be a catalytically inactive pseudo-methyltransferase that may be the relic of an ancestral methyltransferase. The other appears to be an essential structural component of the ketoreductase, the catalytic domain of which lies further down the polypeptide chain. The ACP and TE domains, which would have been expected to be located in the upper arm section, are absent from the crystal structure, presumably because of extreme flexibility of the linker connecting these domains to the body of the complex. Nevertheless, crystal structures of isolated ACP and TE domains have been obtained independently.

Unexpectedly, the crystal structure also revealed an asymmetry in the two sides of the molecule with the arm and leg sections positioned closer together in one of the reaction chambers. Since the upper and lower sections of the molecule are connected by a single linker region, it appears that flexibility of this linker would permit tilting of the molecule back and forth to facilitate interaction of the ACP domain in the top section with the chain-extending domains in the bottom section alternatively in the two reaction chambers Fig. 3.12. This possibility is consistent with the earlier proposal, based on biochemical and electron microscopy studies, that biosynthesis of FAs in the two reaction chambers functions asynchronously. Thus while the ACP in one reaction chamber is engaged with the acyl-chain elongating domains in the lower section of the molecule, the ACP in the other reaction chamber engages the adjacent enzymes in the upper section to facilitate β-carbon processing reactions. It is also likely that flexibility of the linker could permit rotation of the upper section relative to the lower section of the molecule, a possibility that would escape detection by crystallography. Such an event would effectively allow functional contacts between the ACPs in the top section of the molecule with either of the two pairs of MAT-KS domains in the lower section of the molecule, as predicted earlier by results of mutant complementation studies in vitro (Fig. 3.11).

One of the limitations of crystallographic analyses is that they typically provide a snapshot of the molecule in a

(a)

(b)

(c)

(d)

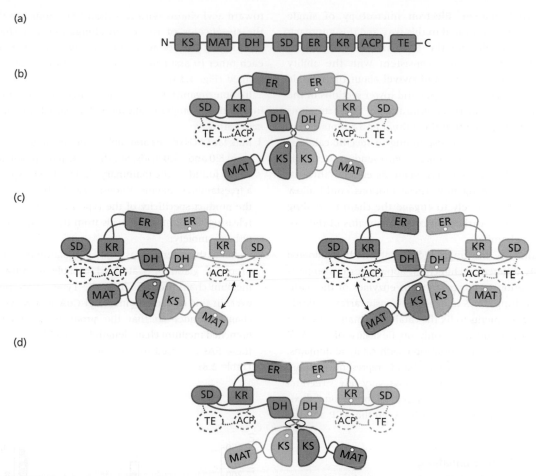

Fig. 3.12 Structure and conformational dynamics of the animal FAS. (a) Domain map of the 2505-residue polypeptide. Abbreviations: KS, β-ketoacyl synthase; MAT, malonyl/acetyl transferase; DH, dehydratase; SD, structural domain; ER, enoylreductase; KR, β-ketoacylreductase; ACP, acyl carrier protein; TE, thioesterase. The SD is required for KR activity and also contains an inactive pseudo-methyltransferase. (b) Structural organization of the FAS dimer. The two subunits are distinguished by black and blue colouring. Extreme mobility of the ACP and thioesterase (TE) domains has precluded their precise location and they are shown as dashed lines. The dimeric nature of the ER, DH and KS domains stabilizes the overall dimeric structure of the FAS, which contains two complete reaction chambers for fatty acid synthesis. The locations of the active sites of catalytic domains contributing to the reaction chamber on the right side are shown as white dots. Those contributing to the left chamber are positioned on the back face of the structure. The upper section of the molecule, which contains the domains required for β-carbon processing and chain termination, and the lower section, which contains the domains required for chain extension, are connected only by a narrow, coiled, flexible junction. (c) Tilting of the molecule back and forth at the flexible junction facilitates interaction of the ACP alternatively with functional domain of the upper and lower sections of the molecule in the two reaction chambers. (d) Swivelling of the two sections of the FAS at the flexible junction also permits the KS and MAT domains of both subunits to function in either of the two reaction chambers, consistent with the head-to-head, coiled subunit model (see Fig. 3.11).

The figure is based on data from electron microscope studies of EJ Brignole, S Smith & FJ Asturias (2009) Conformational flexibility of metazoan fatty acid synthase enables catalysis, *Nat Struct Mol Biol* **17**: 273–9 and the X-ray structure paper of T Maier, M Leibundgut & N. Ban (2008) The crystal structure of a mammalian fatty acid synthase, *Science* **321**: 1315–22.

single conformation. Electron microscopy of single molecules is not limited in this way and this approach was used to identify the FAS in a wide range of different conformations consistent with the ability of the molecules to tilt and swivel about the flexible linker connecting the upper and lower sections of the molecule. The near continuum of conformations observed suggested that the animal FAS adopts different conformations corresponding to specific catalytic events. The gallery of images representing these different conformations has been assembled into an animation showing how these motions could allow the ACP alternatively to engage the chain extending domains and ß-carbon processing domains of the two reaction chambers.

In summary, the fungal and animal FASs have adopted radically different architectural approaches in compartmentalization of the component enzymes of the pathway. The fungal FAS utilizes an extensive array of purely structural elements to fix the catalytic domains in a rigid arrangement and relies only on flexibility of the ACP tether to effect communication with catalytic domains. Animal FAS, on the other hand, represents a more parsimonious design that employs very few structural elements and relies on substantial conformational flexibility to facilitate interaction between the ACP and the catalytic domains.

3.1.3.3 Chain termination

The typical end-product of animal FAS enzymes is unesterified palmitic acid. The cleavage of this acid from the complex is catalysed by a thioesterase which, as discussed above, is a covalently linked domain of the complex. Several factors combine to achieve this specificity. First, the reversible transferase that catalyses the loading of substrate moieties from CoA-esters to the enzyme-bound thiolester group has a high specificity for acetyl and butyryl groups. Thus, once the acyl chain has grown longer than 4C it cannot readily escape the FAS. Second, the condensation reactions are much faster for medium-chain acyl substrates thereby ensuring that, once elongation has started, it rapidly proceeds to longer chain lengths. By contrast, chain lengths of 16C and longer are not readily transferred to the condensing enzyme domain and tend to dwell on the 4′-phosphopantetheine of the ACP domain, where they are targets for the thioesterase. Third, the thioesterase domain exhibits little activity

toward acyl chains with less than 16C atoms, so that the specificities of the chain-elongation and chain-terminating components of the FAS complement each other in assuring that C16 is the major product formed (Fig. 3.13).

In some organisms, FAs shorter than C16 are produced in quantity through modification of the specifity of the *de novo* pathway.

1 Milk fat TAGs of rats and rabbits contain large quantities of 8:0 and 10:0 acids. Smith and Knudsen independently found, in the mammary glands of these species, a freestanding enzyme, thioesterase II, able to modify the product specificity of the type I cytosolic FAS by releasing medium-chain acids from the phosphopantetheine moiety.

2 A similar thioesterase was found by Kolattukudy in the uropygial gland of waterfowl capable of releasing medium chain products from the Type 1 FAS. However, in this case methylmalonyl-CoA is used as the chain extender so that the products are methyl-branched medium chain-length FAs. Lipids containing these FAs are used to waterproof the bird's feathers (Table 2.8).

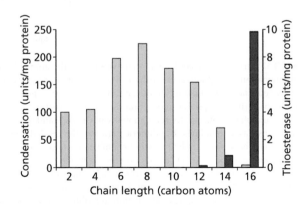

Fig. 3.13 Substrate selectivities for condensation and thioester cleavage determine the chain-length of fatty acid products of animal FAS. Activities were measured with model substrates but show that the capacity of the condensing enzyme (β-ketoacyl-ACP synthase: light blue bars) to transfer acyl chains to its active site cysteine reduces to very low levels with palmitate. In contrast, the thioesterase activity (dark blue) which is negligible up to 14C, increases dramatically for palmitate.

3 In contrast, goat mammary gland, which also produces short- and medium-chain FAs, lacks a thioesterase II enzyme. In this case, specificity of the readily reversible transacylase component of the FAS (MAT) is broadened to include 2C–10C acyl chains enabling the transfer of medium chain-length acyl chains directly to CoA. These acyl-CoAs are then rapidly incorporated into TAGs by microsomal acyltransferases.

4 A medium-chain thioesterase has also been isolated from seeds the California bay tree. This enzyme functions as part of a type II FAS that produces medium-chain FAs for storage in seed lipids. This thioesterase has been introduced into oilseed rape, to produce laurate (12:0)-enriched oils that have industrial utility.

5 For *E. coli*, early experiments showed that the specificity of the 3-ketoacyl-ACP synthase was such that palmitoyl- and vaccenoyl-ACP could not act as primers. However, additional evidence from Cronan's laboratory suggests chain elongation can continue in cells if they are starved of G3P so that the FA products are not transferred into membrane phospholipids. FA biosynthesis in *E. coli* produces 16C and 18C products. These are substrates for acyltransferases involved in phosphoglyceride formation (see Section 4.5.4). The first of those enzymes is G3P acyltransferase (*plsB* gene product), which can work with either acyl-ACP or acyl-CoA substrates. The former come from FAS, while the latter can be produced by activation of exogenous FAs. Thus, the activity of the acyltransferases acts as a chain termination mechanism for *de novo* FA biosynthesis.

6 In *Mycobacterium smegmatis* termination involves transacylation of 16C–24C FAs to CoA. This transacylation is stimulated by polysaccharides, which seem to act by increasing the diffusion of the acyl-CoA esters from FAS rather than promoting acyl transfer from ACP to CoA.

There may be other mechanisms for controlling the chain length of the FAs produced by the various FAS complexes. Certainly, there are plenty of other theories that have been proposed and evidence has been obtained, in some cases, in vitro. Moreover, there are numerous cases where unusual distributions of FAS products are found but about which we know very little of the mechanism of termination.

3.1.4 Mitochondrial fatty acid synthase

Among recently discovered features of mitochondrial biochemistry is their ability to synthesize FAs via an ACP-dependent pathway. The biosynthetic route uses a Type II FAS. The possibility that mitochondria could make FAs *de novo* was reported over 50 years ago but it was assumed that the mechanism was a reversal of β-oxidation. However, in 1988 Brady showed that *Neurospora crassa* mitochondria contained ACP and, since then, other components of the Type II pathway have been isolated. The individual enzymes from eukaryotic mitochondria have sequence similarities to bacterial FAS components.

The mitochondrial pathway produces mainly octanoyl-ACP (although long chain acids can be produced), which is used for lipoic acid biosynthesis and can be transferred directly to the lipolylation sites of various enzymes modified by lipoic acid synthase. Links between mitochondrial FAS and RNA processing have been reported in both yeast and mammals. However, the main role of the pathway seems to be to provide lipoyl moieties for the glycine cleavage enzyme and the α-ketoacid dehydrogenases of the Krebs cycle.

3.1.5 Elongation

Although, as discussed above, the major product of FAS is often palmitate, many tissues contain longer-chain FAs in their (membrane) lipids. For example, in the myelin of nervous tissues, FAs of 18C or greater make up two-thirds of the total, while in many sphingolipids, FAs of 24C are common. In plants, the surface waxes contain mainly very long-chain products in the 28C–34C range.

The formation of these very long-chain FAs is catalysed by the Type III synthases, which are commonly termed elongases because they chain lengthen preformed FAs (either produced endogenously or originating from the diet). Most eukaryotic cells have the capacity to carry out elongation reactions.

Early studies with liver, brain and other mammalian tissues, indicated that there are two elongation systems located in the mitochondria and ER respectively. The mitochondrial system, discovered by Wakil in rat liver, adds 2C units from acetyl-CoA rather than malonyl-CoA. Monoenoic acyl-CoAs are generally preferred to saturated substrates. In tissues such as liver or brain, both NADPH and NADH are needed, whereas heart or skeletal muscles require only NADH. The German biochemist, Seubert, showed the virtual reversal of β-oxidation (see Section 3.2.1) for mitochondrial elongation. However, the enzyme FAD-dependent acyl-CoA dehydrogenase in β-oxidation is replaced by the thermodynamically more favourable enoyl-CoA reductase. The enoyl-CoA reductase isolated from liver mitochondria is different from that of the ER and kinetic studies suggest that its activity largely controls the speed of overall mitochondrial elongation.

The principal reactions for the elongation of longer-chain FAs are found in the membranes of the ER (typically present in microsomal fractions after ultracentrifugation). The reactions involve acyl-CoAs as primers, malonyl-CoA as the donor of 2C units and NADPH as the reducing coenzyme. An example of microsomal elongation is in the nervous system, where large amounts of 22C and 24C SFA are constituents of myelin sphingolipids. Stearoyl-CoA elongase is hardly measureable before myelination begins but rises rapidly during myelination. The mutant 'quaking mouse' is deficient in myelination and has proved to be a useful model for studies of the elongation process. The rate of elongation of 18:0-CoA to 20:0-CoA is normal and that of 16:0-CoA to 18:0-CoA is rather lower than normal in this mutant. The elongation of 20:0-CoA, however, is very much reduced, suggesting that there are at least three elongases in this tissue.

Elongation of FAs occurs in four steps and their chemistry is analogous to that of FAS (see Section 3.1.3.2). However, the elongation reactions are performed by separate proteins which are located on the ER. Three of these are on the cytosolic face of the ER membranes while the dehydratase is embedded in the membrane. The enzymes use acyl-CoA substrates and, preferentially, NADPH as a source of reducing equivalents (Fig. 3.14). For the first reduction, the flow of electrons from NADPH involves two other ER proteins, cytochrome b_5 and cytochrome P450 reductase.

The first enzyme in the four reactions is present as several isozymes, which differ in their substrate selectivity. These condensing enzymes are known as ELOVLs (elongation of very long-chain FAs). Rather less is known of the other three proteins but they seem to have catalytic activity with a variety of fatty acyl chains (although 4 isoforms of the dehydratase have been found in mammals). The seven ELOVLs are listed in Table 3.4, together with some notes on their characteristics. They can be divided into three groups: (a) those that elongate SFA and monounsaturated fatty acids [MUFA] (ELOVL 1, 3, 6); (b) ELOVL isoforms 2 and 4, which are involved in polyunsaturated fatty acid (PUFA) biosynthesis; (c) ELOVL5 which can use a broad range of 16C–22C

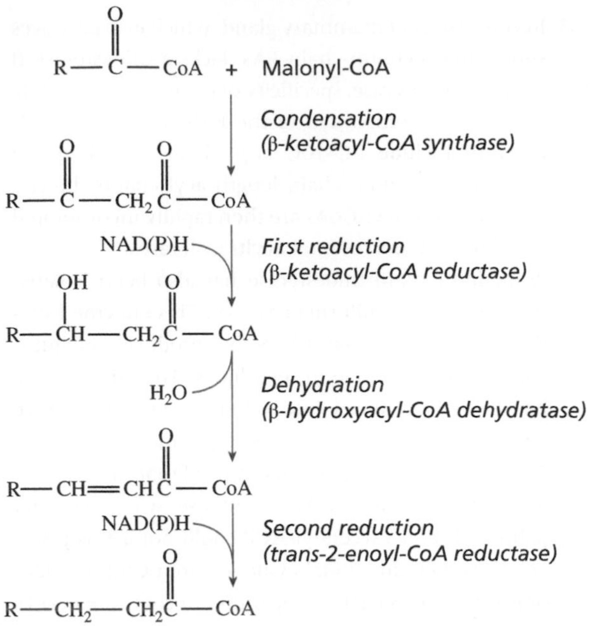

Fig. 3.14 The reactions of fatty acid elongation.

substrates. All ELOVL proteins contain several stretches of amino acids, which are fully conserved in different animals. This includes the HXXHH motif needed for catalytic activity.

Yeast homologues of many of the ELOVLs have been identified and, indeed, yeast mutants have proved to be useful in elucidating the functions and characteristics of many of the isozymes.

One of the most important functions of elongation is in the transformation of the dietary essential fatty acids (EFA), linoleic and α-linolenic acids into longer-chain PUFAs. A series of desaturations and elongations is needed for these processes and the role of specific EVOVLs is shown in Fig. 3.15. Further aspects and details of the generation of very long-chain PUFAs in animals are in Section 3.1.8.3.

Very long-chain FAs in plants are also made by membrane-bound enzyme systems utilizing malonyl-CoA as the source of 2C units in similar fashion to the animal elongases. Acyl-CoAs have been shown to be the substrates in some of these systems (and are presumed to be used by the others) and various elongases have been demonstrated that have different chain-length specificities. Moreover, the individual partial reactions involved have been demonstrated and some purifications achieved. For example, genes coding for the condensing

Table 3.4 Characteristics of the different condensing enzymes (ELOVLs) involved in mammalian fatty acid elongation.

	Substrate selectivity	Notes
ELOVL 1	SFA, MUFA	Deficient in Quaking and Jimpy mice. Important for membrane functions including those for sphingolipids. Ubiquitous tissue expression but especially high in myelin fractions.
ELOVL 2	PUFA	Overlapping function with ELOVL5 but broader selectivity. 20 and 22C PUFA preferred. High activity in testis where high levels of very long-chain PUFAs are present.
ELOVL 3	SFA, MUFA	Glycoprotein. Induced in brown adipose tissue on cold exposure. Important in skin where its deficiency causes impairment of barrier function.
ELOVL 4	PUFA	Glycoprotein. High expression in rod and cone photoreceptor cells probably because it is involved in DHA production. Point mutations cause macular and other dystrophies.
ELOVL 5	PUFA and others	Elongation of 18-22C substrates. Highest expression in testis and adrenal gland where DHA and adrenic (22:4n-6) acid are needed. Regulated by PPAR-α during development.
ELOVL 6	SFA, MUFA (12-16C)	Highly expressed in lipogenic tissues (liver, adipose). Regulated by SREBP and in response to dietary changes.
ELOVL 7	SFA, MUFA	High levels in adrenal glands, kidney, pancreas, prostate. Induced in cancer when it is cytosolic.

enzymes have recently been identified. The production of very long-chain (>18C) FAs is required for the formation of the surface-covering layers, cutin and suberin (see Sections 4.4 & 5.6.1), as well as for seed oil production in commercially important crops such as rape and jojoba (see Chapter 11).

3.1.6 Branched-chain fatty acids

The formation of branched-chain FAs by the Type I FAS of the sebaceous (uropygial) glands of waterfowl has already been mentioned (see Section 3.1.3.2 & Table 3.3). These acids arise because of the use of methylmalonyl-CoA rather than malonyl-CoA, which is rapidly destroyed by a very active malonyl-CoA decarboxylase. The utilization of methylmalonyl-CoA results in the formation of products such as 2,4,6,8-tetramethyldecanoic acid or 2,4,6,8-tetramethyl-undecanoic acid (undecanoic is 11:0) as major products when acetyl-CoA or propionyl-CoA, respectively, are used as primers.

A high proportion of odd-chain and of various polymethyl-branched FAs occurs in the adipose tissue TAGs of sheep and goats, when they are given diets based on cereals such as barley. Cereal starch is fermented by bacteria in the rumen to form propionate, and when the animal's capacity to metabolize propionate via methylmalonyl-CoA to succinate is overloaded, propionyl- and methylmalonyl-CoA accumulate. Garton and his colleagues showed that methylmalonyl-CoA can take the place of malonyl-CoA in FA biosynthesis and that

with acetyl- or propionyl-CoA as primers, a whole range of mono-, di and tri-methyl branched-chain FAs can be produced.

The major FAs in most Gram-positive and some Gram-negative bacterial genera are branched-chain *iso* or *anteiso* FAs. The Type II FAS enzymes present in these bacteria make use of primers different from the usual acetyl-CoA. For example, *Micrococcus lysodeikticus* is rich in 15C acids of both the *iso* type, 13-methyl-C14 or *anteiso* type, 12-methyl-C14. The primers used come from precursors of the amino acids valine and isoleucine, respectively (Fig. 3.16). Thus, isobutyryl-CoA is used as the primer for *iso*-branched-chain FAs and 2-methylvaleryl-CoA for *anteiso* products. These branched-chain FAs increase membrane fluidity in those bacteria that only have low levels of UFA under most growth conditions. As in straight chain FA formation, the primer as its CoA derivative is condensed with malonyl-ACP by condensing enzyme II (FabH). The substrate selectivity of FabH determines whether a particular bacterium can synthesize branched-chain FAs. For example, *E. coli* FabH cannot use branched-chain primers whereas *B. subtilis* has two FabH enzymes both of which prefer branched-chain substrates over acetyl-CoA.

Another common branched-chain FA is 10-methyl-stearic acid (tuberculostearic acid), a major component of the FAs of *Mycobacterium phlei*. In this case, the methyl group originates from the methyl donor S-adenosylmethionine, while the acceptor is oleate esterified in a phospholipid. This is an example, therefore, of FA

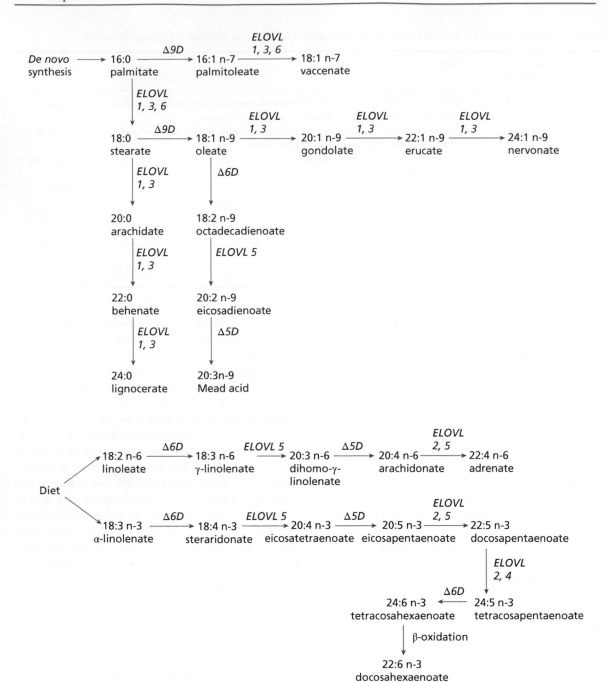

Fig. 3.15 Role of elongases and desaturases in the production of mammalian very long chain fatty acids. Fatty acids are abbreviated with the carbon number followed by the number of double bonds. The positions of the first double bond numbering from the methyl end of the fatty acid is given. Because elongation takes place at the carboxyl end, a series of fatty acids (n-9, n-6 etc.) are formed. However, by convention, the desaturases are usually named from the carboxyl end.

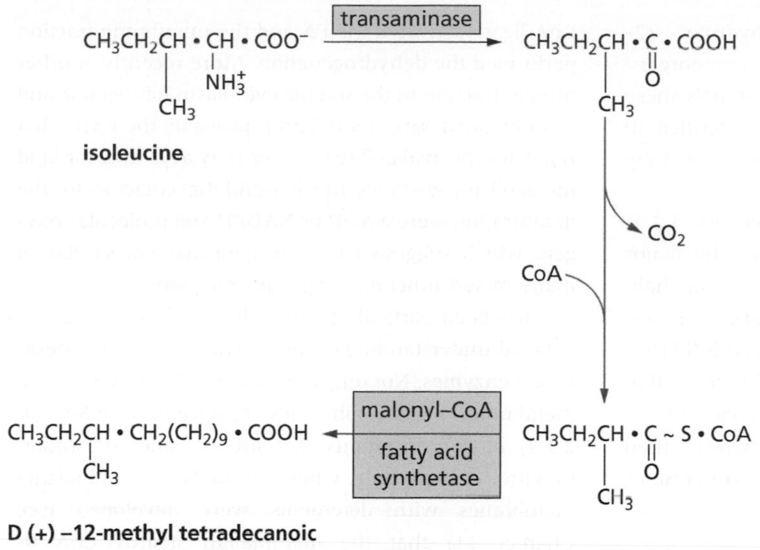

CH$_3$CH$_2$CH • CH • COO$^-$ $\xrightarrow{\boxed{\text{transaminase}}}$ CH$_3$CH$_2$CH • C • COOH
  | | | ||
  | NH$_3^+$ CH$_3$ O
 CH$_3$

isoleucine

CO$_2$

CoA

CH$_3$CH$_2$CH • CH$_2$(CH$_2$)$_9$ • COOH $\leftarrow$ $\boxed{\begin{array}{c}\text{malonyl–CoA}\\ \text{fatty acid}\\ \text{synthetase}\end{array}}$ CH$_3$CH$_2$CH • C ~ S • CoA
  | | ||
 CH$_3$ CH$_3$ O

D (+) –12-methyl tetradecanoic

Fig. 3.16 Production of an *anteiso* branched-chain fatty acid in bacteria.

modification taking place while the acid is in an O-ester rather than as the S-esters of CoA or ACP. The formation of tuberculostearic acid takes place in two steps via a 10-methylenestearic acid intermediate (Fig. 3.17).

3.1.7 The biosynthesis of hydroxy fatty acids results in hydroxyl groups in different positions along the fatty acid chain

Hydroxy FAs are formed as intermediates during various metabolic pathways (e.g. FA biosynthesis, β-oxidation) and are also made by specific hydroxylation reactions. Usually the hydroxyl group is introduced close to one end of the acyl chain. However, mid-chain hydroxlations are also found – a good example being the formation of ricinoleic acid (12-hydroxyoleic acid; Table 2.4). This

acid accounts for about 90% of the TAG FAs of castor oil and about 40% of those of ergot oil, the lipid produced by the parasitic fungus, *Claviceps purpurea*. In developing castor seed, ricinoleic acid is synthesized by hydroxylation of oleate while it is attached to position *sn*-2 of phosphatidylcholine (PtdCho). The 12-hydroxylase accepts electrons from either NADPH or NADH via a cytochrome b$_5$ and uses molecular oxygen. The gene for the enzyme shows many similarities to that for the FA desaturases (see Section 3.1.8.1) and deduced protein structures suggest a common evolutionary origin and reaction mechanism. In contrast to the method of hydroxylation in castor seed, the pathway in *Claviceps* involves hydration of linoleic acid under anaerobic conditions. Thus the hydroxyl group in this case comes from water and not from molecular oxygen.

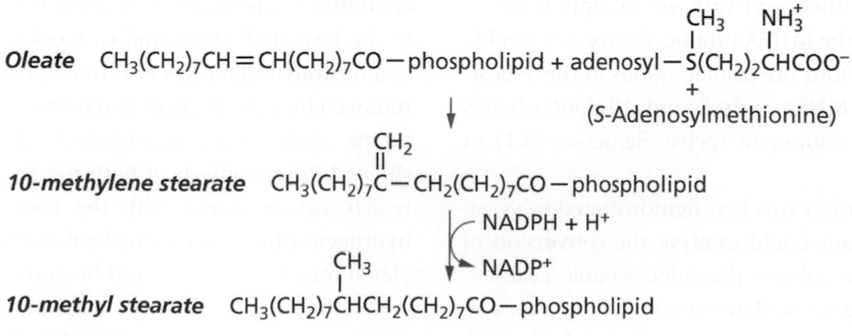

Oleate   CH$_3$(CH$_2$)$_7$CH=CH(CH$_2$)$_7$CO—phospholipid + adenosyl—S(CH$_2$)$_2$CHCOO$^-$
                                                                 | |
                                                               CH$_3$ NH$_3^+$
                                                                     +
                                                          (*S*-Adenosylmethionine)

                                    CH$_2$
                                    ||
10-methylene stearate   CH$_3$(CH$_2$)$_7$C—CH$_2$(CH$_2$)$_7$CO—phospholipid

                                         NADPH + H$^+$
                                    CH$_3$
                                    | NADP$^+$
10-methyl stearate   CH$_3$(CH$_2$)$_7$CHCH$_2$(CH$_2$)$_7$CO—phospholipid

Fig. 3.17 Production of tuberculostearic acid in *Mycobacterium phlei*.

α-Oxidation systems producing α-hydroxy (2-hydroxy) FAs have been demonstrated in microorganisms, plants and animals. In plants and animals these hydroxy FAs appear to be preferentially esterified to certain sphingolipids. α-Oxidation is described more fully in Section 3.2.2.

ω-Oxidation (discussed further in Section 3.2.3) involves a typical mixed-function oxidase. The major hydroxy FAs of plants have an ω-OH and an in-chain OH group (e.g.10,16-dihydroxypalmitic acid). Their biosynthesis seems to involve ω-hydroxylation with NADPH and O_2 as cofactors, followed by in-chain hydroxylation with the same substrates. If the precursor is oleic acid, the double bond is converted into an epoxide, which is then hydrated to yield 9,10-hydroxy groups. These conversions involve CoA esters.

3.1.8 The biosynthesis of unsaturated fatty acids is mainly by oxidative desaturation

3.1.8.1 Monounsaturated fatty acids

UFAs can either be produced anaerobically or in the presence of oxygen, which acts as an essential cofactor. The anaerobic mechanism is rather rare but is used by *E. coli* (and other members of the Eubacteriales) as part of its FAS complex (see Section 3.1.3.2). By far the most widespread pathway is by an oxidative mechanism, discovered by Bloch's team, in which a double bond is introduced directly into the preformed (e.g. saturated) long-chain FA with O_2 and a reduced compound (such as NADH) as cofactors. This pathway is almost universal and is used by yeasts, algae, higher plants, protozoa, animals and most bacteria. Apparently, the two pathways are usually mutually exclusive i.e. organisms usually use one mechanism or the other. However, *Pseudomonas aeruginosa* has three mechanisms! This bacterium can use an anaerobic pathway (like *E. coli*) as well as aerobic mechanisms that use either acyl-CoA or phospholipids as substrates. Most of the MUFA produced have a *c*9-double bond. Some exceptions are double bonds in the 7-position in many algae, MUFA with *c*5- and *c*10-double bonds in Bacilli and a *c*6-monoene (petroselenic; *c*6-18:1) in some plants.

Aerobic desaturation was first demonstrated in yeast. Cell-free preparations could catalyse the conversion of palmitate into palmitoleate (hexadec-9-enoic acid, *c*9-16:1) only if both a particulate microsomal fraction and the supernatant fraction were present. A soluble acid:

CoA ligase activated the FA and the membrane fraction performed the dehydrogenation. More recently another protein fraction in the soluble cytoplasm has been found to stimulate desaturation. This is probably the FABP that regulates the availability of FA or fatty acyl-CoA for lipid metabolizing enzymes. Bloch found that cofactors for the desaturation were NADH or NADPH and molecular oxygen, which suggested to him a mechanism similar to many mixed-function oxygenase reactions.

It has been particularly difficult and slow to obtain a detailed understanding of the biochemistry of the desaturase enzymes. Not only are they usually located in the membranes, but the substrates are micellar (see Section 2.1.7) at concentrations that are suitable for studies in vitro. It was only when methods for solubilizing membranes with detergents were developed (see Chapter 11) that the mammalian stearoyl-CoA 9-desaturase was purified and a better understanding of the enzymic complex emerged. Work by Sato's group in Japan and Holloway in the USA has identified three component proteins of the complex: a flavoprotein, NADH-cytochrome b_5 reductase; a haem-containing protein, cytochrome b_5; and the desaturase itself which, because of its inhibition by low concentrations of cyanide, is sometimes referred to as the cyanide-sensitive factor (CSF). Strittmatter's group in the USA purified the desaturase (and identified its gene), which is a single polypeptide chain of 53 kDa containing one nonhaem iron atom. The essential iron can be reduced in the absence of stearoyl-CoA, by NADH and the electron transport proteins.

Although the stearoyl-CoA desaturation reaction has all the characteristics of a mixed-function oxygenation, nobody has ever successfully demonstrated a hydroxylation as an intermediate step in double bond formation. In spite of our lack of knowledge of the mechanism, certain details have emerged concerning the stereochemistry of the dehydrogenation. Schroepfer and Bloch in the USA and James and co-workers in the UK have demonstrated that the 9 and 10 (*cis*) hydrogen atoms are removed by animal, plant and bacterial systems. Experiments with deuterium-labelled stearate substrates showed isotope effects at both the 9- and 10-positions, which are consistent with the concerted removal of hydrogens rather than a mechanism involving a hydroxylated intermediate followed by dehydration.

The three essential components of the animal 9-desaturase are thought to be arranged in the ER in a manner shown in Fig. 3.18. The cytochrome b_5 is a small haem-containing protein (16–17 kDa), which has a major hydrophilic region and a hydrophobic carboxy terminal anchor of about 40 amino acids. The CSF component is largely within the membrane with only its active centre exposed to the cytosol. Most animal 9-desaturases work well with saturated acyl-CoAs in the range 14C–18C. Two isoforms (SCD1, SCD5) of the stearoyl-CoA desaturase are found in humans but four (SCD1-4) in mice. All these 9-desaturases contain four transmembrane domains and three regions of catalytically essential histidine domains. Of the two human stearoyl-CoA desaturases, SCD1 is ubiquitously expressed whereas SCD5, which is distinct from any of the mouse SCDs, is expressed at high levels in the brain and pancreas.

In yeast, the 9-desaturase is expressed by the *OLE-1* gene. In deficient mutants, the rat liver 9-desaturase

gene can effectively substitute for the missing activity. A general scheme for aerobic FA desaturation is shown in Fig. 3.19 and this applies to the animal 9-desaturase. However, although the animal enzyme clearly uses NADH and cytochrome b_5, evidence suggests that the 9-desaturase of other organisms may not. Interestingly, the enzyme from plant chloroplasts is soluble. Moreover, it uses stearoyl-ACP as substrate instead of stearoyl-CoA (a sensible choice of substrate for a desaturase that is located in the choloroplast) and reduced ferredoxin as a source of reductant.

The availability of the plant 9-desaturase as a soluble protein allowed it to be purified in Stumpf's laboratory and, later, with the availability of a gene coding for the enzyme, it has been possible to obtain a lot of important information about its reaction mechanism using point mutations. Much of this work has come from Shanklin's

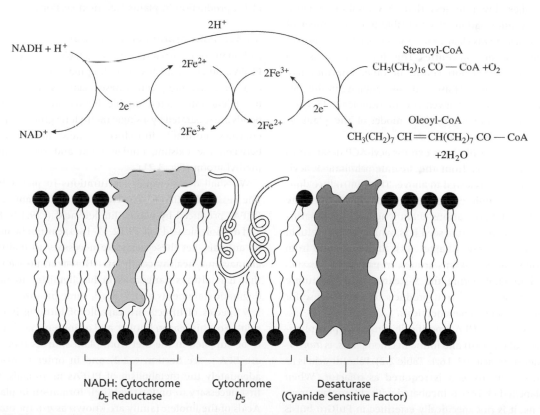

Fig. 3.18 Diagrammatic representation of the animal Δ9-fatty acid desaturase complex. Adapted from HW Cook (1996) Fatty acid desaturation and elongation in eukaryotes. In DE Vance & J Vance, eds., *Biochemistry of Lipids, Lipoproteins and Membranes*, pp. 129–52, Elsevier, Amsterdam, with kind permission of the author and Elsevier Science. Note that the nature of the hydrogen transfer from NADH (which is depicted in the diagram as H⁺) has not been proven. In some Δ5 or Δ6 polyunsaturated fatty acid desaturases, the cytochrome b_5 component is part of the same protein as the desaturase.

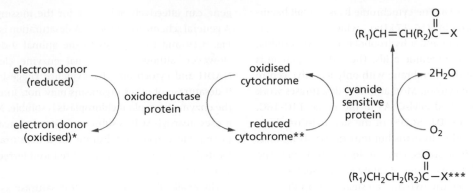

Fig. 3.19 A generalized scheme for aerobic fatty acid desaturation: *e.g. NADH, NADPH, reduced ferredoxin; **e.g. cytochrome b_5;***e.g. acyl-ACP (stearoyl-ACP Δ9-desaturase in plants); acyl-CoA (stearoyl-CoA Δ9-desaturase in animals); oleoyl-phosphatidylcholine (Δ12-desaturase in yeast or plants); linoleoyl-monogalactosyldiacylglycerol (Δ15-desaturase in plant chloroplasts).

laboratory at Brookhaven and he and his co-workers have succeeded in obtaining structural information by X-ray studies down to less than 3 Å resolution. Once sequence information was available from a number of desaturases, it was clear that they belonged to a group of proteins with di-iron centres. This di-iron cluster is also found in the reaction centre of other enzymes like methane monooxygenase and, interestingly, in the oleate hydroxylase that gives rise to ricinoleate (Section 3.1.7). A computer-generated model of the plant 9-desaturase is shown in Fig. 3.20.

Several species of plants express acyl-ACP desaturases which are distinct from the stearate/palmitate 9-acyl-ACP desaturase referred to above. These introduce double bonds into different positions (e.g. 4- and 6-) and are active with FAs of various chain lengths. Such desaturases give rise to unusual seed FAs, which can be useful in biotechnology (see Chapter 11). Indeed, work from Cahoon's and Shanklin's groups, in particular, has succeeded in engineering the 9-stearoyl-ACP desaturase itself to produce different products of potential use for industry as renewable chemicals.

An unusual MUFA, specifically linked to phosphatidylglycerol (PtdGro) and found in chloroplasts, is *trans*-3-hexadecenoic acid (*t*3-16:1; Table 2.2). Palmitic acid is its precursor and oxygen is required as cofactor. When radiolabelled *t*3-16:1 is incubated with chloroplast preparations, it is not specifically esterified in PtdGro but is either randomly esterified in all chloroplast lipids or reduced to palmitic acid. These results suggest that the direct precursor of *t*3-16:1 is palmitoyl-PtdGro rather than palmitoyl-S-CoA or palmitoyl-S-ACP. This is an example of a complex lipid acting as a desaturase substrate and often such substrates seem to be important for PUFA production in plants (see next section).

3.1.8.2 Polyunsaturated fatty acids

Although most bacteria are incapable of producing PUFAs, other organisms, including many cyanobacteria and all eukaryotes, can. These acids usually contain methylene-interrupted double bonds, i.e. the double bonds are separated by a single methylene group. Animal enzymes normally introduce a new double bond between the existing double bond and the terminal methyl group (Fig. 3.21).

We shall describe polydesaturations in plants first. The reason for this is that two of the most abundant PUFAs produced by plants, linoleic acid (*c*,*c*9,12-18:2) and α-linolenic acid (all-*c*9,12,15-18:3) cannot be made by animals, yet these acids are necessary to maintain animals in a healthy condition. (However, please note that it is always dangerous in biochemistry to make dogmatic statements and, indeed, some protozoa and a few species of insects are capable of forming linoleic acid!) For this reason linoleic and α-linolenic acids have to be supplied in the diet from plant sources (i.e. they are EFA – see Section 6.2.2) and in order to discuss adequately the metabolism of PUFAs in animals, it is first necessary to understand their formation in plants. Acids of the linoleic family are known as *n*-6 (or omega (ω)-6) FAs and those derived from α-linolenic acid belong to the *n*-3 (omega(ω) -3) family. Both types are essential for good health and are discussed in Section 6.2.2.

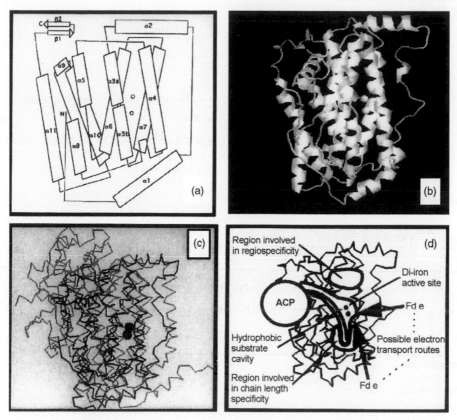

Fig. 3.20 Models of the plant Δ9-stearoyl-ACP desaturase: (a) schematic of secondary structural elements; (b) structural elements; (c) overlay of the desaturase with methane mono-oxygenase; (d) cartoon of functional regions of the desaturase. Reproduced from Fig. 1 of J Shanklin EB Cahoon, E Whittle *et al.* (1997) Structure-function studies on desaturases and related hydrocarbon hydoxylases. In JP Williams, MU Khan & NK Lem, eds., *Physiology, Biochemistry and Molecular Biology of Plant Lipids*, pp. 6–10, Kluwer, Dordrecht, with kind permission of the authors and Kluwer Academic Publishers.

The precursor for PUFA formation in plants and algae is oleate. The next double bond is normally introduced at the 12,13 position (12-desaturase) to form linoleate followed by desaturation at the 15,16 position (15-desaturase) to form α-linolenic acid (all-*c*9,12,15-18:3) as summarized in Fig. 3.22.

α-Linolenic acid is the most common FA found in plants and most fresh-water algae. In marine algae, highly unsaturated 20C acids are predominant, the principal of which (arachidonic and eicosapentaenoic [EPA] acids) are made by the pathways also shown in Fig. 3.22. Because of its high concentration in leaf tissue (Tables 2.3 and 2.13), α-linolenate is the most abundant FA on land.

The possibility that desaturation could occur on fatty acyl chains esterified in complex lipids was first suggested by experiments with the phytoflagellate, *Euglena gracilis*.

These organisms can live either photosynthetically or heterotrophically and can synthesize both plant and animal types of PUFAs. The animal types of FAs accumulate in the phospholipids and the plant types in the galactolipids. It proved impossible to demonstrate plant-type desaturations in vitro when acyl-CoA or acyl-ACP thioesters were incubated with isolated cell fractions. Either the desaturase enzymes were labile during the fractionation of the *Euglena* cells, or the substrates needed to be incorporated into the appropriate lipids before desaturation could take place. The next series of experiments (in A.T. James's laboratory) was done with *Chlorella vulgaris*, a green alga that produces a very simple pattern of plant-type lipids. When cultures of the alga were labelled with ^{14}C-oleic acid as a precursor of the 18C PUFAs, labelled linoleic and α-linolenic acids were

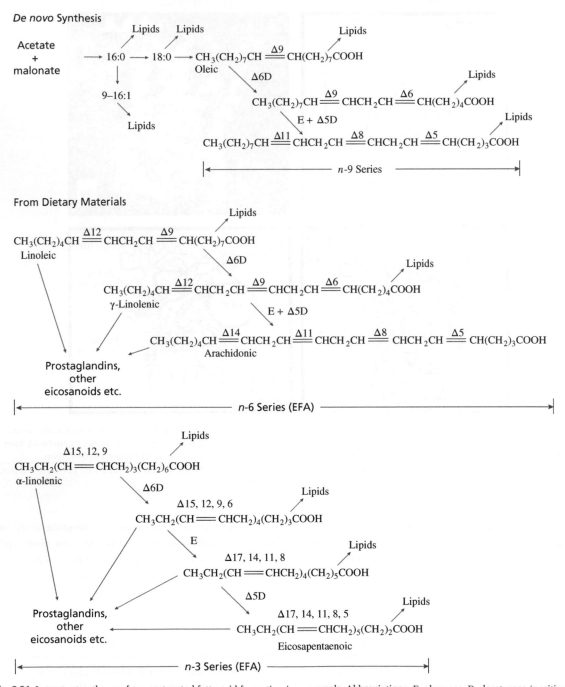

Fig. 3.21 Important pathways for unsaturated fatty acid formation in mammals. Abbreviations: E, elongase; D, desaturase (positional specificity indicated).

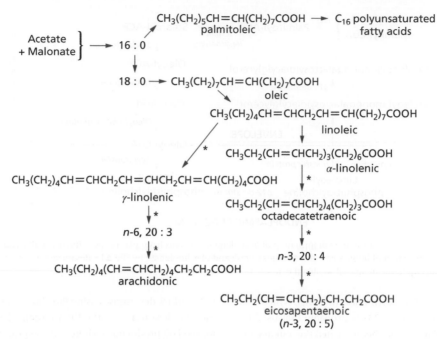

Fig. 3.22 Major pathways for polyunsaturated fatty acis synthesis in plants and algae. *Indicates a pathway found in high levels in marine algae and mosses, but less commonly in other algae or plants.

produced and the label was located only in the PtdCho fraction. Next, synthetic ^{14}C-labelled oleoyl-PtdCho was tested as a substrate for desaturation: the only product formed was linoleoyl-PtdCho. An important loophole that had to be closed was the possibility that during the incubation, labelled oleic acid might be released, activated to the CoA or ACP thiolester, desaturated as a thiolester and then esterified to the same complex lipid very rapidly. Appropriate control experiments eliminated this possibility.

Soon after this discovery, desaturations involving lipid-bound FAs were shown to occur in the mould *Neurospora crassa*, various yeasts such as *Candida utilis* and *Candida lipolytica*, higher plants and several animal tissues.

One of the best studied systems has been that in the leaves of higher plants. It will be recalled that biosynthesis of FAs *de novo* in plants occurs predominately in the plastids. FAS forms palmitoyl-ACP, which is elongated to stearoyl-ACP and then desaturated to oleoyl-ACP. The latter can then be hydrolysed and re-esterified by ACSs on the chloroplast envelope to oleoyl-CoA. Although oleoyl-CoA can be used as a substrate for in vitro systems and will be desaturated rapidly to

linoleate in the presence of oxygen and NADH, both substrate and product accumulate in PtdCho. Indeed, careful experiments by Roughan and Slack in New Zealand, Stymne and Stobart (Sweden and UK) and others have provided considerable evidence that the actual desaturase substrate is 1-acyl-2-oleoyl-PtdCho. Although PtdCho may also be the substrate for linoleate desaturation in a few systems, for leaf tissue the final desaturation appears usually to utilize another lipid, monogalactosyldiacylglycerol (MGDG), as a substrate. These pathways are indicated in Fig. 3.23.

One difficulty with the scheme (shown in Fig. 3.23) is the necessity for movement of oleate out of, and linoleate back into, the plastid. The transport of oleate and its esterification into PtdCho is solved by the participation of oleoyl-CoA, which is water-soluble. In fact, acyl-CoAs can be bound to acyl-CoA binding proteins (see Section 3.1.1) that help to prevent any adverse effects from raised acyl-CoA levels having detergent activity. How the linoleate returns to the chloroplast is not known at present, although the diacylglycerol (DAG) part of the PtdCho molecule may be recycled intact. Certain phospholipid exchange proteins, which have been isolated from plants by Yamada and co-workers, may play a role here.

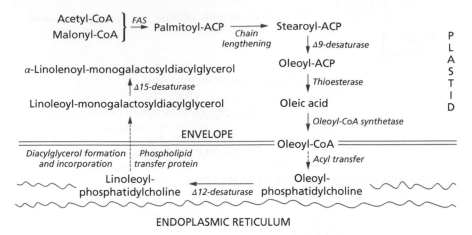

Fig. 3.23 A simplified depiction of the overall formation of α-linolenate in leaves from plants operating the 'eukaryotic pathway' of lipid synthesis. Note formation of linoleate by the Δ12-desaturase and of α-linolenate by the Δ15-desaturase is membrane-associated whereas the reactions up to oleate are all soluble (i.e. in the plastid stroma).

The pathway for PUFA formation in plants, which was discussed above (see Fig. 3.23) is probably that used by the majority of plants. It has been termed the 'eukaryotic pathway' because it involves the participation of extra-chloroplastic compartments and particularly because 18C FAs are esterified in position *sn*-2 of participating lipids as they would be in other eukaryotes, like animals. By contrast, desaturation (and formation of chloroplast lipids) continues within the chloroplast in some plants and such mechanisms are termed 'prokaryotic'. For the latter desaturations, MGDG is used as substrate – allowing the formation of α-linoleate and, also, hexadecatrienoate (16:3) at position *sn*-2. An example of a plant operating the prokaryotic pathway would be spinach (see Table 2.13). However, the most important point to stress is that for all plants, PUFAs are made on complex lipid substrates.

A small, scruffy, plant weed named *Arabidopsis thaliana* has proved to be exceptionally useful to geneticists and biochemists because of its small genome and short generation time. Browse and Somerville made mutants of *Arabidopsis* and discovered a number with modified FA and/or lipid patterns. They then used these mutants to reveal the genes coding for desaturase expression. By such experiments we now know the identity of genes for desaturations used in the 'prokaryotic' and 'eukaryotic' pathways referred to above.

The genes for the main desaturases used by plants have been identified in a variety of species. Seven different desaturases are used with five in the plastid and two

(12- and 15-desaturases using PtdCho as substrate) on the ER. The desaturases in the ER are particularly important for seed oil production, whereas those producing UFAs in the plastid are more important in nonseed tissues. However, even in leaves the relative importance of the different subcellular compartments for overall PUFA formation differs between different plants and is also affected by development and environmental stresses.

A few plant species accumulate PUFA which have a 6-double bond (e.g. γ-linolenic acid; all-*c*6,9,12-18:3). The 6-desaturase introducing the double bond at this position belongs to a group of 'front-end' desaturases, so-called because, they (and many animal desaturases; see later), introduce the double bond on the carboxyl side of existing double bonds (rather than towards the methyl end, like the 12- and 15-desaturases). Plant 6-desaturases plants also contain the donor of reduced equivalents, cytochrome b$_5$, physically fused to the N-terminal end of the CSF moiety. These desaturases also contain a modified residue (glutamine for histidine) in the third histidine box characteristic of the membrane-localized plant desaturases.

Although complex lipid substrates have been studied best in plants, the conversion of eicosatrienoyl-PtdCho into arachidonoyl-PtdCho is an example of a similar reaction in animals (e.g. rat liver). Interestingly, some yeasts have been shown to contain oleoyl-CoA as well as oleoyl-PtdCho desaturases.

In Fig. 3.22 some desaturase reactions involved in the production of 20C or 22C PUFAs by algae are shown.

Because marine or freshwater algae are often at the base of food chains, the production of arachidonic, eicosapentaenoic (EPA; all-*c*5,8,11,14,17-20:5) and docosahexaenoic (DHA; all-*c*4,7,10,13,16,19-22:6) acids is particularly important for the health of higher organisms which often have limited ability to synthesize them. The importance of these PUFAs for good health is covered in Section 6.2.2 and their metabolism is described in Section 3.5.

3.1.8.3 Formation of polyunsaturated fatty acids in animals

Double bonds are found mainly at the 9-, 6-, 5- and 4-positions in animal PUFAs and it has been assumed for a long time that four desaturations were involved. However, although 9-, 6- and 5-desaturase activities have been measured in a variety of animal tissues, the 4-desaturase has not been convincingly demonstrated. Moreover, Sprecher found an alternative pathway to the major 4-PUFA, DHA, (22:6,*n*-3). Thus, 22:5(*n*-3) is elongated instead of being desaturated directly (Fig. 3.24). The 24:5(*n*-3) product is then desaturated at the 6-position and shortened by two carbons through β-oxidation (see Section 3.2.1), in peroxisomes. This alternative to 4-desaturation is known as the Sprecher Pathway and a similar mechanism is involved in the creation of 22:5 (*n*-6) as shown in Fig. 3.24. (This contrasts with the direct 4-desaturation of 22:5(*n*-3) which algae are able to do, as shown in Fig. 3.22.)

Both the animal 5- and 6-desaturases are membrane-bound (as are all the PUFA-forming desaturases) and localized in the ER. They are both 'front end' desaturases with a fused cytochrome b₅ domain at the N-terminus, have the three histidine boxes (characteristic of membrane desaturases) and two membrane-spanning regions. They are widely expressed in human tissues and, in some, they produce specialized products. For example, the 6-desaturase in the sebaceous gland of skin can use palmitate to synthesize sapienoic acid (*c*6-16:1) a major component of sebum.

The animal desaturases share several characteristics. Their reaction mechanism is that depicted in Fig. 3.19. They use molecular oxygen, a reduced pyridine nucleotide and an electron transport system consisting of a cytochrome and a related reductase enzyme. Almost invariably the first double bond introduced is in the 9-position and subsequent bonds are methylene

interrupted (i.e. separated by a methylene group) rather than being conjugated. They are all in the *cis* configuration.

The most important substrates for the first polydesaturation are oleic acid (either produced by the animal or coming from the diet), linoleic and α-linolenic acids (only from the diet). The structural relationships between the families of FAs that arise from these three precursors are most easily recognized by using the system that numbers the double bonds from the methyl end of the chain. Hence, oleic acid gives rise to a series of (*n*-9) FAs, the linoleic family is (*n*-6) and the α-linolenic family is (*n*-3). The first desaturation is at position 6 and the sequence is one of alternate elongations (see Section 3.1.5) and desaturations (Fig. 3.24).

Under some circumstances, the 6-desaturase may exert significant control over the overall desaturation and elongation process with a detectable build-up of its substrate FAs. This can have physiological consequences as discussed in Section 6.2.2.2. The regulation of enzymes involved in PUFA biosynthesis is not well defined, but it is known that development (age), diet and diabetes all affect 6- and 5-desaturase activities (see Section 3.1.11.5). The role of UFA biosynthesis in the formation of prostaglandins and other signalling molecules will be discussed further in Section 3.5.

3.1.9 Biohydrogenation of unsaturated fatty acids takes place in rumen microorganisms

Desaturation of acyl chains is widespread in Nature. The reverse process, namely the hydrogenation of double bonds, is found in only a few organisms. These organisms are commonly found in the rumens of cows, sheep and other ruminant animals. Linoleic acid, for example, can be hydrogenated by rumen microbes (anaerobic bacteria and protozoa) to stearic acid by a series of reactions shown in Fig. 3.25.

First of all, the substrate FAs must be released from leaf complex lipids by the action of acyl hydrolases (see Section 4.6.5). The first reaction of the unesterified linoleic acid involves isomerization of the *cis*-12,13 double bond to a *trans*-11,12 bond, which is then in conjugation with the *cis*-9,10 double bond. The enzyme responsible for this isomerization has been partially purified from the cell envelope of *Butyrivibrio fibriosolvens* and can act on α-linolenic as well as linoleic acid. Next, hydrogen is added across the *cis*-9,10 bond to form *trans*-vaccenic acid (*t*11-18:1), which is further reduced

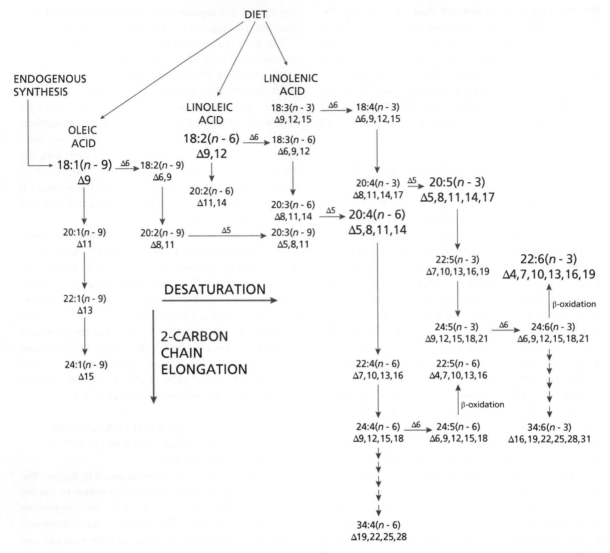

Fig. 3.24 Major pathways for polyunsaturated fatty acid synthesis in animals. Note the alternating sequence of desaturation in the horizontal direction and chain elongation in the vertical direction in the formation of polyunsaturated fatty acids from dietary essential fatty acids. Type size for individual fatty acids reflects, in a general way, relative accumulation in tissues. Adapted from HW Cook (1996) Fatty acid desaturation and elongation in eukaryotes. In DE Vance & J Vance, eds., *Biochemistry of Lipids, Lipoproteins and Membranes*, pp. 129–52, Elsevier, Amsterdam, with kind permission of the author and Elsevier Science.

to stearic acid. Analogous reactions occur with FAs other than linoleic acid, but the positions of the *cis* and *trans* bonds will, of course, be different.

In spite of the high activity of rumen microorganisms (including the further breakdown of the FAs by oxidation) ruminants do not appear to suffer from EFA deficiency. The amount of unchanged EFA passing through the rumen (up to 4% of dietary intake) is sufficient for

the needs of the animal. Hydrogenation can, however, be reduced by giving ruminants 'protected fats', thereby enriching their tissues with PUFAs (see Section 6.1.2.2).

Another example of UFA hydrogenation occurs in *Bacillus cereus*, which can reduce oleic to stearic acid. This reductase is induced by an increase in growth temperature and seems to be involved in the regulation of membrane fluidity (see also Section 5.1.3).

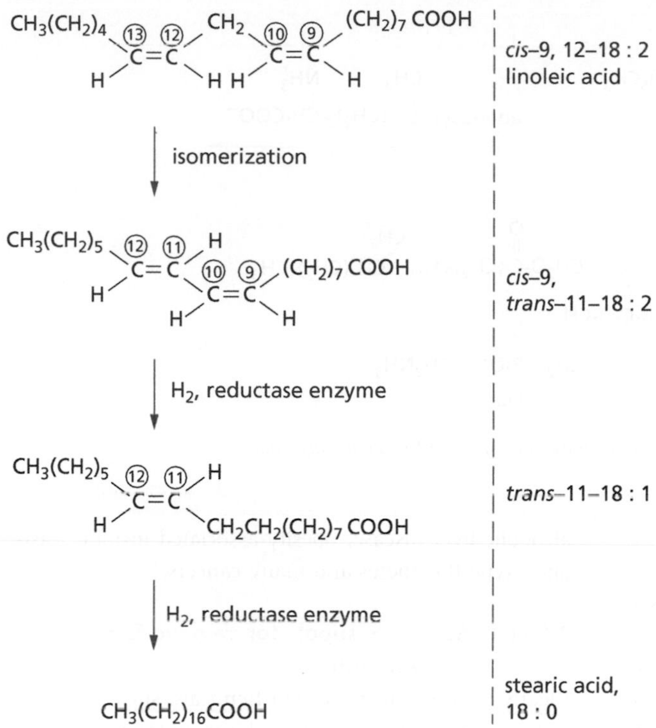

CH$_3$(CH$_2$)$_4$ —(13)(12)— CH$_2$ —(10)(9)— (CH$_2$)$_7$ COOH

cis–9, 12–18 : 2
linoleic acid

↓ isomerization

CH$_3$(CH$_2$)$_5$ —(12)(11)— (10)(9)— (CH$_2$)$_7$ COOH

cis–9, $trans$–11–18 : 2

↓ H$_2$, reductase enzyme

CH$_3$(CH$_2$)$_5$ —(12)(11)— CH$_2$CH$_2$(CH$_2$)$_7$ COOH

$trans$–11–18 : 1

↓ H$_2$, reductase enzyme

CH$_3$(CH$_2$)$_{16}$COOH

stearic acid, 18 : 0

Fig. 3.25 Biohydrogenation (by rumen microorganisms).

3.1.10 The biosynthesis of cyclic fatty acids provided one of the first examples of a complex lipid substrate for fatty acid modifications

The only FA ring structures we shall discuss are those containing cyclopropane or cyclopropene rings. The methylene group in cyclopropane acids originates from the methyl group of methionine in S-adenosyl methionine ('active methionine'). This is the same methyl donor involved in the formation of 10-methylene stearic acid and 10-methyl stearic acid from oleic acid (see Section 3.1.6). The acceptor of the methyl group is likewise an UFA. Thus, cis-vaccenic acid (c11-18:1) gives rise to lactobacillic acid, while oleic acid yields dihydrosterculic acid, the saturated derivative of sterculic acid (Table 2.4). These reactions occur in a number of bacteria and in certain families of higher plants, e.g. Malvaceae and Sterculaceae.

When Law and his colleagues purified cyclopropane synthase from *Clostridium butyricum*, they found that it would catalyse the formation of cyclopropane FAs from ^{14}C-labelled methionine only if micellar solutions of phospholipids were added. They discovered that the real acceptor for the methylene group was not the unesterified MUFA or its CoA or ACP thioester, but phosphatidylethanolamine (PtdEtn) – the major lipid of the organism (see Fig. 3.26).

The biosynthesis of cyclopropane and the related cyclopropane acids in higher plants has been studied by experiments with radioactive precursors. In this method, a proposed precursor for the compounds being studied is supplied to the plant and its incorporation into more complex molecules and/or conversion to products is studied at successive time intervals. The sequence in which the radiolabel appears in different compounds can be used to deduce the pathways by which they are made. Thus it has been shown that oleic acid gives rise to the cyclopropane derivative of stearic acid (dihydrosterculic acid). The latter can either be shortened by α-oxidation (see Section 3.2.2) or desaturated to give sterculic acid (Fig. 3.27). Desaturation of the α-oxidation product (dihydromalvalic acid) similarly yields malvalic acid (8, 9-methylene-8-17:1).

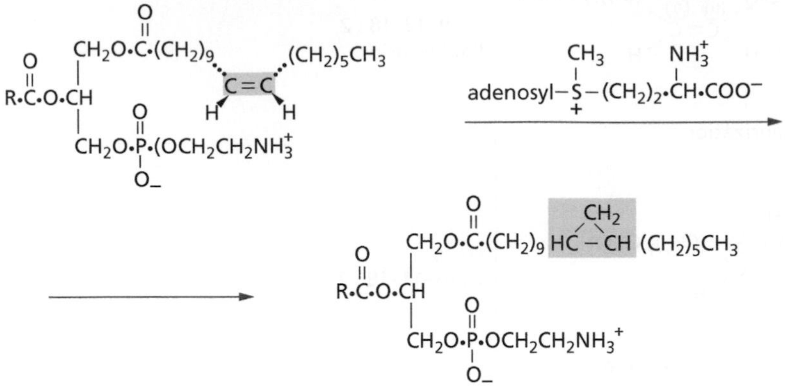

Fig. 3.26 Formation of a cyclopropane fatty acid using phosphatidylethanolamine in *Clostridium butyricum*.

3.1.11 Control of fatty acid biosynthesis in different organisms

Since FAs (and acyl lipids) are major constituents of microorganisms, plants and animals it is not surprising that their biosynthesis should be under strong regulation. In bacteria, factors such as nutrient supply and temperature are major players while in plants growth rate, environmental stress and developmental stages are important.

Genetic, dietary and other factors may influence *de novo* biosynthesis in animals. For humans with a relatively high fat proportion (30–40% energy: see Section 9.1.2.2) in the diet, *de novo* lipogenesis (DNL) plays less importance than in many other animals. Nevertheless, it is higher in obese and hyperinsulinaemic individuals.

DNL is especially important during development in the foetus and young infants and it is also implicated in alcoholic liver disease, obesity-associated insulin resistance, type II diabetes and many cancers.

3.1.11.1 Substrate supply for *de novo* fatty acid biosynthesis

Clearly, in order to make biochemicals, such as FAs, a source of substrates is needed. This includes not only the carbon primers but also reducing equivalents, ATP etc.

For bacteria, FA biosynthesis is coordinated with phospholipid formation and, hence, growth. This can easily be seen when phospholipid formation ceases at the end of a growth phase when substrates diminish. Feedback inhibition takes place to inhibit early steps in the FA biosynthetic pathway, probably mediated by long-chain acyl-ACPs.

In plants, two spectacular examples are light-stimulated FA biosynthesis and lipid accumulation in

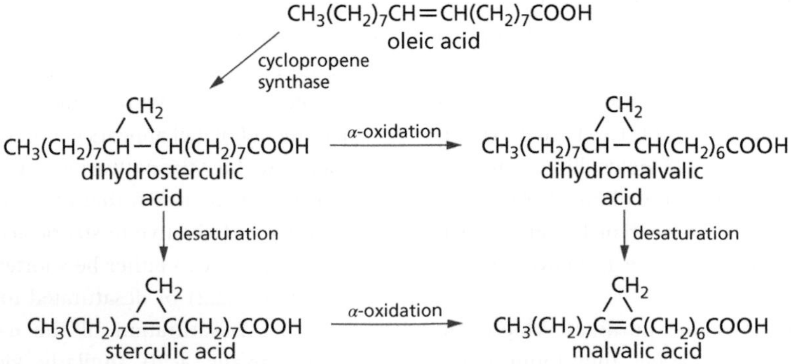

Fig. 3.27 Synthesis of sterculic and malvalic acids by plants.

oil seeds. In leaves, light stimulates FA biosynthesis about 20-fold partly by enzyme regulation (see Section 3.1.11.3) but also because the availability of NADPH and ATP is much increased. For oilseeds, the enzymes involved in FA formation are active during the oil deposition phase (see Section 4.1.3) after which they become inactive in the mature seed.

In animals, when acetyl-CoA production from pyruvate is high, citrate becomes elevated in turn. It is transported from mitochondria by the tricarboxylate anion carrier to the cytosol where it activates ACC and (by the action of ATP:citrate lyase) provides acetyl-CoA substrate.

3.1.11.2 Acetyl-CoA carboxylase and its regulation in animals

It is often thought that enzymes near the beginning of metabolic pathways may play an important role in controlling the carbon flux down the pathway. (Note that the term 'regulatory enzyme' is incorrect, since all enzymes in a pathway are potentially regulating and, therefore, all of them play a role in overall control – see Fell's (1997) book in **Further reading** for a good discussion.) Because ACC catalyses the first committed step for FA (and lipid) biosynthesis, it has been well studied to see whether it could be important for flux control. Indeed, it is now accepted generally that ACC activity plays a key role in the overall control of FA biosynthesis, especially in animal tissues.

Both the activity of ACC and the rate of FA biosynthesis fluctuate in response to various internal or external factors which affect lipogenesis. These include diet, hormones and developmental or genetic factors. The major control that ACC exerts has been clearly demonstrated recently by experiments in which an ACC mRNA-specific ribozyme gene was expressed. The rate of FA biosynthesis in these experiments was shown to be proportional to the amount of ACC mRNA and ACC protein. Following demonstration of two isoforms of animal ACCs (see Section 3.2.1.3), it should be borne in mind that most experiments on this enzyme's regulation were carried out before such knowledge was available. Thus, some detailed interpretations may have to be modified in the future. For the time being, however, we will assume that when control of FA biosynthesis depends on ACC, the isoform mainly involved is ACC1. However, in liver (where FA oxidation and biosynthesis are both important) ACC2 is doubly critical. ACC2-null

mice, lacking the malonyl-CoA pool which it normally generates, continually oxidize FAs and accumulate little fat. On the other hand, liver-specific deletion of ACC1 reduces FA biosynthesis and hepatic TAG1 accumulation but does not affect FA oxidation. (The role of ACC2 in β-oxidation of FAs is discussed in Section 3.2.1.6.) The activity of ACC can be regulated rapidly or over a longer time course in several ways (see Table 3.5).

Activation and inhibition

Because ACC catalyses the first committed step in lipid biosynthesis, and because its substrate lies at a crossroads between carbohydrate and lipid metabolism, then its acute regulation is clearly important. Hydroxytricarboxylic acids, such as citrate, have long been thought to be physiologically important because a rise in their concentrations could indicate a constraint in the TCA cycle and, hence, a need to convert excess carbon into lipid stores. Because citrate is a precursor of acetyl-CoA (see Section 3.1.3) it also acts as a positive feed-forward activator.

Citrate promotes conversion of the inactive protomer into the catalytically active polymer. Activation by citrate is unusual in that it increases the rate of reaction (V_{max}) without affecting the K_m of its substrates. Both partial reactions are stimulated. Although (as mentioned above) citrate has a logical role in controlling carbon flux into lipid stores as opposed to ATP generation, its exact physiological role is still unclear. Citrate also prevents binding of the ACC inhibitor, acyl-CoA (below).

Other classical effectors of ACC activity are acyl-CoAs, which can be regarded as end products of FA biosynthesis in mammals. Inhibition by long-chain (16C–20C) acyl-CoAs is competitive towards citrate and noncompetitive towards the three substrates acetyl-CoA,

Table 3.5 Relative acetyl-CoA carboxylase (ACC) levels and rate of enzyme synthesis in different conditions.

	Protein content	Synthesis of ACC protein
Rat		
Normal	1	1
Fasted	0.25	0.50
Re-fed	4	4
Alloxan-diabetic	0.50	0.50
Mouse		
Normal	1	1
Obese mutant	4	3

ATP and bicarbonate. Saturated fatty acyl-CoAs of 16C–20C are most effective and unsaturated acyl-CoAs are considerably less inhibitory.

It is, of course, important to consider whether the cellular concentration of the putative regulator molecules – citrate and acyl-CoAs – are sufficient to allow the physiological control of ACC. Indeed, cellular concentrations of these molecules under various metabolic conditions are consistent with their role in regulation. For example, it has been found that in liver between 50% and 75% of the cellular citrate is in the cytoplasm with the ACC. The estimated concentration of 0.3–1.9 mM is close to the concentration needed for half maximal activation. Furthermore, glucagon or (dibutyryl)cAMP, which lowers the rate of FA biosynthesis, also reduces the cytoplasmic concentration of citrate. Recently, a 22 kDa cytosolic protein (MIG 12) has been discovered which lowers the threshold for citrate activation of ACC. Changes in expression of MIG 12 were found both in vitro and in vivo to change activity of ACC and alter TAG accumulation in the liver significantly. The effects were mostly found for ACC 1. The K_i for free long-chain acyl-CoAs is about 5 nM. Although cytosolic levels of acyl-CoAs can reach over 1000 nM, most are bound to acyl-CoA binding protein. However, the free acyl-CoA concentration is in the range 2–10 nM which is sufficient to have an important regulatory effect.

Phosphorylation/dephosphorylation

ACC1 and ACC2 are both phosphoproteins and control by covalent phosphorylation is well established. Purified ACCs contain as many as nine phosphorylation sites and these are concentrated either at the N-terminus or between the biotin-binding and carboxyltransferase functional domains (Fig. 3.28).

Phosphorylation by AMP-activated protein kinase (AMPK) causes a large decrease in V_{max} and, of the four important phosphorylation sites, its phosphorylation of serine 79 (in ACC1) seemed to be critical for inactivation. In contrast, the cAMP-dependent protein kinase caused mainly an increase in the K_a for citrate (the concentration for half-maximal activation) and a more modest decrease in V_{max}. Serine 1200 was critical for this inactivation although Ser77 is also phosphorylated. Other kinases may also play a role in phosphorylating the carboxylase and, hence, inactivating it (see Fig. 3.28). Because protein kinases are affected by various signalling pathways (such as glucagon and adrenaline) then indirectly the latter will influence the activity of ACC.

Role of the hypothalamus

Malonyl-CoA concentrations in the hypothalamus are thought to be important for regulating food intake and energy expenditure. Both ACC1 and ACC2, as well as FAS, are present at high levels in hypothalamic neurons

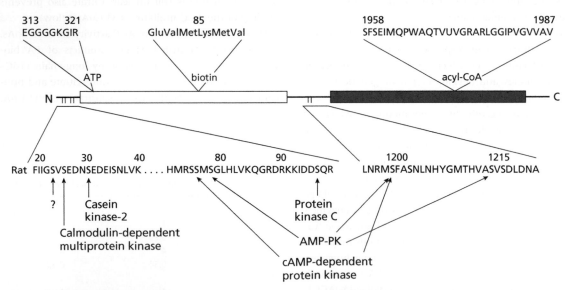

Fig. 3.28 Domain map of animal acetyl-CoA carboxylase showing locations of functional and phosphorylation sites. AMP-PK= 5'-AMP-dependent protein kinase. Taken from KH Kim (1997) Regulation of mammalian acetyl-CoA carboxylase. *Ann Rev Nutr* **17**: 77–99, with kind permission of the author and Annual Reviews Inc.

and increases in malonyl-CoA concentration there cause a rapid suppression of food intake. This was first shown pharmacologically but later confirmed in knockout mice. Conversely, when viral overexpression of malonyl-CoA decarboxylase reduced malonyl-CoA concentrations, food intake was stimulated. Nutritional conditions (fasted, fed) affect hypothalamic malonyl-CoA concentrations and, hence, change food intake. Most of these effects, via ACC mediation of malonyl-CoA concentrations, appear to be due to changes in AMPK activity.

Synthesis and degradation

Long-term regulation of ACC in animals can be due to changes in enzyme amounts. The tissue concentration of the carboxylase protein has been shown to vary with the rate of FA biosynthesis under a variety of nutritional, hormonal, developmental or genetic conditions. Measurements with specific antibodies have shown that, for example, fasted rats have only one-quarter the normal levels of the liver enzyme, while genetically obese mice show a four-fold increase. The amount of an enzyme protein accumulating is due to the net rates of synthesis and degradation. Depending on the specific trigger for changes in ACC concentration, one or both of these factors may be altered. For example, the increases in enzyme content in fasted/re-fed rats is due only to changes in the rates of its synthesis (Table 3.5). By contrast, the decreases in enzyme content in fasted animals is due both to diminished synthesis and to accelerated breakdown.

The effects of hormones and dietary status are interrelated. Lipogenesis is higher when there are elevated circulating glucose or insulin concentrations. Furthermore, the nature of the diet is important. For example, glucose (or its carbohydrate precursors) stimulates lipogenesis, while PUFAs reduce it. Regulation is at the protein level with both transcriptional and post-translational mechanisms involved.

Both isoforms of ACC are transcribed from multiple promoters. The ACC1 gene has at least three promoters (PI, PII, PIII). While PI and PII give rise to the same product, PIII gives rise to an N-terminal variant in some tissues (e.g. liver, mammary gland) but not in others (e.g. muscle, adipose tissue). ACC2 has at least two promoters.

Role of transcription factors

A number of transcription factors can affect expression of ACC. These include sterol regulatory element (SRE) binding protein (SREBP; see Section 7.3.1), liver X receptor (LXR; see Section 7.3.3) and carbohydrate response element binding protein (ChREBP; see Section 7.3.1). Because such transcription factors usually affect both ACC and FAS, we will consider them in the context of overall lipogenesis later (see Section 3.1.11.4).

3.1.11.3 Acetyl-CoA carboxylase regulation in other organisms

Yeast ACC is also inhibited by acyl-CoAs but, unlike the animal enzyme, is unaffected by citrate. In some mutant strains, an activation by fructose-1,6-bisphosphate has been demonstrated. Like the mammalian carboxylase, the enzyme from yeast is also regulated by changes in its amount. An interesting example of the role of fatty acyl-CoA in mediating the rate of biosynthesis of acetyl-CoA carboxylase has been demonstrated in FA mutants of *C. lipolytica*. These mutants contained no apparent ACS and hence, were unable to grow on exogenous FAs, when their own FA biosynthesis was blocked by inhibitors. However, further examination showed that the mutants did have one type of ACS (ACS II) in common with normal cells but lacked ACS I, which was needed for membrane lipid biosynthesis. ACS II is used for activating FAs destined for β-oxidation (see Section 3.2.1). Thus, these two ACSs are responsible for generating two pools of acyl-CoAs in different parts of the cell. The acyl-CoAs formed by ACS I are in the cytosol and these cause repression of ACC (Fig. 3.29).

Transcriptional regulation of *E. coli* ACC appears to be quite complex. As mentioned before, there are four genes that code for subunits of the complex (*accA-D*). While transcription of all four genes is under growth rate control, there are several promoters involved. One promotes the *accBC* operon and there are separate promoters for *accA* and for *accD*.

In addition to transcriptional regulation, *E. coli* ACC is also inhibited by feedback inhibition by the end-product of fatty acid synthesis, acyl-ACP.

In plants, ACC has been identified as an important site for flux control during FA biosynthesis in leaves. FA biosynthesis in such tissues is very much stimulated by light (about 20-fold) and this increase is accompanied by changes in the pool sizes of acetyl and malonyl thioesters consistent with large alterations of ACC activity. This was confirmed by the use of specific inhibitors that showed that most of the regulation of carbon flux into FAs and lipids was by ACC. However, the mechanism of

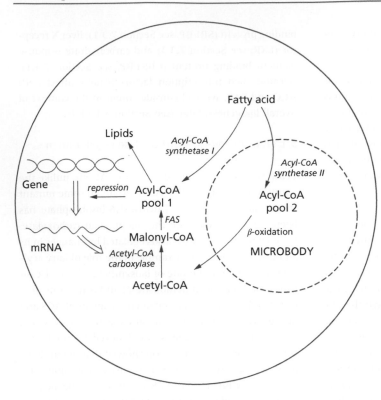

Fig. 3.29 Interaction of acyl-CoA pools with fatty acid metabolism in the hydrocarbon-utilizing yeast *Candida lipolytica*. Reproduced from S Numa (1981) Two long-chain acyl coenzyme A synthetases: their different roles in fatty acid metabolism and its regulation. *Trends Biochem Sci*, Fig. 2, p. 115 with kind permission of Dr S Numa and Elsevier.

control is poorly understood. In light stimulation experiments, the activation of ACC may be secondary to stromal solute changes (ATP, ADP, Mg^{2+} and pH) that accompany photosynthesis. In addition, reversible phosphorylation and redox modulation (both also caused by chloroplast changes during photosynthesis) seem to alter plant ACC activity.

3.1.11.4 Regulation of fatty acid synthase
In the same way as the actual levels of ACC protein can be changed in animals under dietary influence, so can that of FAS (Table 3.6). Up to 20-fold differences have been observed in the concentrations of, for example, liver FAS between starved and carbohydrate re-fed animals. Alterations in both synthesis and degradation of the enzyme seem to be involved.

It should be noted that dietary factors that affect FAS levels do not affect all tissues equally. While the liver is highly influenced by such regulation, the FAS of brain is unaffected. This is just as well, for it would be extremely disadvantageous for a young animal to have its brain development influenced dramatically by its day-to-day nutritional state!

While refeeding a high carbohydrate diet causes an increase in hepatic FAS protein concentrations, fat feeding reduces them. However, the effects of dietary fat are complex, PUFAs being particularly effective.

In addition, a number of hormones, including insulin, glucocorticoids, glucagon and oestradiol, have also been found to produce acute alterations in concentrations of various mammalian FASs (Table 3.6). Longer term factors can also play a role in determining enzyme levels. Thus, for example, FAS increases in mammary gland

Table 3.6 Effect of nutritional state or hormones on liver fatty acid synthase levels.

Cause increase	Cause decrease
Re-feeding	Starvation
Insulin	Alloxan-diabetes
β-Oestradiol	Glucagon[a]
Hydrocortisone	
Growth hormone	

[a] Glucagon can induce enzyme synthesis in embryonic liver, but will reduce levels of fatty acid synthase in adult animals.

during mid to late pregnancy and early in lactation. In brain development, on the other hand, the synthase is highest in foetal and neonatal animals and decreases with maturity. Like ACC, the concentration of FAS is also much higher in tissues of certain genetic mutants such as the obese hyperglycaemic mouse. The enzyme in the latter mutant is less subject to control by dietary factors, such as fasting, when compared to normal mice.

These changes in animal FAS have now been studied at the subcellular level. FAS mRNA has been isolated from different tissues after various hormonal (insulin, glucagon and thyroid hormone are the most important) or dietary manipulations and translated in vitro. Recombinant plasmids have been used to ascertain the size of FAS mRNA and also the amounts of this mRNA in tissues. By these means it has been shown that differences in FAS activity are caused by changes in enzyme concentration rather than its intrinsic activity. These alterations are themselves changed by the balance of enzyme synthesis and degradation.

In recent years it has become clear that several transcription factors are involved in the overall control of biogenesis and in modulating transcription of both ACC and FAS. FAS is coded by a single copy gene with one promoter in most animals. In humans, however, it is transcribed from two promoters. Promoter PI is involved important when demand for lipogenesis is high and contains TATA and CAAT boxes. Promoter PII contains no TATA or CAAT boxes and is involved in lower levels of FAS expression at a constitutive level.

One transcription factor is called upstream stimulating factor (USF). Although USF is expressed ubiquitously, it is involved in the expression of several developmentally regulated or tissue-specific genes. For example, in adipocytes it is involved in insulin-mediated activation of the FAS promoter. Full activation also needs SREBP binding (see below).

SREBP itself binds to the sterol regulatory element and activates genes involved in the uptake and biosynthesis of lipids. These are three isoforms (1a, 1c, 2) which are formed as transmembrane protein precursors embedded in the ER. For them to interact with the nucleus needs proteolytic cleavage of the cytosolic N-terminal domain. SREBP plays a critical role in inducing transcription of FAS and other lipogenic enzymes in response to fasting/refeeding.

While isoforms 1a and 2 play important roles in regulating cholesterol metabolism, SREBP-1c is mainly

involved in genes for FA biosynthesis. These include ACC and FAS as well as other lipogenic enzymes. In most cases, synergistic binding of USF and SREBP is needed to activate the FAS promoter.

LXR is in a specific subclass of the nuclear hormone receptor family. It acts as a sensor for cellular cholesterol concentrations, is activated by oxysterols and inactivated by some UFAs. There are two isoforms – LXRα is predominant in adipose tissue and liver while LXRβ is important in brain. LXR seems to be an important regulator of SREBP-1c expression. Indeed there are two LXR binding sites in the SREBP-1c promoter. In LXR knockout mice, lipogenic enzyme induction is insensitive to insulin. Moreover LXRα is induced strongly by this hormone.

During feeding, glucose itself (as well as insulin) causes glycolytic and lipogenic gene induction. The main transcription factor involved is ChREBP. This appears to be particularly important in liver where it can bind to the PI promoter of the ACC1 gene as well as a FAS promoter and similar elements of other glycolytic and lipogenic genes. ChREBP phosphorylation appears to be involved in its activation.

DNL is strongly reduced in fasting but, on feeding a high carbohydrate diet, is dramatically increased. In contrast, a high-fat diet supresses lipogenesis. PUFAs are particularly effective and act via the above transcription factors as well as PPARs.

A common way in which the rate of a particular metabolic reaction can be controlled is through the supply of substrate. In the case of animal FAS, this regulation has been examined with regard to NADPH. However, it appears that NADPH production is adjusted to cope with the altering demands of FA biosynthesis rather than the other way around. In contrast, supply of malonyl-CoA by ACC activity (see Section 3.1.3.1) is considered to be a major factor regulating overall FA (and lipid) formation under many conditions.

The above mechanisms probably apply also for organisms other than animals. In developing seeds, where there is a spectacular rise in fat accumulation during a particular state of maturation (see Section 4.1.3 & Fig. 4.2), the increase in FA biosynthesis has been correlated with increases in concentrations of various synthase proteins including ACP. Conversely, the rise in activity of FAS in photosynthetic tissues on illumination undoubtedly requires the simultaneous production of NADPH substrate by photosystem I.

The nature of plant FAS products can be altered by the relative activity of the component enzymes. Experiments in vitro with specific inhibitors have shown that the different condensing enzymes are responsible for the chain length of the final products. This fact has now been exploited in transgenic crops where the ratio of 16C to 18C products can be changed. In addition, introduction of foreign genes such as the medium-chain acyl-ACP thioesterase (from the Californian bay tree or from coconut) can be used to terminate chain lengthening prematurely and allow laurate accumulation (see also Section 11.6.1).

E. coli adjusts its FA composition in response to growth temperature. This adaptive response is the result of a change in the activity of 3-ketoacyl-ACP synthase II (FabF). Such alterations in its activity occur very rapidly (within 30 s of temperature downshift) and are therefore independent of transcriptional control. Moreover, mutants lacking synthase II (FabF) are unable to adapt to lower temperatures. Modifying the activity of other condensing enzymes, such as ketoacyl-ACP synthase I does not affect the temperature response.

FA biosynthesis is coordinately regulated with membrane lipid formation in *E. coli*. After phospholipid biosynthesis ceases, acyl-ACP concentrations build up. This feeds back to reduce *de novo* biosynthesis by inhibiting ACC and the FabH and FabI components of FAS.

FA formation in *E. coli* is also regulated by an unusual nucleotide, guanosine 5′-diphosphate-3′-diphosphate (ppGpp), which causes a marked reduction in stable RNA biosynthesis. ppGpp directly inhibits phospholipid biosynthesis and the build-up of acyl-ACP which in turn, reduces FA formation. Overexpression of an acyl-thioesterase relieves this by hydrolysing acyl-ACPs.

3.1.11.5 Control of animal desaturases

UFAs are present in all living cells. In complex lipids they are important in regulating the physical properties of lipoproteins and membranes (Chapters 5 & 7). As NEFAs, they can act as regulators of metabolism in cells (see above) or as precursors for physiologically active compounds (see Sections 3.5, 6.2.2 & 10.3.3). The control of UFA production must therefore be important, yet our knowledge of the subject is sparse. Research has lagged behind that on FAS, partly because of the difficulties of isolating and studying membrane-bound enzymes, as discussed earlier.

A good example of the importance of UFAs comes from experiments on temperature-induced change in membrane fluidity (see Chapter 5). The adaptive response of *E. coli* has already been referred to above. Some protists that used to be referred to as simple animals (for example the protozoa *Tetrahymena* and *Acanthamoeba*), are able to produce linoleic acid, in contrast to mammals. The 12-desaturase responsible in *Acanthamoeba* has been studied in some detail. When *Acanthamoeba* is cooled, its membranes lose their fluidity and the organism is unable to phagocytose (it feeds on soil bacteria). The 12-desaturase is then induced and the membranes become more fluid, as oleate is converted into linoleate, and phagocytosis can commence again. The induction of gene expression is caused by low temperatures or, independently, by oxygen (which is more soluble at low temperatures).

Mammalian 9-stearoyl-CoA desaturase (SCD-1 isoform) shows extreme responses to dietary alterations. When fasted animals are re-fed a fat-free diet, there can be a 'super-induction' with levels of desaturase activity increased more than 100-fold. Likewise, when rat pups are nursed by mothers on an EFA-deficient diet, their SCD-1 mRNA is nearly 100-fold that of control levels. In contrast, brain 9-desaturase concentrations are hardly affected in keeping with the essential nature of UFAs in nervous tissue development. Dietary PUFA, particularly linoleic and arachidonic acids, inhibit 9-desaturase activity more than they do *de novo* FA biosynthesis. The SCD-2 gene undergoes coordinate transcriptional down-regulation in response to these PUFAs.

As with ACC and FAS, several transcription factors are involved in regulating desaturase activity in mammals (see Sections 7.3.1–7.3.3 for more detailed descriptions of the transcription factors). These include SREBP-1c, LXR, ChREBP and a peroxisome proliferator activated receptor (PPAR-α). SREBP-1c generally activates FA formation and that includes the three desaturases (SCD, 5- and 6-desaturases). LXR is involved in cholesterol and acyl lipid homeostasis. A LXR response element has been identified in SCD and is probably present in the other two desaturases. PPAR-α is found in the promoters of both SCD and the 6-desaturase. Like SREBP-1c, PPAR-α is involved in the feedback induction of 6-desaturase when highly unsaturated PUFA (like DHA) levels are low. Finally, SCD is a target of ChREBP. Indeed dietary glucose, independent of insulin, increases expression of desaturases (and elongases). It is thought that ChREBP plays an important role in the PUFA-mediated repression of desaturases when both its mRNA stability and nuclear translocation are decreased.

3.2 Degradation of fatty acids

The main pathways of FA degradation involve oxidation at various points on the acyl chain or 'lipoxidation' at certain double bonds of specific UFAs.

There are three main types of FAs oxidation, termed α, β and ω. They are named depending on which carbon on the acyl chain is attacked:

$$CH_3(CH_2)_n \underset{\omega}{\uparrow} CH_2 \underset{\beta}{\uparrow} CH_2 \underset{\alpha}{\uparrow} COO^-$$

Of the oxidations, β-oxidation is the most general and prevalent.

3.2.1 β-Oxidation is the most common type of biological oxidation of fatty acids

Long-chain FAs, combined as TAGs, provide the long-term storage form of energy in the adipose tissues of animals (see also Section 9.1). In addition, many plant seeds contain TAG stores (see Section 9.3). Once FAs have been released from TAGs, they are degraded principally by the liberation of 2C (acetyl-CoA) fragments in β-oxidation. The mechanism was originally proposed over a century ago (1904) by Knoop. He synthesized a series of phenyl-substituted FAs with odd-numbered or even-numbered carbon chains and found that the odd-numbered substrates were metabolized to phenylacetate. Knoop was using the phenyl group in the way that modern biochemists would use a ^{14}C-radiolabel. At around the same time the proposed intermediates were isolated by Dakin and, therefore, the basic information about β-oxidation was available 50 years before the enzymic reactions were demonstrated.

When, in 1944, Leloir and Munoz showed that β-oxidation could be measured in cell-free preparations from liver, it was not long before several of its important features were revealed. Lehninger found that ATP was needed to initiate the process, which seemed particularly active in mitochondria. Following the isolation of coenzyme A by Lipmann, Lynen was able to demonstrate that the 'active' acetate was acetyl-CoA and Wakil and Mahler showed that the intermediates were CoA-esters. With the availability of chemically synthesized acyl-CoA substrates, Green in Wisconsin, Lynen in Munich and Ochoa in New York could study in detail the individual enzymes involved.

3.2.1.1   Cellular site of β-oxidation

It was originally thought that β-oxidation was confined to mitochondria. Although animal mitochondria do contain all the enzymes necessary and are a major site for β-oxidation, other subcellular sites, such as the microbodies, are implicated. Peroxisomes or glyoxysomes together are often referred to as microbodies. They contain a 'primitive' respiratory chain where energy released in the reduction of oxygen is lost as heat. The presence of an active oxidation pathway in microbodies was first detected in the glyoxysomes from germinating seeds by de Duve in 1976. Since that time the various enzymes involved have been purified and characterized for microbodies from animals as well as from plants. Glyoxysomes are like peroxisomes in containing flavin oxidases and catalases instead of the energy-coupled electron transport chain of mitochondria. In addition, they contain enzymes of the glyoxylate pathway that enables carbon from fat to be used to make carbohydrate instead of being lost as CO_2.

Microbodies occur in all major groups of eukaryotes including yeasts, protozoa, plants and animal but their contribution to total β-oxidation differs considerably between tissues. In animals, microbodies are particularly important in liver and kidney. In fact, in liver it seems that mitochondria and microbodies collaborate in overall FA oxidation. Thus, microbodies oxidize very long-chain FAs to medium-chain products, which are then transported to mitochondria for complete breakdown. In this way very long-chain FAs, such as erucate (c13-22:1), which are poor substrates for mitochondria, can be catabolized. Microbodies (which also contain α-oxidation enzymes) are important for the degradation of other lipid substrates such as some xenobiotics and the eicosanoids.

In contrast, the glyoxysomes from germinating seeds are capable of the complete breakdown of FAs to acetyl CoA. They integrate this metabolism with the operation of the glyoxylate cycle, which allows plants (in contrast to animals) to synthesize sugars from acetyl-CoA. Leaf tissues also contain peroxisomes and recent work indicates that β-oxidation in leaves is always found in peroxisomes with significant activity in mitochondria in some circumstances.

3.2.1.2   Transport of acyl groups to the site of oxidation: the role of carnitine

In mammals, FAs are transported between organs in the circulation either in the form of NEFAs bound to albumin or as TAGs in the cores of plasma lipoproteins (especially chylomicrons and very low density lipoproteins (VLDL – see Section 7.2.3). TAGs are hydrolysed on the outer surface of cells by LPL and the liberated NEFAs have been shown to enter liver, adipose and heart tissue cells by saturable and nonsaturable mechanisms. Several FATPs have been identified and they are involved in the saturable mechanism. When FAs are present at high concentrations, nonsaturable uptake can take place due to passive diffusion.

Once inside cells, FAs can be activated to acyl-CoAs by various ligases. Most of the activating enzymes are ATP-dependent ACSs, which act in a two-step reaction (see Section 3.1.1.1).

Both NEFAs and acyl-CoAs are capable of binding to distinct cytosolic proteins known as FABPs. The best known of these is the so-called 'Z-protein' of liver. These low molecular mass (about 14 kDa) FABPs have been suggested to function for intracellular transport or to provide a temporary binding site for potentially damaging compounds such as acyl-CoAs (see Section 3.1.1). Specific acyl-CoA binding proteins have also been discovered recently. The importance of FABP is illustrated in knock-out mice lacking heart FABP which show exercise intolerance and, later, die of cardiac hypertrophy. In plants FABPs have various functions that include helping plants cope with stress.

Because the inner mitochondrial membrane is impermeable to CoA and its derivatives, fatty acyl-CoAs formed in the cytosol cannot enter the mitochondria

directly for oxidation. The observation, by Bremer and others, that carnitine could stimulate the oxidation of fatty acids in vitro led to the idea that long-chain FAs could be transported as carnitine esters.

$$(CH_3)_3\overset{+}{N}CH_2\underset{\underset{\displaystyle OH}{|}}{CH}CH_2\,COO^-$$

Carnitine

The theory really began to take shape when an enzyme (carnitine:palmitoyl transferase, CPT) was discovered, which would transfer long-chain acyl groups from CoA to carnitine. Two isoforms of the enzyme have been identified. In mitochondria CPT1 is located within the outer mitochondrial membrane, while CPT 2 is on the inner membrane (Fig. 3.30).

Peroxisomes and the ER also have CPTs. Examination of these enzymes shows that they are very similar to the mitochondrial CPT 1 and are also severely inhibited by malonyl-CoA (see below). These transferases are collectively termed CPTo.

The acyl-carnitines cross the outer mitochondrial membrane through a porin channel and the inner membrane via a carnitine:acylcarnitine translocase, which causes a one-to-one exchange, thus ensuring that the mitochondrial content of carnitine remains constant (Fig. 3.30). FAs of fewer than ten carbons can be taken up by mitochondria

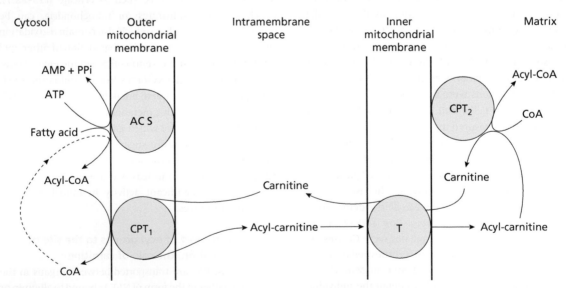

Fig. 3.30 Movement of acyl residues into mitochondria via carnitine. Abbreviations: ACS, acyl-CoA synthase; T, translocase; CPT, carnitine:palmitoyltransferase.

as NEFAs, independent of carnitine. This is in spite of the fact that a third carnitine acyltransferase, which transfers acyl groups of 2C–10C, has been found in mitochondria. The function of this enzyme is uncertain although it has been suggested to be useful in regenerating free CoA within the mitochondrial matrix.

3.2.1.3 Control of acyl-carnitine is very important

Because mitochondria, peroxisomes and the ER have very active acyl-carnitine translocases, this means that as soon as acyl-carnitine is made by CPTo, it will be taken into these compartments. Thus, any modulation of CPTo activity will rapidly alter the utilization of long-chain acids within these organelles or the ER lumen or, alternatively, lead to a rise in acyl-CoAs in the cytosol. If the latter occurs then a host of effects can result including, for example, increased DAG or ceramide formation leading in turn to protein kinase C (PKC) activation or apoptosis, respectively (see Sections 8.1.3 & 8.5).

Activity of CPTo is controlled by malonyl-CoA concentrations. These CPTs are integral membrane proteins, but both their catalytic and regulatory sites are exposed on the cytosolic side of the outer mitochondrial (peroxisomal and ER) membranes. Malonyl-CoA is generated by the ACC2 isoform of acetyl-CoA carboxylase. CPT_1 (the mitochondrial form of CPTo) is extremely sensitive to malonyl-CoA, the Ki for malonyl-CoA being in the micromolar range.

The idea that ACC2 plays a key role in the regulation of β-oxidation is supported by three facts. First, CPT1, an essential component for mitochondrial β-oxidation, is very sensitive to malonyl-CoA, which is only produced by ACC. Second, tissues such as heart muscle, which are nonlipogenic, contain large amounts of ACC activity. The main form in these mitochondria-rich tissues is ACC2. Thirdly, physiological conditions that decrease ACC activity and malonyl-CoA concentrations are accompanied by accelerated β-oxidation.

3.2.1.4 Enzymes of mitochondrial β-oxidation

The reactions of β-oxidation are shown in Fig. 3.31 and essentially involve four enzymes working in sequence (acyl-CoA dehydrogenase, enoyl hydratase, a second dehydrogenase and a thiolase), which results in the cleavage of two carbons at a time from the acyl chain. The 2C product, acetyl-CoA, is then used by, for example, the TCA cycle to yield energy and the acyl-CoA (two carbons shorter) is recycled.

Each of the four enzymes involved exist as isoforms with different chain-length selectivities (see Table 3.7). Thus, for the efficient degradation of typical FA substrates, the different isoforms must cooperate. Three of the enzymes (hydratase, 3-hydroxacyl dehydrogenase and thiolase) have their long-chain isoforms present as a trifunctional complex on the inner mitochondrial membrane. All the other isoforms of the component

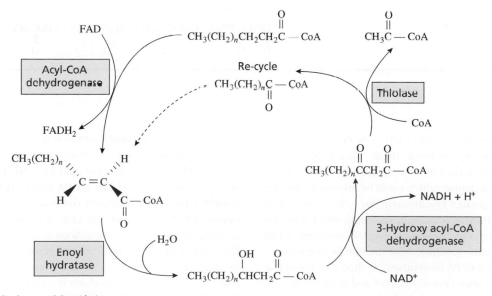

Fig. 3.31 The fatty acid β-oxidation cycle.

Table 3.7 Enzymes of the mitochondrial β-oxidation cycle in animals (reaction scheme in Fig. 3.31).

Full name	Trivial name	Description
Acyl-CoA dehydrogenase (EC 1.3.2.2)		Four acyl-CoA dehydrogenases with overlapping chain-length selectivity (short-chain, medium-chain, long-chain and very long chain). First three are soluble in mitochondrial matrix and are homotetramers. FAD is tightly but non-covalently bound. Very long chain acyl-CoA dehydrogenase is a hornodimer in the inner membrane. All form a *trans* 2,3-double bond.
Enoyl-CoA hydratase (EC 4.2.1.17)	Enoyl hydrase Crotonase	Two enzymes have been identified in heart mitochondria; one is crotonase or short-chain enoyl-CoA hydratase; The second is a long-chain enoyl-CoA hydratase. Crotonase activity is so high in some tissues that the second enzyme may have little function in β-oxidation there (e.g. in liver). However, in most tissues they probably cooperate in fatty acid degradation. The long-chain enoyl-CoA hydratase is a component enzyme of the trifunctional β-oxidation complex that also contains long-chain activities of L-3-hydroxyacyl-CoA dehydrogenase and thiolase. The complex is located in the inner mitochondrial membrane. Both hydratases are specific for the *trans* enoyl substrate and form the L(+) stereo-isomer.
L-3-Hydroxyacyl-CoA dehydrogenase		Three dehydrogenases have been found in mitochondria. All use NAD* but differ in their acyl specificities. The main enzyme is soluble and can use a variety of acyl chain lengths. However, it has poor activity for long chains, which are metabolized by a second dehydrogenase located in the inner membrane as part of the trifunctional β-oxidation complex. A third enzyme is soluble and can work with short-chain substrates with or without 2-methyl substituents but is mainly involved in leucine metabolism.
Acyl-CoA; acetyl-CoA S-acyltransferase (EC2.3.1.16)	Thiolase	Three thiolases are present in mitochondria. The first is soluble, specific for acetoacetyl-CoA and probably mainly functions in ketone body and isoleucine metabolism. Another soluble thiolase has a broad specificity while a long-chain thiolase is part of the membrane-bound trifunctional β-oxidation complex. All the enzymes function in thiolytic cleavage:

$$RCH_2CCH_2C-SCoA + HS-E \rightleftharpoons RCH_2C-S-E + CH_3C-SCoA$$

and in acyl transfer:

$$RCH_2C-S-E + CoASH \rightleftharpoons RCH_2C-SCoA + HS-E$$

Thus the overall thiolase reaction is:

$$RCH_2CCH_2C-SCoA + CoASH \rightleftharpoons RCH_2C-SCoA + CH_3C-SCoA$$

enzymes are soluble, but there is evidence of metabolite channelling between them. This improves the efficiency of β-oxidation and prevents intermediates from accumulating, which could be inhibitory and also lead to a greater requirement for scarce CoA. As a generalization, it is thought that long-chain acids are oxidized by the membrane-located isozymes while medium-chain substrates use the matrix enzymes.

In contrast to FA biosynthesis, it should be noted that β-oxidation uses CoA-derivatives and gives rise to a 3-hydroxy intermediate with the L(+) configuration.

3.2.1.5 Other fatty acids containing branched-chains, double bonds and an odd number of carbon atoms can also be oxidized

So far we have assumed that the FA being oxidized is a straight chain, fully saturated compound. This is not necessarily the case and the ease with which other compounds are oxidized depends on the position along the chain of the extra group and the capacity of the cell for dealing with the end-products. From acids of odd chain length, one of the products is propionic acid and the ability of an organism or tissue to oxidize such FAs is governed by its ability to oxidize propionate. Liver, for example, is equipped to oxidize propionate and therefore deals with odd-chain acids quite easily; heart, on the other hand, cannot perform propionate oxidation and degradation of odd-chain acids grinds to a halt. The end-product of propionate oxidation, succinyl-CoA, arises by a mechanism involving the B_{12} coenzyme and the biotin-containing propionyl-CoA carboxylase:

$$CH_3CH_2COSCoA + CO_2 + ATP \xrightarrow{\text{Biotin propionyl-CoA carboxylase}}$$
Propionyl-CoA

$$CH_3CH(COOH)COSCoA + ADP + Pi \xrightarrow{B_{12}} HOOCCH_2CH_2COSCoA$$
Methymalonyl-CoA   Succinyl-CoA

Similarly, branched-chain FAs with an even number of main-chain carbon atoms may eventually yield propionate, while the oxidation of the odd-numbered branched-chain acids proceeds by a different route involving 3-hydroxy-3-methylglutaryl-CoA (HMG-CoA) (Fig. 3.32).

Most natural FAs are unsaturated. In addition, as discussed before (see Section 2.1.3) most double bonds are *cis* and, in PUFAs, are methylene interrupted (three carbons apart). When UFAs are β-oxidized, two problems may be encountered – the UFAs have *cis* double bonds and these may be at the wrong position for β-oxidation. A typical UFA might be linoleic acid and its oxidation is illustrated in Fig. 3.33. At first, β-oxidation proceeds normally (by three cycles of oxidation) but then a *cis* double bond is encountered at the wrong position (position 3). The German biochemist, Stoffel, has shown that an isomerase exists to convert the *cis*-3 compound into the necessary *trans*-2-fatty acyl-CoA. The isomerase will also act on *trans*-3 substrates, though at lower rates. Once over this obstacle, β-oxidation can again continue to eliminate a further two carbons and then dehydration produces a 2-*trans*,4-*cis*-decadienoyl-CoA from linoleoyl-CoA. The discovery of a 2,4-dienoyl-CoA reductase by Kunau and Dommes showed that this enzyme could use

NADPH to yield 3-*trans*-decenoyl-CoA as its product. The enoyl-CoA isomerase then moves the double bond to position 2 and β-oxidation can proceed again as normal (Fig. 3.33).

There are additional isomerases or reductases present in mitochondria which can cope with double bonds at unusual positions such as intermediates produced during β-oxidation of conjugated linoleic acid (*c9,t*11-18:2). Otherwise, such intermediates might accumulate and inhibit β-oxidation.

3.2.1.6 Regulation of mitochondrial β-oxidation

There are two major products of β-oxidation. Complete oxidation of an even-numbered FA yields acetyl-CoA, which can be fed into the TCA cycle. In some tissues, however, notably liver and the rumen epithelial cells of ruminant animals, acetoacetate accumulates. This compound, with its reduction product, 3-hydroxbutyrate, and its decarboxylation product, acetone, make up a group of metabolites known as ketone bodies. Free acetoacetic acid may accumulate in liver in two ways. The CoA derivative may be enzymically hydrolysed to the free acid and CoA and the liver tissue lacks the thiokinase to reconvert the acid back into its thioester. Alternatively, acetoacetyl-CoA may be converted into HMG-CoA, which is subsequently cleaved to free acetoacetic acid:

$$CH_3COCH_2COSCoA + CH_3COSCoA \rightleftharpoons CH_3CCH_2-COOH + CoA$$
Acetoacetyl-CoA   Acetyl-CoA

with side chains:
$$\begin{array}{c} CH_2COSCoA \\ | \\ | \\ OH \end{array}$$

Hydroxymethyl-
glutaryl-CoA

$$\updownarrow$$

$$CH_3COCH_2COOH + CH_2COSCoA$$
Acetoacetate   Acetyl-CoA

HMG is an important intermediate in cholesterol biosynthesis and this pathway provides a link between

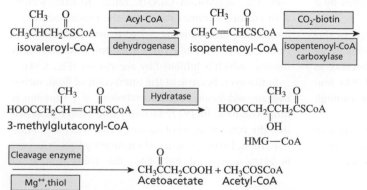

Fig. 3.32 Oxidation of odd-numbered branched-chain fatty acids.

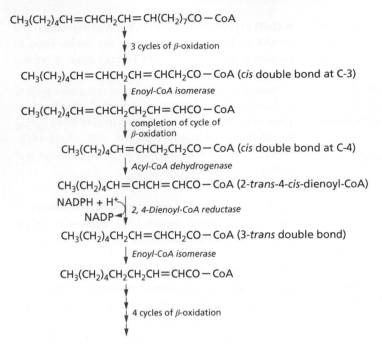

$$CH_3(CH_2)_4CH=CHCH_2CH=CH(CH_2)_7CO-CoA$$

    ↓ 3 cycles of β-oxidation

$$CH_3(CH_2)_4CH=CHCH_2CH=CHCH_2CO-CoA\ (cis\ \text{double bond at C-3})$$

    ↓ Enoyl-CoA isomerase

$$CH_3(CH_2)_4CH=CHCH_2CH_2CH=CHCO \cdot CoA$$

    completion of cycle of
    ↓ β-oxidation

$$CH_3(CH_2)_4CH=CHCH_2CH_2CO-CoA\ (cis\ \text{double bond at C-4})$$

    ↓ Acyl-CoA dehydrogenase

$$CH_3(CH_2)_4CH=CHCH=CHCO-CoA\ (2\text{-}trans\text{-}4\text{-}cis\text{-dienoyl-CoA})$$

NADPH + H⁺
     2, 4-Dienoyl-CoA reductase
NADP⁺

$$CH_3(CH_2)_4CH_2CH=CHCH_2CO-CoA\ (3\text{-}trans\ \text{double bond})$$

    ↓ Enoyl-CoA isomerase

$$CH_3(CH_2)_4CH_2CH_2CH=CHCO-CoA$$

    ↓ 4 cycles of β-oxidation

Fig. 3.33 The oxidation of linoleic acid.

FA and cholesterol metabolism. Ketone bodies are also excellent fuels for the brain during starvation, as the brain cannot utilize long-chain FAs as fuels directly.

Acetyl-CoA is therefore the substrate for two competing reactions: with oxaloacetate to form citrate or with acetoacetyl-CoA to form ketone bodies (ketogenesis). Which reaction predominates depends partly on the rate of β-oxidation itself and partly on the redox state of the mitochondrial matrix, which controls the oxidation of malate to oxaloacetate and, hence, the amount of oxaloacetate available to react with acetyl-CoA. The proportion of acetyl groups entering the TCA cycle relative to ketogenesis is often referred to as the 'acetyl ratio'. The overall rate of β-oxidation may be controlled by a number of well-known mechanisms:

1 *The availability of FAs.* There are two aspects to this availability. First, FAs have to be made available by hydrolysis of lipid stores or from the diet. Second, as discussed in Section 3.2.1.3, the entry of FAs into mitochondria is under careful regulation through the activity of CPT_1.

2 *The rate of utilization of β-oxidation products,* which in turn can either lead to specific inhibition of particular enzymes or to 'feedback inhibition' of the whole sequence.

The concentration of NEFAs in plasma is controlled by catecholamines (which stimulate) and insulin (which inhibits) breakdown of TAGs in adipose tissue stores (see Sections 4.2 & 9.1.2 for more information). Once the FAs enter cells they can be degraded to acetyl-CoA or used for lipid biosynthesis. The relative rates of these two pathways depend on the nutritional state of the animal, particularly on the availability of carbohydrate.

In muscle, the rate of β-oxidation is usually dependent on both the NEFA concentration in the plasma as well as the energy demand of the tissue. A decrease in energy demand by muscle will lead to a build-up of NADH and acetyl-CoA. Increased NADH/NAD⁺ ratios lead to inhibition of the mitochondrial TCA cycle and increase further the acetyl-CoA/CoASH ratio. Kinetic studies with purified enzymes show that the major sites of the β-oxidation inhibition are 3-hydroxacyl-CoA dehydrogenase, which is inhibited by NADH, and 3-ketoacyl-CoA thiolase, which is inhibited by acetyl-CoA (Fig. 3.34).

In the liver, because of the interaction of lipid, carbohydrate and ketone body metabolism, the situation is more complex. In 1977, McGarry and Foster proposed that the concentration of malonyl-CoA was particularly important. In the fed state, when carbohydrate (glucose) is being converted into FAs, the concentration of

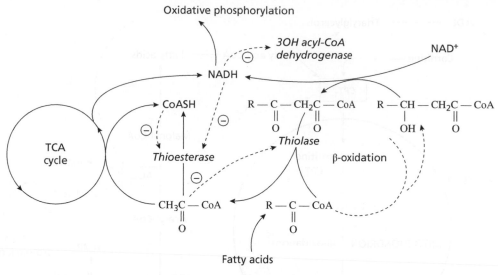

Fig. 3.34 Regulation of β-oxidation in muscle. Inhibition of enzymes is indicated.

malonyl-CoA is raised, inhibiting β-oxidation. In a fasting state, lowered malonyl-CoA allows CPT_1 to function at high rates and this stimulates β-oxidation and ketogenesis (Fig. 3.35). For lipogenic tissues such as liver and adipose tissue ACC1 is the main form, while in heart and skeletal muscle it is ACC2. In the latter tissues the levels of malonyl-CoA are controlled not only by ACC but also by a cytosolic malonyl-CoA decarboxylase.

As mentioned in Section 3.2.1.3, it seems that ACC controls β-oxidation through the inhibition of CPT_1, by malonyl-CoA. In the fasted condition, a high glucagon/insulin ratio elevates cellular cAMP, thus allowing short-term inhibition of ACC (see Section 3.1.11.2) by phosphorylation. Reduction of the glucagon/insulin ratio on feeding reverses this effect. Thus, hepatic FA biosynthesis and degradation are co-regulated by hormonal levels.

3.2.1.7 Fatty acid oxidation in *E. coli*

When *E. coli* is grown on a medium containing FAs rather than glucose, a 200-fold induction of the enzymes of β-oxidation is seen. By the use of deficient mutants, it has been possible to identify a number of genes responsible for coding the relevant enzymes. These genes are named FA degradation (*fad*) genes and there are seven of them, at six different sites on the *E. coli* chromosome. Together they form a regulon, which is under the control of the *fadR* gene (see Section 3.1.1.1 & Fig. 3.2). In the presence of medium-chain or long-chain FAs, the *fad* regulon is

coordinately induced as acyl-CoAs bind to a small repressor protein (produced by *fadR*) and stop it interacting with the *fad* gene promoters. All the genes except the electron transport protein (which is constitutive) are induced.

Of the different genes, one (*fadL*) codes for a permease (which is a specific pore that allows NEFA passage across the outer membrane), thus permitting rapid uptake of FAs into *E. coli*. FadD codes for a single ACS, while *fadE* codes for an acyl-CoA dehydrogenase.

Perhaps the most interesting feature of β-oxidation in *E. coli* was the discovery of a homogeneous protein (FadB) that contained five enzyme activities (enoyl-CoA hydrolase, 3-hydroxyacyl-CoA dehydrogenase, thiolase, *cis*-3,*trans*-2-enoyl-CoA epimerase, 3-hydroxyacyl-CoA epimerase). The protein is coded by the *fadAB* operon. The protein exists as an $\alpha_2\beta_2$ heterotetramer. The β chain is smaller, contains the thiolase and (confusingly) is coded for by the *fadA* gene. The α-chain contains the other activities and is coded by the *fadB* gene.

3.2.1.8 β-Oxidation in microbodies

It was mentioned in Section 3.2.1.1 that microbodies are an important (in plants, the main) site of FA oxidation. Microbodies do not have an electron transport system coupled to energy production as in mitochondria. Moreover, the flavin-containing oxidase, which catalyses the first reaction, transfers electrons to oxygen to produce

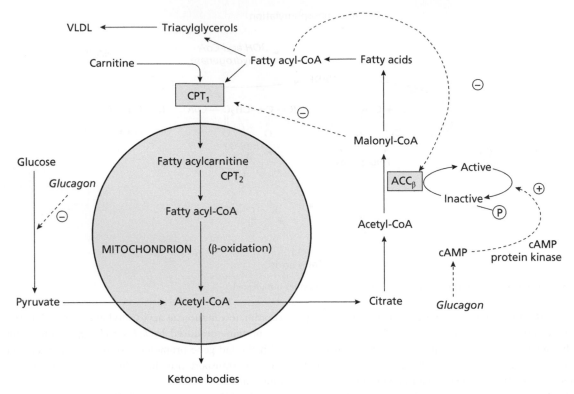

Fig. 3.35 Regulation of fatty acid metabolism in liver. Abbreviations: CPT$_1$, CPT$_2$, carnitine palmitoyltransferase 1 and 2; ACC, acetyl-CoA carboxylase; VLDL, very low density lipoproteins.

H$_2$O$_2$ (which is rapidly destroyed by catalase). Thus, this energy is lost as heat. Nevertheless, microbodies produce chain-shortened acyl-CoAs, acetyl-CoA and NADH, all of which can exit the organelle. The initial uptake of FAs into the organelle is most probably via a malonyl-CoA-sensitive CPT$_o$ (see Section 3.2.1.3) or, possibly, NEFAs or acyl-CoAs can cross the microbody membrane.

As mentioned above, β-oxidation in microbodies differs from that in mitochondria (Fig. 3.36). The first step, catalysed by acyl-CoA oxidase, introduces a *trans*-2 double bond and produces H$_2$O$_2$, which is destroyed by catalase. In some tissues (e.g. rat liver) there are several oxidases with different substrate selectivities. The next two reactions (hydration, dehydrogenation) are catalysed by a multifunctional protein (which also may contain the *cis*-3, *trans*-2-enoyl-CoA epimerase). NADH is a product of the dehydrogenase reaction. No epimerase is contained in yeast and fungi, so that in these organisms the protein is bifunctional. The last reaction uses a thiolase and, in liver, there are two enzymes, one of which is constitutively expressed and the other is

induced by peroxisomal proliferators. In mammals, there are two multifunctional enzymes (see above). One (MFE1) contains three enzymes including the epimerase. The other (MFE2) contains only the first two enzymes. They also differ in some features of their reactions.

As with entry of FAs into microbodies (above) there is controversy about the exit of chain-shortened acyl-CoAs

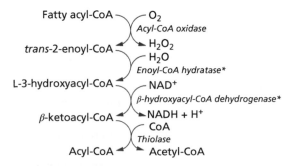

Fig. 3.36 Peroxisomal β-oxidation of fatty acids. *Catalysed by a trifunctional protein in liver and a bifunctional protein in yeast or fungi.

from animal microbodies. Although it was believed previously that such compounds could cross the membrane relatively easily, it is now increasingly thought that a carnitine transport system is used. Because of the absence of an electron transport chain in microbodies there is no internal means of regenerating NAD$^+$. The reoxidation of NADH is thought to occur either by a G3P shuttle or by movement of NADH to the cytosol and NAD$^+$ back.

In animal tissues, such as liver, the microbody system seems incapable of oxidizing long-chain acyl-CoAs completely, owing to the limited substrate specificity of the acyl-CoA oxidases. The medium-chain products are transferred to carnitine by a peroxisomal medium-chain carnitine acyltransferase. The acyl-carnitines can then be moved to the mitochondria for further oxidation. Studies on this system and on peroxisomal β-oxidation in general have been facilitated by the fact that hypolipidemic drugs like clofibrate, high fat diets, starvation or diabetes all increase peroxisomal oxidation considerably. Under average conditions peroxisomes are thought to contribute up to 50% of the total FA oxidative activity of liver. A disease known as neonatal adrenoleukodystrophy is characterized by the accumulation of very long-chain FAs in the adrenals. In Zellweger Syndrome, peroxisome formation is severely affected. In addition to these inherited diseases (where there is a general deficiency in peroxisomal β-oxidation), there are other diseases where one or other individual enzyme is missing. These include X-linked adrenoleukodystrophy, where mutations in the *ABCD1* gene (a gene coding for a protein involved in the peroxisomal import of FAs and/or fatty acyl-CoAs) causes accumulation of very long-chain FAs (see also Section 10.1.2).

In contrast to the chain length limitation of β-oxidation in mammalian peroxisomes, the glyoxysomes of fatty seedlings and the peroxisomes of plant leaves, fungi or yeasts are capable of completely oxidizing FAs to acetate. As mentioned before, no cooperation with mitochondria is necessary (apart from regeneration of NAD$^+$) and, indeed, many plant mitochondria do not appear to contain β-oxidation enzymes.

3.2.2 α-Oxidation of fatty acids is important when substrate structure prevents β-oxidation

α-Oxidation is important in animals for the formation of α-hydroxy FAs and for chain shortening, particularly during catabolism of molecules that cannot be metabolized directly by β-oxidation. Brain cerebrosides and other sphingolipids (see Table 2.4, Table 2.14 & Section 2.3.3) contain large amounts of α-hydroxy FAs and a mixed-function oxidase, which requires O_2 and NADPH to form them, has been identified in microsomal fractions. Breakdown of α-hydroxy FAs also takes place in such fractions but is probably localized in peroxisomes.

The peroxisomal α-oxidation system has been well-characterized, partly because it is very important for the breakdown of branched-chain FAs – in particular, phytanic acid (3, 7, 11, 15-tetramethyl palmitic acid), which is formed in animals from dietary phytol, the side-chain of chlorophyll. In phytanic acid the methyl group at C3 prevents β-oxidation. Phytanic acid is activated to its CoA ester, hydroxylated at the *sn*-2 position (to give an α-hydroxy intermediate), followed by cleavage to yield pristanal (an aldehyde) with the release of formyl-CoA. The latter gives rise to CO_2. Pristanal is then oxidized (using NAD) to pristanic acid, which as pristanoyl-CoA, is a substrate for β-oxidation, which continues with the release of propionyl-CoA. (The 2-methyl group on pristanic acid does not inhibit β-oxidation provided it has the correct stereochemistry. The initial metabolism of phytanic acid is shown in Fig. 3.37.

An inborn error of metabolism where α-oxidation is impaired is Refsum disease (see also Section 10.1.2). In this primarily neurological disorder, phytanoyl-CoA hydroxylase is deficient and huge amounts of phytanic acid accumulate. In blood, this may represent 30% of the total FAs. Ataxic neuropathy develops and the disease is normally fatal. To survive, patients must have a low phytol diet.

Plants have an active α-oxidation system that is concerned, amongst other functions, with the turnover of the phytol moiety of chlorophyll as described above. The mechanism has been studied in several tissues. Although there is still some controversy as to whether O_2 or H_2O_2 is the substrate and which cofactors are involved, it seems probable that O_2 is activated by a reduced flavoprotein and that a hydroperoxide intermediate is the active species that reacts with the FA.

Until 1973 it seemed that there must be at least two different pathways for α-oxidation in plants, depending on the source of the enzymes. The pathways that had been studied in pea leaves and in germinating peanut cotyledons apparently had different cofactor requirements and different intermediates seemed to be

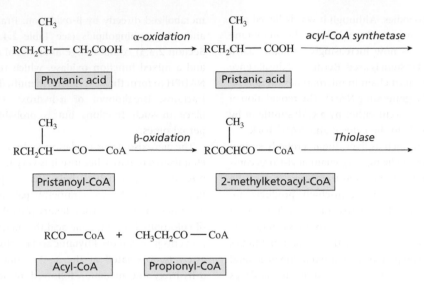

Fig. 3.37 Initial metabolism of phytanic acid in animals.

involved. The discrepancies have now been resolved and a unified pathway proposed by Stumpf and his team in California (Fig. 3.38). They showed that molecular oxygen was a requirement and, in the peanut system, that a hydrogen peroxide-generating system was involved. Enzymes catalysing the reduction of peroxides, such as glutathione peroxidase, reduced α-oxidation and increased the production of D-hydroxpalmitate from palmitate. This pointed to the existence of a peroxide (2-hydroperoxypalmitate) intermediate. Accumulation of hydroxy acids in some experiments was due to the intermediates being channelled into a dead-end pathway. After the hydroperoxy intermediate, CO_2 is lost and a fatty aldehyde produced. This is oxidized by an NAD^+-requiring system to yield a FA, one carbon less than the original (palmitic acid) substrate. α-Oxidation in plants is extra-mitochondrial and may be in the cytoplasm or associated with the ER (depending on the tissue).

3.2.3 ω-Oxidation uses mixed-function oxidases

ω-Oxidation of straight-chain FAs yields dicarboxylic acid products. Common FAs are only slowly catabolized by such a system, since β-oxidation is usually highly active. However, in substituted derivatives, ω-oxidation is often an important first step to allow subsequent β-oxidation to take place. ω-Oxidation has an ω-hydroxy FA as an

intermediate and the enzyme is a mixed-function oxidase that uses O_2 and NADPH as co-substrates. The enzyme is very similar to, or identical with, the cytochrome P450-dependent drug-hydroxylating enzyme system.

An ω-hydroxylase has been studied in *Pseudomonas* by Coon and co-workers. They showed that the enzyme had nonhaem iron as, apparently, the only prosthetic group involved directly in the hydroxylation reactions. The system was fractionated into three components: a nonhaem iron protein (similar to rubredoxin), a flavoprotein and a final component needed for hydroxylase activity.

In plants, the ω-hydroxylase system is responsible for synthesis of the ω-hydroxy FA components of cutin and suberin (see Section 5.6.1). Kolattukudy has studied the reactions in preparations from *Vicia faba*. NADPH and O_2 were cofactors and the enzyme was a typical mixed-function oxidase. However, the involvement of P450 in the plant system is unproven since, although the hydroxylation is inhibited by CO_2, the inhibition is not reversed by 420–460 nM light (a property typical for cytochrome P450 systems). Where dihydroxy FAs are being synthesized for cutin or suberin, the ω-hydroxy FA is the substrate for the second hydroxylation. Like ω-oxidation, mid-chain hydroxylation also requires NADPH and O_2 and is located in the ER.

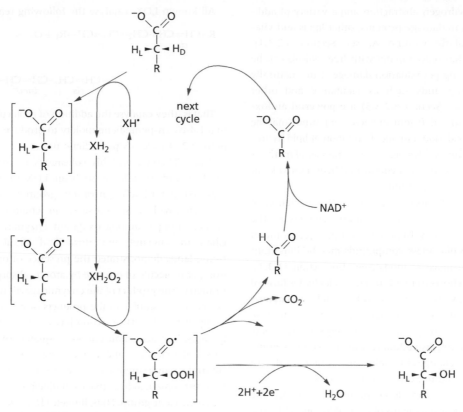

Fig. 3.38 Proposed α-oxidation pathway in plants.

3.3 Chemical peroxidation is an important reaction particularly of polyunsaturated fatty acids

One of the characteristic reactions of lipids that are exposed to oxygen is the formation of peroxides. Indeed, among nonenzymic chemical reactions taking place in the environment at ambient temperatures, the oxidation of unsaturated compounds is, perhaps, the most important both from an industrial and a medical point of view. In biological tissues, uncontrolled lipid peroxidation causes membrane destruction and is increasingly regarded as an important event in the control or development of diseases (see Sections 10.2.1 & 10.3). In food, oxidation (either enzymically or chemically catalysed) can have desirable as well as adverse consequences (see also Sections 6.1.3.4 & 10.5.1).

In common with other radical chain reactions, lipid peroxidation can be divided into three separate processes: initiation, propagation and termination. During initiation a very small number of radicals (e.g. transition metal ions or a radical generated by photolysis or high-energy irradiation) allow the production of $R^\bullet$ from a substrate RH:

$$X^\bullet + RH \rightarrow R^\bullet + XH$$

Propagation then allows a reaction with molecular oxygen:

$$R^\bullet + O_2 \rightarrow ROO^\bullet$$

and this peroxide radical can then react with the original substrate:

$$ROO^\bullet + RH \rightarrow ROOH + R^\bullet$$

Thus, the events form the basis of a chain-reaction process.

Free radicals such as $ROO^\bullet$ (and $RO^\bullet$, $OH^\bullet$ etc. which can be formed by additional side-reactions) can react at random by hydrogen abstraction and a variety of addition reactions to damage proteins, other lipids and vitamins (particularly vitamin A, see Section 6.2.3.1). Compounds that react rapidly with free radicals can be useful in slowing peroxidation damage. Thus, naturally occurring compounds such as vitamin E and other tocopherols (see Section 6.2.3.3) are powerful antioxidants and tissues deficient in such compounds may be prone to peroxidation damage. Formation of lipid hydroperoxides can be readily detected by a number of methods of which the absorption of conjugated hydroperoxides at 235 nm is particularly useful.

Termination reactions may lead to the formation of both high and low molecular mass products of the peroxidation reaction. Depending on the lipid, some of the low molecular mass compounds may be important flavours (or aromas) of foods (see also Section 11.3.1). For example, short- to medium-chain aldehydes formed from UFAs may give rise to rancidity and bitter flavours on the one hand or more pleasant attributes such as those associated with fresh green leaves, oranges or cucumbers on the other hand. Fish odours are attributed to a ketone. Some relevant changes in quality or nutritional value of food are indicated in Sections 6.1.2.3 & 11.4.2.

In general, it is considered desirable to reduce the initiation reaction as a means of controlling peroxidation. Apart from natural antioxidants, BHT (3,5-di-t-butyl-4-hydroxytoluene) is often used as a food additive. Metal binding compounds and phenolic compounds (present in olive oil, a key component of the 'Mediterranean diet') may also be inhibitory as well as the endogenous superoxide dismutase and glucose oxidase-catalase enzyme systems.

3.4 Peroxidation catalysed by lipoxygenase enzymes

The second kind of peroxidation is catalysed by a family of enzymes known as lipoxidases (or lipoxygenases, LOX). These enzymes were originally thought to be present only in plants, but it has now been realized that they catalyse very important reactions in animals (see Sections 3.5.9–3.5.11). Important sources of the enzyme are peas and beans (especially soybean), cereal grains and oil seeds. LOX was originally detected by its oxidation of carotene and has been used extensively in the baking industry for bleaching carotenoids in dough.

All known LOXs catalyse the following reaction:

$$R{-}CH{=}CH{-}CH_2{-}CH{=}CH{-}R_1 + O_2 \rightarrow$$

$$\underset{cis}{} \qquad \underset{cis}{}$$

That is, they catalyse the addition of molecular oxygen to a 1,4-cis, cis-pentadiene moiety to produce a 1-hydroperoxy-2, 4-$trans$, cis-pentadiene unit.

When Theorell and his colleagues in Sweden first purified and crystallized soybean LOX in 1947, they reported that it had no prosthetic group or heavy metal associated with it. Such a situation would make lipoxygenase unique among oxidation enzymes. However, Chan in England and Roza and Franke in The Netherlands demonstrated the presence of one atom of iron per molecule of enzyme by atomic absorption spectrometry. The product of the enzymic reaction – a hydroperoxide – is similar to the products of purely chemical catalysis, but the LOX reaction has several distinguishing features. The activation energy is smaller than that for chemical reactions, and the enzyme has very specific substrate requirements. In order to be a substrate, the FA must contain at least two cis double bonds interrupted by a methylene group. Thus, linoleic (18:2n-6) and α-linolenic (18:3n-3) acids are good substrates for the plant enzymes while arachidonic acid (20:4n-6) and DHA (22:6n-3), the major PUFAs in mammals, are attacked by different LOXs in their tissues. Like chemically catalysed peroxidation, the LOX reaction involves free radicals and can be inhibited by radical trapping reagents such as the tocopherols. The reaction sequence shown in Fig. 3.39 represents the currently accepted pathway.

LOXs are widespread in plants. One effect of the enzymes in plant tissues is to yield volatile products with characteristic flavours and aromas – either desirable or undesirable. This aspect is dealt with in more detail in the next section, where the physiological importance of plant LOX is covered. In general it seems that plant LOXs work best with NEFA substrates. These are released from storage TAGs by lipases or, more importantly, from membrane lipids by nonspecific acylhydrolases or phospholipases (PL)A.

Several LOXs are found in mammals, being distinguished from one another by the position of oxygen insertion. Their role in producing leukotrienes and other oxy-lipids is discussed in Sections 3.5.9 and 3.5.10. As for

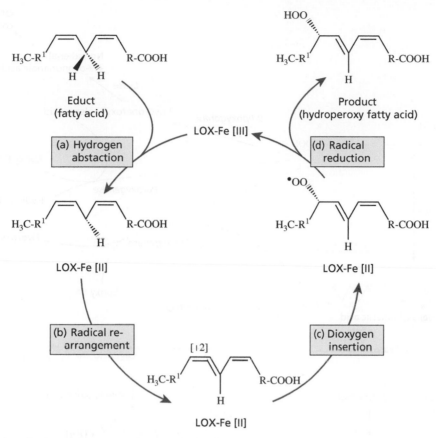

Fig. 3.39 Detailed mechanism of the LOX-reaction. LOX catalysed oxygenation of fatty acids consists of four consecutive elementary reactions, the stereochemistry of which are tightly controlled. (a) Stereoselective hydrogen abstraction from a *bis*allylic methylene. The hydrogen atom is removed as a proton and the resulting electron is picked up by the ferric nonheme iron that is reduced to the ferrous form. (b) Radical rearrangement: During this elementary reaction the radical electron is dislocated either in the direction of the methyl end of the fatty acid ([+2] rearrangement) or in the direction of the carboxylate [−2] rearrangement). (c) Oxygen insertion: Molecular dioxygen is introduced antarafacially (from opposite direction of the plane determined by the double bond system) related to hydrogen abstraction. If the hydrogen located above the double bond is removed, dioxygen is introduced from below this plane. (d) peroxy radical reduction: The peroxy radical formed via oxygen insertion is reduced by an electron from the ferrous nonheme iron converting the radical to the corresponding anion. Thereby the iron is reoxidized to its ferric form. Finally, the peroxy anion is protonated.

the plant enzymes, animal LOXs use NEFAs that must first be released from complex lipids to provide sufficient substrate.

3.4.1 Lipoxygenases are important for stress responses and development in plants

Whereas the major membrane PUFAs in animals have chain lengths of 20C and 22C, which are thus the most important substrates for LOX attack, in plants α-linolenate is the most abundant fatty acid. Breakdown of this FA

is known as the α-linolenic acid cascade (or Lipoxygenase pathway), which gives rise to a variety of important molecules (Figs. 3.40 & 3.41).

Plant LOXs are stable enzymes which are present in high concentrations in many plant tissues especially leaves. They are multifunctional and can catalyse three different types of reaction: (1) dioxygenation of lipids forming hydroperoxides (above), (2) secondary conversion of the latter to keto lipids (hydroperoxidase reaction), (3) formation of epoxy leukotrienes (leukotriene synthase). In plants, dioxygenation is the most

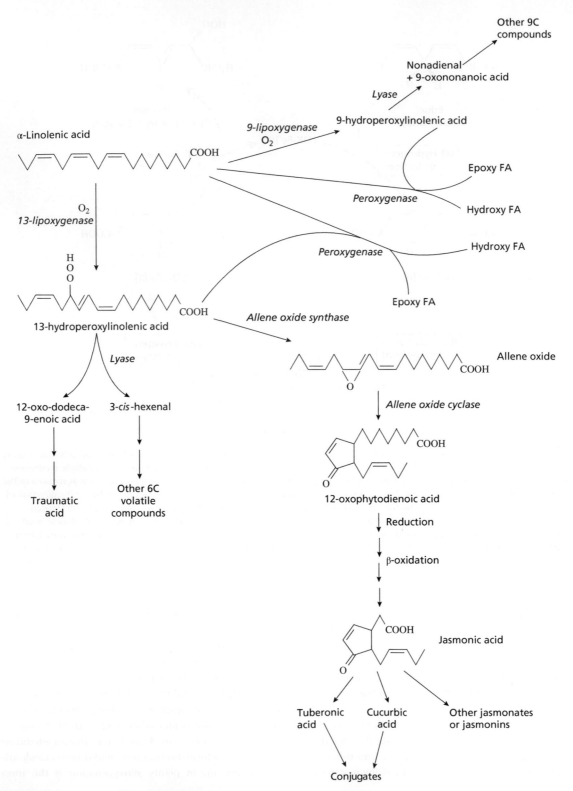

Fig. 3.40 The lipoxygenase pathway and oxylipin formation in plants.

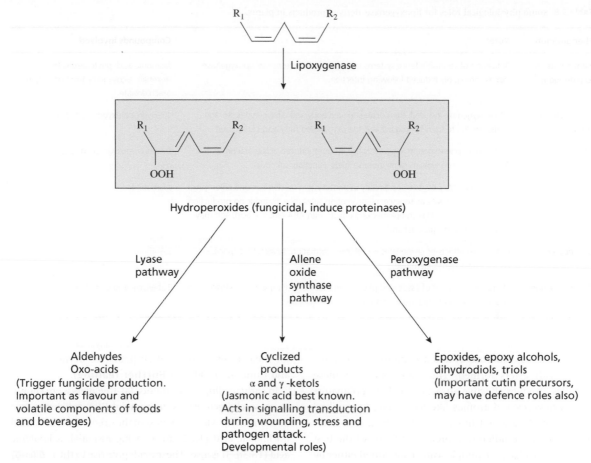

Fig. 3.41 Lipoxygenase products and physiological roles in plants.

prevalent. There are two classes of LOXs, the 9-LOXs and the 13-LOXs which differ in the position of the peroxide moiety introduced onto the carbon chain.

The FA hydroperoxides produced by plant LOXs can be converted into three main products (Fig. 3.40) with important known functions:

1 Cooxidative reactions with peroxygenase give a mixture of epoxy and hydroxy FAs (depending on the nature of the acceptor FA for the mono-oxygenation). The products have roles in the formation of cutin (the surface covering of leaves – see also Section 5.6.1) and in responding to pest attack.

2 Hydroperoxide lyase cleaves the hydroperoxide into an aldehyde and an oxo-UFA. Their metabolic products have a role in pest defence and appear to act as pollinators and herbivore attractants, especially for flowers and fruits. They are also important flavour and aroma components of foods, drinks and perfumes.

3 Allene oxide synthase gives rise to the precursor of jasmonic acid. The latter and associated derivatives are commonly known as the jasmonates, although Hildebrand has proposed the term 'jasmonins' to include both classes of compounds. The jasmonins have important effects on plant growth, development and senescence.

Some of the important roles for LOX-derived plant products are listed in Table 3.8.

Under item (2) above, hydroperoxide lyase cleavage is listed. This usually occurs with the 13-hydroperoxide derivative, thus giving rise to a 6C aldehyde. If the aldehyde is the unsaturated hexenal, the products will be rather volatile. The aldehyde can be converted into its alcohol (by alcohol dehydrogenase), the double bond is

Table 3.8 Some physiological roles for lipoxygenase-derived products in plants.

Phenomenon	Notes	Compounds involved
Plant resistance to pathogens	Activation of several defence systems in 'hypersensitive response' lipoxygenase expression rapidly induced following infection.	Jasmonic acid; phytoalexins (e.g. hexenal); eicosanoids from pathogen arachidonate
Mobilization of storage lipids	13-Lipoxygenase induced, translocates to lipid body and dioxygenates storage lipids, which are then hydrolysed and the oxygenated fatty acid catabolized.	13-Hydroperoxygenated lipids
Drought stress	Widespread phenomenon in plants is the drought-induced accumulation of 9-hydroperoxy derivatives of membrane lipids. Function unknown.	9-Hydroperoxygenated lipids
Senescence	Lipoxygenase may be involved directly in photosystem inactivation and chlorophyll oxidation. Jasmonate as senescence-promoting substance, by inducing production of certain proteins. Has complementary effects with ABA (abscisic acid) and may influence ethylene production.	Jasmonins
Fruit ripening	Differential effects of jasmonates on ethylene formation. Several fruit-specific lipoxygenase genes identified.	Jasmonins
Tuber induction	Stimulate a number of associated phenomena such as cell expansion, cytoskeleton structure and carbohydrate accumulation.	Tuberonic acid and other jasmonins

isomerized (*cis*-3 to *trans*-2) and acetate (or other short-chain acid) esters formed. The combination of these volatiles provides effective attractants to both pollinators and herbivores and aromas like 'cut grass', 'cucumber' and those of various fruits arise from them. It is noteworthy that the different aromas and tastes of the best quality olive oils, for example, can be identified either by human tasters or via sophisticated gas chromatographic (GC) analysis of the lipoxygenase-derived volatiles.

3.5 Essential fatty acids and the biosynthesis of eicosanoids

Some types of PUFAs, the so-called essential FAs (EFAs – see Section 6.2.2.1), are converted into oxygenated FAs with potent physiological effects. These include effects on muscle contraction, cell adhesion, immune function and vascular tone.

An 'essential' dietary component is one that is needed for the normal development and function of an animal. There is usually a minimum amount that must be supplied. However, this requirement varies with gender, age, species and whether there are particular stresses. EFAs are such components but the situation is complex because there are many such structures and the evidence

for their 'essentiality' is often poor (see Section 6.2.2.4 and Cunnane (2003) in **Further reading**).

In early feeding experiments, fat-free diets were shown to cause characteristic symptoms of EFA deficiency (see Section 6.2.2). Although most of the original work focused on linoleic acid (18:2*n*-6) as being essential, α-linolenic acid (18:3*n*-3) is also. These acids give rise to the *n*-6 (ω-6) and *n*-3 (ω-3) series of PUFAs, which produce different physiological effects (see Section 3.5.13).

The reason that animals need dietary linoleic and α-linolenic acids is that they lack the 12- and 15-desaturases to synthesize them from oleic acid (see Section 3.1.8.2). These acids are essential for growth, reproduction and good health and must, therefore, be obtained in the diet directly from plant sources or indirectly by consuming animals which have eaten algae or plants. During certain developmental stages (e.g. pregnancy, infancy) there may be an additional need for the very long chain metabolites, arachidonate and DHA because they may not be formed in adequate amounts from the precursor EFAs. Cunnane has termed this situation as being 'conditionally indispensable' for arachidonate and DHA.

The first important oxygenated products to be identified as having potent biological activities were 20C derivatives known collectively as the eicosanoids.

3.5.1 The pathways for prostaglandin biosynthesis are discovered

Although prostaglandins (PG) were the first biologically active eicosanoids to be identified, it is now known that the EFAs are converted into several different types of eicosanoids. (Eicosanoid is a term meaning a 20C FA derivative.) The various eicosanoids are important examples of local hormones. That is, they are generated in situ and, because they are rapidly metabolized, only have activity in the immediate vicinity.

A summary of the overall pathway for generation of eicosanoids is shown in Fig. 3.42. EFAs can be attacked by LOXs, which give rise to leukotrienes (LTs) or hydroxy FAs and lipoxins. Alternatively, metabolism by cyclo-oxygenase (COX) gives cyclic endoperoxides from which the classic PGs, thromboxanes (TX) and prostacyclin (PGI) can be synthesized. A third possibility is via cyto-chrome P450 oxygenation (see Section 3.5.12), where atomic oxygen is introduced, leading to FA hydroxyl-ation or epoxidation of double bonds. Whereas both the LOX and COX reactions arise from the formation of a FA radical, in P450-oxygenation, activation of molecular oxygen is involved. After this, one oxygen atom is trans-ferred to the FA substrate and one is reduced to water. Because, historically, PG biosynthesis was elucidated before that of the other eicosanoids, we shall describe PG formation first.

Once the structures of PGE and PGF had been defined, two research teams led by van Dorp in The Netherlands and Bergström and Samuelsson in Sweden, drove much of the rapid development of the field, including decipher-ing the biosynthetic pathways. Both realized that the most likely precursor of PGE_2 and $PGF_{2\alpha}$ was arachidonic acid. This was then demonstrated by both groups simul-taneously (with the same preparation of tritiated arachi-donic acid) by incubation with whole homogenates of sheep vesicular glands. An important point noted was that the NEFA was the substrate and not an activated form (see later for discussion about release of NEFAs from tissues in Section 3.5.14). The ability of the arachi-donic acid chain to fold allows the appropriate groups to come into juxtaposition for the ring closure to occur (Figs. 3.43 & 3.44). The reactions take place in the microsomal fraction, but a soluble heat-stable factor is required. This cofactor can be replaced by reduced glu-tathione. The reaction also requires molecular oxygen. Labelling with $^{18}O_2$ demonstrated that all three oxygen atoms in the final prostaglandin (Fig. 3.44) are derived from molecular oxygen.

3.5.2 Prostaglandin biosynthesis by cyclo-oxygenases

The key initial reaction in PG formation is catalysed by prostaglandin endoperoxide synthase, usually known as COX (cyclo-oxygenase).

Once the arachidonate substrate has been released from membrane lipids, (Fig. 3.42) COX converts it to a cyclic product (Fig. 3.44), PGH_2. Similar reactions will

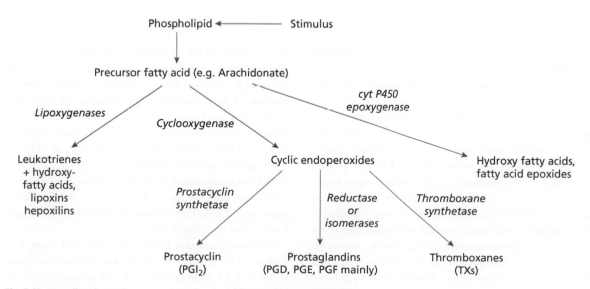

Fig. 3.42 Overall pathway for conversion of essential fatty acids into eicosanoids.

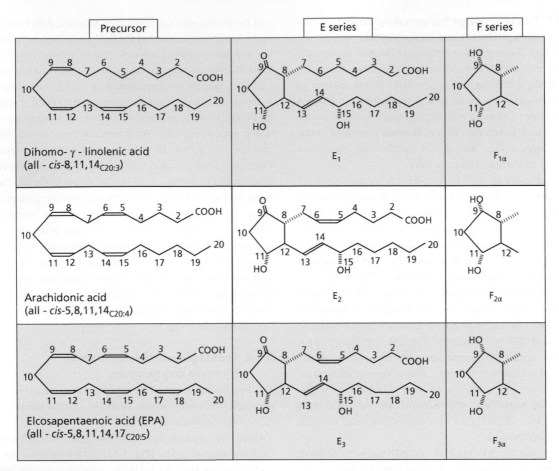

Fig. 3.43 Structures of prostaglandins E and F and their precursors.

occur with other substrates (e.g. EPA) to yield different endoperoxides.

There are two major COX isoforms and crystal structures for both have been obtained. The COXs are haemoproteins and they exhibit both COX and peroxidase activities. COX-1 is constitutively expressed in many mammalian cells and tissues and appears to be responsible for the formation of PGs involved in the general regulation of physiological events. On the other hand, COX-2 is present at low basal levels in inflammatory cells. It is strongly induced by inflammatory stimuli such as cytokines, endotoxins, tumour promoters and some lipids. Both isoforms have similar V_{max} and K_m values for arachidonate, undergo suicide inactivation and their reactions are initiated by hydroperoxide. (Suicide inactivation (or mechanism-based inhibition) refers to

irreversible enzyme inhibition that occurs when an enzyme binds a substrate analogue and forms an irreversible covalent-bound complex during normal catalysis. A good example is the inhibition of COXs by aspirin, described below.) However, COX-2 needs much lower levels of hydroperoxide and has a substrate selectivity that is somewhat different from that of COX-1. Both enzymes will, nevertheless, utilize a broad spectrum of PUFA substrates such as linoleate, γ-linolenate (all-*cis*-6,9,12-18:3), α-linolenate (all-*cis*-9,12,15-18:3), arachidonate and EPA. In turn, the resultant endoperoxides can form a host of different PGs.

The overall reaction is shown in Fig. 3.44. First, PG endoperoxide synthase inserts two molecules of oxygen to yield a 15-hydroperoxy-9,11-endoperoxide with a substituted cyclopentane ring (PGG_2; Fig. 3.44). This is

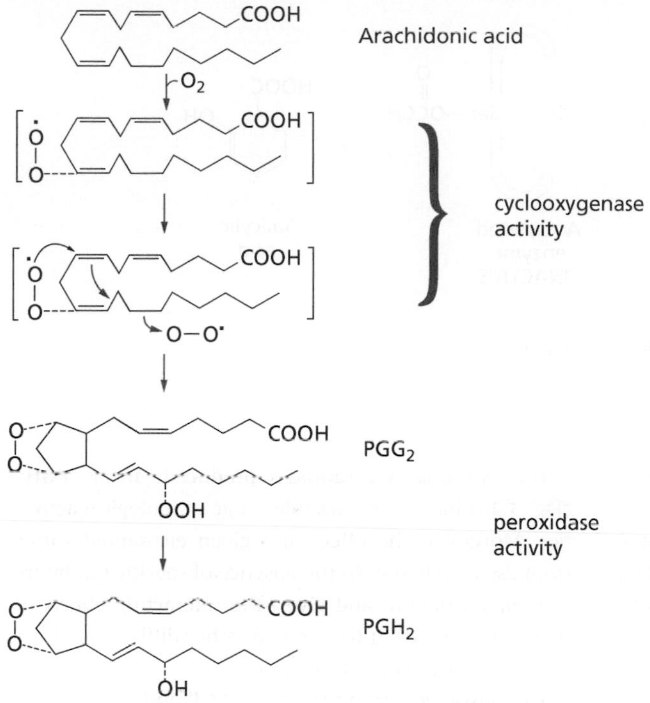

Fig. 3.44 Mechanism of biosynthesis of the cyclic endoperoxide, PGH$_2$.

the COX activity of the enzyme. The peroxidase activity then reduces PGG$_2$ to its 15-hydroxy analogue, PGH$_2$.

As mentioned above, the COX activity requires a hyperoxide activator to remove a hydrogen atom from position 13 on the incoming FA, and therefore allow attack by oxygen. This is an unusual mechanism because oxygenases usually work by activating the oxygen substrate. There are other interesting features of the enzyme. For example, the free radical intermediates generated can also inactivate the enzyme. This self-deactivation of the COX occurs in vivo as well as in purified preparations. It may ensure that only a certain amount of endoperoxide is generated even when large quantities of precursor FAs are available, since minute amounts have profound physiological effects. Both the COXs have a dual localization in the cell, being present in both the ER and the nuclear envelope.

3.5.3 Nonsteroidal anti-inflammatory drugs are cyclo-oxygenase inhibitors

A classic inhibitor of both COXs is aspirin (used for over 100 years) and, indeed, much of our knowledge of these enzymes has come from studies with this and other

nonsteroidal anti-inflammatory drugs (NSAIDs). In addition, gene disruption and overexpression of COX isoforms have confirmed that, as a generalization, the therapeutic anti-inflammatory action of the NSAIDs is due to inhibition of COX-2 while simultaneous inhibition of COX-1 causes most of the unwanted side-effects such as gastric ulceration.

Of the various NSAIDs (such as aspirin, ibuprofen, indomethacin and diclofenac) aspirin is the best known. Aspirin competes with arachidonate for binding to the COX active site. Although arachidonate binds about 10,000 times better than aspirin, once bound, aspirin acetylates a serine residue (serine 530) at the active site and causes irreversible COX inactivation (Fig. 3.45). Thus, the cell can restore COX activity only by making fresh enzyme. In COX-1 the serine 530 residue, although at the active site, is not needed for catalysis and acetylation by aspirin results in steric hindrance to prevent arachidonate binding. For COX-2, acetylation by aspirin still permits oxygenation of arachidonate, but the usual product PGH$_2$ is not made.

Not only is aspirin used for pain relief but low-dose therapy is used to reduce platelet TX formation, which reduces platelet aggregation and, hence, blood clotting in

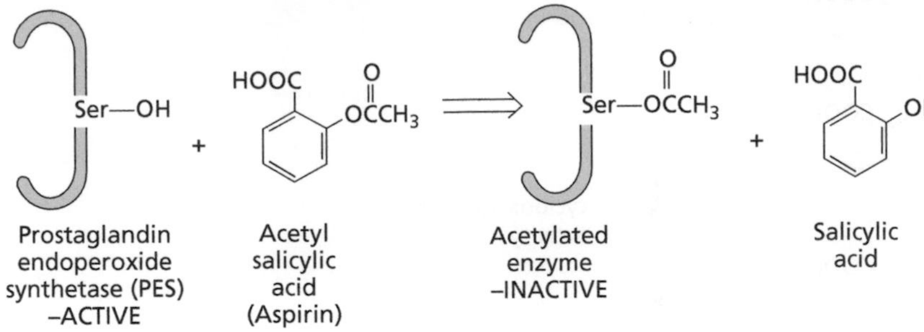

Fig. 3.45 Mechanism for the inactivation of the cyclooxygenase-1 by aspirin.

patients at risk from cardiovascular disease (see also Section 10.5.3).

Although the other NSAIDs inhibit COX activity, most of them cause reversible enzyme inhibition by competing with arachidonate for binding. A well-known example of a reversible NSAID is ibuprofen. The use of NSAIDs is a huge part of the pharmaceutical market and currently accounts for around £8 billion in annual sales – or to put it another way, approximately 10^{11} tablets of aspirin are consumed annually! Because NSAIDs inhibit both COXs, there has been active development of COX-2 specific drugs, which should possess anti-inflammatory properties without the unwanted side-effects of common NSAIDs. However, the first such drug (rofecoxib, marketed as Vioxx) had to be withdrawn because of unwanted cardiovascular effects. This has now been explained because it also affected formation of PGI_2 (an antithrombotic compound). Some competing actions of different eicosanoids are discussed in Section 3.5.13 and below.

3.5.4 Cyclic endoperoxides can be converted into different types of eicosanoids

The enzymes responsible for the further metabolism of PGH are present in catalytic excess to COX and, hence, are not regulatory except in the sense that the balance of their activities determines the pattern of PGs and TXs, which are formed in a given tissue. Thus, although Fig. 3.46 depicts some possible conversions of PGH_2 into various biologically active eicosanoids, prostanoid biosynthesis is cell-specific. For example, platelets form mainly TxA_2; endothelial cells produce PGI_2 as their major prostanoid; while kidney tubule cells synthesize predominately PGE_2.

The various eicosanoids produced from PGH_2 (Fig. 3.46) have a remarkable range of biological activities. Moreover, the effect of a given eicosanoid varies from tissue to tissue. In the absence of specific inhibitors of their formation, and given the interacting effects of many eicosanoids, it has proved rather difficult to elucidate all their biological functions.

The classic PGs are in the PGD, PGE and PGF groups. These, and other prostanoids, bind to cell receptors and there are pharmacologically distinct receptors for each of the known prostanoids. Moreover, there may be multiple receptors for a given PG. For example, in the case of PGE_2, four distinct receptors have been identified. Several receptors have been cloned and shown to have typical membrane-spanning regions. So far they all appear to belong to the G-protein-coupled receptor (GPCR) family and will, therefore, modify cellular reactions through classic G-protein signalling pathways (see Section 7.3.4 for more information).

PGD synthase (which is an isomerase) has high activity in brain and the spinal cord and PGD_2 is also the main COX product of mast cells. Glutathione-dependent and -independent forms of PGD synthase have been found (Fig. 3.46). PGD_2 inhibits platelet aggregation, increases platelet cAMP content and has a membrane receptor distinct from that for PGI_2 (see Section 3.5.7). It can act as a peripheral vasoconstrictor, pulmonary vaso-constrictor and bronchoconstrictor. The latter activity can be demonstrated when PGD_2 is inhaled. It is thought to have various neuromodulatory actions (e.g. it can decrease noradrenaline (US: norepinephrine) release from adrenergic nerve terminals) and overproduction of PGD_2 may be involved in the hypotensive attacks of patients with mastocytosis (a rare condition caused by

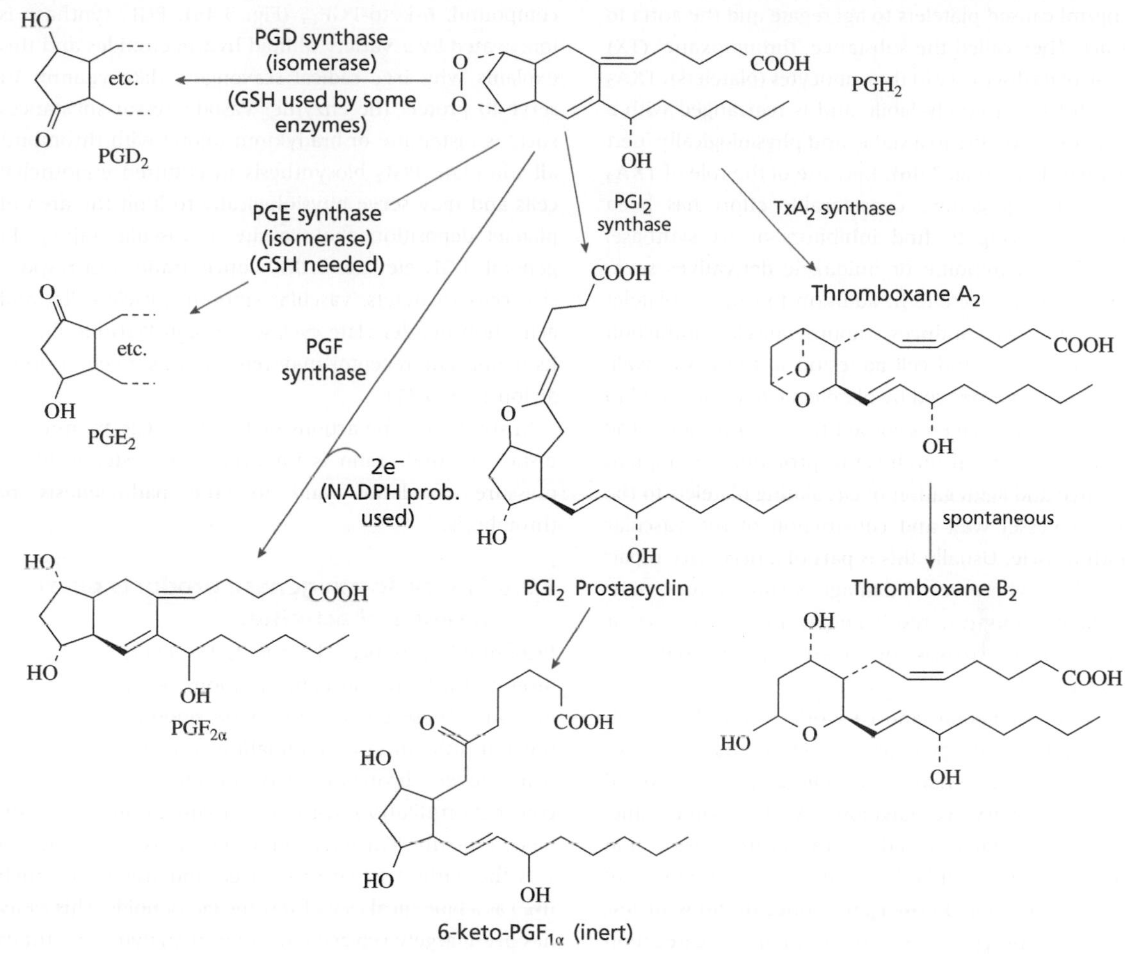

Fig. 3.46 Conversion of PGH$_2$ into various eicosanoids.

the accumulation of mast cells that release histamine and other substances into the blood).

PGE$_2$ is the main arachidonate metabolite in kidneys, where it reduces vasopressin (antidiuretic hormone)-stimulated water reabsorption and may help to control renin release. It may also help to mediate the metabolism and interaction of macrophages with other cells. PGE synthases (again isomerases) are unique in that all of them use glutathione as a cofactor.

PGF$_{2\alpha}$ seems to be the so-called luteolytic factor produced by mammalian uteri. When it is injected into cows it causes regression of the *corpus luteum* and induces ovulation and is used commercially as a regulator of ovulation in dairy cows. PGF$_{2\alpha}$ is useful for the induction of abortions in women in midtrimester. In addition, if

injected as a slow-releasable form, it will act as a birth control agent probably by inhibiting implantation of the fertilized ovum in the uterine wall. In contrast to the formation of PGD or PGE, the biosynthesis of PGF is a two-electron reduction. A PGF synthase, which uses NADPH as reductant, has been purified from lung.

3.5.5 New eicosanoids are discovered

Apart from the classic PGs, two other products of endo-peroxide metabolism have important cardiovascular actions: thromboxanes and prostacyclins.

In 1973, Hamberg and Samuelsson in Sweden were studying the role of PGs in platelet aggregation. They discovered two new COX products, one of which was highly active in stimulating aggregation. As little as 5 ng/ml caused platelets to aggregate and the aorta to contract. They called the substance 'thromboxane' (TX) because of its discovery in thrombocytes (platelets). TXA_2 (Fig. 3.46) is extremely labile and is rearranged with a half-life of about 30 s to a stable and physiologically inert derivative, TXB_2 (Fig. 3.46). Because of the role of TXA_2 in platelet aggregation, considerable effort has been devoted to trying to find inhibitors of TX synthase. Imidazole and pyridine or imidazole derivatives seem to be the most effective. In addition to causing platelet aggregation, TXA_2 induces smooth muscle contraction (vasoconstriction) and cell adhesion to the vessel wall. Thus TXA_2 is synthesized by platelets when they bind to subendothelial collagen exposed by micro-injuries. The newly synthesized thromboxane promotes subsequent adherence and aggregation of circulating platelets to the damaged vessel wall and constriction of the vascular smooth muscle. Usually this is part of a necessary repair process but, where vessel damage is chronic (or extensive), blood clots can result and cause arteriovascular crises such as strokes or heart attacks (see also Section 10.5.3).

In 1976, Needleman and co-workers found an enol-ether (later termed prostacyclin -PGI_2) that was produced in vascular endothelial cells and which would provoke coronary vasodilation. At the same time, Vane showed that PGI had an anti-aggregatory action on platelets. In fact, it is the most powerful inhibitor of platelet aggregation known and concentrations of less than 1 ng/ml prevent arachidonate-induced aggregation in vitro. The physiological effects of PGI are essentially the opposite to those of TX (Table 3.9). Thus, for healthy blood vessels, production of PGI_2 by the vascular endothelial cells counteracts the effects of TX on platelet aggregation. Like TX, PGI_2 has a short half-life and rearranges to a stable and physiologically inert

compound, 6-keto-$PGF_{1\alpha}$ (Fig. 3.46). PGI_2 synthase is inactivated by a variety of lipid hydroperoxides and this explains why free radical scavengers (like vitamin E) serve to protect the enzyme. Vasodepressor substances such as histamine or bradykinin (along with thrombin) all stimulate PGI_2 biosynthesis in cultured endothelial cells and may serve physiologically to limit the area of platelet deposition about a site of vascular injury. In general, PGI_2 elevates cAMP concentrations in responsive cells (platelets, vascular smooth muscle cells) and activation of adenylate cyclase through PGI_2 binding to its membrane receptor may represent its mechanism of action (Fig. 3.47).

In summary, the actions of TXs and PGIs are antagonistic and their ratio is important for systemic blood pressure regulation and for the pathogenesis of thrombosis.

3.5.6 The cyclo-oxygenase products exert a range of activities

Eicosanoids produced by the COXs exert a range of profound activities at concentrations down to 10^{-9} g/g of tissue. These include effects on smooth muscle contraction, inhibition or stimulation of platelet stickiness, pain, fever, bronchoconstriction/dilation and vasoconstriction/dilation with a consequent influence on blood pressure. The effect on smooth muscle contraction was the earliest to be recognized and was once much used as a biological assay for some eicosanoids. This assay has been largely replaced by modern analytical methods such as GC-mass spectrometry (MS) or a combination of high performance liquid chromatography (HPLC) separations followed by radioimmunoassay.

The effects of PGs, in particular, on smooth muscle contraction, are utilized in both medicine and agriculture. PGE_2 is known to have a cytoprotective effect in the upper gut while PGI_2 is used in dialysis and the treatment of peripheral vascular disease. Continuous perfusion of very small amounts (e.g. 0.01 mg PGE_1/min) into pregnant women causes uterine activity similar to that encountered in normal labour without any effect on blood pressure. PGs and synthetic analogues are now being used for induction of labour and for therapeutic abortion in both humans and domestic animals. The exact time of farrowing of piglets, for example, can be precisely controlled by PG treatment. For abortions in women, a solution of PGF_2 at a concentration of 5 mg/ml is administered by intravenous infusion at a rate of 5 mg/

Table 3.9 Opposing effects of prostacyclin and thromboxanes on the cardiovascular systems.

Thromboxanes produced in platelets	Physiological effect	Prostacyclin synthesized in arterial wall
Stimulates	Platelet aggregation	Inhibits
Constricts	Arterial wall	Relaxes
Lowers	Platelet cAMP levels	Raises
Raises	Blood pressure	Lowers

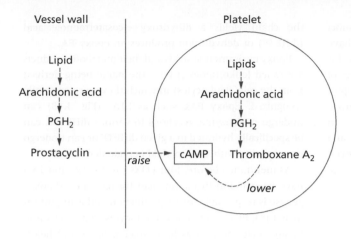

Fig. 3.47 Interaction of thromboxanes and prostacyclins in regulating platelet cAMP levels.

min until abortion is complete (about **5** h). The method appears, so far, to be safe with few side-effects and no further surgical intervention is required.

PGE (PGE$_2$ is the best studied) is also an effective antagonizer of the effect of a number of hormones on NEFA release from adipose tissue (see Sections 4.3.3 & 9.1.2.3). It is likely that its mode of action, and that of TXs and PGIs, in smooth muscle and platelets involves interaction with the relevant GPCR in the cell membrane coupled to adenylate cyclase, so that their physiological activities are largely expressed through regulation of cellular cAMP concentrations (see Section 7.3.4 for more information on GPCRs). The sequence of events is probably:

1 release of arachidonic acid from a membrane phospholipid by the activation of phospholipase A (PLA, see 4.6.2);
2 the formation of the active compound via the cyclic endoperoxide;
3 the reaction of the active molecules with specific receptor sites on the target cell membrane;
4 the positive or negative regulation of adenylate cyclase.

In fat cells (adipocytes), PGE$_2$ acts to reduce cAMP levels; in platelets PGIs tend to raise cAMP concentrations, while TXs have the reverse effect (Table 3.9). The TXA$_2$ receptor acts primarily through activation of the phosphoinositidase C pathway and consequent elevation of cytosolic Ca^{2+} and activation of protein kinase C-β (see Section 8.4). The interactive effects of TXA$_2$ and PGI with regard to platelet cAMP concentrations are depicted in Fig. 3.47.

In healthy blood vessels prostacyclin is produced, which counteracts the TX effect on cAMP concentrations and, therefore, reduces platelet adhesion. In artificial or damaged vessels, the TX effect is dominant and platelet activation and adhesion can take place initiating a blood clot (and, in extreme cases, leading on to a stroke or heart attack; see also Section 10.5.3).

Another area of PG action that has been studied in detail is their role in water retention by the kidney. The final event in urine production occurs in the terminal portion of the renal tubule where vasopressin (antidiuretic hormone) interacts with V$_2$-vasopressin receptors on the tubule to cause movement of water back into the tissue. Vasopressin does this by activating adenylate cyclase and increasing cellular cAMP concentrations. (cAMP itself has been shown to have a hydro-osmotic effect also). Increased cellular cAMP triggers insertion of aquaporin-2 channels into the plasma membrane and thus causes the reabsorbtion of water from the glomerular filtrate. Low concentrations of PGE$_2$ inhibit these events by counteracting vasopressin-induced cAMP formation.

3.5.7 Prostanoids have receptors that mediate their actions

In the last decade, several prostanoid receptors have been identified and partly characterized. Their use, together with 'knockout' mice, has enabled a more rational consideration of earlier results on the physiological action of prostanoids.

For PGs there are ten receptors, each encoded by a separate gene. For example, for PGE$_2$ there are four different receptors called EP1-4. Each has a distinct molecular mechanism of action. Two receptors have been identified for PGD_2 and there are others for $PGF_{2\alpha}$, PGI_2 and TxA_2. To elucidate the precise roles of receptors, knockout mice lacking a functional receptor have been bred. Using these animals it has been shown, for example, that EP1 receptors are involved in colon cancer, EP2 and EP4 function in allergy and bone resorption while EP3 receptors have a role in fever. The receptors have seven transmembrane segments and belong to the GPCR family (see Section 7.3.4).

3.5.8 Prostaglandins and other eicosanoids are rapidly catabolized

We have already mentioned that TXA_2 and PGI have very short half-lives in vivo. In addition, it was shown by Vane and Piper in the late 1960s that PGs like PGE_2 or $PGF_{2\alpha}$ are rapidly catabolized and did not survive a single pass through the circulation. The lung plays a major role in this inactivation process, which is usually initiated by oxidation of the hydroxyl group at C15. The 13-double bond is next attacked and further degradation involves β- and ω-oxidation (see Sections 3.2.1 & 3.2.3) in peroxisomes. Concentrations of major active prostaglandin products in blood are less than 10^{-10} M and, because of their rapid catabolism, they can only act as local hormones (also called autocoids), which modify biological events close to their sites of biosynthesis. Moreover, in contrast to typical circulating hormones, prostanoids are produced by practically every cell in the body. They exit the cell via carrier-mediated transport before being inactivated rapidly in the circulation.

3.5.9 Instead of cyclo-oxygenation, arachidonate can be lipoxygenated or epoxygenated

For over 40 years it has been known that plant tissues contain LOXs, which catalyse the introduction of oxygen into PUFAs (see Section 3.4). In 1974 Hamberg and Samuelsson found that platelets contained a 12-LOX and, since that time, 5-, 8- and 15-LOXs have also been discovered in animals. The immediate products are hydroperoxy FAs which, for the arachidonate substrate, are hydroperoxy eicosatetraenoic acids (HPETEs).

The HPETEs can undergo three reactions (Fig. 3.48). The hydroperoxy group can be reduced to an alcohol, thus forming a hydroxyeicosatetraenoic acid (HETE). Alternatively, a second lipoxygenation elsewhere on the chain yields a dihydroxy eicosatetraenoic acid (diHETE) or dehydration produces an epoxy FA.

Epoxy eicosatrienoic acids and their metabolic products are called leukotrienes (LTs) – the name being derived from the cells (leukocytes) in which they were originally recognized. Epoxy FAs, such as LTA_4 (Fig. 3.48) can undergo nonenzymic reactions to various diHETEs, can be specifically hydrated to a given diHETE or can undergo ring-opening with GSH to yield peptide derivatives.

As mentioned before, four LOXs (5-, 8-, 12- and 15-) have been found in mammalian tissues. The 5-lipoxygenase is responsible for LT production and is important in neutrophils, eosinophils, monocytes, mast cells and keratinocytes as well as lung, spleen, brain and heart. Products of 12-LOX activity also have biological activity, e.g. 12-HPETE inhibits collagen-induced platelet aggregation and 12-HETE can cause migration of smooth muscle cells in vitro at concentrations as low as one femtomolar (10^{-15} M). In contrast, few biological activities have been reported for 15-HETE, although this compound has a potentially important action in inhibiting the 5- and 12-LOXs of various tissues. The physiological role of the 8-LOX, the most recently discovered, is not yet clear.

The opening of the epoxy group of LTA_4 by the action of glutathione S-transferase attaches the glutathionyl residue to the 6-position to form LTC_4 (Fig. 3.48). Unlike most glutathione S-transferases, which are soluble, the LTC_4 synthase is a microsomal protein with a high preference for its LTA_4 substrate. LTC_4 can then lose a γ-glutamyl residue to give LTD_4 and the glycyl group is released in a further reaction to give LTE_4 (Fig. 3.49).

Prior to their structural elucidation, LTs were recognized in perfusates of lungs as slow-reacting substances (SRS) after stimulation with cobra venom (a source of PLs) or as slow-reacting substances of anaphylaxis (SRS-A) after immunological challenge. LTC_4 and LTD_4 are now known to be major components of SRS-A.

Arachidonic acid and other oxygenated derivatives of the arachidonate cascade can also be metabolized by cytochrome P450-mediated pathways, of which the most important in animals appears to be the epoxygenase pathway (see Section 3.5.12).

3.5.10 Control of leukotriene formation

Unlike the COX products, the formation of LTs is not solely determined by the availability of free arachidonic acid in cells (see Section 3.5.14). Thus for example,

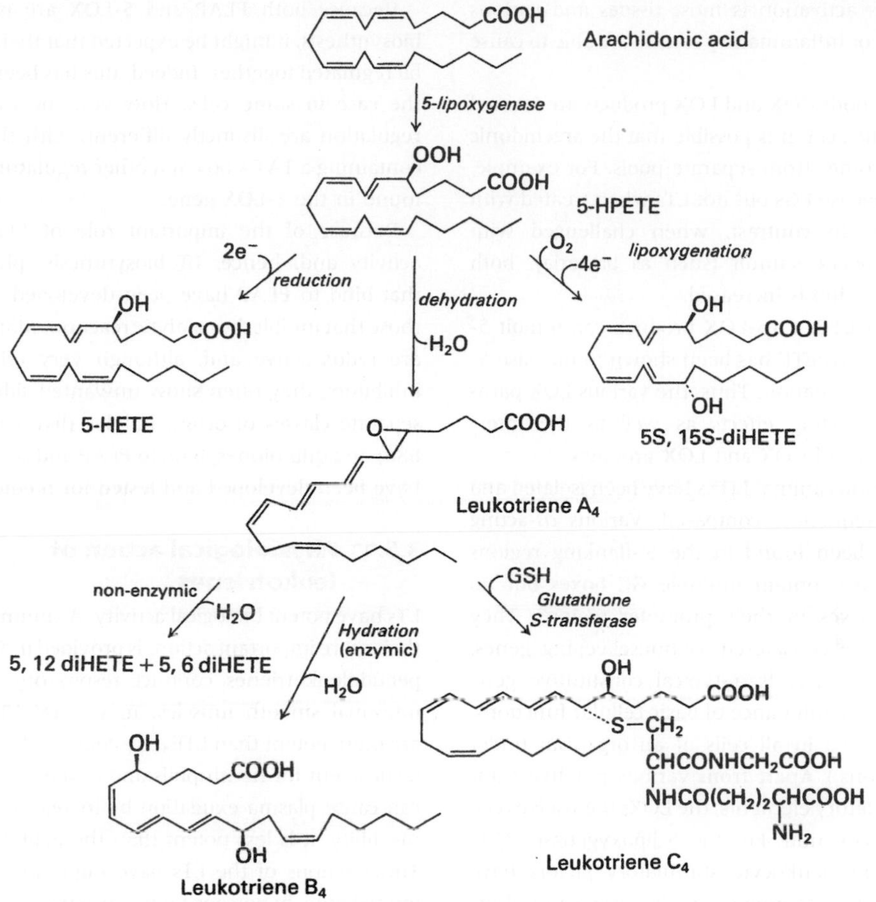

Fig. 3.48 Formation of leukotrienes from arachidonic acid.

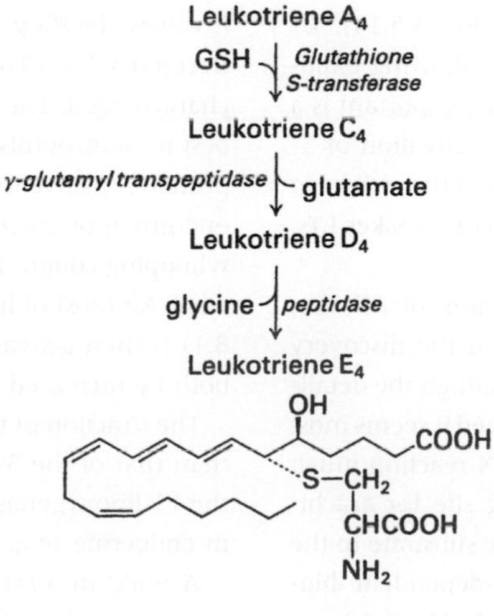

Fig. 3.49 Formation of peptidoleukotrienes.

5-LOX requires activation is most tissues and various immunological or inflammatory stimuli are able to cause this.

In some cells both COX and LOX products are formed from arachidonic acid. It is possible that the arachidonic acid substrate comes from separate pools. For example, macrophages release PGs but not LTs when treated with soluble stimuli. In contrast, when challenged with insoluble phagocytic stimuli (such as bacteria), both PG and LT formation is increased.

Whereas 15-HETE, a 15-LOX product, can inhibit 5- and 12-LOXs, 12-HPETE has been shown to increase 5-HETE and LTB_4 formation. Thus, the various LOX pathways have interacting effects as well as the inter-dependent actions of COX and LOX products.

Genes for various animal LOXs have been isolated and their primary sequences compared. Various *cis*-acting elements have been found in the 5'-flanking regions and all the genes contain multiple GC boxes but no typical TATA boxes in their promoter regions. They can, therefore, be considered as housekeeping genes. (A housekeeping gene is a typical constitutive gene required for the maintenance of basic cellular functions. They are expressed in all cells of an organism under normal conditions.) Apart from various putative transcriptional regulatory elements, the LOXs themselves can be subject to activation. For the 5-lipoxygenase, ATP, Ca^{2+} and various leukocyte stimulatory factors have been demonstrated. Moreover, in the absence of phospholipid, the purified enzyme has poor activity, suggesting that it may need to act at a membrane interface.

Whereas the key event for PG production is the release of substrate (NEFA) by PLA_2 (see Section 3.5.14), LT biosynthesis occurs only in intact cells following exposure to certain stimuli, of which the most important is a rise in intracellular calcium. Moreover, activation of 5-LOX seems to involve its translocation from the cytosol to membrane surfaces, where it becomes active, makes LTs, and undergoes suicide inactivation.

Studies with a highly potent inhibitor of LT biosynthesis, MK-886, in intact cells led to the discovery of 5-LOX activating protein (FLAP). Although the details of FLAP's function are still being elucidated it seems most likely that the protein makes the 5-LOX reaction much more efficient by containing a binding site for arachidonic acid and facilitating transfer of the substrate to the enzyme. A possible model for FLAP-dependent biosynthesis and release of LTC_4 is shown in Fig. 3.50.

Because both FLAP and 5-LOX are needed for LT biosynthesis, it might be expected that their genes would be regulated together. Indeed, this has been shown to be the case in some cells. However, the mechanisms of regulation are distinctly different, with the FLAP gene containing a TATA box and other regulatory motives not found in the 5-LOX gene.

Because of the important role of FLAP in 5-LOX activity and, hence, LT biosynthesis, pharmaceuticals that bind to FLAP have been developed in addition to those that inhibit the catalytic reaction. Many of the latter are redox-active and, although very effective 5-LOX inhibitors, they often show unwanted side-effects. Two separate classes of drugs, indoles (like compound HK-886) and quinolones, bind to FLAP and several members have been developed and tested for medical use.

3.5.11 Physiological action of leukotrienes

LTs have potent biological activity. A summary of some of their more important actions is provided in Table 3.10. The peptidoleukotrienes contract respiratory, vascular and intestinal smooth muscles. In general LTC_4 and LTD_4 are more potent than LTE_4. By contrast, LTB_4 is a chemotactic agent for neutrophils and eosinophils. Although it can cause plasma exudation by increasing vascular permeability, it is less potent than the peptidoleukotrienes. These actions of the LTs have implications for asthma, immediate hypersensitivity reactions, inflammatory reactions and myocardial infarction (see Section 10.3.1 for discussion of lipids and the immune system).

The action of LTs at the molecular level first of all involves binding to specific high-affinity receptors. Receptors for LTB_4, LTC_4, LTD_4 and LTE_4 have been characterized. The detailed mechanism has been studied best in neutrophils where binding of LTB_4 involves a G-protein sensitive to pertussis toxin. (Pertussis toxin is the endotoxin produced by *Bordetella pertussis*, which causes whooping cough. Its role in respiratory function is obvious.) An inositol lipid-specific PLC (see Sections 4.6.4 & 8.1) is then activated and intracellular Ca^{2+} is elevated both by increased influx and by release from stores.

The function of the 12- and 15-LOXs is much less clear than that of the 5-LOX. Nevertheless, it is thought that the 15-lipoxygenase products (15-HETE) may play a role in endocrine (e.g. testosterone) secretion.

A build-up of 12-HETE (46 times normal) has been reported in lesions of patients suffering from the skin

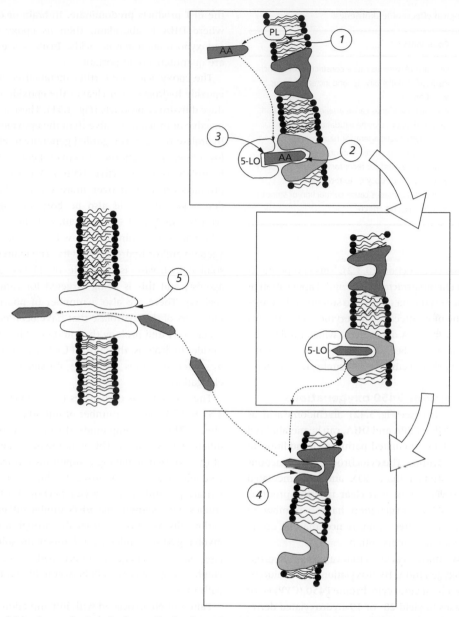

Fig. 3.50 A model for the synthesis and release of leukotriene LTC$_4$. FLAP and LTC$_4$ synthase are shown as integral nuclear membrane proteins, with a phospholipase (PL) associating with this membrane to release AA(1). Following the release of AA, 5-LOX (5-LO) translocates to the nuclear membrane(3) in a process regulated by FLAP(2). 5-LOX then converts AA into LTA$_4$, which is subsequently converted into LTC$_4$ synthase(4). LTC$_4$ is exported from the cell by a membrane carrier(5). Points at which this process can be inhibited are indicated (1–5; blockade of PL, FLAP, 5-LOX, LTC$_4$ synthase and the carrier protein, respectively). Reproduced from PJ Vickers (1998) FLAP: structure, function and evolutionary aspects. In AF Rowley, H Kuhn, T Scewe, eds., *Eicosanoids and Related Compounds in Plants and Animals*, pp. 97–114, Portland Press, London, with kind permission of the author and Portland Press.

Table 3.10 Biological effects of leukotrienes.

Effect	Comments
Respiratory	Peptidoleukotrienes cause constriction of bronchi especially smaller airways and increase mucus secretion
Microvascular	Peptidoleukotrienes cause arteriolar constriction, venous dilation, plasma exudation
Leukocytes	LTB_4 chemotactic agent for neutrophils, eosinophils, e.g. increase degranulation of platelets, cell-surface receptors and adherence of polymorphonucleocytes to receptor cells
Gastrointestinal	Peptidoleukotrienes cause contraction of smooth muscle (LTB_4 no effect)

disease psoriasis (see Section 10.3.2). This compound is known to be a chemoattractant and topical application of 12-HETE causes erythema (similar to sunburn). Interestingly, one of the most effective anti-erythema treatments is etretinate (a vitamin A derivative), which inhibits the 12-LOX. Etretinate has been used for psoriasis patients, but it is particularly effective for antisunburn treatment.

3.5.12 Cytochrome P450 oxygenations

As mentioned earlier (see Fig. 3.42), arachidonic acid, as well as the n-3 PUFAs, EPA and DHA, can be metabolized by cytochrome P450-mediated pathways in addition to COX and LOX reactions. The cytochrome P450 reactions are much less studied than COX and LOX-mediated metabolism. Nevertheless, it is clear that the products of cytochrome P450 oxygenation have a number of extremely important effects and, hence, may be useful targets for therapeutic intervention.

There are two initial reactions that can occur, hydroxylation or epoxygenation. Hydroxylation of arachidonic acid is catalysed by several cytochrome P450 (CYP) ω- or ω-1 hydroxylases to yield 20- or 19-hydroxylated derivatives, respectively. However, the main reactions involve CYP epoxygenases. The epoxy group that is formed can occur at any of the double bond positions of the substrate PUFA. Thus, from arachidonate there are four epoxy products, five from EPA and six from DHA. The individual epoxyeicosatrienoic acids (EETs) from arachidonate are shown in Fig. 3.51 while the general pathway is in Fig. 3.52. The epoxygenases work well with both n-3 and n-6 PUFAs (including 18C PUFAs like linoleic acid) but, because arachidonate is the main PUFA in most tissues,

the EET products predominate. In brain or optic tissue, where DHA is abundant, then its epoxy derivatives (epoxydocosapentaenoic acids, EDPs – see Fig. 3.53) are quantitatively important.

The epoxy acids are further metabolized by a soluble epoxide hydrolase that cleaves the epoxide ring to produce dihydroxy products (Fig. 3.51). These are generally inactive or much less active than the epoxides. Thus, the hydrolase reaction is regarded generally to give rise to a loss of beneficial effects. Moreover, because the soluble hydrolase is more active with n-3 epoxy FAs, such products are turned over more quickly than the n-6 derivatives. It should also be born in mind that, as with the COX and LOX enzymes, different tissues (or different cells within tissues) will show variable epoxygenase and/or hydrolase activity. The expression of the soluble hydrolase is inducible by PPARα and PPARγ agonists and this is being targeted for cancer therapy (below). There are also a number of pharmacological inhibitors of the hydrolase that have been developed in order to try and prolong the life time of beneficial EETs. Besides hydrolysis, the epoxy PUFAs can be metabolized by other reactions including β-oxidation and chain elongation.

The epoxy PUFA derivatives (EETs, EEQs, EDPs: Figs. 3.51 & 3.53) have a number of important physiological effects. The n-6 compounds (EETs) have potent anti-inflammatory effects. This is noteworthy because most of the oxidized signalling compounds formed from arachidonic acid (such as prostaglandin E_2 or leukotriene B_4) are pro-inflammatory (see Section 3.5.13). Animal studies have shown that EETs inhibit inflammation in various disease models probably through a mechanism involving NF-κB. Indeed, inhibitors of the soluble hydrolase have a synergistic effect with COX inhibitors, emphasizing the cross-talk between different eicosanoid pathways.

Pain is often associated with inflammation – including neuropathic pain. Such pain seems to be relieved significantly by epoxy PUFAs including EETs. Considering that in the USA alone some 30% of the population complain of chronic pain which costs $635 billion to treat (inadequately), this is a very important subject for medical advances.

While the above effects of EETs are of obvious benefit, their effect on angiogenesis (formation of new blood vessels from preexisting ones) is important in cancer development. Nevertheless, there are beneficial

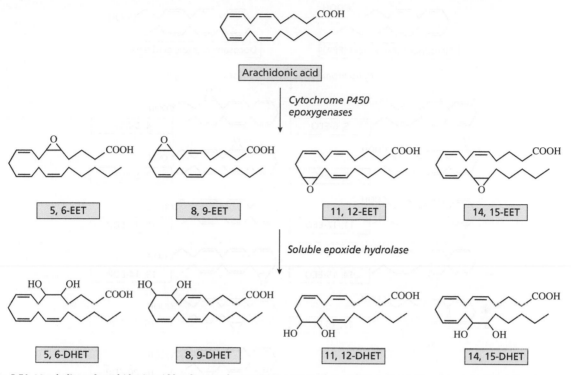

Fig. 3.51 Metabolism of arachidonic acid by the cytochrome P450 epoxygenase pathway. Metabolism by epoxygenases produces four isomers of epoxyeicosatrienoic acids (EETs) which are converted to dihydroxy eicosatrienoic acids (DHETs) by a soluble epoxide hydrolase.

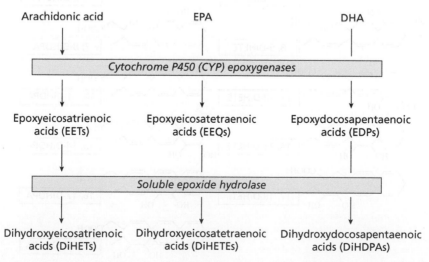

Fig. 3.52 Overview of metabolism of major PUFAs by cytochrome P450 oxygenation. Epoxygenase action can generate epoxy groups at any of the double bonds to give 4 EETs but 6 EDPs. The epoxy group is hydrolysed by a soluble hydrolase to give dihydroxy products. The reactions are shown in more detail for arachidonic acid in Fig. 3.51 and for DHA in Fig. 3.53.

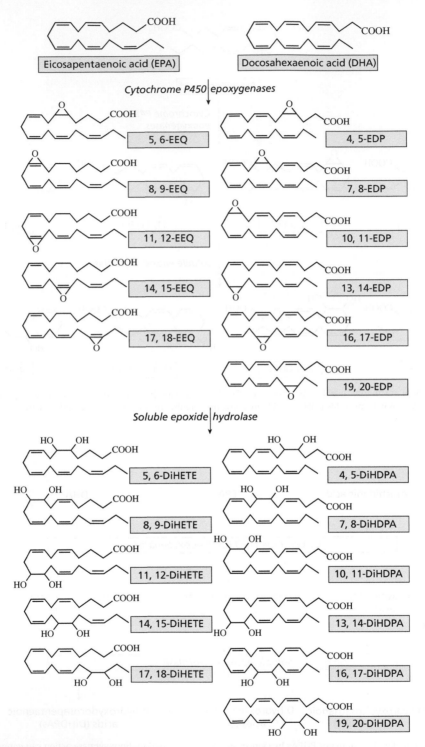

Fig. 3.53 Metabolism of the n-3 PUFAs, EPA or DHA, by the cytochrome P450 epoxygenase pathway. Epoxygenation of EPA gives rise to 5 regioisomers of epoxyeicosatetraenoic acids (EEQs) while that of DHA yields 6 regioisomers of epoxydocasapentaenoic acids (EDPs). The soluble epoxide hydrolase converts these to the diol derivatives, dihydroxyeicosatetraenoic acids (DiHETEs) or dihydroxy-docasapentaenoic acids (DiHDPAs), respectively.

opportunities for the stimulation of angiogenesis such as in hepatic insufficiency, diabetic wound healing and for burn treatments.

Angiogenesis is needed for tumour growth and metastasis in almost all types of cancer. The connection of EETs with pro-angiogenesis was demonstrated when up-regulation of CYP epoxygenation increased cancer cell proliferation. Moreover, CYP epoxygenase expression is increased in a variety of cancer cell lines.

In contrast, cell culture experiments with the epoxy-derivatives of the n-3 PUFA, DHA (EDPs), while having little effect on cancer cell proliferation, strongly inhibited invasion. Overall EDPs appeared to inhibit tumour growth and metastasis through their reduction of angio-genesis. EDPs represent a class of lipid mediators having striking anticancer properties. In fact, this may explain some of the reported beneficial effects of dietary n-3 PUFAs against certain types of cancer (see Section 10.2).

EDPs have been shown to be the most potent epoxy-PUFAs for the dilation of blood vessels. In some test systems, they had potency up to 1000-fold more than EETs. Moreover, like EETs the EDPs were potent anti-inflammatory compounds and also reduced pain.

Thus, although the cytochrome P450-mediated path-way for PUFA metabolism has been somewhat neglected compared to those of COX and LOX, it may offer many future opportunities for the treatment of inflammation, hypertension and pain. The DHA-derived EDPs may also provide new avenues for cancer therapy as well as explaining some of the noted differences between the effects of n-3 and n-6 PUFAs in our diet.

3.5.13 Important new metabolites of the n-3 PUFAs, eicosapentaenoic and docoasahexaenoic acids have recently been discovered

We have already mentioned the important effects of n-3 PUFAs in addition to those of the n-6 series, like arach-idonate. One of the important implications of these contrasting types of PUFA is that they are in competition for many of the important enzymes that metabolize them. This includes not only the synthetic enzymes that convert dietary linoleic and α-linolenic acids into their 20C products (arachidonic acid and EPA, respec-tively; Fig. 3.54) but also those that produce the biologi-cally effective metabolites, like COXs and LOXs. This has important implications for the ratio of dietary n-3/n-6 PUFAs and has led to the generally held assumption that

Western diets have too much n-6 PUFAs or that the ratio of n-6/n-3 PUFAs is too high. A healthy ratio of 4 has been recommended whereas current UK and US diets have a ratio of 10–20 (see Box 10.1).

The main reason for this concern is that, although both n-3 and n-6 PUFAs are essential in the diet (see Section 6.2.2) their oxygenated metabolites (PG etc.) have con-trasting effects (Table 3.11). Thus, the n-6 PUFA arachi-donic acid gives rise to products which are usually pro-inflammatory, whereas metabolites of EPA and DHA are often anti-inflammatory (see Section 10.3.3).

Clearly, the 20C PUFA, EPA, could be expected to be metabolized by COX isoforms to produce a range of eicosanoids which have an additional double bond. Thus, PGE_3 is produced instead of the PGE_2 product from arachidonic acid. The 5-LOX also metabolizes EPA and gives rise to LTs like LTB_5. These n-3 PUFA metabolites usually produce the opposite effects to their n-6 PUFA counterparts. This provides another important reason for getting an appropriate balance of n-6/n-3 PUFAs in the diet.

In the early 2000s, new classes of compounds derived from the n-3 PUFAs, EPA and DHA were discovered. Research by Charlie Serhan's group at Harvard first identified compounds now referred to as resolvins (Ser-han *et al.* 2008). The name derives from the ability of such metabolites to terminate acute inflammatory epi-sodes or 'resolve' them (resolution). Depending on whether EPA or DHA is the starting material, then an initial oxidation can be either by a cytochrome P450 enzyme or 15-LOX (Figs. 3.55 & 3.56). However, both EPA and DHA can be metabolized by COX-2 after it has been acetylated by aspirin (see 3.5.3). Indeed, this was how the resolvins were first discovered. 5-LOX and epoxidation reactions are then involved in producing the resolvins, two of which are shown in Figs. 3.55 and 3.56. Resolvins derived from EPA are in the E series, while those from DHA are in the D_1 (PD_1) series.

Once DHA has been converted into a hydroxylated derivative, 5-LOX activity gives resolvins, while epoxy-genation produces protectins. A significant contribution to our knowledge about one protectin, neuroprotectin D_1, has come from Nicholas Bazan's laboratory in New Orleans. The production of 22C open-chain compounds directly from DHA solves a question about the prevalence of DHA in nervous tissue when, previously, all the PUFA-derived biological signalling molecules were 20C com-pounds (i.e. eicosanoids).

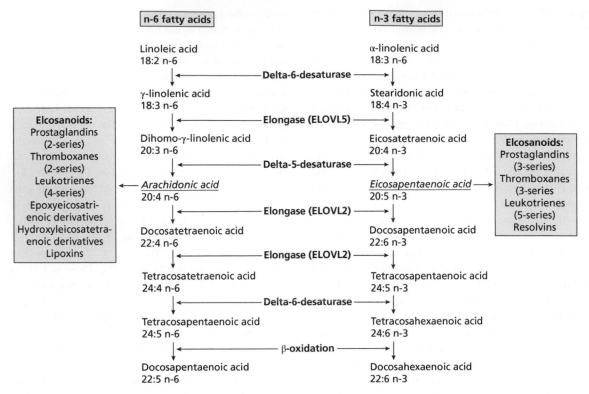

| n-6 fatty acids | | n-3 fatty acids |

Fig. 3.54 n-3 and n-6 fatty acid metabolism. From G Schmitz & J Ecker (2008) The opposing effects of n−3 and n−6 fatty acids. *Prog Lipid Res* **47**:147–55. Reproduced with permission of Elsevier.

In 2008, yet another type of LOX-derived metabolites of DHA – the maresins – was discovered. The maresins have potent anti-inflammatory and pro-resolving properties. Like the resolvins, they are hydroxyl-derivatives of DHA but their biosynthesis is not yet elucidated.

Resolution is an active process that ends an acute inflammatory episode. The inflamed tissue reverts to its basal state as inflammatory cells are cleared and any further neutrophil recruitment ceases. Simultaneously, tissue neutrophils undergo apoptosis and are

Table 3.11 Comparison of the physiological effects of n-3 and n-6 PUFA-derived compounds.

Class	n-6 PUFA (arachidonic acid-derived compound)	Physiological effects		n-3 PUFA (EPA/DHA-derived compound)
Prostaglandin	PGE_2	Pro-arrhythmic	Anti-arrhythmic	PGE_3
	PGI_2	Pro-inflammatory	Anti-arrhythmic	PGI_3
Leukotriene	LTB_4	Pro-inflammatory	Anti-inflammatory	LTB_5
Epoxy-derivatives	11,12-EET	Anti-inflammatory	Anti-inflammatory	17,18-EEQ
Resolvins	–		Anti-inflammatory	RVE_1
Neuroprotectin	–		Anti-inflammatory	RVD_1 -RVD_4
Thromboxane	TXA_2	Platelet activator	Vasodilation	TXB_3
	TXB_2	Vasoconstriction	Platelet inhibitor	

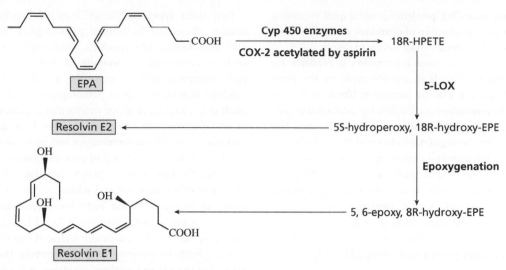

Fig. 3.55 E-series resolvin metabolism. From MJ stables & DW Gilroy (2011). Old and new generation lipid mediators in acute inflammation and resolution. *Prog Lipid Res* **50**:35–51. Reproduced with permission of Elsevier.

removed by phagocytic macrophages. Any dysregulation of the process can lead to prolonged inflammation, which is important in the pathology of several chronic inflammatory diseases (see Section 10.3). Conditions where

resolvins have been shown to be efficacious include allergic asthma, colitis, inflammatory pain, peritonitis, pneumonia and sepsis. The resolvins act through specific receptors and have potent regulatory effects on

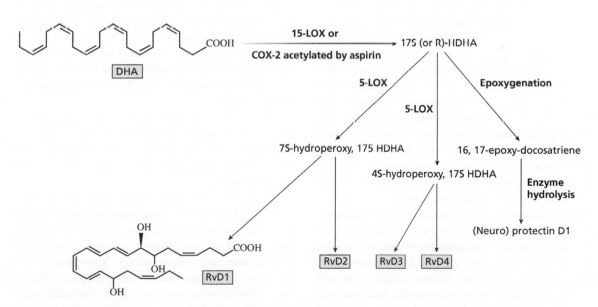

Fig. 3.56 D-series resolvin and protectin metabolism. From MJ stables & DW Gilroy (2011). Old and new generation lipid mediators in acute inflammation and resolution. *Prog Lipid Res* **50**:35–51. Reproduced with permission of Elsevier.

leukocytes, including preventing neutrophil swarming and decreasing production of cytokines, chemokines and reactive oxygen species from inflamed tissues.

Of the protectins, most information is available for neuroprotectin, PD_1. It is actively made by the brain and microglia but also by peripheral blood monocytes in which its production is stimulated by activation of Toll-like receptor 7 (TLR7) a sensor of tissue damage. The resulting immune-regulatory effects of PD_1 include suppression of inflammatory cytokines and pro-inflammatory lipid mediators and promotion of T-cell apoptosis (see Section 10.3). PD_1 is protective in experimental models of ischaemic stroke, oxidative stress, asthma and Alzheimer's disease. Indeed, Alzheimer's patients given dietary DHA supplements have lower levels of several cytokines known to be reduced by PD_1.

3.5.14 For eicosanoid biosynthesis, an unesterified fatty acid is needed

As discussed in Section 3.5.13, there are two series of biologically active lipids formed from fatty acids – the n-3 and n-6 series. The main relevant acids in these groups are EPA or DHA and arachidonate, respectively. These very long-chain PUFAs are esterified in phosphoglycerides, from which they have to be released to generate the NEFA substrates for COXs, LOXs etc. (see Fig. 3.42).

We will begin by discussing arachidonate (the main 20C n-6 PUFA). The concentration of nonesterified arachidonic acid in cells is normally far below the K_m for PGH synthetase. Thus, the first stage for eicosanoid formation will normally be an activation of the release of arachidonate from position sn-2 of phosphoglycerides that contain it. In animals, two classes of phospholipids are thought to play major roles as sources of arachidonate in cells: PtdCho (the major membrane constituent) and the phosphoinositides (by virtue of the high enrichment of arachidonate at position 2). Thus, hydrolysis of phosphoinositides not only produces two second messengers directly (see Section 8.1) but lipase attack on the released DAG may also initiate an arachidonate cascade. Other potential sources of arachidonic acid are the plasmalogens (see 2.3.2.5) that also have a high enrichment at the sn-2 (acyl) position. Because plasmalogens are poor substrates for PLA_2, they are usually hydrolysed by a plasmalogenase first. Thus, release of arachidonic acid from plasmalogens could be controlled independently from that for diacylphosphoglycerides.

Two main types of stimuli increase arachidonate release. These can be termed physiological (specific) and pathological (nonspecific). Physiological stimuli, such as adrenaline, angiotensin II and certain antibody-antigen complexes, cause the selective release of arachidonic acid. In contrast, pathological stimuli, such as mellitin (a component of bee venom that activates PLA_2) or tumour promoters like phorbol esters, have generalized effects on cellular membranes and promote release of all FAs from position sn-2 of phosphoglycerides.

Of the various cellular lipases stimulated by hormones or other effectors and which, in theory, could give rise to arachidonate hydrolysis, cytosolic PLA_2 is the most important. Cytosolic PLA_2 ($cPLA_2$) is stimulated by Ca^{2+} and hormonally induced mobilization of Ca^{2+} leads to movement of the enzyme from the cytosol to the ER and nuclear envelope. It is relatively specific for arachidonate and is stimulated by phosphorylation.

Secretory PLA_2 ($sPLA_2$) is also stimulated by Ca^{2+}, but at the higher concentrations found outside the cell. It is relatively nonspecific towards different phospholipids and towards the FA at the sn-2 position. Its involvement in PG biosynthesis has been shown in endothelial cells by the use of antibodies that prevent it binding to the cell surface. It has been suggested that $cPLA_2$ produces the initial burst of PG biosynthesis, whereas the secretory enzyme is involved in late-phase PG formation after cells have been stimulated further by cytokines, inflammatory mediators, or growth factors. In fact, the initial burst is usually thought to be due to COX-1 activation by $cPLA_2$, whereas cytokines and inflammatory factors (together with $cPLA_2$) induce $sPLA_2$, which releases arachidonate used by COX-2.

It is also of interest that phospholipase A_2, the activity of which is needed to initiate eicosanoid production, is also needed to produce another type of biologically active lipid: platelet activating factor (PAF – see Sections 4.5.11 & 10.3.7.2).

PLs are, of course, also involved in the release of n-3 very long-chain PUFAs. Most mammals contain higher concentrations of DHA than EPA and DHA is especially important in nervous or optic tissues. In contrast to arachidonate (see above), different phosphoglycerides are important sources of DHA. In particular, PtdEtn (especially its plasmalogen form) and phosphatidylserine (PtdSer) contain high amounts of DHA.

3.5.15 Essential fatty acid activity is related to double bond structure and to the ability of such acids to be converted into physiologically active eicosanoids

Work from van Dorp's lab in the Netherlands and from Holman's laboratory at the Hormel Institute, Minnesota, originally showed that only those FAs (including new synthetic odd-numbered acids) that act as precursors for biologically active eicosanoids have EFA activity. This and other results superseded the old dogma that EFA all had the *n*-6, *n*-9 double bond system. Indeed, some synthetic acids without this structure had EFA activity (Table 3.12). Van Dorp and his colleagues (at the Unilever Laboratories at Vlaardingen) postulated that, of the *n*-6 PUFA, only those FAs capable of being converted into the 5,8,11,14-tetraenoic FAs of chain lengths 19C, 20C and 22C would show EFA activity because only these tetraenoic acids can give rise to physiologically active eicosanoids. However, it is now known that columbinic acid (*t*5,*c*9,*c*12-18:3), when given to EFA-deficient rats, normalizes growth and cures their dermatitis, yet it is unable to form an eicosanoid.

Moreover, although there is a correlation between EFA activity and the potential to be converted into eicosanoids, one cannot cure EFA deficiency by infusion of eicosanoids

because they are rapidly destroyed and because different cells produce their own special pattern of eicosanoids. However, one of the first organs to show EFA deficiency is the skin, whose water permeability is very much increased in this condition. Topical application of EFAs to skin can reverse the deficiency symptoms and there is now evidence that topical application of prostaglandins can also be therapeutic (see Sections 6.2.2 & 10.3.2 for other comments about lipids and skin disease).

We are still a long way from being able to account for the function of all the EFAs that enter the body and the fate of the eicosanoids that may be formed from them. Eicosanoids are metabolized very rapidly and their metabolites are excreted in the urine or bile. One approach to studying daily eicosanoid production is to isolate and analyse such metabolites. In this way it has been estimated that 1 mg of PG metabolites is formed each day in humans – considerably less than the 10 g of EFA that are thought to be necessary daily. Part of this contrast is due to the fact that PUFAs are β-oxidized rather rapidly. Moreover, the demonstration that EFA like linoleic acid may have additional functions (see Section 6.2.2) apart from being eicosanoid precursors may account for this apparent discrepancy and raises the possibility that new roles for EFAs may be discovered in the future.

Table 3.12 Relationship between fatty acid structure and EFA activity.

Fatty acid chain length	Position of double bonds		EFA potency (unit g^{-1})
	From carboxyl end (Δ)	From methyl end (ω)	
18:2	9, 12	6, 9	100
18:3	6, 9, 12	6, 9, 12	115
18:3	8, 11, 14	4, 7, 10	0[a]
18:4	6, 9, 12, 15	3, 6, 9, 12	34
18:4	5, 8, 11, 14	4, 7,10, 13	0[a]
19:2	10, 13	6, 9	9
20:2	11, 14	6, 9	46
20:3	8, 11, 14	6, 9, 12	100
20:3	7, 10, 13	7, 10, 13	0[a]
20:4	5, 8, 11, 14	6, 9, 12, 15	139
21:3	8, 11, 14	7, 10, 13	56
22:3	8, 11, 14	8, 11, 14	0[a]
22:5	4, 7, 10, 13, 16	6, 9, 12, 15, 18	139

[a] Cannot give rise to any prostaglandins.

KEY POINTS

- Of the many hundreds of FAs found in Nature, the most common are even chain compounds of 16C–22C in length. The even numbers of carbon atoms in most FA chains are a consequence of the method of biosynthesis. Nevertheless, mechanisms are in place to produce and catabolize FAs of different structures.

Biosynthesis of fatty acids

- FAs are usually converted into thioesters (coenzyme A or ACP derivatives) to aid their metabolism.

- Both FAs and acyl-CoAs can be bound to specific proteins (e.g. FABPs) in cells. This aids their utilization but also prevents undesirable physiological consequences if the concentrations of NEFAs or acyl-CoAs become excessive.

- *De novo* biosynthesis of FAs uses ACC and FAS to produce mainly saturated even-chain FAs of 16C or 18C (i.e. palmitate and stearate).

- ACC is a biotin-containing enzyme that catalyses the first committed step of FA biosynthesis. ACC can be a multi-enzyme complex of four individual proteins as in *E. coli* or a multifunctional protein as in mammals. The mechanism of ACC in both forms is similar. First, biotin is carboxylated by an ATP-dependent reaction. Second, the carboxyl group is transferred to acetyl-CoA to produce malonyl-CoA.

- As befits its importance, ACC activity is subject to precise regulation in all organisms. In mammals, acute regulation involves allosteric and covalent mechanisms including phosphorylation. Long-term regulation by diet causes changes in the amount of carboxylase protein. Different control mechanisms are used by *E. coli* and by higher plants.

- FASs are divided into Type I (multifunctional proteins) or Type II (multi-enzyme complexes). The mammalian Type I FAS is a homodimer that releases end products as NEFAs. The yeast FAS produces acyl-CoAs, contains two dissimilar proteins and operates as an $\alpha_6 \beta_6$ complex.

- Type II FASs are found in many bacteria (such as *E. coli*), cyanobacteria and in the chloroplasts of algae and higher plants. All the enzymes catalysing the partial reactions (condensation, reduction, dehydration, second reduction) as well as ACP can be purified and studied individually. In *E. coli* the FAS can also produce MUFAs (mainly *cis*-vaccenate) as well as long-chain saturated products.

- Once FAs have been made *de novo*, they can be modified by desaturation, elongation or other reactions. Elongation involves addition of two carbons from malonyl-CoA and takes place using acyl-CoA substrates in the ER. The 4 consecutive reactions are similar to FAS but ACP is not involved. Different, chain-length selective, elongases are found in both mammals and plants.

- Desaturation usually takes place by an aerobic mechanism in which hydrogen atoms are removed from the FA and a reductant is required to convert oxygen into water. The desaturases differ from each other in the position at which the double bond is introduced and the nature of the reductant and the acyl substrate. Those desaturases producing PUFAs usually introduce the additional double bond three carbons from an existing double bond.

Degradation of fatty acids

- Oxidation of SFA or UFA can take place by α-, β- or ω-oxidation. α-Oxidation is common in plants and brain tissue and releases CO_2 to create a FA one carbon shorter than the substrate. An α-(2-)hydroxy intermediate is formed which, in some tissues, is put to use.

- β-Oxidation is the most important mechanism for FA degradation. It uses a cycle of four reactions (oxidation, hydration, oxidation, thiolysis) to produce acetyl-CoA units and is the main way organisms release the energy stored in TAG. β-Oxidation takes place in both mitochondria as well as microbodies. These two locations have different features and, in addition, there are special mechanisms to deal with the β-oxidation of UFAs.

- ω-Oxidation uses mixed-function oxidases and is useful for dealing with substituted acyl chains where β-oxidation cannot work initially. The oxidases are similar to drug-hydroxylating systems and utilize cytochrome P450, NADPH and molecular oxygen.

Peroxidation systems

- Chemical oxidation (peroxidation to form peroxides and free radicals) can be important in the destruction of PUFAs, when it can lead to the development of diseases or food spoilage. Antioxidants (e.g. vitamin E) can slow this process.

- Peroxidation can also be catalysed by LOX enzymes. In plants the prevalent linoleic and α-linolenic acids are converted into oxidized products important as flavours and aromas in foods or as signalling compounds for plant development or in stress responses. The initial hydroperoxides produced by LOXs are further metabolized by several different pathways to yield aromas (such as 'cucumber') or to yield jasmonate hormones.

Essential fatty acids, eicosanoids and other bioactive compounds in mammals

- Certain PUFAs, synthesized by algae and plants, of both the n-3 and n-6 series, are EFAs. The core EFAs (linoleic and α-linolenic acids) cannot be made by mammals and their longer chain PUFA products can often be produced only in inadequate amounts. So mammals need a dietary source of such PUFAs.

- Almost all mammalian cells make biologically active eicosanoids (20C compounds) from arachidonate, EPA and other 20C PUFAs by the action of COXs, LOXs or cytochrome P450 epoxygenation. COXs produce PGs, PGI and TXs while LOXs form LTs. The eicosanoids have potent receptor-mediated actions near their site of production.

- Depending on the substrate FA the eicosanoids produced have different activity. In general, those derived from the n-6 PUFA arachidonate tend to be pro-inflammatory while those produced from the n-3 PUFA, EPA, are non-or anti-inflammatory.

- Some signalling molecules (resolvins, protectins, maresins) have been recently discovered to be produced particularly from n-3 PUFAs including the 22C DHA. These are thought to be important in resolving inflammatory episodes (resolvins) and in protecting sensitive tissues, like brain, from oxidative damage (protectins).

Further reading

General

Gunstone FD, Harwood JL, Dijkstra AJ, eds. (2007) *The Lipid Handbook*, 3rd edn. CRC Press, Boca Raton.

Harwood JL, ed. (1998) *Plant Lipid Biosynthesis*. Cambridge University Press, Cambridge.

The Lipid Library (http://lipidlibrary.aocs.org).

Murphy DJ, ed. (2005) *Plant Lipids: Biology, Utilisation and Manipulation*. Blackwell Science, Oxford.

Ratledge D & Wilkinson SG, eds. (1989) *Microbial Lipids*, **2** vols. Academic Press, London.

Vance DE & Vance JE, eds. (2008) *Biochemistry of Lipids, Lipoproteins and Membranes*, 5th edn. Elsevier, Amsterdam.

Fatty acid synthesis

Bernlohr DA, Simpson MA, Hertzel AV & Banaszak LJ (1997) Intracellular lipid-binding proteins and their genes. *Annu Rev Nutr* **17**:277–303.

Brownsey RW, Boone AN, Elliott JE, Kulpa JE & Lee WM (2006) Regulation of acetyl-CoA carboxylase. *Biochem Soc Trans* **4**:223–7.

Cahoon E, Lindqvist Y, Schneider G & Shanklin J (1997) Redesign of soluble fatty acid desaturases from plants for altered substrate specificity and double bond position. *Proc Natl Acad Sci, USA* **94**:4872–7.

Cronan JE & Waldrop GL (2002) Multi-subunit acetyl-CoA carboxylases. *Prog Lipid Res* **41**:407–35.

Fell D (1997) *Understanding the Control of Metabolism*. Portland Press, London.

Frohnest BI & Bernlohr DA (2000) Regulation of fatty acid transporters in mammalian cells. *Prog Lipid Res* **39**:83–107.

Glatz JFC & van der Vusse GJ (1996) Cellular fatty acid-binding proteins: their function and physiological significance. *Prog Lipid Res* **35**:243–82.

Hannerland NH & Spener F (2004) Fatty acid-binding proteins – insights from genetic manipulations. *Prog Lipid Res* **43**:328–49.

Harwood JL (1996) Recent advances in the biosynthesis of plant fatty acids. *Biochim Biophys Acta* **1301**:7–56.

Jakobsson A, Westerberg R & Jacobsson A (2006) Fatty acid elongases in mammals: their regulation and roles in metabolism. *Prog Lipid Res* **5**:237–49.

Joshi AK, Witkowski A & Smith S (1998) The malonyl/acetyl transferase and β-ketoacyl synthase domains of the animal fatty acid synthase can cooperate with the acyl carrier protein domain of either subunit. *Biochemistry* **37**:2515–23.

Kim KH (1997) Regulation of mammalian acetyl-CoA carboxylase. *Annu Rev Nutr* **17**:77–99.

Knowles JR (1989) The mechanism of biotin-containing enzymes. *Annu Rev Biochem* **58**:195–221.

Leonard AE, Pereira SL, Sprecher H & Huang YS (2004) Elongation of long-chain fatty acids. *Prog Lipid Res* **43**:36–54.

Lomakin IB, Xiong Y & Steitz TA (2007) The crystal structure of yeast fatty acid synthase, a cellular machine with eight active sites working together. *Cell* **129**:319–32.

Miyazaki M & Ntambi JM (2008) Fatty acid desaturation and chain elongation in mammals. In DE Vance & JE Vance, eds., *Biochemistry of Lipids, Lipoproteins and Membranes*, 5th edn. Elsevier, Amsterdam, pp. 191–211.

Ohlrogge JB & Jaworski JG (1997) Regulation of fatty acid synthesis. *Annu Rev Plant Physiol Plant Mol Biol* **48**:109–36.

Parsons JB & Rock CO (2013) Bacterial lipids: metabolism and membrane homeostasis. *Prog Lipid Res* **52**:249–76.

Rock CO (2008) Fatty acid and phospholipid metabolism in prokaryotes. In DE Vance & JE Vance, eds., *Biochemistry of Lipids, Lipoproteins and Membranes*, 5th edn. Elsevier, Amsterdam, pp. 59–96.

Schweizer E & Hofmann J (2004) Microbial type I fatty acid synthases (FAS): major players in a network of cellular FAS systems. *Microbiol Mol Biol R* **68**:501–17.

Smith S & Tsai SC (2007) The type I fatty acid and polyketide synthases: a tale of two megasynthases. *Nat Prod Rep* **24**:1041–72.

Smith S, Witkowski A & Joshi AK (2003) Structural and functional organization of the animal fatty acid synthase. *Prog Lipid Res* **42**:289–317.

Sul HS & Smith S (2008) Fatty acid synthesis in eukaryotes. In DE Vance & JE Vance, eds., *Biochemistry of Lipids, Lipoproteins and Membranes*, 5th edn. Elsevier, Amsterdam, pp. 155–90.

Wallis JG & Browse J (2002) Mutants of Arabidopsis reveal many roles for membrane lipids. *Prog Lipid Res* **41**:254–78.

Watkins PA (1997) Fatty acid activation. *Prog Lipid Res* **36**:55–83. ve fpuinve.

White SW, Zheng J, Zhang YM & Rock CO (2005) The structural biology of type II fatty acid biosynthesis. *Annu Rev Biochem* **74**:791–831.

See also specific chapters in multiauthor volumes in the *General* section above.

Fatty Acid Degradation

Baker A, Graham IA, Holdsworth M, Smith SM & Theodoulou FL (2006) Chewing the fat: β-oxidation in signalling and development. *Trends Plant Sci* **11**:124–32.

Eaton S, Bartlett K & Pourfarzam M (1996) Mammalian mitochondrial β-oxidation. *Biochem J* **320**:345–57.

Gerhardt B (1992) Fatty acid degradation in plants. *Prog Lipid Res* **31**:417–46.

Kunau WH, Dommes V & Schulz H (1995) β-Oxidation of fatty acids in mitochondria, peroxisomes and bacteria: a century of continued progress. *Prog Lipid Res* **34**:267–342.

Schultz H (2008) Oxidation of fatty acids in eukaryotes. In DE Vance & JE Vance, eds., *Biochemistry of Lipids, Lipoproteins and Membranes*, 5th edn. Elsevier, Amsterdam, pp. 131–54.

Wanders RJA, Vreken P, Ferdinandusse S *et al.* (2001) Peroxisomal fatty acid α and ß-oxidation in humans: enzymology, peroxisomal metabolite transporters and peroxisomal diseases. *Biochem Soc Trans* **29**:250–67.

Zammit VA (1999) Carnitine acyltransferase: functional significance of subcellular distributions and membrane topology. *Prog Lipid Res* **38**:199–224; see also Zammit (1999) under Fatty acid synthesis – Supplementary reading.

See also specific chapters in multiauthor volumes in the *General* section above.

Eicosanoids and related compounds

Cunnane SC (2003) Problems with essential fatty acids: time for a new paradigm? *Prog Lipid Res* **42**:544–68.

Gaffney BJ (1996) Lipoxygenase: structural principles and spectroscopy. *Annu Rev Biochem* **25**:431–59.

Haeggstrom JZ & Wetterholm A (2002) Enzymes and receptors in the leukotriene cascade. *Cell Mol Life Sci* **59**:742–53.

Schmitz G & Ecker J (2008) The opposing effects of *n*-3 and *n*-6 fatty acids. *Prog Lipid Res* **47**:147–55.

Smith WL & Murphy RC (2008) The eicosanoids: cyclooxygenase, lipoxygenase and epoxygenase pathways. In DE Vance & JE Vance, eds., *Biochemistry of Lipids, Lipoproteins and Membranes*, 5th edn. Elsevier, Amsterdam, pp. 331–62.

Smith WL, DeWitt DL & Garavito RM (2000) Cyclooxygenases: structural, cellular and molecular biology. *Annu Rev Biochem* **69**:145–82.

Spector AA, Fang X, Snyder GD & Weintraub NL (2004) Epoxyeicosatrienoic acids (EETs): metabolism and biochemical function. *Prog Lipid Resh* **43**:55–90.

Stables MJ & Gilroy DW (2011) Old and new generation lipid mediators in acute inflammation and resolution. *Prog Lipid Res* **50**:35–51.

Zeldin DC (2001) Epoxygenase pathways of arachidonic acid metabolism. *J Biol Chem* **276**:36059–62.

Zhang G, Kodani S & Hammock BD (2014) Stabilized epoxygenated fatty acids regulate inflammation, pain, angiogenesis and cancer. *Prog Lipid Res* **53**:108–23.

See also specific chapters in multiauthor volumes in the *General section* above.

Supplementary reading
Fatty acid synthesis

Browse J & Somerville CR (1991) Glycerolipid metabolism, biochemistry and regulation. *Annu Rev Plant Phys* **42**:467–506.

Campbell JW & Cronan JE (2001) Bacterial fatty acid biosynthesis: targets for antibacterial drug discovery. *Annu Rev Microbiol* **55**:305–32.

Cronan JE Jr (2003) Bacterial membrane lipids: where do we stand? *Annu Rev Microbiol* **57**:203–24.

Du ZY & Chye ML (2014) Interactions between Arabidopsis acyl-CoA-binding proteins and their protein partners. *Planta* **238**:239–45.

Girard J, Ferré P & Foufelle F (1997) Mechanisms by which carbohydrates regulate expression of genes for glycolytic and lipogenic enzymes. *Annu Rev Nutr* **17**:325–52.

Guillou H, Zadravec D, Martin PG & Jacobsson A (2010) The key roles of elongases and desaturases in mammalian fatty acid metabolism: insights from transgenic mice. *Prog Lipid Res* **49**:186–200.

Hildebrand DF, Yu K, McCracken C & Rao SS (2005) Fatty acid manipulation. In D Murphy, ed., *Plant Lipids: Biology, Utilisation and Manipulation*. Blackwell, Oxford, pp. 67–102.

Nakamura MT & Nara TY (2004) Structure, function and dietary regulation of delta 6, delta 5 and delta 9 desaturases. *Annu Review Nutr* **4**:345–76.

Ntambi JH & Miyazaki M (2003) Recent insights into stearoyl-CoA desaturase-1. *Curr Opin Lipidol* **14**:255–61.

Ohlrogge JB (1999) Plant metabolic engineering: are we ready for phase two? *Current Opin Plant Biol* **2**:121–2.

Riederer K & Möller C, eds. (2006) *Biology of the Plant Cuticle*. Blackwell, Oxford.

Shanklin J & Cahoon EG (1998) Desaturation and related modifications of fatty acids. *Annu Rev Plant Phys* **49**:611–41.

Tocher DR, Leaver MJ & Hodgson PA (1998) Recent advances in the biochemistry and molecular biology of fatty acid desaturases. *Prog Lipid Res* **37**:73–118.

Voelker TA, Worrell AC, Anderson L *et al.* (1992) Fatty acid biosynthesis redirected to medium chains in transgenic oilseed plants. *Science* **257**:72–3.

Xiao S & Chye ML (2011) New roles for the acyl-CoA-binding proteins in plant development, stress responses and lipid metabolism. *Prog Lipid Res* **50**:141–51.

Zammit VA (1999) The malonyl-CoA-long chain acyl-CoA axis in the maintenance of mammalian cell function. *Biochem J* **343**:505–15.

Fatty acid degradation

Graham IA & Eastmond PJ (2002) Pathways of straight and branched chain fatty acid catabolism in higher plants. *Prog Lipid Res* **41**:156–81.

McGarry JD (1995) The mitochondrial carnitine palmitoyltransferase system: its broadening role in fuel homeostasis and new insights into its molecular features. *Biochem Soc Trans* **23**:321–4.

Tanaka K & Coates PM, eds. (1990) *Fatty Acid Oxidation: Chemical, Biochemical and Molecular Aspects*. Liss, New York.

Eicosanoids and related compounds

Blée E (1998) Phytooxylipins and plant defense reactions. *Prog Lipid Res* **37**:33–72.

Cheng Y, Austin SC, Rocca B *et al.* (2002) Role of prostacyclin in the cardiovascular response to thromboxane A2. *Science* **296**:539–41.

Diczfalusy U (1994) β-Oxidation of eicosanoids. *Prog Lipid Res* **33**:403–28.

Grechkin A (1998) Recent developments in biochemistry of the plant lipoxygenase pathway. *Prog Lipid Res* **37**:317–52.

Hui Y, Ricciotti E, Crichton I *et al.* (2010) Targeted deletions of cyclooxygenase-2 and atherogenesis in mice. *Circulation* **121**:2654–64.

Ji R-R, Xu Z-Z, Strichartz G & Serhan CN (2011) Emerging roles of resolvins in the resolution of inflammation and pain. *Trends Neurosci* **34**:599–609.

Kahn H & O'Donnell V (2006) Inflammation and immune regulation by 12/15 lipoxygenases. *Prog Lipid Res* **45**:334–56.

Kang TJ, Mbonye UR, DeLong CJ, Wada M & Smith WL (2007) Regulation of intracellular cyclooxygenase levels by gene transcription and protein degradation. *Prog Lipid Res* **46**:108–25.

Marnett LJ & Kalgutkar AS (1999) Cyclooxygenase 2 inhibitors: discovery, selectivity and the future. *Trends Pharmacol Sci* **20**:465–9.

Oliw EH (1994) Oxygenation of polyunsaturated fatty acids by cytochrome P450 monooxygenases. *Prog Lipid Res* **33**:329–54.

Rouzer C & Masnett L (2003) Mechanism of free radical oxygenation of polyunsaturated fatty acids by cyclooxygenases. *Chem Rev* **103**:2239–2304.

Serhan CN, Chiang N & Van Dyke TE (2008) Resolving inflammation: dual anti-inflammatory and pro-resolution lipid mediators. *Rev Immunol* **8**:349–61.

Smith WL, DeWitt DL & Garavito RM (2000) Cyclooxygenases, structural, cellular and molecular biology. *Annu Rev Biochem* **69**:149–82.

Spector AA, Fang X, Snyder GD & Weintraub NL (2004) Epoxyeicosatrienoic acids (EETs): metabolism and biochemical function. *Prog Lipid Res* **43**:55–90.

Sugimoto Y & Narumiya S (2007) Prostaglandin E receptors. *J Biol Chem* **282**:11613–17.

Uddin M & De Levy BD (2011) Resolvins: Natural agonists for resolution of pulmonary inflammation. *Prog Lipid Res* **50**:75–81.

Vane JR, Bakhle YS & Botting RM (1998) Cyclooxygenases 1 and 2. *Annu Rev Pharmacol* **38**:97–120.

CHAPTER 4

The metabolism of complex lipids

As described in Chapter 2, the major storage and membrane lipids contain fatty acids (FAs) attached to other molecules such as glycerol or sphingosine. In this chapter we will describe the biosynthesis and catabolism of such molecules and, importantly, how these processes can be regulated.

We will start by discussing the metabolism of storage lipids which, in most organisms, are the triacylglycerols (TAGs). There are several different biosynthetic pathways to TAGs. Animals and plants can transfer two acyl groups to glycerol 3-phosphate (G3P). The product is then dephosphorylated and the remaining hydroxyl is acylated. Another pathway in animals resynthesizes TAGs from the absorbed digestion products of dietary TAGs. There are also reactions that modify the acyl distributions on the glycerol of preexisting TAGs.

4.1 The biosynthesis of triacylglycerols

There is an important difference between animals and plants with respect to TAG composition and metabolism. Plants must (of necessity) synthesize their TAGs from simple starting materials according to their requirements, since they have no dietary source of preformed lipids. Unlike animals, they have the ability to synthesize linoleic (c,c-9,12-18:2,n-6) and α-linolenic (all-c9,12,15-18:3, n-3) acids that they normally possess in abundance: animals rely on plants for these essential nutrients (see Sections 3.5 & 6.2.2). The FA composition of animal TAGs is greatly influenced by their diet and, therefore, ultimately by the food materials they eat. The way in which dietary TAGs are modified by animals may differ between species and from organ to organ within a species. Such modifications are not necessarily always effected by the animal's own cells. For example, in ruminants such as cows, sheep or goats, the microorganisms present in the rumen hydro-

genate the double bonds of dietary polyunsaturated fatty acids (PUFAs) like linoleic and α-linolenic acids to form a mixture of mainly saturated fatty acids (SFAs) and *cis* and *trans*-monounsaturated (monoenoic) acids (MUFAs, see Section 3.1.9). An outstanding and significant feature of TAG FA composition is that it is usually quite distinctly different from that of the phospholipids or the nonesterified fatty acid (NEFA) pool. An understanding of this cannot be obtained simply from analyses of FA distributions but must depend upon the study of different metabolic pathways and of the individual enzymes in those pathways (that are involved in the biosynthesis of each lipid class).

4.1.1 The glycerol 3-phosphate pathway in mammalian tissues provides a link between triacylglycerol and phosphoglyceride metabolism

Historically, the *de novo* pathway, now usually known as the G3P pathway for TAG biosynthesis (Fig. 4.1), was worked out first. It was first proposed by the American biochemist, Eugene Kennedy in 1956, based on the earlier work of Kornberg and Pricer, who first studied reactions 1 and 4 (Fig. 4.1), the formation of phosphatidic acid (PtdOH) by stepwise acylation of G3P. Kennedy also demonstrated the central role of PtdOH in both phosphoglyceride (see Section 4.5) and TAG biosynthesis and one of his outstanding contributions was to point out that diacylglycerols (DAGs) derived from PtdOH form the basic building blocks for TAGs as well as phosphoglycerides.

Steps 1 and 4 in Fig. 4.1, the stepwise transfer of acyl groups from acyl-coenzyme A to G3P, are catalysed by two distinct enzymes specific for positions 1 and 2. For animals, the enzyme that transfers acyl groups to position 1 (acyl-CoA:glycerol 3-phosphate 1-O-acyltransferase, GPAT) exhibits marked selectivity for saturated acyl-CoA thioesters whereas the second enzyme (acyl-CoA:1-acyl glycerol

Lipids: Biochemistry, Biotechnology and Health, Sixth Edition. Michael I. Gurr, John L. Harwood, Keith N. Frayn, Denis J. Murphy and Robert H. Michell.
© 2016 John Wiley & Sons, Ltd. Published 2016 by John Wiley & Sons, Ltd.

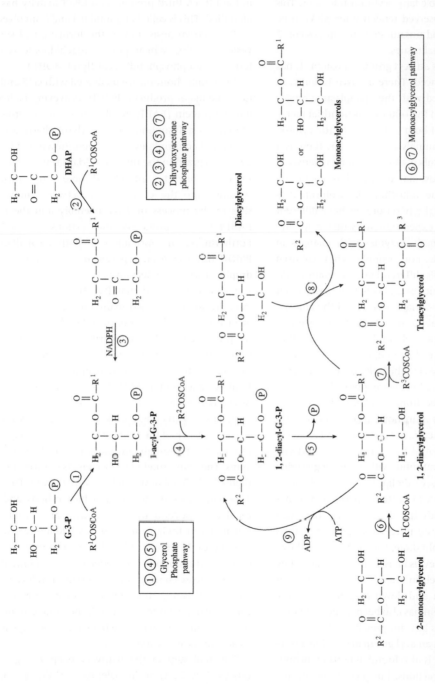

Fig. 4.1 The biosynthesis of triacylglycerols in mammals. The figure illustrates the three main pathways: the glycerol 3-phosphate (G3P) pathway (reactions 1, 4, 5, 7), the dihydroxyacetone phosphate (DHAP) pathway (reactions 2, 3, 4, 5, 7) and the monoacylglycerol pathway (reactions 6, 7). The enzymes involved are: (1) glycerol 3-phosphate acyltransferase (GPAT); (2) dihydroxyacetone phosphate acyltransferase; (3) acyl-dihydroxyacetone phosphate reductase; (4) lysophosphatidate acyltransferase (LPAT); (5) phosphatidate phosphohydrolase (PAP); (6) monoacylglycerol acyltransferase; (7) diacylglycerol acyltransferase (DGAT). A minor route from diacylglycerols may proceed via reaction 8: (DAG transacylase). In some tissues, DAG can be reconverted into phosphatidic acid by DAG kinase (9).

phosphate 2-*O*-acyltransferase, usually referred to as lyso-phosphatidate acyltransferase, LPAT) shows selectivity towards mono- and dienoic fatty acyl-CoA thioesters. This is in accord with the observed tendency for SFAs to be enriched in position 1 and unsaturated ones in position 2 in the lipids of most animal tissues.

GPAT, which transfers an acyl group to position 1, has been cloned and sequenced. There are two isoforms in mammals, one associated with the endoplasmic reticulum (ER), the other with the outer mitochondrial membrane. The active sites of both isoforms face the cytosol. In the liver of most mammals studied so far, there is a similar activity at both subcellular locations, whereas in other tissues, the microsomal enzyme has about ten times the activity of the mitochondrial isoform. It is not yet known whether the two isoforms have different functions but there is experimental evidence to link changes in activity of both enzymes with changes in overall TAG biosynthesis. For example, when cultured 3T3-L1 preadipocyte cells differentiate into mature adipocytes, the activity of microsomal GPAT increases about 70-fold and the mitochondrial activity (and the amount of its mRNA) increases about tenfold. It is interesting to speculate whether questions about the function of GPAT isoforms might be resolved by manipulating 3T3-L1 cells to produce knockout lines in which the genes coding the isoforms are either deleted or overexpressed.

GPAT was first purified from *E. coli* membranes and, more recently, a rat liver mitochondrial enzyme has also been purified. The purified *E. coli* enzyme, with a molecular mass of 83 kDa, proved to be inactive, but activity was restored by reconstitution with phosphoglyceride preparations, principally cardiolipin and phosphatidylglycerol (PtdGro). A 90 kDa protein ('p90'), which has about 30% identity with the *E. coli* GPAT, has been cloned in mice. When the mice were fasted and then given a diet rich in carbohydrate, the mRNA for p90 was rapidly induced at high levels in liver, muscle and kidney but at relatively low levels in brain. Such induction was up-regulated by insulin and down-regulated by cyclic AMP (cAMP). This protein has also been detected in fully differentiated adipocytes but not preadipocytes.

LPAT, which transfers an acyl group to position 2 of 1-acylG3P to form PtdOH, is also found in both mitochondrial and microsomal fractions, but predominantly the latter. It has proved more difficult to purify this activity but two human isoforms, α and β with 46% homology, have been cloned. LPAT-α is present in all tissues (but predominantly skeletal muscle), whereas the β-isoform is found primarily in heart, liver and pancreas. It is localized in the ER. (A third protein with LPAT activity has been identified. This is called endophilin-I and is involved, not in TAG biosynthesis, but in the formation of synaptic vesicles, during which process arachidonate is transferred to lysophosphatidic acid [lysoPtdOH].)

When mitochondria are incubated with G3P and acyl-CoA, the major product is PtdOH. However, if a liver FA binding protein (FABP; an old name was 'Z' protein) is present as a lysophosphatidate (lysoPtdOH) acceptor, lysoPtdOH accumulates. It has been shown that lysoPtdOH synthesized in the mitochondria is transferred to the ER and is converted into PtdOH there. More research is required into the cooperativity between intracellular sites in the process of TAG assembly and the role of FABPs in this process (see also Sections 3.1.1 & 7.1.3). Furthermore, the formation of both lysoPtdOH and PtdOH has also been reported on the surface of lipid droplets (LDs – see Section 5.5).

It should be noted that PtdOH, the product of the esterification of the two hydroxyls of G3P, can also be formed by phosphorylation of DAGs by DAG kinase in the presence of ATP. The contribution of this reaction to overall biosynthesis of PtdOH is ill defined but it is unlikely to be important in overall TAG biosynthesis. Its importance is more likely to be in the phosphorylation of DAGs released from PtdIns(4,5)$P2$ in the phosphoinositidase C signalling pathway (see Section 8.1.3).

The next step of the G3P pathway (step 5 in Fig. 4.1) is catalysed by phosphatidate phosphohydrolase (PAP) and lipins. The lipin protein family consists of three members (lipin-1, -2, -3) all of which have PAP activity. They have important roles in glycerolipid biosynthesis and gene regulation. Several human diseases are attributed to mutations in their genes (e.g. myoglobinurea, Majeed syndrome). Moreover, *LPIN-1* polymorphisms are associated with numerous metabolic traits including insulin and/or glucose concentrations and statin-induced myopathy (see Chapter 10 for more information about these conditions). PAP is found in both membrane-bound and soluble forms, which may be relevant to the regulation of its activity (see Section 4.3.2).

The final step in the pathway (step 7, Fig. 4.1) is catalysed by DAG acyltransferase (DGAT), an enzyme unique to TAG biosynthesis. DGAT plays a key role in determining the carbon flux into TAGs. For example, the rate of TAG formation in liver hepatocytes is controlled by the affinity of DGAT for acyl-CoA. Although DGAT activity was measured in chicken liver in the 1950s, it was only recently that genes encoding two isoforms of DGAT were cloned. DGAT1 and DGAT2 have been identified in numerous organisms. In addition, several DGAT-related genes that code proteins also using acyl-CoAs in similar reactions (such as monoacylglycerol (MAG) acyltransferase (MGAT) and wax synthases) have been identified.

The first DGAT1 gene was cloned from mouse in 1998 based on its sequence homology with acyl-CoA: cholesterol acyltransferase (ACAT) (see Section 7.2.4.2). However, expression of various DGAT1 genes showed that none of them could form cholesteryl esters. DGAT1 orthologues have been found in numerous animal and plant tissues. It is not present in most yeasts, including *Saccharomyces cerevisiae*, although it has been detected in *Yarrowia lipolytica* which is a commercially important oleaginous yeast.

The second DGAT, DGAT2, was originally found in a fungus. Surprisingly, the gene coding the protein shows no sequence homology with DGAT1 or ACAT. DGAT2 is widely distributed in eukaryotes – animals, plants, algae, fungi. Knockout experiments in mice suggest that DGAT2 plays a more important role than DGAT1 in overall TAG biosynthesis. On the other hand, in plants DGAT1 is usually more active, with DGAT2 being especially important in species that accumulate TAGs with unusual FAs.

Both DGAT isoforms are very hydrophobic proteins with DGAT1 predicted to have 8–10 regions that can form transmembrane domains. Both mouse isoforms will accept a broad range of acyl-CoA substrates but DGAT1 can utilize a wider range of acyl acceptors, i.e. it also has MGAT, wax synthase and acyl-CoA: retinol acyltransferase activities. Based on their sequences, the DGAT1 isoforms isolated from different tissues belong to a large family of proteins called membrane-bound *O*-acyltransferases (MBOAT). The MBOAT family includes ACAT and LPAT. DGAT2, with its distinct sequence, does not belong to the MBOAT family.

DGAT1 in animals is associated with the ER but DGAT2 is also found on LDs. On stimulation of TAG biosynthesis, DGAT2 activity can be seen associated with distinct regions of the ER, at or near the surface of LDs and also with mitochondria. Since an isoform of GPAT has also been found in mitochondria, it is possible that this organelle may have latent capacity for TAG formation.

For DGAT, which is the best studied of the acyltransferases of the Kennedy pathway, it is clear that the substrate selectivity is sufficiently wide to permit a variety of FAs to be esterified. Thus, it is not surprising that numerous examples have been found of TAGs that contain unnatural or 'xenobiotic' acyl groups. Xenobiotic carboxylic acids may arise from many sources including herbicides, pesticides and drugs of various kinds. Many are substrates for acyl-CoA synthetase (ACS) and the resulting xenobiotic acyl-CoAs may be incorporated into TAGs, cholesteryl esters, phospholipids and other complex lipids. An example of a xenobiotic TAG is the one formed from ibuprofen, a commonly used nonsteroidal anti-inflammatory drug (NSAID). Such compounds are normally stored in adipose tissue, where they may have relatively long half-lives since the ester bonds of many xenobiotic acyl groups are poor substrates for lipases. Much needs to be learned of their metabolism and potential toxicity.

4.1.2 The dihydroxyacetone phosphate pathway in mammalian tissues is a variation to the main glycerol 3-phosphate pathway and provides an important route to ether lipids

In the 1960s, the American biochemists Hajra and Agranoff discovered that radio-phosphorus was incorporated from ^{32}P-ATP into a hitherto unknown lipid, which they identified as acyl dihydroxyacetone phosphate. Further research demonstrated that dihydroxyacetone phosphate (DHAP) could provide the glycerol backbone of TAGs without first being converted into G3P (Fig. 4.1). The first reaction of this so-called 'DHAP pathway' is the acylation of DHAP at position 1, catalysed by the enzyme DHAP acyltransferase. Although first studied in a microsomal fraction, the activity has been found in peroxisomes. The latter contain many enzymes of lipid metabolism including those of the DHAP pathway.

DHAP acyltransferase has two separate roles. The first is in providing an alternative route to TAGs as illustrated in Fig. 4.1. (The second role, in ether lipid biosynthesis, is described later in Section 4.5.10.)

DHAP acyltransferase has been purified after detergent solubilization of peroxisomal membranes from guinea pig liver and human placenta. The purified protein has a molecular mass of 65–69 kDa. Certain chemical substances, including the hypolipidaemic drug clofibrate, cause peroxisomes to proliferate (see Section 7.3.2).

Under such conditions, the activity of DHAP acyltransferase increases 2–3-fold.

The next step in the pathway is the reduction of the keto group in 1-acyl-DHAP to form 1-acylG3P, linking once more into the main TAG biosynthetic pathway. This enzyme is located on the cytosolic side of the peroxisomal membrane and is notable in that it requires NADPH rather than NADH. NADPH is normally associated with reactions of reductive synthesis such as FA biosynthesis (see Section 3.1.3). There is some evidence to support the view that the activity of the DHAP pathway is enhanced under conditions of increased FA biosynthesis and relatively reduced in conditions of starvation or when the animal is given a high fat diet that is particularly rich in unsaturated fatty acids (UFA). Evidence from Amiya Hajra's laboratory in Ann Arbor, published in 2000, suggests that the DHAP pathway may indeed make a significant contribution to TAG accumulation when 3T3-L1 preadipocytes differentiate towards adipocytes in culture. However, we do not yet have a clear picture of the quantitative significance of the DHAP pathway in overall TAG assembly.

Peroxisomes do not contain the enzymes catalysing the final steps of acylglycerol biosynthesis. The end-products of peroxisomal biosynthesis, acyl or alkyl (see below) DHAP or 1-*O*-acyl- or 1-*O*-alkylG3P must be exported to the ER before acylation at position 2, dephosphorylation to DAG or its ether analogue and the final acylation at position 3 by DGAT can occur. The details of the transport mechanism from peroxisomal to microsomal membranes are unclear, but there is some evidence for the participation of a FABP.

It is now generally agreed that the second and more important role for DHAP acyltransferase is to catalyse the first step in the biosynthesis of ether lipids, a process that occurs on the luminal side of peroxisomal membranes. The peroxisome is now regarded as the principal site of biosynthesis of the alkyl (ether) lipids. The mechanism of formation of the alkyl linkage at position 1 of DHAP is described in more detail in Section 4.5.10 because the alkyl phospholipids are more widespread and of greater physiological importance than the nonpolar ('neutral') alkyl acylglycerols. However, as described in Section 2.2.2, the nonpolar alkyl lipids are found in significant quantities in the liver oils of sharks. The vital importance of ether lipids is illustrated by a number of serious neurological diseases resulting from lack of peroxisomes or peroxisomal enzymes. Thus, one piece of research, in

which human cDNA for DHAP acyltransferase was cloned, showed that absence of the enzyme causes severe neurological impairment and skeletal deformities but no alteration in overall TAG biosynthesis. Moreover, for the all too prevalent Alzheimer's disease, where there are significant decreases in plasmalogen lipids in patients (that correlate with cognitive impairment), recent results show reduced peroxisomal activity.

4.1.3 Formation of triacylglycerols in plants involves the cooperation of different subcellular compartments

The primary pathway for the biosynthesis of TAGs in plants that use lipids as their major energy store, is the G3P (or 'Kennedy') pathway. However, there are sufficient differences from animals in terms of subcellular location and sources of substrates to merit a separate discussion of plant TAG biosynthesis.

To study the different enzymes in the pathway, seeds or fruits need to be harvested at the time when the rate of lipid accumulation is most rapid, since there are distinct phases of development as illustrated in Fig. 4.2. In phase 1, cell division is rapid but there is little deposition of storage material, whether it be protein, lipid or carbohydrate. In phase 2, there is a fast accumulation of storage material. Moreover, if the TAGs in the lipid stores contain unusual FAs, the special enzymes needed for their biosynthesis are active only at this stage. Finally, in phase 3, desiccation takes place with little further metabolism.

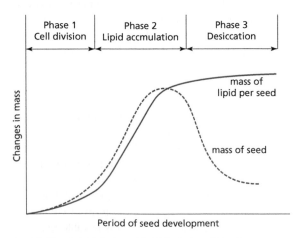

Fig. 4.2 The accumulation of lipid during seed development.

FA biosynthesis *de novo* is concentrated in the plastids of plant cells (see Section 3.1.3). By a combination of acetyl-CoA carboxylase (ACC) and fatty acid synthase (FAS), palmitoyl-acyl carrier protein (ACP) is produced. This is chain-lengthened by a specific condensing enzyme, called KAS II, to stearoyl-ACP, which is desaturated to oleoyl-ACP by a 9-desaturase (see Section 3.1.8.1). Under most conditions, palmitate and oleate are the main products of biosynthesis in plastids (chloroplasts in leaves). These acyl groups can be transferred to G3P within the organelle or hydrolysed, converted into CoA-esters and exported outside the plastid (Fig. 4.3). If FAs are esterified to G3P within the plastid, then 16C acids (mainly palmitate) are attached at position *sn*-2, whereas 18C acids (mainly oleate) are attached at position *sn*-1. This is sometimes referred to as a 'prokaryotic pathway' because it resembles the arrangement of 16C and 18C acids on complex lipids in cyanobacteria.

The biosynthesis of TAGs for energy storage purposes takes place in the ER, not in the plastid. Therefore, during the relatively short period of oil accumulation (a few days

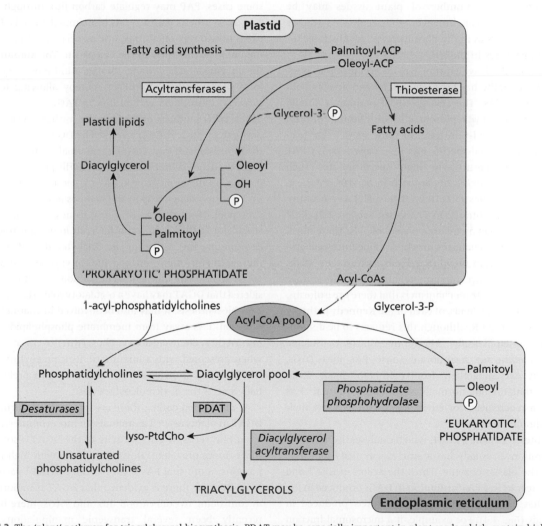

Fig. 4.3 The 'plant' pathway for triacylglycerol biosynthesis. PDAT may be especially important in plant seeds which contain highly unsaturated oils because PtdCho is the substrate for linoleate and α-linolenate formation (Section 3.1.8.2) although experimental evidence is lacking. Abbreviations: lyso-PtdCho, l-acylphosphatidylcholine; PDAT, phospholipid: diacylglycerol acyltransferase. This enzyme may account for over half of the total TAG made in some plants.

only) mechanisms must be in place to export acyl groups from the plastid to ER membranes. The first requirement is a thioesterase to hydrolyse ACP derivatives to NEFAs. During oil accumulation, the activity of this enzyme needs to be much higher than that of plastid GPAT, LPAT and PAP as there is no need for a rapid production of DAGs in plastids when the carbon flux is mainly to the ER.

After NEFAs are attached to CoA in the plastid envelope they move rapidly to the ER (Fig. 4.3). It is not known how this transport takes place, although acyl-CoA-binding transport proteins which have been purified from a number of plant tissues may be involved. It is clear, however, that acyl-CoAs (rather than acyl-ACPs as in the plastid) serve as substrates for acyltransferases in the ER.

The ER GPAT, which transfers an acyl group to position 1, generally has broad specificity and accepts both SFAs and UFAs. The acyl group at position 1 should therefore broadly represent the distribution of acyl groups in the acyl-CoA pool. In some species that produce unusual seed-specific FAs (e.g. *Cuphea spp*), GPAT may exhibit particularly high specificity for such FAs. (TAGs of *Cuphea spp* are rich in 8C–10C FAs; e.g. *C. koehneana* oil has 95% 10:0.) 'Unusual' FAs are almost invariably concentrated in TAGs (see Section 2.2), little being esterified in membrane phospholipids. There must, therefore, be mechanisms for channelling 'unusual' acyl groups away from membrane lipids (where they might adversely influence membrane function) into storage lipids. One possible mechanism is that there are isoforms, specific to storage lipids, of each of the Kennedy pathway enzymes in the ER, although this remains to be demonstrated unequivocally. Microsomal fractions isolated from *Cuphea spp* incorporate those FAs into TAGs, actively excluding them from membrane lipids. Even seeds that do not normally synthesize 'unusual' FAs have a mechanism for excluding such FAs from their membrane lipids.

In the ER of plants, LPAT, which catalyses the acylation at position 2, usually has a stricter specificity for UFAs than the plastid enzyme. Thus, the action of GPAT and LPAT results in a FA distribution in DAGs produced in the ER that is different from that in DAGs produced in plastids (Fig. 4.4). For example, if a microsomal fraction of developing safflower seeds is incubated with 16:0-CoA alone, lysoPtdOH accumulates, with little formation of PtdOH. Incubation with 18:1-CoA or 18:2-CoA, which

are accepted efficiently by both GPAT and LPAT, results in the formation of PtdOH. Likewise, accumulation of lysoPtdOH is also seen when a microsomal fraction from developing rapeseed is incubated with erucoyl-CoA. In rapeseed oil, erucic acid (*c*13-22:1) is exclusively located at positions 1 and 3. As erucic acid is a valuable industrial commodity (for use as a lubricating oil) there is currently much interest in either modifying the substrate specificity of LPAT or introducing into rapeseed a gene for LPAT from another plant species that does not discriminate against erucic acid.

A PAP located in the ER then acts to generate DAGs. In some cases, PAP may regulate carbon flux through the TAG biosynthesis pathway. As in mammalian cells, PAP exists in two separate forms, one associated with the ER membranes, the other in the cytoplasm. The amount of the enzyme attached to membranes is influenced by the local concentration of NEFAs, thereby allowing feed-forward control over carbon flux to TAG.

The DAG products of PAP can be further acylated at position 3 with acyl-CoA, catalysed by DGAT to complete the Kennedy pathway. This enzyme usually has less specificity than the acyltransferases that esterify positions 1 and 2. Thus, in many plants, the FAs that accumulate at the *sn*-3 position depend upon the composition of the acyl-CoA pool. Nevertheless, in a few tissues, the substrate specificity of DGAT may also have an important role in determining the nature of the final stored lipid. From measurements of enzyme activities in vitro and because DAGs accumulate during lipid deposition, it is often considered that DGAT may have a regulatory role in the rate of TAG biosynthesis. It may also be involved in channelling 'unusual' FAs away from membrane phospholipids and into TAGs as discussed above. Thus, DGAT of castor bean, when presented with a mixture of di-ricinoleoyl (*c*9,12-OH-18:1) and di-oleoyl glycerol species in vitro, preferentially selects the di-ricinoleoyl species.

As mentioned earlier, there are two major isoforms of DGAT in plants as well as animals. In most commercially important seeds, it is the activity of the DGAT1 isoform that is more prevalent. However, species with high proportions of unusual FAs seem to utilize DGAT2 particularly. Because there is evidence that DGAT is important in controlling overall carbon flux into TAGs, there have recently been several transgenic manipulations to increase its activity in plants. These have resulted in significant increases in oil yields in crops like oilseed rape and soybean.

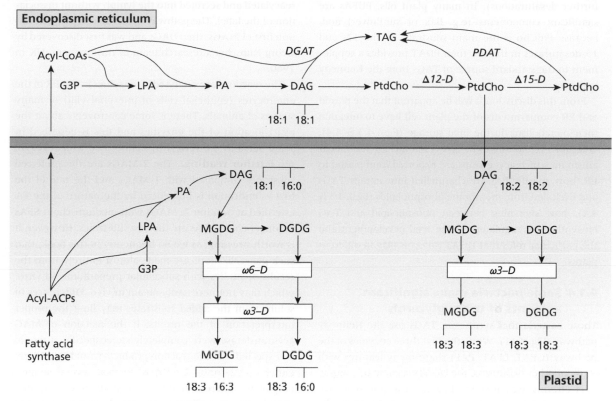

Fig. 4.4 Pathways for storage and plastid lipid synthesis in '16:3' and '18:3' plants showing location of desaturase enzymes which form PUFA. This is a simplified diagram of the one by JG Wallis & J Browse (2002) Mutants of Arabidopsis reveal many roles for membrane lipids. *Prog Lipid Res* **41**:254–78. Both types of plants produce TAG in the endoplasmic reticulum via the Kennedy Pathway and ancillary enzymes (e.g. PDAT). In 16:3 plants, the DAG used for galactosylglyceride synthesis is generated within the plastid whereas in 18:3 plants, at least, the bulk of the DAG comes from PtdCho (containing linoleate) in the ER. * the palmitate at position sn-2 on MGDG is first desaturated at the ω-9 (Δ-7) position. Abbreviations: LPA, lysophosphatidic acid, PA, phosphatidic acid.

During active oil accumulation, carbon from DAGs is preferentially channelled into TAG formation. However, at other phases of the seed's life cycle, the utilization of DAGs for the biosynthesis of membrane phospholipids, such as phosphatidylcholine (PtdCho), (catalysed by the enzyme CDPcholine:diacylglycerol cholinephosphotransferase), is more important. In some plants, the latter enzyme may also play a role in TAG biosynthesis (by equilibrating DAG with PtdCho), as illustrated in Fig. 4.3. The reaction catalysed by this enzyme is approximately in equilibrium and can, therefore, allow the rapid exchange of DAGs between their pool and that of the newly synthesized PtdCho. Since PtdCho is the substrate for oleate (and linoleate) desaturation in seeds (see Section 3.1.8.2), the reversible nature of the cholinephosphotransferase allows the DAG pool to become

enriched in PUFAs. This process has been studied in particular detail in safflower (which has an oil rich in PUFAs, comprising 75% linoleate). It has been shown that the acyl-CoA pool can also be utilized through the activity of an acyl-CoA:lysophosphatidylcholine acyltransferase (see below). In contrast, for plants where less unsaturated oils accumulate (e.g. avocado) the subsidiary flux of DAGs through PtdCho and their consequent desaturation, is much less important.

There are other enzyme reactions which may be of importance during TAG biosynthesis in plants. In particular, phospholipid: diacylglycerol acyltransferase (PDAT) catalyses the formation of TAGs in an acyl-CoA-independent manner (Fig. 4.3). This enzyme also produces lysophosphatidylcholine (lysoPtdCho), which is the substrate for another acyltransferase to regenerate PtdCho (for further desaturations). In many plant oils, PUFAs are significant components (e.g. flax or sunflower) and, because PtdCho is the main substrate for the 12- and 15-desaturases in the ER, then PDAT provides a supplement to the standard source of TAGs from the Kennedy pathway.

From this discussion it will be apparent that the plastid and ER compartments of the plant cell have to integrate their metabolism during lipid storage (Figs. 4.3 & 4.4). Questions remaining to be answered include the mechanism by which acyl groups are exported from plastid to ER, how specific FAs are channelled into storage TAGs and excluded from membrane phospholipids (Figs. 4.3 & 4.4), how switching between phospholipid and TAG biosynthesis is regulated during seed development and the particular roles that DGAT isoforms play in different plants.

4.1.4 Some bacteria make significant amounts of triacylglycerols

Those bacteria that synthesize TAGs use the Kennedy pathway (Fig. 4.11). While the first three enzymes of the pathway (GPAT, LPAT, PAP) share many features with those of other organisms, the DGATs seem to be unique. This enzyme was first purified from a gram-negative bacterium *Acinetobacter calcoaceticus* and found to be bifunctional (with wax synthase activity also). In vitro it also has MGAT activity. The bifunctional enzyme has no sequence relation to either wax synthases from plants or animals or to either of the DGAT isoforms from eukaryotes. Similar bifunctional enzymes have also been found in other bacteria such as the important pathogens *Streptomyces spp.* or *Mycobacterium tuberculosis*.

4.1.5 The monoacylglycerol pathway

The main function of the monoacylglycerol (MAG) pathway (Fig. 4.1) is to resynthesize TAGs from the MAGs formed during the digestion of fats in the small intestine. Since the MAG pathway usually uses partial acylglycerols derived from food, it is a mechanism for modifying TAGs rather than one that makes new TAGs (see Section 7.1.2).

During the hydrolysis of dietary TAG in the intestinal lumen by pancreatic lipase, the FAs from positions 1 and 3 are preferentially removed. The remaining 2-MAGs are relatively resistant to further hydrolysis. When 2-MAGs, radiolabelled in both FA and glycerol moieties, were given in the diet, the molecules were absorbed intact,

reacylated and secreted into the lymph without dissociation of the label. The pathway involves a stepwise acylation first of MAGs, then DAGs and was first discovered by Georg Hübscher's research team in Birmingham, UK in 1960.

The reactions are catalysed by enzymes in the ER of the enterocytes (epithelial cells of intestinal villi) of many species of animals. There is some controversy about the exact location of the enzymes and this is described in some detail in the review by Lehner and Kuksis (1996; see **Further reading**). The 2-MAGs are the preferred substrates compared with 1-MAGs and the rate of the first esterification is influenced by the nature of the FA esterified at position 2. MAGs with medium-chain SFAs or longer chain UFAs are the best substrates. However, it is worth stressing, as we do frequently in this book, that such generalizations are made almost entirely from the results of studies with subcellular preparations in vitro, which may not necessarily obtain in vivo. Differences in solubility of the added substrates may limit the proper interpretation of the results. If the reaction of MAG acyltransferase were completely stereospecific, it would be expected that the reaction products would be entirely either *sn*-1,2- or *sn*-2,3-DAGs. In fact, several studies, employing different analytical methods, have indicated that the reaction mixture contains about 90% 1,2- and about 10% 2,3-DAGs, so the reaction may not be completely stereospecific.

DGAT is specific for *sn*-1,2-DAGs and will not acylate the *sn*-2,3- or *sn*-1,3-isomers. Diunsaturated or mixed-acid DAGs are better substrates than disaturated compounds when measured in vitro. Again, we have to be cautious when we interpret results of this kind. Lipids containing UFAs are more easily emulsified than saturated ones, so that we may not be observing differences in selectivity of the enzyme for FA composition, but differences in solubility of the substrates when such enzyme assays are performed. There is little information on which DGAT isoform is used later in the MAG pathway.

Attempts to purify the individual enzymes involved in the MAG pathway have met with only partial success. Frequently (but not always!) the partly purified preparation has contained all three enzyme activities: MGAT, DGAT and ACS. This has led to the concept that there is a complex of enzymes acting in concert, now generally referred to as 'triacylglycerol synthase' or 'triacylglycerol synthase complex'. A molecular mass of 350 kDa has been proposed by the Canadian, Kuksis. A more purified preparation that migrated as a 37 kDa band on sodium dodecyl sulphate-polyacrylamide gel electrophoresis (SDS-PAGE) had MGAT activity, but whether it was a genuine subunit of the complex or a proteolytic fragment was not clear.

The MAG pathway has also been demonstrated in the liver and adipose tissue of the hamster and rat and these enzymes have also been partly purified. The activity is particularly high in pig liver. Under certain conditions it appears to compete with the Kennedy pathway for acyl groups and may serve to regulate the activity of the latter pathway. The origin of the MAG substrate in tissues other than intestine is not known and the role of the pathway in these other tissues is far from clear.

The liver and intestinal enzymes differ in substrate specificity, thermolability and response to different inhibitors, suggesting the existence of separate isoforms. Because the liver isoform has a preference for 2-MAGs that contain PUFAs, it has been suggested that one of its roles may be to prevent excessive degradation of PUFAs under conditions of high rates of β-oxidation. Consistent with this proposal is the high activity of liver MGAT in neonatal life and in hibernating animals. The normal neonatal rise in MGAT is attenuated in rat pups given an artificial high-carbohydrate diet compared with those sucking mother's milk.

An alternative route to TAG assembly in intestinal tissue was discovered by Lehner and Kuksis in the early 1990s. This involves the transfer of acyl groups between two DAG molecules without the intervention of acyl-CoA, catalysed by DAG transacylase (Fig. 4.1). The activity is located in the ER and has been solubilized and purified to homogeneity as a 52 kDa protein. Its precise function is unknown, but because its activity can be as much as 15% of that of DGAT, it could supply significant amounts of TAG. The substrates are 1,2- and 1,3-DAGs and the MAG product can be fed into the MAG pathway.

4.2 The catabolism of acylglycerols

Catabolism refers to the metabolic breakdown of complex biological molecules. One of the main themes running through this book is that, with very few exceptions, lipids in biological tissues are in a dynamic state: they are continually being synthesized and broken down. This is known as turnover. Complete catabolism of acylglycerols

takes place in two stages. Hydrolysis of the ester bonds that link fatty acyl chains to the glycerol backbone is brought about by enzymes known as lipases. Following the action of a lipase, releasing FAs, the latter may be further catabolized by oxidation (see Section 3.2). Acylglycerols themselves are not substrates for oxidation. Alternatively, however, the FAs released from acylglycerols may follow other metabolic pathways, including re-esterification with glycerol to make new acylglycerols. Therefore the lipases may be seen as 'gatekeepers', releasing FAs from the acylglycerol energy stores for oxidation or for further metabolism.

There are many lipases, some related in families, and they differ in respect to their substrates and the positions in substrates of the bonds that they hydrolyse. There are also many enzymes called esterases that hydrolyse ester bonds in general, but lipases form a distinct class and the distinction lies in the physical state of their substrates. The milieu in which a lipase acts is heterogeneous: the lipid substrate is dispersed as an emulsion in the aqueous medium, or is present as a fat droplet (e.g. within the mammalian adipocyte), and the enzyme acts at the interface between the lipid and aqueous phases. If, by some means, a single-phase system is obtained, for example when the TAG contains short-chain FAs (as in triacetoylglycerol – sometimes called 'triacetin') or when a powerful detergent is present, then the lipid may be hydrolysed by a less specific esterase, rather than a more specialized lipase.

Our understanding of lipase action has increased markedly in recent years with the use of X-ray crystallography to determine the three-dimensional structure of one such enzyme, pancreatic lipase (see also Section 7.1.1). This work, which has involved the crystallization of pancreatic lipase in the presence of emulsified fat, has provided a mechanistic basis for the phenomenon of interfacial activation, the activation of the enzyme that occurs on the surface of an oil-in-water droplet. Most importantly, in the presence of emulsified lipid, conformational changes in the enzyme lead to the opening of a 'lid' that allows access of the substrate to the active site. This lid-opening mechanism appears to be common to other members of the same family of lipases.

4.2.1 The nature and distribution of lipases

Lipases are widespread in nature and are found in animals, plants and microorganisms. The initial step in the hydrolysis is generally the splitting of the FAs esterified at positions *sn*-1 and *sn*-3 (see Box 2.2). TAG lipases in general are ineffective in hydrolysis of the *sn*-2 position ester bond. Therefore 2-MAGs tend to accumulate, at least in artificial models of lipase action. In vivo, 2-MAGs may be removed by cellular uptake, by nonenzymic isomerization to the 1(3)-form so that complete hydrolysis can occur, or by the action of a more specific MAG lipase.

Some lipases not only hydrolyse FAs on the primary positions of acylglycerols, but will also liberate the FA esterified at position 1 of phosphoglycerides. In most cases, the rate of hydrolysis is independent of the nature of the FAs released. There are, however, several exceptions to this rule. The ester bonds of FAs with chain lengths less than 12 carbon atoms, especially the shorter-chain lengths of FAs in milk TAGS, are cleaved more rapidly than those of the more common chain-length FAs (14C–18C), while the ester bonds of the very long-chain PUFAs, eicosapentaenoic acid, 20:5*n*-3 (EPA) and docosahexaenoic acid, 22:6*n*-3 (DHA), found in algae or in the oils of fish and marine mammals are more slowly hydrolysed. The lipases of some microorganisms have a marked FA selectivity. For example, the fungus *Geotrichum candidum* possesses a lipase that seems to be specific for oleic acid in whichever position it is esterified.

4.2.2 Animal triacylglycerol lipases play a key role in the digestion of food and in the uptake and release of fatty acids by tissues

Animal TAG lipases may be broadly classified into extracellular and intracellular. Extracellular lipases are secreted from their cells of synthesis and act upon TAGs that are present in the extracellular environment, releasing NEFAs that may be taken up into cells. Therefore, they are concerned mainly with cellular uptake of FAs. There are four major extracellular TAG lipases (in adults), which are members of one family and have sequence and structural similarities to each other. Pancreatic lipase is, as its name suggests, secreted from the exocrine tissue of the pancreas into the small intestine, where it acts to hydrolyse dietary TAGs so that the constituent FAs and MAGs may be absorbed from the intestine into enterocytes. Its action is described in more detail in Section 7.1.1.

Within the body, TAGs circulate in the plasma in the form of macromolecular complexes, the lipoproteins (see Section 7.2). Some lipoprotein TAGs are taken up directly into cells that express receptors, which bind and internalize complete particles, but this is mainly a route for the clearance of 'remnant' particles, once they have lost much of their TAG. Most of the TAGs are hydrolysed within the vascular compartment by lipoprotein lipase (LPL) to release FAs that may again be taken up into cells. (This enzyme is called 'clearing factor lipase' in older literature, because of its ability to clear the turbidity of the plasma caused by the presence of large TAG-rich lipoprotein particles.) LPL is related to pancreatic lipase. It is synthesized within the parenchymal cells of tissues (e.g. adipocytes, muscle cells, milk-producing cells of the mammary gland), but then exported to the endothelial cells that line blood capillaries. Here it is bound to the luminal membrane of the endothelial cell (facing into the blood) where it can act upon the TAGs in lipoprotein particles as they pass by. The FAs it releases may then diffuse into the adjacent cells for re-esterification and storage (e.g. in adipose tissue) or for oxidation (in muscle). (See Section 7.2.5 for more information on the physiological role of LPL.)

Hepatic lipase is the third member of this family. As its name suggests, it is synthesized within the liver and, just like LPL, exported to bind to the endothelial cells that line the hepatic sinusoids – the tiny vessels that are the liver's equivalent of capillaries. It plays a role particularly in hydrolysis of TAGs in smaller lipoprotein particles, and also assists in binding these particles to the receptors that may remove them from the circulation (see Section 7.2.4.3).

The fourth member of the family is endothelial lipase. It was discovered only in 1999 and its function is not yet clear. It is expressed by the endothelial cells in several tissues including liver, lung, kidney and placenta, and some endocrine tissues including thyroid, ovary and testis. It is more active as a phospholipase (with A_1-type activity, see Section 4.6.2) than as a TAG lipase. It has been suggested that it plays a role in lipoprotein metabolism and vascular biology.

Of the intracellular lipases, one has been particularly well studied: the so-called 'hormone-sensitive lipase' (HSL) expressed in adipocytes and in some cells with an active pattern of steroid metabolism (adrenal cortex cells, macrophages, testis). Its name reflects the fact that its activity is rapidly regulated by a number of hormones (see Section 4.3.3 & Box 9.1) although this can be confusing because other lipases, especially LPL, are also regulated by hormones. HSL acts on the surface of the TAG droplet, stored within adipocytes, to release FAs that may be delivered into the circulation for transport to other tissues where they may be substrates for oxidation or for re-esterification to glycerol. HSL is also a cholesteryl esterase and acts with equal efficiency to hydrolyse cholesteryl esters: this is presumably its role in cells that metabolize steroids. Adipose tissue also contains a lipase that is much more active on MAGs than is HSL.

Lipases are still being discovered and it may be a long time before the whole complex jigsaw puzzle of acylglycerol breakdown can be pieced together. For instance, it is known that an intracellular lipase is involved in hydrolysis of TAGs within hepatocytes, releasing FAs that are then re-esterified before incorporation into very low density lipoprotein particles (VLDL; the lipoprotein particles secreted by the liver; see Section 7.2.3). This lipase has not yet been characterized (see Section 7.2.3 for further discussion).

4.2.3 Plant lipases break down the lipids stored in seeds in a specialized organelle, the glyoxysome

Seeds that contain lipid may have as much as 80% of their dry weight as TAGs. Plants such as soybean face two particular problems in using such energy reserves. First, these plants have to mobilize the lipid rapidly and break it down to useful products. This overall process involves the synthesis of degradative enzymes as well as the production of the necessary membranes and organelles that are the sites of such catabolism. Secondly, plants with lipid-rich seeds must be able to form water-soluble carbohydrates (mainly sucrose) from the lipid as a supply of carbon to the rapidly elongating stems and roots. Animals cannot convert lipid into carbohydrate (Fig. 4.5) because of the

decarboxylation reactions of the Krebs (tricarboxylic acid) cycle (isocitrate dehydrogenase and 2-oxoglutarate dehydrogenase). Thus, for every two carbons entering the Krebs cycle from lipid as acetyl-CoA, two carbons are lost as CO_2. In plants, these decarboxylations are avoided by a modification of the Krebs cycle, which is called the glyoxylate cycle. This allows lipid carbon to contribute to the biosynthesis of oxaloacetate, which is an effective precursor of glucose and, hence, sucrose.

When water is imbibed into a dry seed, there is a sudden activation of metabolism once the total water content has reached a certain critical proportion. So far as lipid-storing seeds are concerned, TAG lipase activity is induced and studies by Huang in California, and others, suggest that this lipase interacts with the outside of the oil droplet through a specific protein (an oleosin; see Sections 5.5.2 & 9.3.1.2) that aids its binding. This interaction has some similarities to the binding of LPL to VLDL in animals (see Section 7.2.5). The lipases in seeds hydrolyse the 1 and 3-positions of TAG and because the acyl group of the 2-MAG products can migrate rapidly to position 1, the lipases can completely degrade the lipid stores.

In general, plant lipases show rather little selectivity. Only when the seed oil contains large amounts (>75%) of SFA or unusual FAs are lipases with an appropriate selectivity needed. For example, *Cuphea procumbens* contains an oil that is highly enriched in medium-chain FAs and its seed lipase has a 20-fold preference for capric acid (10:0) compared to, for example, palmitic acid.

The liberated FAs are activated to CoA-esters and broken down by a modification of β-oxidation, which takes place in specialized microbodies. Because these microbodies also contain the enzymes of the glyoxylate cycle (see above), they have been termed glyoxysomes by the Californian biochemist, Harry Beevers. Beevers and his group worked out methods for the isolation of glyoxysomes from germinating castor bean seeds and showed that FA β-oxidation was confined to this organelle. The glyoxysomes have only a temporary existence. They are formed during the first two days of germination and, once the lipid stores have disappeared (after about 6 days), they gradually break down. Nevertheless, in leaf tissues the role of glyoxysomes in the β-oxidation of FAs is replaced by other microbodies (see Section 3.2.1.1).

Most plant lipases appear to be membrane-bound although soluble enzymes are present in some tissues.

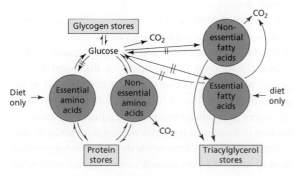

Fig. 4.5 Permitted and forbidden interconversions of fats, carbohydrates and proteins in animals.

There has been some interest in using the latter in the food industry in order to carry out the transesterification of TAGs (see Section 11.2.4) and, thus, modify their composition for certain purposes (e.g. to make fats that resemble cocoa butter ('cocoa butter equivalents') for chocolate). These transesterification reactions are favoured by a very low water content (usually less than 1%) of the reaction mixture.

Plant lipase activities also have implications for food quality, when the release of NEFAs can cause deterioration. For example, during olive oil production, significant endogenous FA levels, themselves, determine whether an oil can be considered high quality ('extra virgin') or not. Alternatively, in poorly stored wholemeal flour, endogenous lipases give rise to FAs that can be broken down by lipoxygenases (see Section 3.4) and cause off-flavours and a reduction in baking quality.

4.3 The integration and control of animal acylglycerol metabolism

The major function of TAGs in animals is as a source of FAs to be used as metabolic fuel. A full description of the overall fuel economy of the body requires an understanding of the metabolism (storage, transport, biosynthesis, oxidation) of the FAs as well as of the integration of fat and carbohydrate metabolism.

4.3.1 Fuel economy: the interconversion of different types of fuels is hormonally regulated to maintain normal blood glucose concentrations and ensure storage of excess dietary energy in triacylglycerols

The maintenance of fuel reserves within fairly narrow limits is referred to as fuel homeostasis. Glycogen is stored within muscle and liver cells (and in small amounts elsewhere, e.g. in the brain). Because glycogen is stored in a hydrated form, the amount that can be stored in a cell is limited, and also it would be disadvantageous for the body to carry around excessive amounts of this fuel source. In adult humans the liver glycogen store is typically about 100 g, and it is probably no coincidence that this is approximately the amount of glucose needed by the brain in 24 h. Glycogen stores are for immediate, 'emergency' use in maintaining carbohydrate supply for tissues that require it, such as the brain, and their total amount is held within fairly narrow limits. Protein is potentially a major energy source (the adult human body has about 12.5 kg of protein). However, there is no specific storage form of protein, and so all body protein plays some specific role (enzymes, structural, etc.). Therefore, it is not generally used as a fuel store, and indeed its degradation during starvation is specifically spared. Again, therefore, its amount appears to be controlled within fairly narrow limits. In contrast, humans and other mammals appear to have an almost infinite capacity for storage of TAGs, achieved both by expansion of existing adipocytes (so that a mature, fat-filled human adipocyte may be 0.1 mm in diameter) and by an increase in the number of adipocytes. (This will be discussed further in Section 9.1.)

The various fuels can be interconverted to some extent. Excess carbohydrate and protein can be converted into fat, and amino acids from protein can be converted into carbohydrate (as happens during starvation as a means of supplying new glucose). However, FAs cannot be converted into carbohydrate (see Section 4.2.3; Fig. 4.5) in animals and do not appear to be significant precursors of amino acids. The constituent glycerol from TAGs that have been hydrolysed to release FAs may, however, be an important precursor of glucose during starvation.

When the energy in the diet exceeds immediate requirements, excess carbohydrate is preferentially used to replenish glycogen stores. Excess protein tends to be oxidized after satisfying the tissues' needs for protein synthesis. Any remaining excess of either fuel, or of fat, then tends to be converted into TAGs for storage in adipose tissue. Adipose tissue TAG is the ultimate repository for excess dietary energy (Chapter 9 gives further details).

The control of these interconversions is a composite function of the amount of energy in the diet, the nature of the dietary constituents and the concentrations of the relevant hormones in the blood. A dominant role is played by insulin. Its concentration in the blood helps to coordinate the flow of fuel either into storage or from the stores into various tissues as required. High concentrations of insulin characterize the fed state when ample dietary fuel is available from the diet; low levels signal the starved state when the animal's own reserves need to be called upon. However, it has recently become clear that dietary FAs themselves may play an important role in regulating the expression of various genes involved in these processes (see Section 7.3 for more detail).

4.3.2 The control of acylglycerol biosynthesis is important, not only for fuel economy but for membrane formation, requiring close integration of storage and structural lipid metabolism

An important concept of metabolism is that of turnover, which describes the continual renewal, involving biosynthesis and breakdown, of body constituents. The rates of turnover may differ widely between different biological molecules in different tissues. The rates of forward and backward reactions may also differ. Thus, when there is net synthesis of a tissue component, the rate of synthesis is faster than the rate of breakdown, but both may be proceeding simultaneously. When metabolic control is exercised, by whatever means, it may affect the rate of synthesis, the rate of degradation, or both. Turnover allows a fine degree of control of metabolic pathways. Control may be exercised on the synthesis or degradation of enzymes catalysing metabolic reactions, at the level of gene transcription or translation, or on the allosteric control of enzymes by small molecules or cofactors.

In animals, net TAG accumulation occurs when energy supply exceeds immediate requirements. Most diets contain both fat and carbohydrates. When there is an excess of energy as carbohydrates, the body switches mainly to a pattern of carbohydrate oxidation so that these are used as the preferential metabolic fuel. When there is a considerable excess of carbohydrates for some time, then the tissues convert them into FAs that are esterified into acylglycerols. Conversely, when there is a preponderance of fat in the diet, fat biosynthesis from carbohydrate is depressed in tissues and fat oxidation and fat storage will tend to predominate. This involves conversion of the products of fat digestion into lipoproteins (see Section 7.2.3), which circulate in the bloodstream. When the lipoproteins reach the tissues, FAs are released from the acylglycerols at the endothelial lining of the capillaries, a process catalysed by the enzyme LPL (see Sections 4.2.2 & 7.2.5). The FAs are taken up into the cells. Once inside, they are either oxidized or esterified into acylglycerols (see Fig. 4.6). In discussing the control of acylglycerol biosynthesis, we shall be discussing both the esterification of FAs synthesized *de novo* or released from

circulating lipoproteins. The control of FA biosynthesis itself is discussed in Section 3.1.11.

Although acylglycerols may be synthesized in many animal tissues, the most important are: the small intestine, which resynthesizes TAGs from components absorbed after digestion of dietary fats; the liver, which is concerned mainly with synthesis from carbohydrates and redistribution; adipose tissue, which is concerned with longer term storage of fat; and the mammary gland, which synthesizes milk fat during lactation. There is also turnover of a relatively smaller pool of TAGs within muscle cells.

In the enterocytes of the small intestine, the G3P pathway seems to supply a basal rate of TAG biosynthesis between meals but the postprandial influx of dietary FAs appear to be the main factors controlling the rate of TAG biosynthesis via the 2-MAG pathway. In other tissues, in which the G3P pathway is predominant, TAG biosynthesis is more tightly regulated by the prevailing nutritional state.

The nutritional regulation of acylglycerol biosynthesis has been studied mostly in liver and adipose tissue. In neither tissue, however, is our understanding complete, mainly because of the difficulty of isolating the relevant enzymes, which are insoluble and membrane-bound. Early attempts to purify the enzyme DGAT, for example, were described as 'masochistic enzymology'. This has changed as modern methods of cloning by homology searching (screening cDNA libraries for related sequences) have been applied.

The cellular concentrations of the substrates for the acyltransferases that catalyse acylglycerol esterification, acyl-CoA and G3P, are influenced by nutritional status. Since G3P is produced from DHAP, an intermediate in the pathways of glycolysis and gluconeogenesis, the main factors influencing the amounts of G3P available for acylglycerol biosynthesis are those that regulate the levels and activities of the enzymes of glucose metabolism. Starvation reduces the intracellular concentration of G3P severely, whereas carbohydrate feeding increases it. Intracellular concentrations of acyl-CoA increase during starvation.

In adipose tissue, there is evidence that the supply of G3P may regulate TAG biosynthesis. Insulin stimulates glucose uptake by adipocytes and, by implication, G3P production. Certainly TAG biosynthesis in adipose tissue increases in the period following a meal, as is necessary to accommodate the influx of FAs from LPL

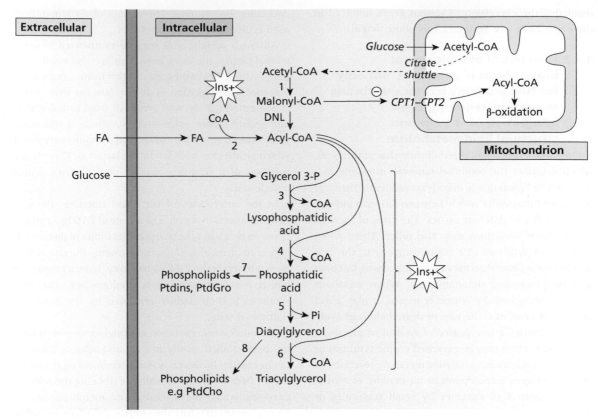

Fig. 4.6 Central role of acyl-CoA in hepatic lipid metabolism. Acetyl-CoA is formed either from esterification (sometimes called activation) of a fatty acid with CoA, or from the pathway of *de novo* lipogenesis (DNL). It may be utilized for β-oxidation in the mitochondria (catabolism) or for glycerolipid biosynthesis. The relative rates of these pathways are determined by the nutritional state, mediated largely by variations in plasma insulin concentration. When insulin concentrations are high (well-fed state) (shown as Ins+), generation of malonyl-CoA through the activity of acetyl-CoA carboxylase (1) inhibits entry of acyl-CoA into the mitochondrion by inhibition of carnitine palmitoyl transferase-1 (CPT_1). At the same time, the pathway of glycerolipid synthesis is stimulated by insulin. Other enzymes/pathways shown are: (2) acyl-CoA synthase (also known as acid:CoA ligase); (3) acyl-CoA:glycerol phosphate 1-*O*-acyl transferase (GPAT); (4) acyl-CoA:1-acylglycerol phosphate 2-O-acyl transferase (LPAT); (5) phosphatidate phosphohydrolase (PAP); (6) diacylglycerol acyltransferase (DGAT); (7) pathway of phospholipid biosynthesis via CDP-DAG (see Figs. 4.11 and 4.13) leading to phosphatidylinositol (PtdIns), phosphatidylglycerol (PtdGro) and others; (8) 1,2-diacylglycerol:choline phosphotransferase, leading to phosphatidycholine (PtdCho) and other phospholipids (see Figs. 4.11 and 4.12). Glucose metabolism is also shown in outline; glucose (via glycolysis) leads to the provision of glycerol 3-phosphate, and via pyruvate dehydrogenase in the mitochondrion to acetyl-CoA. For acetyl-CoA produced by this route to be a substrate for *de novo* lipogenesis, it must be exported to the cytoplasm using the citrate shuttle.

(itself stimulated by insulin) acting on circulating TAGs in the capillaries (see Section 9.1.2.1). This might equally, though, reflect stimulation of the enzymes of TAG biosynthesis by insulin and other factors. One such factor is the 76-amino acid peptide known as acylation stimulating protein (ASP). The production of ASP from adipocytes will be described in more detail in Section 7.3.5.

In the liver, however, there is little evidence that intracellular concentrations of G3P or acyl-CoA are important in regulating acylglycerol biosynthesis. Instead, the rate of acylglycerol biosynthesis in the liver appears to reflect the relative activities of the competing pathways for FA utilization, acylglycerol biosynthesis and β-oxidation (Fig. 4.6). Entry of FAs into the mitochondrion for oxidation is mediated by carnitine palmitoyl transferase 1 (CPT1; see Section 3.2.1.2), and this enzyme is powerfully suppressed by an increase in the cytosolic concentration of malonyl-CoA, the first committed intermediate in FA biosynthesis (see Sections 3.1.3 & 3.2.1). The formation of malonyl-CoA is stimulated under well-fed conditions (when insulin levels are high; see Section 3.1.11) and so FA oxidation will be suppressed and acylglycerol biosynthesis favoured under these conditions. The opposite will be true in starvation or energy deficit. An important question is whether there is also coordinated regulation of the enzymes of acylglycerol biosynthesis. The answer appears to be yes, but pinpointing the locus of control has proved difficult.

The enzymes that may be involved in regulation of mammalian acylglycerol biosynthesis include the acyltransferases that link successive molecules of acyl-CoA to the glycerol backbone, and PAP (Fig. 4.1). The first of these enzymes is GPAT (see Section 4.1.1). It has been suggested that GPAT and CPT1 (see above), both expressed on the outer mitochondrial membrane, represent an important branch point in FA metabolism, leading acyl-CoA into esterification or oxidation respectively. GPAT expression is generally regulated in parallel with the rate of TAG biosynthesis in different nutritional states. This regulation is brought about by a number of factors including insulin, for which there is a specific response element (sterol regulatory element, SRE), mediated by its binding protein (SREBP) in the promoter region of the mitochondrial GPAT gene. GPAT expression is regulated almost exactly in parallel with the expression of FAS.

Nutritional and hormonal factors appear to influence the mitochondrial GPAT activity more than the microsomal GPAT. Thus, when fasted rats are given a diet low in fat and rich in carbohydrate, mitochondrial GPAT activity increases six-fold with little change in the microsomal GPAT activity. Similarly in perfused rat liver, inclusion of insulin in the perfusion fluid increases the mitochondrial GPAT activity four times more than the microsomal. During starvation, hepatic mitochondrial GPAT activity decreases, but the overall capacity for TAG biosynthesis in the liver remains unchanged provided that β-oxidation is inhibited. It seems that in starvation, the decrease in GPAT activity is due primarily to competition by CPT1 for acyl-CoA (see Section 3.2.1.6).

The activity of the next enzyme in the pathway, LPAT, is increased 2.5-fold in liver postnatally and about 60-fold during the differentiation of 3T3-L1 preadipocytes.(see Section 9.1.2.2) Changes in mRNA for the enzyme in response to dietary changes have not been reported.

The activity of PAP, like that of GPAT, generally runs parallel to the potential for overall acylglycerol biosynthesis in that tissue, and it has been suggested that PAP is the major locus for regulation of TAG biosynthesis. However, it now appears more likely that control is 'shared' by a number of enzymes. PAP activity in liver is increased by high levels of dietary sucrose and fat, by ethanol and by conditions, such as starvation, that result in high concentrations of plasma NEFAs. It is also increased in obese animals. It is decreased in diabetes and by administration of drugs that result in a reduction of circulating lipid concentrations. The factors that tend to increase the activity of PAP are also those that result in an increased supply of SFA and MUFA to the liver, namely those FAs normally esterified in simple acylglycerols. If PAP activity is low, its substrate – PtdOH, the central intermediate in lipid metabolism – does not accumulate, but becomes a substrate for the biosynthesis of acidic membrane phosphoglycerides such as phosphatidylinositol (PtdIns; see Sections 4.5.6 & 8.1). Phosphoglyceride metabolism makes more extensive demands on a supply of UFAs than does simple acylglycerol metabolism and the activities of enzymes that divert PtdOH into phosphoglycerides rather than TAG metabolism tend to be elevated in conditions where UFA predominate (Fig. 4.6).

PAP exists in the cytosol and in the ER. In mammals, up to three PAPs, called 'lipins', have been reported. There is another distinct isoform associated with the cell membrane but this is thought to be involved more with signal transduction than with TAG biosynthesis. Most PAPs also hydrolyse ceramide 1-phosphate, sphingosine 1-phosphate and lysoPtdOH. Consequently they are preferably called lipid phosphatases. The cytosolic enzyme is physiologically inactive but translocates to the membranes of the ER on which PtdOH is being synthesized. The translocation process seems to be regulated both by hormones and substrates: cAMP, for example, displaces the enzyme from membranes, whereas increasing concentrations of NEFAs and their CoA esters promote its attachment. Insulin, which has the effect of decreasing intracellular concentrations of cAMP, ensures that the translocation is more effective at lower FA concentrations. The mechanisms that cause these changes are not yet established but may involve the reversible phosphorylation of PAP. It seems, therefore, that the membrane-associated PAP is the physiologically active form of the enzyme and the cytosolic form represents a reservoir of potential activity. This phenomenon of translocation is seen with some other enzymes involved in lipid metabolism. Enzymes that exist in different locations in the cell and can regulate metabolism by moving from one location to another are called 'ambiquitous enzymes'. Other examples in lipid metabolism are the CTP:phosphorylcholine cytidylyltransferase, which is important for regulation of the biosynthesis of PtdCho (see Section 4.5.5), and HSL in adipocytes (see Section 4.3.3).

Many attempts to purify the mammalian PAP and clone its cDNA were unsuccessful. However, comparisons with the yeast enzyme allowed the isolation of the mammalian orthologue. This was called lipin-1. In deficient mice, lack of lipin-1 causes fatty liver dystrophy, deficiency of adipose tissue (lipodystrophy), insulin resistance and increased β-oxidation (see also Section 10.4.2). There are two other lipins (2 and 3) with PAP activity in mammals (see Section 4.1.1). Lipins-1 and -3 are both needed for full PAP activity in liver and adipose tissue, possibly because of heterodimerization.

The DAGs formed by the action of PAP can be used either for TAG biosynthesis (Fig. 4.1), or for phosphoglyceride formation (see Fig. 4.11 in Section 4.5.4). The biosynthesis of phospholipids takes precedence over that of TAGs when the rate of biosynthesis of DAG is relatively low. This ensures the maintenance of membrane turnover and bile secretion, which are more essential processes in physiological terms than the accumulation of TAGs. The mechanisms that achieve preferred biosynthesis of phosphoglycerides are not certain but probably include the high affinity of choline phosphotransferase for DAG. The supply of choline may eventually limit PtdCho biosynthesis and a major effect of choline deficiency is to drive DAG into TAG formation, leading to development of a fatty liver.

The last enzyme in the biosynthesis of TAG, DGAT, is strongly expressed in tissues that have a high rate of TAG biosynthesis, including small intestine and adipocytes. However, the level of expression is relatively low in the liver, which is surprising and leaves open the possibility that yet another enzyme is still awaiting identification. As yet, little is known of its regulation, although it has been claimed to be an important locus of control of TAG

biosynthesis by ASP (see Section 7.3.5) in adipocytes. There is evidence for short-term regulation of its activity by reversible phosphorylation, but this has not yet been shown conclusively.

Further research into the hormonal and nutritional control of these processes may give insights into how to control the common diseases of lipid metabolism (see Chapter 10).

4.3.3 Mobilization of fatty acids from the fat stores is regulated by hormonal balance, which in turn is responsive to nutritional and physiological states

In physiological states demanding the consumption of fuel reserves, the resulting low concentrations of insulin turn off the biosynthetic pathways and release the inhibition of HSL within adipocytes. The activity of HSL is regulated in the short term by a cascade mechanism illustrated in Fig. 4.7. In the longer term it is also regulated by control of transcription; its expression is up-regulated, for instance, during prolonged fasting.

Short-term regulation is brought about by reversible phosphorylation. The phosphorylated enzyme is active and the dephosphorylated form inactive. Changes in activity in vivo, or in intact cells, are much greater than can be achieved when the purified enzyme is phosphorylated in vitro and acts on a synthetic lipid emulsion. This has led to the realization that phosphorylation of HSL is associated with translocation of the enzyme from a cytosolic location to the surface of the intracellular LDs (see Section 5.5.4). This may involve 'docking' with a protein, perilipin, that is associated with the LD surface, and is itself a substrate for phosphorylation under similar conditions to HSL, as discussed further in Box 9.1.

Phosphorylation of HSL is mediated by the enzyme protein kinase A (PKA), or cAMP-dependent protein kinase. This in turn is activated by the binding of cAMP when cellular cAMP concentrations are elevated through increased activity of the enzyme adenylate cyclase (sometimes termed 'adenylyl cyclase'). The activity of the cyclase is under the control of the catecholamines adrenaline (US: epinephrine – a true hormone, released from the adrenal medulla) and noradrenaline (US: norepinephrine – a neurotransmitter released from sympathetic nerve terminals in adipose tissue). Binding

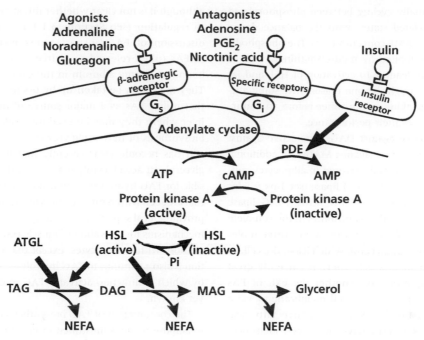

Fig. 4.7 Short-term regulation of the activity of hormone-sensitive lipase (HSL) in adipocytes. HSL is regulated in the short term by reversible phosphorylation brought about the enzyme protein kinase A (also known as cyclic AMP-dependent protein kinase). Protein kinase A is activated by the binding of cyclic AMP (cAMP), generated by the action of adenylate cyclase on ATP. In turn, adenylate cyclase is regulated by membrane-associated heterotrimeric guanine-nucleotide binding proteins, known as G-proteins. These link cell-membrane hormone receptors with adenylate cyclase. These are stimulatory (G_s) and inhibitory (G_i) G-proteins that link with the appropriate receptors (see Section 7.3.4 for more information on G protein-coupled receptors). Insulin suppresses the activity of HSL by causing its dephosphorylation. This reflects activation of a particular phosphodiesterase (PDE), that breaks down cAMP and therefore reduces the activity of protein kinase A. There are constitutively expressed protein phosphatases that return HSL to its inactivated state under these conditions. HSL is mainly active against diaglycerols (DAG), less so against triacylglycerols (TAG). TAG hydrolysis is mainly brought about by the enzyme commonly called adipose triglyceride (TAG) lipase (ATGL). There is a specific, highly active monoacylglycerol (MAG) lipase in adipose tissue. The overall reaction is the liberation from stored TAG of three non-esterified fatty acids and a molecule of glycerol. Regulation of the whole pathway is described in more detail in Box 9.1.

of catecholamines to β-adrenergic receptors in the cell membrane activates adenylate cyclase through the intermediary G-proteins (proteins that bind GTP, and couple receptors to adenylate cyclase). The situation is complex because catecholamines can also bind to α-adrenergic receptors that act through inhibitory G-proteins to *reduce* adenylate cyclase activity, although the overall effect in vivo is usually towards *activation*. There are several locally produced mediators, such as adenosine and prostaglandins (formed from dietary essential fatty acids (EFAs; see Sections 3.5 & 6.2.2) that inhibit adenylate cyclase activity, again via specific

cell-surface receptors and inhibitory G-proteins. This complex system undoubtedly exists to regulate fat mobilization extremely precisely (see Section 9.1.2 for further discussion of the physiological importance of this regulation in mammals).

Counter-regulation is achieved mainly by insulin, which acts via cell-surface insulin receptors, to activate a specific form of phosphodiesterase that catalyses the breakdown of cAMP. Under these conditions, protein phosphatases in the cell (which do not appear to be regulated but are present in high activity) dephosphorylate HSL, which is thus inactivated. Presumably HSL in vivo is continually cycling between phosphorylated and dephosphorylated states, with the balance determined by cellular cAMP concentrations. The phosphodiesterase can be inhibited by methylxanthines, such as caffeine, therefore leading to activation of HSL and fat mobilization. This might be one reason why some athletes claim that drinking strong coffee before an endurance event improves their performance.

HSL is most active against DAGs, less active against TAGs and has little activity against MAGs. An additional TAG lipase has also been discovered in adipocytes, adipose triglyceride (triacylglycerol) lipase (see Box 9.1 for further details). There is a highly active MAG lipase expressed in adipocytes that contributes to the complete hydrolysis of stored TAGs, with release of three molecules of FA and one of glycerol, which leave the cell, as discussed further in Box 9.1. There is currently great interest in the question of whether the exit of FAs occurs by diffusion across the cell membrane or via a specific transport protein. Several putative fatty acid transport proteins (FATP) have now been identified,

although it is not clear whether this step might be open to regulation (see Sections 3.1.1 & 7.1.2 for further discussion of FA transport across membranes). The NEFAs (often referred to as free fatty acids, FFA) are bound to plasma albumin in the circulation (Fig. 4.8). They may then be taken up by tissues such as muscle that utilize FAs as a major source of fuel, and by the liver where they may be used for oxidation or for the biosynthesis of new acylglycerols.

It has become clear recently that both adipose triglyceride [triacylglycerol] lipase and HSL are also responsible for TAG hydrolysis within skeletal muscle. The LDs that are seen within muscle fibres on electron microscopy provide a local supply of FAs during intense exercise. The mechanisms for activation of muscle HSL may be similar to those within adipocytes, except that muscle contraction is also a stimulus to TAG hydrolysis. The mechanism by which contraction activates TAG mobilization is not yet known.

The coordination of TAG biosynthesis and breakdown in adipose tissue is illustrated in Fig. 4.8.

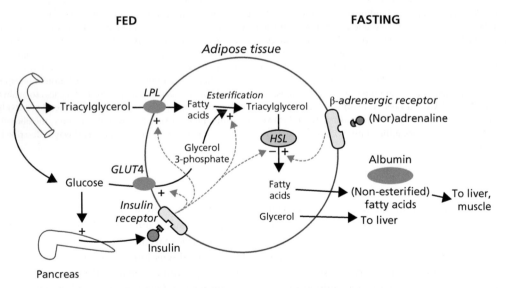

Fig. 4.8 The coordination of triacylglycerol synthesis and breakdown in adipose tissue. In the fed state, dietary fatty acids are delivered as chylomicron-triacylglycerol, which is hydrolysed by lipoprotein lipase (LPL) in adipose tissue capillaries to liberate fatty acids, that may be taken up into adipocytes and esterified for storage as triacylglycerol. Glycerol 3-phosphate for this process is provided by glycolysis, following glucose uptake by the insulin-regulated glucose transporter GLUT4. High circulating glucose concentrations in the fed state stimulate pancreatic secretion of insulin, which stimulates all aspects of the pathway of triacylglycerol storage via the cell-surface β-adrenergic receptors. This process delivers nonesterified fatty acids to the circulation, for use as a fuel by other tissues including liver and skeletal muscle. In addition, the liver secretes triacylglycerol in the form of very low-density lipoprotein (Section 7.2.3), and this is also a substrate for adipose tissue LPL, thus creating a metabolic cycle between liver and adipose tissue.

4.3.4 Regulation of triacylglycerol biosynthesis in oil seeds

TAG deposition in oil seeds is very important from an agricultural and economic perspective. These seed oils are important for food, animal feeds and as renewable sources of industrial chemicals. Recently, there has been considerable interest in using plant oils for bio-diesel (see Section 11.2.3) but this raises the serious consequence of 'food versus fuel' and, thus, should be treated cautiously.

Because of its importance, the control of biosynthesis of plant oils has become a 'hot topic' for research over the past decade. There have been numerous attempts to increase oil yields by manipulating various enzymes in the overall biosynthetic pathway, often using the 'model' plant Arabidopsis for the research. In many cases, the genes chosen for up-regulation have been based on rather flimsy evidence that they are appropriate choices. It is only recently that analyses of metabolism based on metabolic flux or metabolic control analysis have been employed. These methods give an overall picture of metabolic pathways and thus highlight areas which may be useful for manipulation. The overall use of metabolic control in examining regulation of pathways is discussed in the excellent book by Fell (see **Further reading**).

Recently there has been a number of encouraging results in increasing oil yields through breeding or through genetic manipulation. In the latter case, it is clear (as in many cases with different organisms) that results obtained with one species cannot, necessarily, be applied directly to others. Because metabolic control analysis predicts that control over a pathway is shared by all the constituent enzymes then, first, there is no such thing as 'the rate controlling enzyme' and, second, that in manipulating gene expression it is usually necessary to increase expression of more than one gene. In that regard, there has been interest recently in a transcription factor, *Wrinkled*, that was first discovered in a mutant variant of Arabidopsis which had wrinkled seeds. This transcription factor changes expression of a number of genes, including increasing that for FA biosynthetic enzymes. Where supply of FAs is limiting for TAG formation, then *Wrinkled* can be of considerable benefit (see also Section 5.5.2).

A good example of the targeted benefit of knowing whether a particular enzyme exerts strong flux control comes from oilseed rape. Biochemical experiments in the 1990s indicated that DGAT activity could be important and this was confirmed by experiments in the related brassica Arabidopsis. Flux control analysis showed that DGAT exerted significant control and, when the enzyme's activity was increased in transgenic plants, the manipulated oilseed rape gave significantly raised oil yields.

As mentioned above, because regulation of a metabolic pathway needs several enzymes or substrate supplies to be changed, then current research uses changed expression of several genes (in a 'cassette'). Recent successes include the use of 'push/pull' strategies where increased supply of substrates at the beginning of a pathway is supplemented by more active final reactions (e.g. increased expression of *Wrinkled* and DGAT). Moreover, since accumulation of any lipid is a balance between synthesis and degradation (and it has been shown that lipase activity can reduce oil yields) then manipulation of both synthetic and degradative enzymes can be of benefit.

Details of plant TAG biosynthesis are shown in Fig. 4.3 and a comparison of the process in different organisms is given in Fig. 4.9.

4.4 Wax esters

These esters of long-chain FAs and long-chain fatty alcohols provide energy sources, as an alternative to TAG, in some animals, plants and microorganisms.

4.4.1 Occurrence and characteristics

Wax esters have the general formula R^1COOR^2 and are esters of long-chain FA (R^1COOH) with long-chain fatty alcohols (R^2OH). They occur in some species of bacteria, notably the mycobacteria and corynebacteria. However, they are mainly important for providing a form of energy reserve that is an alternative to TAG in certain seed oils (e.g. jojoba), the oil of the sperm whale, the flesh oils of several deep-sea fish (e.g. orange roughy) and in zooplankton. Thus, a major part of jojoba oil, 70% of sperm whale oil and as much as 95% of orange roughy oil consists of wax esters. In global terms, however, the zooplankton of the oceans are of greatest importance. At certain stages of their life cycles, the zooplankton may synthesize and store massive amounts of wax esters, as much as 70% of their dry weight. These then become a significant component of the marine food chain.

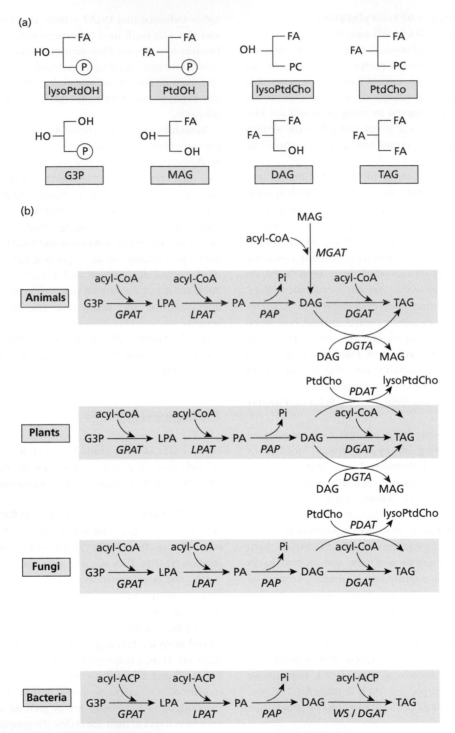

Fig. 4.9 A comparison of triacylglycerol synthesis in different organisms. The four enzymes of the Kennedy Pathway (Fig. 4.7) (GPAT, LPAT, PAP, DGAT) are used by all organisms except that, in those bacteria that produce TAG, the DGAT has dual functionality as a wax ester synthase (WS). Most, if not all, bacteria use acyl-ACP substrates. In the upper part (a) schematic depictions of key structures are shown. Part (b) compares pathways in different organisms. Abbreviations: MGAT, acyl-CoA: monoacylglycerol acyltransferase; DGTA, diacylglycerol transacylase; PDAT, phospholipid: diacylglycerol acyltransferase. Based on a diagram in Q Liu, RMP Siloto, R Lahner, SJ Stone & RJ Weselake (2012) Acyl-CoA: diacylglycerol acyltransferase: molecular biology, biochemistry and biotechnology, *Prog Lipid Res* **51**: 350–77 with permission of the authors and Elsevier.

The component FAs and alcohols of wax esters usually have an even number of between 10 and 30 carbon atoms. Normally, they are straight-chain saturated or monounsaturated chains of 16C–24C. Branched and odd-numbered chains are rare in both constituents except for the bacterial waxes. Wax esters from cold-water organisms exist as oils down to about 0 °C and their physicochemical properties differ from those of the TAGs in that they are more hydrophobic and less dense, which may help to provide greater buoyancy for marine animals.

4.4.2 Biosynthesis of wax esters involves the condensation of a long-chain fatty alcohol with fatty acyl-CoA

Zooplankton must synthesize their own wax esters *de novo* since their primary food, the phytoplankton, are devoid of these lipids. Indeed, all marine animals so far studied (mammals and fish as well as crustaceans), can synthesize wax esters *de novo*.

The fatty acyl components of wax esters are synthesized by the malonyl-CoA pathway (see Section 3.1.3), although they can also be derived from dietary lipids. Because wax esters are often characterized by very long chain acyl groups, further elongation of the products of the FAS may be necessary, as described in Section 3.1.5.

Waxes sometimes contain one or more keto groups that have escaped removal after the condensation stage of FA biosynthesis (see Section 3.1.3.2).

The fatty alcohol components are formed from acyl-CoAs by reduction, first by acyl-CoA reductase to form a fatty aldehyde, and then by an aldehyde reductase to form the alcohol (Fig. 4.10). The enzymes are membrane-bound and require NADPH for the reduction; the aldehyde is normally a transient intermediate and does not accumulate. The final esterification is catalysed by acyl-CoA: alcohol transacylase (Fig. 4.10).

4.4.3 Digestion and utilization of wax esters is poorly understood

The wax esters of fish and of jojoba seed oils are poorly hydrolysed by the pancreatic lipases of the human digestive system so that these lipids have poor nutritive value for humans. However, fish, such as salmon and herring grow rapidly when feeding on zooplankton rich in wax esters, yet do not contain these lipids themselves. Their digestive systems are adapted to the efficient hydrolysis of wax esters, most of the products being absorbed and resynthesized into TAG. The tissue breakdown of wax esters (so that the products can be utilized as metabolic fuels) presumably involves lipases or esterases analogous to the lipases in adipose tissue of

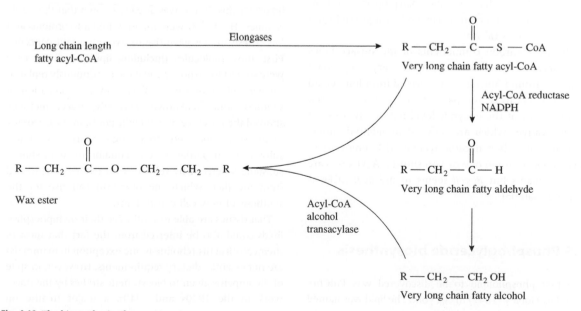

Fig. 4.10 The biosynthesis of wax esters.

organisms that store TAG, but these reactions have been little studied.

Some authorities now believe that wax esters have to be considered as key compounds in the transmission of carbon through the marine food chain. Wax esters seem to be found in greatest amounts in places where animals experience short periods of food plenty followed by long periods of food shortage, for example in polar regions, where the short summer limits the period of phytoplankton growth, or in deep waters with a low biomass. On a mass basis, somewhat more energy can be stored as wax esters than as TAGs because the oxygen content of wax esters is proportionately less.

4.4.4 Surface lipids include wax esters and a wide variety of other lipids

Although true waxes are esters of long-chain FAs and fatty alcohols, the term wax is often used to refer to the whole mixture of lipids found on the surfaces of leaves, or on the skin or the fur of animals. These surface lipids comprise a complex mixture of wax esters, long-chain hydrocarbons, nonesterified long-chain FAs, alcohols and sterols. They are responsible for the water-repellent character of the surface and are important in conserving the organism's water balance and providing a barrier against the environment. Some components, notably the long-chain NEFAs, have antimicrobial properties and this also contributes to their protective role in a physiological as well as a physical way.

The hydrocarbon components of the surface lipids (see Sections 3.1.5 & 5.6.1) have very long chain lengths (around 30C) and are formed from long-chain FAs. The pathway (or pathways) has not yet been defined fully at the enzymic level but involves use of FA derivatives which are reduced to aldehyde intermediates and then decarbonylated (CO removed) to form a hydrocarbon one carbon shorter. Acyl-CoAs are used in plants but *E. coli* seems to use acyl-ACPs as starting material.

4.5 Phosphoglyceride biosynthesis

The first phospholipid to be discovered was PtdCho, found by Gobley (1847) in egg yolk. The lipid was named

'lecithin' after the Greek lekithos (egg yolk). Interest in the chemistry of phospholipids increased significantly with the extensive investigations of Thudicum, who isolated and analysed lipids from many animal tissues (particularly the brain) and published his results in *A Treatise on the Chemical Constitution of the Brain* (1884). It began to be appreciated later that the difficulties involved in handling these substances and obtaining a pure product were enormous. Another factor tending to discourage research into phospholipids was the very prevalent but erroneous idea that they were metabolically inert and that, once laid down during the initial growth of the tissue, their turnover was very slow: they were assumed to be purely structural.

All phosphoglycerides are formed from a basic parent compound – PtdOH – which is also a key intermediate in TAG biosynthesis. PtdOH itself arises from acylation of G3P. There is considerable variation in the way in which the polar head groups are attached to the phosphate of PtdOH.

4.5.1 Tracer studies revolutionized concepts about phospholipids

The myth that phospholipids were slowly turning over structural molecules was exploded by the Danish chemist Hevesy who, in 1935, demonstrated that a radioactive isotope of phosphorus (^{32}P) could be rapidly incorporated as inorganic orthophosphate into tissue phospholipids. By this time it was already known that the stable isotopes ^{2}H and ^{15}N were incorporated into proteins and fats. These studies gave rise to two important concepts. First, most molecules (including lipids) in living cells were subject to turnover and were continuously replaced (or parts of them were replaced) by a combination of synthesis and breakdown. Secondly, tracer methods showed the presence of metabolic pools. In these experiments, such pools were circulating mixtures of chemical substances, in partial or total equilibrium with similar substances derived by release from tissues or absorbed from the diet, which the organism can use for the synthesis of new cell constituents.

That tissues are able to synthesize their own phospholipids could also be inferred from the fact that most of their constituents (choline is one exception in mammals) are not essential dietary requirements. However, in spite of the impetus given to biosynthetic studies by the tracer work in the 1930s and 1940s, a major treatise on phospholipids in the early 1950s (Wittcoff's *Phosphatides*) could give virtually no information about their biosynthesis.

4.5.2 Formation of the parent compound, phosphatidate, is demonstrated

The different parts of phosphoglycerides – FAs, phosphate and headgroup – are capable of turning over independently. Thus, to study phospholipid biosynthesis we must first learn the origin of each constituent and then how they are welded together. Eighteen years after Hevesy's demonstration of the rapid rate of phospholipid turnover, came the first real understanding as to how complete phospholipids are built up.

Two American biochemists, Kornberg and Pricer, found that a cell-free enzyme preparation from liver would 'activate' FAs by forming their coenzyme A thioesters. They then went on to demonstrate that these activated FAs could be used in acyl transfer reactions to esterify *sn*-G3P forming *sn*1,2-diacyl-G3P (PtdOH). We now know that there are two distinct acyl-CoA:glycerol phosphate-*O*-acyltransferases, specific for positions 1 and 2 (see Section 4.1.1). Furthermore, organisms containing the Type II FAS (see Section 3.1.3.2), which form acyl-ACP products, can use these as substrates for the acyltransferases. Such organisms are bacteria like *Escherichia coli*, cyanobacteria, algae and higher plants. In fact, plant cells produce PtdOH within their plastids using acyl-ACPs but also contain acyl-CoA:G3P acyltransferases on the ER for the biosynthesis of extra-chloroplast lipids such as PtdCho or TAGs.

PtdOH is the precursor molecule for all glycerophospholipids and was at first thought not to be a normal constituent of tissue lipids. Later studies have shown it to be widely distributed, but in small amounts. In fact, it is very important for the concentration of PtdOH to be carefully regulated because of its function as a signalling molecule (see Section 8.4).

Thus G3P is one of the building units for phosphoglyceride biosynthesis. It is mainly derived from the glycolytic pathway by reduction of DHAP, although other methods are used to various extents by different organisms or tissues (e.g. phosphorylation of glycerol by glycerol kinase). Likewise, PtdOH can also be produced by direct phosphorylation of DAG using DAG kinase. Because this kinase is important in controlling the two critical metabolic intermediates (DAG and

PtdOH), which also serve as signalling molecules, it is not surprising that there are multiple isoforms that have various regulatory domains. The kinases are thought to play important roles in cardiac function, neural plasticity and signalling, T-cell responses and immune reactions (Section 10.3).

4.5.3 A novel cofactor for phospholipid biosynthesis was found by accident

After the problem of PtdOH biosynthesis had been solved, interest then grew in the remainder of the pathways to more complex lipids. The first significant finding in connection with PtdCho biosynthesis was made when Kornberg and Pricer demonstrated that the molecule phosphorylcholine was incorporated intact into the lipid. This they did by incubating phosphorylcholine, labelled both with ^{32}P and ^{14}C in a known proportion, with an active liver preparation and finding that the ratio of ^{32}P to ^{14}C radioactivities remained the same in PtdCho. Choline may come via a number of pathways having their origins in protein metabolism. For instance, the amino acid serine may be decarboxylated to ethanolamine, which may then be methylated thrice to form choline. Phosphorylcholine arises by phosphorylation of choline with ATP by choline kinase. The story of how the base cytidine, familiar to nucleic acid chemists, was found to be involved in complex lipid formation is a good example of how some major advances in science are stumbled upon by accident, although the subsequent exploitation of this finding by the American biochemist, Kennedy, typifies careful scientific investigation at its best. Kennedy proved that cytidine triphosphate (CTP), which was present as a small contaminant in a sample of ATP, was the essential cofactor involved in the incorporation of phosphorylcholine into lipid and later isolated the active form of phosphorylcholine, namely cytidine diphosphocholine (CDP-choline). The adenine analogue had no reactivity.

4.5.4 The core reactions of glycerolipid biosynthesis are those of the Kennedy pathway

Readers will already be familiar with the basic Kennedy pathway in the biosynthesis of TAGs (see Section 4.1.1). A simplified version is shown in Fig. 4.11. Four reaction steps (involving three acyltransferases and PAP) are used to produce storage TAGs. Two of the intermediates, PtdOH and DAG are used for the formation of anionic

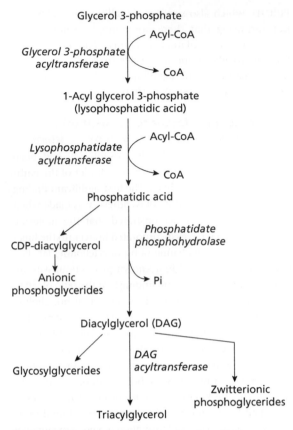

Fig. 4.11 The basic Kennedy pathway for glycerolipid biosynthesis in animals and plants.

phosphoglycerides (PtdGro, PtdIns and diphosphatidyl-glycerol, also called 'cardiolipin') and zwitterionic phosphoglycerides (PtdCho and phosphatidylethanolamine [PtdEtn]), respectively. DAG is also the precursor of the glycosylglycerides in plants (see Section 4.7).

For phosphoglycerides, as well as other glycerolipids, the first two acyltransferases (which esterify positions *sn*-1 and *sn*-2 of glycerol) have typical fatty acyl specificities. Thus, in animals and for the extra-chloroplastic (ER) acyltransferases in plants, SFAs (e.g. palmitate) are preferred for esterification at position 1, whereas UFAs (e.g. oleate) are attached to the 2-hydroxyl. In chloroplasts, by contrast, the acyltransferases that utilize acyl-ACPs typically place a saturated or unsaturated 18C acid at position 1, while palmitate is preferred at position 2. Therefore, in plants, the subcellular origin of the backbone of glycerolipids can be deduced from the FA distribution (see Sections 3.1.8.2 & 4.1.3).

GPAT catalyses the first, committed, step in glycerolipid assembly. In mammals, two isoforms (present in the outer membrane of mitochondria and the ER) have been reported and can be distinguished by their relative sensitivity to N-ethylmaleimide. In most tissues the mitochondrial activity is relatively minor (about 10% total) but, in liver, it is as active as the ER isoform. Furthermore, the liver mitochondrial enzyme is under both nutritional and hormonal control. This property allowed the first isolation of a cDNA for a mammalian enzyme involved in glycerophospholipid synthesis when the cDNA for mouse GPAT was isolated by Sul and co-workers using differential screening. The cDNA was shown to code for the mitochondrial isoform by transfecting cell cultures and measuring an increase in acyltransferase activity in mitochondria, but not in the ER. When the deduced amino acid sequence of the mouse GPAT was compared to other acyltransferases a conserved arginine residue was identified that might be useful for binding the negatively charged substrate G3P or acyl-CoA. The identification of a conserved arginine agreed with the finding that arginine-modifying agents such as phenylglyoxal inhibited the enzyme.

Because the activity of the ER GPAT does not vary with dietary or hormonal changes, it was thought to be primarily important for phosphoglyceride biosynthesis. However, in some circumstances it may play a vital role in TAG formation also. For example, in differentiating adipocytes or in neonatal liver its activity increases more than the mitochondrial isoform when the rate of TAG biosynthesis increases (see Section 4.1.1). Activity of the microsomal GPAT can be regulated by phosphorylation/dephosphorylation.

Transcription of the mitochondrial GPAT is decreased by starvation and glucagon but increased by a high carbohydrate diet. These conditions do not affect the ER isoform. The SREBP (see Section 7.3.1) appears important in the regulation of GPAT.

LPAT in most eukaryotic organisms has a high specificity for UFAs. Indeed, in plant seeds the high specificity of this enzyme has proved to be an obstacle for the engineering of 'designer' oils in transgenic plants (see Section 11.6.1) where the accumulation of unusual FAs at position 2 is severely restricted. Moreover, as noted above, the chloroplasts of plants contain a second isoform of the enzyme that uses acyl-ACP substrate and prefers palmitoyl-ACP rather than oleoyl-ACP as a substrate.

LPAT cDNAs have been cloned from yeast, plants and mammals. In humans, isoforms are present with different tissue distributions (see Section 4.1.1). Intriguingly, the human α-isoform is present at very high levels in testes where it has been suggested that it acts to generate PtdOH for signalling purposes (see Section 8.4). Two LPATs have been cloned from humans and there are another 4 putative sequences. Activity in the ER is much greater than in the mitochondria implying that substrate lysoPtdOH in the latter has to be transported to the ER for further acylation.

PtdOH can also be produced by DAG kinase and there are other proteins distinct from LPAT, such as endophilin, which have particular roles during vesicle fusion and recycling when the conversion of lysoPtdOH to PtdOH alters the curvature of the membrane bilayer (see Section 5.1.2).

PAP catalyses the key dephosphorylation reaction, which yields DAG (Fig. 4.11) and thus directs carbon away from acidic (anionic) phosphoglyceride biosynthesis. Two forms of the enzyme have been identified in animals. One is present in both the cytosol and ER and its activity is altered by translocation from cytosol to ER. The cytosolic form is inactive and its translocation to the ER is under the influence of FAs, fatty acyl-CoAs and PtdOH itself. This makes good sense because a build-up in its substrate (PtdOH) will activate the enzyme. Moreover, when the enzyme is used for TAG production (see Sections 4.1.1 & 4.5.2), the supply of acyl-CoAs (or FAs) will stimulate the Kennedy pathway itself.

A second isoform of animal PAP is present in the plasma membrane where it is thought to have a role in signal transduction. Multiple isoforms have also been detected in yeast and in plants, where they again have different subcellular locations.

4.5.5 The zwitterionic phosphoglycerides can be made using cytidine diphospho-bases

As can be seen in Fig. 4.11, DAG generated by PAP can have several fates, one of which is to be used for generation of the zwitterionic (charged molecules but with no overall charge at physiological pH) phosphoglycerides, namely PtdCho and PtdEtn. In animals and plants the CDP-base pathway is the main pathway for production of PtdCho, which represents 40–50% of the total lipids in most of their membranes.

Three enzymic steps are used (Fig. 4.12). First, the base (either choline or ethanolamine) is phosphorylated by a

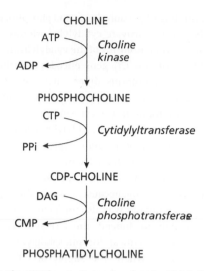

Fig. 4.12 The CDP-base pathway for phosphatidylcholine synthesis in eukaryotes. Equivalent enzymic reactions are used for the conversion of ethanolamine into phosphatidylethanolamine.

kinase that uses ATP. In most tissues, including those from animals, plants and the yeast *Saccharomyces cerevisiae*, separate choline and ethanolamine kinases are found. In fact, the genes for these enzymes in yeast have been mapped to separate chromosomes. However, some choline kinases do have limited activity with the alternative ethanolamine substrate. In mice, there are two separate genes that code for choline kinase α and β. Because the kinase functions as a dimer, three isomers are possible (αα, αβ, ββ). In most tissues, the αβ form predominates.

Cytidylyltransferase, the second enzyme in the pathway, links the phosphorylated base to cytidine monophosphate (CMP) to form CDP-choline or CDP-ethanolamine. In those tissues where PtdCho biosynthesis has been studied in detail, the cytidylyltransferase activity seems to be important in regulating the overall rate of phospholipid formation. This includes the initiation of lung surfactant biosynthesis (see Section 10.3.7.3), indole 3-acetic acid (an auxin or plant hormone) induced plant growth and the formation of Golgi-derived secretory vesicles in yeast. In animals, the cytidylyltransferase contains lipid binding and phosphorylation domains. Near the N-terminus is a nuclear localization signal. Although the enzyme tends to be mainly soluble (cytoplasmic) it will associate with membranes, including the nuclear envelope, when it is activated. Binding of phospholipids and phosphorylation at a number of sites activate cytidylyltransferase.

Two genes encode isoforms of the cytidylyltransferase, α and β. In mouse, the CT_β gene encodes a further two isoforms. The CT_β isoforms differ from the CT_α ones which contain a nuclear localization signal. Thus, the CT_β isoforms are found in the cytosol.

In plants cholinephosphate cytidylyltransferase activity is increased long term by gene expression, as well as short term by activation. Of the factors identified as being important, CTP and AMP concentrations were significant. This was interesting because, in yeast, the supply of CTP (by CTP synthetase) was also found to stimulate the formation of PtdCho. Indeed, the CTP synthetase has been shown to be critical for phosphoglyceride biosynthesis in yeast, where the supply of CTP can be limiting. The enzyme is allosterically regulated by CTP product inhibition and its activity is controlled by phosphorylation. The human equivalent CTP synthetase is also controlled by phosphorylation (which *inhibits*, as distinct from yeast where it *stimulates*).

The CDP-choline or CDP-ethanolamine produced by the cytidylyltransferases are rapidly utilized by phosphotransferase enzymes, which release CMP and transfer the phosphorylated base to DAG. The phosphotransferases are integral membrane proteins and the ethanolaminephosphotransferase has never been purified from any source. However, from selectivity experiments with different DAGs or by examining the molecular species of PtdEtn or PtdCho formed, two separate phosphotransferases seem to be present in animals and plants. In yeast, the genes for these two individual phosphotransferases have been isolated. The cholinephosphotransferase can only use CDP-choline whereas the ethanolaminephosphotransferase can use both CDP-base substrates.

As well as being involved in the production of PtdCho, CDP-choline can be used for the formation of the sphingosine-containing phospholipid, sphingomyelin (see Section 2.3.3). However, the major pathway for the formation of sphingomyelin transfers phosphorylcholine from PtdCho to a ceramide (see Section 4.8.6).

4.5.6 CDP-diacylglycerol is an important intermediate for phosphoglyceride formation in all organisms

In Fig. 4.11 it was seen that PtdOH can be converted into CDP-diacylglycerol (CDP-DAG) using another cytidylyltransferase (CDP-diacylglycerol synthase, CDP-DAG

synthase). The CDP-DAG is an important intermediate for the acidic phosphoglycerides (PtdIns, PtdGro) and diphosphatidylglycerol (cardiolipin) in animals and plants. In yeast it also has a role in phosphatidylserine (PtdSer), PtdEtn and PtdCho production, whereas in *E. coli*, all phosphoglycerides are formed via CDP-DAG. It should be noted that just as CDP-choline provides phosphorylcholine in the final reaction of PtdCho formation (see Section 4.5.5 & Fig. 4.12), CDP-DAG provides PtdOH, leaving CMP as the other product (Fig. 4.13).

We will begin by considering the situation in animals and plants, for which the main reactions are shown in Fig. 4.13. The CDP-DAG intermediate can react with either *myo*-inositol to give rise to the various inositol-containing phospholipids or with G3P. PtdIns is the main inositol-containing phospholipid in all eukaryotes. It can be further phosphorylated to yield the various polyphosphoinositides. The complex interactions of PtdIns kinases and the relevant phosphatases are shown in Fig. 4.14. These allow the rapid interconversions necessary during regulation of the signalling functions of the phosphorylated derivatives of PtdIns (see Section 8.1).

The reaction of G3P with CDP-DAG forms an intermediate that does not accumulate because it rapidly loses a phosphate to yield PtdGro (Fig. 4.13). In animals, almost all of the PtdGro is converted into diphosphatidylglycerol, which accumulates as an important constituent of the inner mitochondrial membrane. An exception is the epithelial Type II cell of the lung that produces pulmonary surfactant (see Section 10.3.7.3) containing significant amounts of PtdGro. A significant constituent of alveolar macrophages is another polyglycerolphospholipid – *bis*monoacyl (glycerol) phosphate.

The formation of PtdGro in plants and algae is most active in the chloroplast where PtdGro is the only significant phosphoglyceride (Table 2.10). A recent study using Arabidopsis has shown that a double deletion mutant (lacking the two PtdGro synthases), which could not therefore produce PtdGro, could not synthesize thylakoid membranes in chloroplasts, indicating a vital role for PtdGro there (see Table 2.10).

In yeast, the CDP-DAG pathway is not only important for acidic phosphoglyceride formation but is also the main way in which the zwitterionic phospholipids are made. A key step in the latter connection is the reaction of CDP-DAG with serine to produce PtdSer. The overall scheme for phosphoglyceride synthesis in yeasts is shown in Fig. 4.15.

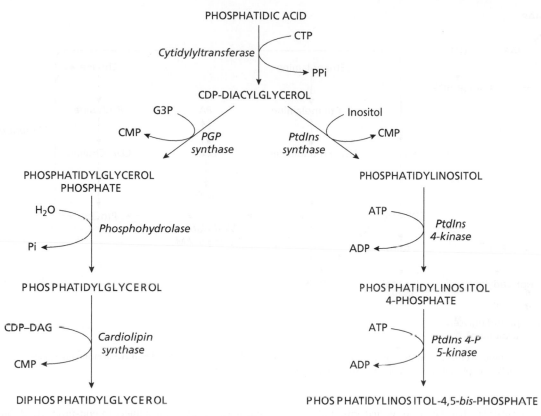

Fig. 4.13 Pathways for the synthesis of acidic phosphoglycerides in animals and plants. Abbreviations: G3P, glycerol 3-phosphate; PGP, phosphatidylglycerol phosphate.

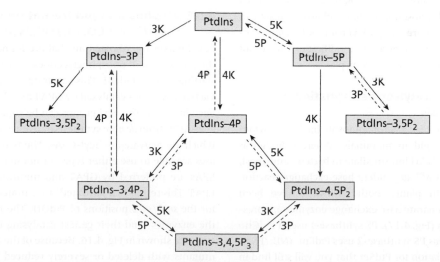

Fig. 4.14 Interconversion reactions for the inositol phosphoglycerides. Kinase reactions are indicated by a K with the pre-fix numeral indicating the position of phosphorylation on the inositol ring. Phosphatases are indicated by a P. Please note also Fig. 8.3 where similar reactions are detailed but from the perspective of lipids as signalling molecules. Based on Y Liu & VA Bankaitis (2010) Phosphoinositide phosphatases in cell biology and disease. *Prog Lipid Res* **49**:201–17.

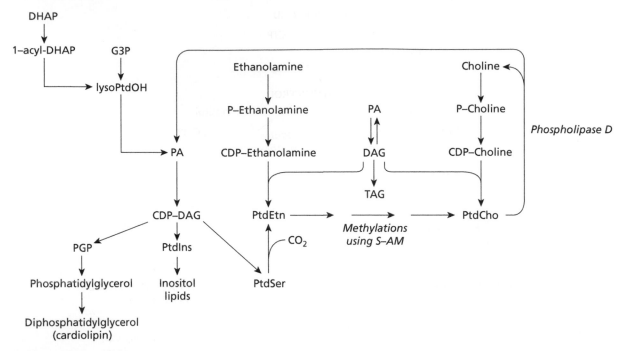

Fig. 4.15 Phosphoglyceride synthesis in yeast (*Saccharomyces cerevisiae*). Abbreviations: DHAP, dihydroxyacetone phosphate; PGP, phosphatidylglycerol phosphate; G3P, glycerol 3-phosphate; S-AM, S-adenosylmethionine.

After CDP-DAG has been converted into PtdSer, a decarboxylation yields PtdEtn, which can then be methylated three times to produce PtdCho. These reactions are described later in more detail and are typical of bacterial systems. Indeed, phospholipid biosynthesis in yeast can be regarded as a more complex version of *E. coli* metabolism (Fig. 4.16), but where the modifications have not gone quite as far as in animals or plants.

4.5.7 Phosphatidylserine formation in mammals

PtdSer is a minor (5–15% total phospholipids) but widespread phospholipid in mammals. Whereas PtdSer is made via a CDP-DAG intermediate in bacteria and yeast, it is made via a Ca^{++}-dependent base exchange reaction in mammals. In plants, both pathways have been reported. Two separate base exchange enzymes are present in mammals (Fig. 4.17). PS synthase-1 uses a PtdCho substrate whereas PS synthase-2 uses PtdEtn. (NB: 'PS' is an older abbreviation for PtdSer that you will still find in some publications.) PS synthase-1 is widely expressed whereas PS synthase-2 is most highly expressed in testes where its deficiency can cause sterility. The activity of PS

synthases seems to be subject to end-product inhibition by PtdSer itself and there is some cross-talk with PtdEtn biosynthesis.

4.5.8 All phospholipid formation in *E. coli* is via CDP-diacylglycerol

E. coli makes use of reactions that are found in yeast for the production of phosphoglycerides via CDP-DAG. The starting point is PtdOH. This substrate is made from G3P via two acylations catalysed by GPAT and then LPAT (see Section 4.1.1). The acyl groups can be supplied either as acyl-ACPs from *de novo* synthesis or from exogenous FAs, which are activated to acyl-CoAs. The two acyltransferases are able to use either type of thioester substrate but SFAs are preferred by GPAT and unsaturated ones by LPAT. This results in a saturated/unsaturated enrichment for the *sn*-1/*sn*-2 positions of PtdOH. The reactions and the enzymes (and their genes) catalysing the following steps are shown in Fig. 4.16. Because of the availability of mutants with deleted or severely reduced gene expression, we have been able to learn much about the function of specific phosphoglycerides recently from experiments with *E. coli*. This bacterium has one of the simplest

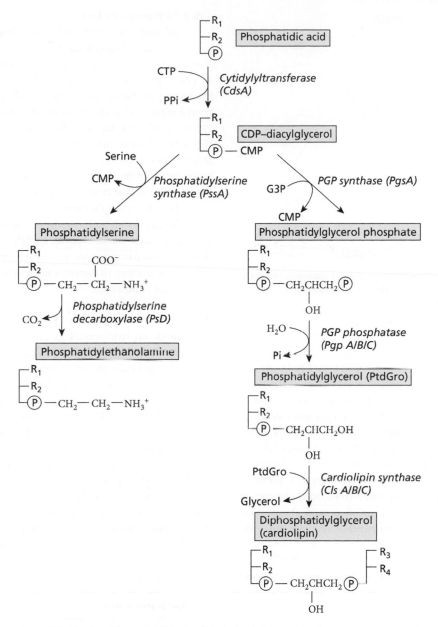

Fig. 4.16 Major reactions of phoshoglyceride synthesis in *E. coli* (gene nomenclature for the enzymes concerned are shown in parenthesis).

phospholipid compositions (approximately 75% PtdEtn, 15% PtdGro, 10% diphosphatidylglycerol) and so is relatively straightforward to study. Of note is that two molecules of PtdGro are used to make diphosphatidylglycerol contrast to the pathway in mammals (Fig. 4.13).

Also it does not methylate PtdEtn and cannot produce PtdCho.

PtdCho is found in about 10% of bacteria and is usually present in small amounts. However, there is one exception, where it comprises about three-quarters of the total

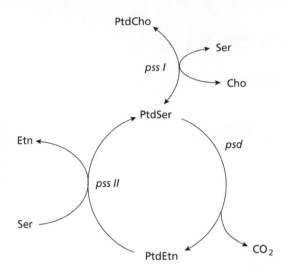

Fig. 4.17 Phosphatidylserine formation in mammals. *pss1* codes for PtdSer synthase 1; *pss2* codes for PtdSer synthase 2; *psd* codes for PtdSer decarboxylase.

membrane lipid in *Acetobacter aceti*. It used to be thought that PtdCho was made only by methylation of PtdEtn but many of the PtdCho-containing bacteria have a PtdCho synthase where choline can accept a phosphatidyl group from CDP-DAG. Since many of such bacteria are pathogenic, such a reaction may be able to make use of choline from the host.

For the methylation pathway, successive methylations of PtdEtn occur in three steps using the methyl donor, S-adenosylmethionine. In general, the bacterial phospholipid *N*-methyltransferase catalyses all three methylations.

4.5.9 Differences between phosphoglyceride biosynthesis in different organisms

A major difference between phosphoglyceride formation in yeasts and bacteria compared to animals is in the production of PtdSer. In the simpler organisms, PtdSer plays a major role as an intermediate in the production of PtdEtn (and PtdCho) and is made from CDP-DAG (Fig. 4.16). However, in animals it is made via an exchange reaction (see Section 4.5.7). Further fine-tuning of the proportions of different phosphoglycerides can be made by the presence of PtdSer decarboxylase (Fig. 4.17).

For the other phospholipids, the importance of different pathways in different organisms is listed in Table 4.1. With the exception of diphosphatidylglycerol production, the same pathways are used by different organisms, but it is the number of enzymes present or their relative activity that is different.

4.5.10 Plasmalogen biosynthesis

Ether lipids are widespread in animal tissues and, of these, the plasmalogens are the most abundant. In mammals, ether links are predominately found in the choline and ethanolamine glycerolipid classes.

Although the pathway for the formation of ethanolamine and choline plasmalogens uses CDP intermediates, there are differences from the biosynthesis of diacylphospholipids. First, the ethanolamine and choline phosphotransferases use a so-called 'plasmalogenic DAG' (1-alkenyl-2-acyl-*sn*-glycerol – not strictly a DAG but a term that has been widely used by researchers). Hajra and his group showed that the starting point for its biosynthesis is DHAP (see also Section 4.1.2). This compound is first acylated at positions *sn*-1 and -2. The acyl

Table 4.1 Major differences in phosphoglyceride synthesis in different organisms.

	Main pathway used		
Lipid	Animals	Yeast	*E. coli*
Phosphatidylcholine	CDP-choline	CDP-DAG	Not present
Phosphatidylethanolamine	CDP-Etn	CDP-DAG	CDP-DAG
Phosphatidylinositol	CDP-DAG	CDP-DAG	Not present
Phosphatidylglycerol	CDP-DAG	CDP-DAG	CDP-DAG
Cardiolipin	CDP-DAG[a]	CDP-DAG[a]	CDP-DAG[a]
Phosphatidylserine	Exchange	CDP-DAG	CDP-DAG

[a] Cardiolipin synthase uses CDP-DAG substrate in animals and yeast but phosphatidylglycerol in *E. coli*. CDP-Etn: CDP-ethanolamine.

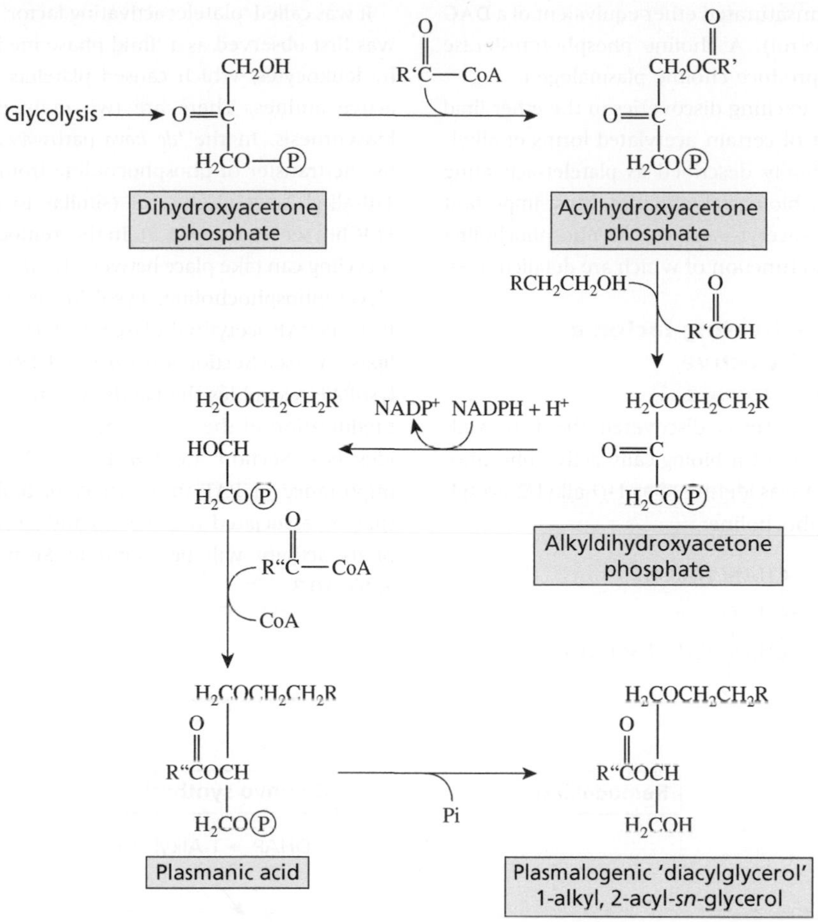

Fig. 4.18 Generation of substrate for plasmalogen synthesis.

group at position 1 is then substituted by a long-chain alcohol and, finally, the keto group at position 2 is reduced. The newly created hydroxyl at position 2 is then acylated and the phosphate group removed to produce a 'plasmalogenic DAG' (Fig. 4.18).

It is interesting to note that the acyl-DHAP at the start of the pathway can also be reduced to form lysoPtdOH, thus providing another potential source of PtdOH and a link between acyl- and alkyl-glycerolipid formation.

Further work on the pathway, which particularly involved the laboratories of Snyder in the USA and Paltauf in Austria, demonstrated that the 1-alkyl, 2-acyl-*sn*-glycerol was used by phosphotransferase enzymes to give saturated ether products.

The desaturation of positions 1 and 2 of the alkyl chain to form an alkenyl chain in plasmalogens is catalysed by cell-free extracts of intestinal epithelial cells, tumour cells

and brain. The desaturase is present in the microsomal fraction but its reaction is stimulated by a high molecular weight, heat-labile factor in the soluble cytosol. The fact that a reduced pyridine nucleotide and molecular oxygen are absolute requirements and that the reaction is inhibited by cyanide but not by CO suggests that this enzyme is very similar to the FA desaturases described in Section 3.1.8. This provides another interesting example of an enzyme catalysing a modification of a hydrocarbon chain in an intact lipid molecule.

For choline plasmalogen the situation appears even more complicated because the desaturation of the alkyl chain to form the unsaturated alkenyl chain does not take place on a choline alkyl phospholipid. Instead, the ethanolamine plasmalogen appears to be made first and then the head group exchanged for choline or the ethanolamine plasmalogen is converted via a series of reactions into the unsaturated ether equivalent of a DAG (alk-1-enylacylglycerol). A choline phosphotransferase enzyme can then produce choline plasmalogen.

One of the most exciting discoveries in the ether lipid field has been that of certain acetylated forms of alkylglycerolipids (originally described as platelet-activating factor) with potent biological activities. Most important of these is 1-alkyl-2-acetyl-*sn*-glycerol 3-phosphocholine the metabolism and function of which are detailed next.

4.5.11 Platelet activating factor: a biologically active phosphoglyceride

Hanahan's group in Texas discovered the first well-documented example of a biologically active phosphoglyceride in 1980. It was identified as 1-*O*-alkyl-2-acetyl-*sn*-glycero-3-phosphocholine:

It was called 'platelet activating factor' (PAF) because it was first observed as a 'fluid phase mediator', produced by leukocytes, which caused platelets to release vaso-active amines. There are two main pathways for its biosynthesis. In the '*de novo* pathway', PAF is formed by the transfer of phosphocholine from CDP-choline to 1-0-alkyl-2-acetyl-glycerol (similar to the synthesis of PtdCho: see Section 4.5.5). In the 'remodelling pathway', recycling can take place between lysoPAF and alkyl-acyl-glycerophosphocholine. LysoPAF itself is formed from PAF via PAF acetylhydrolase which is a type of phospholipase A_2 (see Section 4.6.2 & Fig. 4.19). The recycling of lysoPAF resembles the Lands-type reactions involved in modification of the acyl composition of phosphoglycerides (see Section 4.6.2 & Fig. 4.23). Because of the importance of PAF in a variety of biological processes that are associated with health and disease, more details of its activity will be found in Section 10.3.7.2 and Table 10.3.

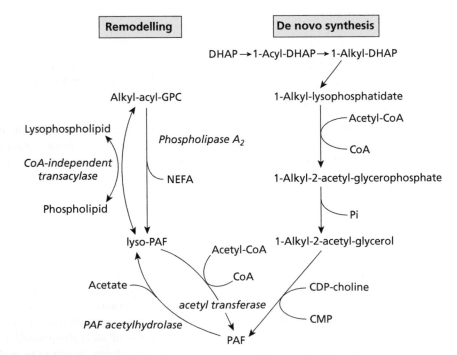

Fig. 4.19 Metabolism of platelet activating factor (PAF). Abbreviations: DHAP, dihydroxyacetonephosphate; GPC, glycerophosphorylcholine. The conversion of DHAP to 1-alkyl-lysophosphate (in the *de novo* synthetic pathway) is shown in Fig. 4.18. The acetylhydrolase is a particular example of phospholipase A_2 enzymes (Section 4.6.2 and Table 4.2).

4.6 Degradation of phospholipids

A variety of hydrolytic enzymes, the phospholipases (PLs), exists to remove selectively the different constituents of the phospholipid molecule: the acyl groups at positions *sn*-1 and *sn*-2 of phosphoglycerides, the phosphorylated headgroup or the headgroup alone.

PLs are classified according to the positions of their attack on the substrate molecule as illustrated in Fig. 4.20. PLs of type A (PLA) yield a monoacyl (lyso) phospholipid while PLsC and D yield a lipid (DAG and PtdOH, respectively) and a water-soluble product. PLD can use a hydroxyl group in an organic molecule instead of water as acceptor and, thus, catalyse *transphosphatidylation* rather than *hydrolysis*. (This activity can result in the production of the artefact phosphatidylmethanol during methanol extraction of tissue extracts if the enzyme is not inactivated!) PLs of type B differ from PLsA$_1$ or A$_2$ in that they can hydrolyse acyl groups at both positions, but they are relatively rare. There are also lysophospholipases and acyl hydrolases (which can act against acyl lipids in general) in some tissues.

All true PLs share the same general property of having relatively low activity against monomeric soluble phospholipids but become fully active against aggregated structures, such as phospholipid solutions above their critical micellar concentration (CMC) or membrane phospholipids in bilayer or hexagonal phase structures (see also Section 5.1). They play an obvious role as digestive enzymes whether it be in the digestive secretions of mammals (see Section 7.1.1) or in bacterial secretions. In addition, some enzymes, such as PLA$_2$, play important roles in remodelling the acyl composition of membrane lipids. Several types of PL are also needed for the production of lipid-derived signalling molecules (see Section 3.5 & Chapter 8) and which are associated with cellular regulation.

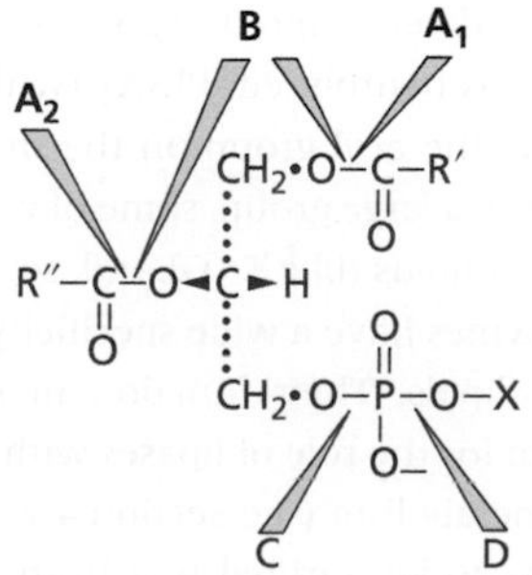

4.6.1 General features of phospholipase reactions

As mentioned above, PLs are distinguished from general esterases by the fact that they interact with interfaces in order to function. The difference in reaction velocity with substrate concentration for these two types of enzymes is illustrated in Fig. 4.21. Whereas esterases show classic Michaelis-Menten kinetics, the PLs show a sudden increase in activity as the substrate (phospholipid) concentration reaches the CMC (Fig. 4.21) and the molecules tend to form aggregates or micelles with polar ends in the aqueous environment (Fig. 4.22). Because of the physical nature of the substrate, enzyme activity is dependent both on the (hydrophobic) interaction with the aggregate and on the formation of a catalytic Michaelis complex. Several factors have been suggested as being responsible for the increased rate of hydrolysis at interfaces:

1 the very high local substrate concentration;
2 substrate orientation and physical state at the interface;
3 increased rates of diffusion of products away from the enzyme;
4 increased enzyme activity due to conformational changes on binding to the interface.

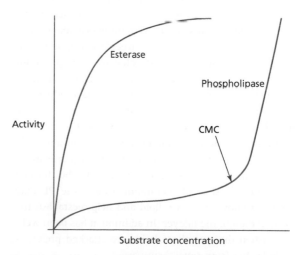

Fig. 4.21 Comparison of the enzyme kinetics of esterases and phospholipases. Abbreviation: CMC, critical micellar concentration.

Fig. 4.20 The sites of action of phopholipases.

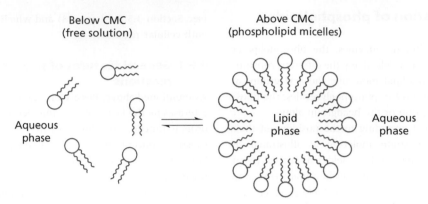

Fig. 4.22 Aggregation of a phospholipid at its critical micellar concentration (CMC).

For a typical phospholipid in solution, formation of a micelle increases its effective local concentration by at least three orders of magnitude – depending on the CMC (factor 1 above). The British biochemist Dawson and also Dutch workers including van Deenen, de Haas and Slotboom have studied how the nature of aggregated lipid (factor 2) influences markedly the activity of PLs. Such activity seems to depend on four parameters – surface charge, the molecular packing within the aggregate, the polymorphism of the aggregate and the 'fluidity' of the phospholipid's acyl chains (see also Section 5.1). The surface charge can be quite different from the bulk pH and is influenced by ionic amphipaths as well as ions in the aqueous environment. This explains why, for example, the PLB from *Penicillium notatum* will not attack pure PtdCho but is active in the presence of 'activators', such as PtdIns, which give PtdCho micelles a net negative charge. Molecular packing is particularly important for the acylhydrolases, but is also relevant to the phosphodiesterases (PLC). Changes in activity due to this parameter can be caused, for example, during diethylether activation (below). Polymorphic states can include micelles, bilayer structures and hexagonal arrays (see Section 5.1). There is some evidence that certain phospholipases attack preferentially hexagonal phase structures and since these are believed to be formed transiently at sites of membrane fusion, the PL may help to remove 'fusative' lipid and help reestablish the normal membrane bilayer. In addition, it has been well-established that gel-phase lipids are attacked preferentially by PLs from several sources.

Factor 3 (above) is particularly important for PLsA and B, where both products from a typical membrane

(natural) phospholipid are hydrophobic. In the case of PLsC and D, where one product is water-soluble, action at an interface is still useful because the water-soluble product can diffuse into the aqueous environment while the other product can diffuse into the hydrophobic phase. Similar physicochemical arguments explain why many PLs show increased activity in the presence of some organic solvents. Thus, when diethylether is used in assays to stimulate PLA activity, it is thought that accumulating FAs are more readily removed. In addition, the solvent may allow more ready access of the enzyme to the hydrocarbon chains of the phospholipid and, by reducing micellar size, increases the effective surface area for reaction. Finally, conformational changes (factor 4 above) have been shown to take place when some digestive PLs interact with substrate and/or Ca^{2+}. Kinetic studies indicate that such a conformation change is needed for maximal activity.

4.6.2 Phospholipase A activity is used to remove a single fatty acid from intact phosphoglycerides

PLA enzymes are divided into two groups, depending on which acyl bond is hydrolysed. PLsA$_1$ (which hydrolyse the bond linking the acyl group on the *sn*-1 position of glycerol) comprise a large group, some of which may also degrade nonpolar lipids (like TAG) and, so, act as lipases. Usually the enzymes have a wide specificity and act well on lysophospholipids. Their function in most cases is obscure – except for the role of lipases with PLA$_1$ activity in lipoprotein metabolism (see Section 4.2.2 and below).

The first PLA$_1$ to be purified was from *E. coli*, which actually has two separate enzymes – a detergent-resistant enzyme in the outer membrane and a detergent-sensitive enzyme in the cytoplasmic membrane and soluble fractions.

In animals, PLA_1 is present in lysosomes and has an appropriately low pH optimum (about 4.0). It does not need Ca^{2+} for activity but Ca^{2+} and charged amphipaths influence hydrolysis by altering the surface charge on the substrate micelle or membrane.

Also present in animals are two enzymes that, while preferentially hydrolysing TAGs, will also hydrolyse the bond at position 1 of phospholipids. These are the extrahepatic LPL (see Section 4.2.2) and the hepatic lipase. The latter enzyme is activated by apolipoprotein E to hydrolyse phospholipids and by phospholipids to act as a lipase.

$PLsA_2$ are widespread in Nature and their activity was noted first by Bokay over a hundred years ago. He studied the degradation of PtdCho by pancreatic secretions and, at the turn of the 20th century, cobra venom PL activity was discovered. These two sources have proved useful for purifying and studying the enzyme.

The PLA_2 family is constantly being added to as more enzymes are discovered. To date more than 30 such enzymes have been discovered in mammals (Table 4.2). They are subdivided based on their structures, catalytic mechanisms, localizations and evolutionary relationships. The largest family comprises the secreted $PLsA_2$ ($sPLA_2$). These are low molecular mass (about 14 kDa), Ca^{2+}-requiring extracellular enzymes and 10 have been identified. The cytosolic family ($cPLsA_2$) characteristically have an N-terminal C2 (a Ca^{2+}-dependent phospholipid-binding) domain that allows association with membranes. Then there are Ca^{2+}-independent $PLsA_2$ ($iPLsA_2$; nine enzymes), which include patatin-like domain lipases. Some of the $iPLsA_2$ work mainly as phospholipases and some as lipases. There are four enzymes identified in the PAF acetylhydrolase (PAF-AH) family

Table 4.2 The phospholipase A_2 superfamily.

Type	Group	Subgroup	Mol. Mass (kDa)	Catalytic residues
Secreted*	GI	A, B	13–15	His/Asp
($sPLA_2$)	GII	A, B, C, D, E, F	13–17	
	GIII		15–18	
	GV		14	
	GIX		14	
	GX		14	
	GXI	A, B	12–13	
	GXII	A, B	19	
	GXIII		<10	
	GXIV		13–19	
Cytosolic ($cPLA_2$)	GIV	A–F	60–114	Ser/Asp
Ca^{**}-independent ($iPLA_2$)	GVI	A–F	84–90	Ser/Asp
PAF acetylhydrolase	GVII	A, D	40–45	Ser/His/Asp
(PAF-AH)	GVIII	A, B	26–40	
Lysosomal ($LPLA_2$)	GXV		45	Ser/His/Asp
Adipose-specific (AdPLA)	GXVI		18	His/Cys

* Group I, A = cobras, kraits, B = pancreas
Group II, A = rattlesnakes, vipers, B = Gaboon viper
Group III, bee, lizard
Groups V, X, XII, human
Group IX, snail venom
Group XI, green rice shoots
Group XIII, parvovirus
Group XIV symbiotic fungus, bacteria
Adapted from EA Dennis, J Cao, Y-H Hsu, V Magrioti & G Kokotos (2011) Phospholipase A2 enzymes: physical structure, biological function, disease implication, chemical inhibition, and therapeutic intervention. *Chem Rev* **111**:6130–85.

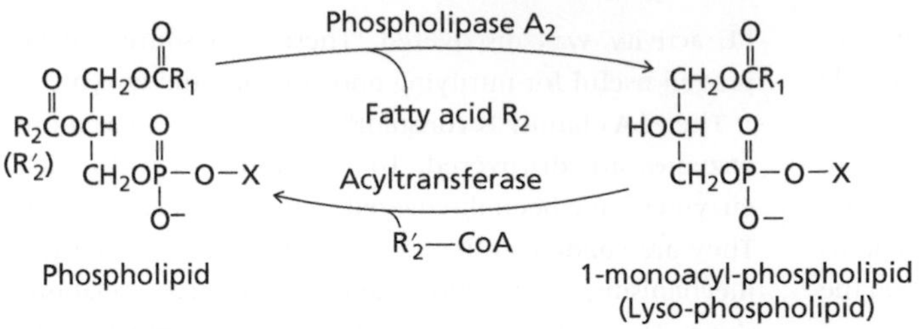

Fig. 4.23 The 'Lands Mechanism' for remodelling the acyl species at the *sn*-2 position of phosphoglycerides.

(see Section 10.3.7.2). In addition, there are two lysosomal PLsA$_2$ and an adipose-specific PLA$_2$.

Pancreatic PLs (sPLsA$_2$, Group IB) are synthesized as zymogens that are activated by the cleavage of a hepta-peptide by trypsin. Cleavage of this peptide exposes a hydrophobic sequence, which then allows interaction of the enzyme with phospholipid substrates. The enzyme is very stable and its seven disulphide bonds no doubt play a key role here. Chemical modification and nuclear magnetic resonance (NMR) studies have shown clearly that the catalytic and binding sites are distinct in both the pancreatic and snake venom PLsA$_2$. In confirmation, it is known that the zymogen form of the pancreatic enzyme is also active even though its binding site is masked. Ca^{2+} is absolutely required for activity and seems to interact with both the phosphate and carbamyl groups of the ester undergoing hydrolysis.

The sequences of pancreatic PLA$_2$ and the enzymes from the venoms of Elapids (cobras, kraits and mambas) are closely related, in contrast to the enzymes from vipers (e.g. *Crotalus* (rattlesnakes)), which are Group II enzymes (Table 4.2). Subtle modifications in the sequence of the Group I-II enzymes cause significant changes in their crystal structure and substrate interaction. For example, pancreatic enzymes (Group IB) crystallize as monomers, rattlesnake enzymes (Group IIA) as dimers and cobra PLsA$_2$ (Group IA) as trimers.

PLsA$_2$ also have other important metabolic functions in addition to the overall destruction of phospholipids as catalysed by digestive pancreatic or venom enzymes. A PLA$_2$ in mitochondrial membranes seems to be inti-mately connected with the energy state of this organelle. Thus, the enzyme is inactive in fully coupled mitochon-dria and only becomes active when ATP and respiratory control drop to low levels. Also, the widespread distribu-tion of PLsA$_2$ allows many tissues to perform retailoring of the molecular species of membrane lipids by the 'Lands mechanism'. In this process, named after Bill Lands, the

American biochemist who first described it, cleavage of the acyl group from the *sn*-2 position yields a lysophos-pholipid that can be reacylated with a different FA from the acyl-CoA pool (Fig. 4.23). This mechanism plays a major role in ensuring that individual membrane lipids have their own particular acyl composition as well as providing a mechanism of removing oxidized acyl groups and replacing them with new FAs.

Some PLsA$_2$ play key roles in signal transduction. The Group IV cPLsA$_2$ translocate to membranes when the intracellular Ca^{2+} concentration rises. They have a very high specificity for arachidonate at the *sn*-2 position of phospholipids and a distinct catalytic mechanism com-pared to the Group I-III enzymes. Ca^{2+} is not involved in catalysis itself but, by promoting enzyme-membrane interaction, increases activity several-fold. Within its catalytic domain a highly conserved sequence around the active-site serine (ser228 in cPLA$_2\alpha$) is identical with that of PLB from *Penicillium* (below) and another serine (ser505) can be phosphorylated by mitogen activated protein (MAP) kinase. Phosphorylation increases the activity significantly and this can account for typical agonist stimulation of PLA$_2$ activity. In addition, a num-ber of cellular agonists (such as the inflammatory cyto-kines interleukin-1 (IL-1) and tumour necrosis factor, TNF) enhance the cellular content of cPLA$_2$ by increased gene expression.

Several PLsA$_2$ are Ca^{2+}-independent. They have cata-lytic serine residues and may be important for the Lands mechanism (above). They also play a role in generating lysoPtdCho, an attractant signal for clearance of apo-ptotic cells by macrophages.

Finally, it is well-recognized that many PLsA have high activity against oxidized or oxygen-containing acyl groups. They can, therefore, play a key role in removing such moieties from oxidized lipids in order to preserve normal lipid function (see above remarks about the Lands mechanism).

4.6.3 Phospholipase B and lysophospholipases

Only one PLB has been highly purified – that from *Penicillium notatum*. This enzyme will, of course, also hydrolyse lysophospholipids and, indeed, under most conditions has much greater activity towards the latter. In the presence of detergents (Triton X-100 is usually used) it shows roughly equal activity towards diacyl- and monoacylphospholipids. The enzyme appears to have two distinct binding sites and the intermediate lysophospholipid must move from one to the other during hydrolysis. Subsequently, PLBs have been characterized from a number of fungi where they appear to be important for pathogenicity.

In contrast to PLB, lysoPLs are widely distributed and are found in microorganisms, bee venoms and mammalian tissues. Many of them have limited activity towards intact phospholipids. Two small lysoPLs (I and II) have been found in mammals. They are Ca^{2+}-independent and have the serine catalytic triad typical of proteases and lipases.

An obvious function of lysoPLs is to prevent a significant build-up of lytic monoacylphospholipids, which would clearly damage membrane function. In addition, the transacylation property of some lysoPLs suggests that they could play a significant role in molecular species remodelling by a process that would be independent of the Lands mechanism (Fig. 4.23).

Recently, a lysoPLD has been purified from serum. It was originally described as a tumour cell mobility-stimulating factor, and hence named autotaxin. It produces lysoPtdOH, an important signalling lipid (see Section 8.3), which functions in cell migration and proliferation.

4.6.4 Phospholipases C and D remove water-soluble moieties

In accordance with the observations showing the independent turnover of phosphate and headgroup moieties, other enzymes named PLsC and D are known to hydrolyse each of the two phosphate links (Fig. 4.20). PLsC have been traditionally associated with secretions from certain pathogenic bacteria. Thus, much of the damage caused by *Clostridium welchii* when giving rise to gas gangrene, is caused by the PLC in its toxins. The enzymes from *Bacillus cereus* are the best studied – three different proteins having been isolated, one having a broad specificity, one being a sphingomyelinase and one attacking PtdIns selectively.

In mammalian tissues, PLsC are found in the cytosol and in lysosomes. In the latter, the enzymes have a broad substrate selectivity, so do not need Ca^{2+} and, as expected, have an acidic pH optimum (about 4.5).

The cPLCs have a neutral pH optimum, require Ca^{2+} and can be divided into three groups. The best studied are those acting on inositol phospholipids, particularly PtdIns $(4,5)P_2$. These are exceptionally important in signal transduction (see Section 8.1.3). Another enzyme hydrolyses PtdCho while there are also specific sphingomyelinases, which are PLC enzymes that release ceramide from sphingomyelin (see Sections 4.8.7 & 8.3).

Several phosphoinositide PLCs have been studied in detail. The PLCδ binds to phosphoinositides in membranes through its pleckstrin homology (PH) domain. This is followed by binding of its C2 (also used for proteins to bind to membranes) and catalytic domains. Ca^{2+} binds to both of the latter.

A two-step catalytic mechanism has been proposed for both bacterial and mammalian PLCs. It involves a cyclic phosphodiester intermediate which, for the bacterial enzyme, can be a significant product.

In plants there are three classes of PLC; nonspecific PLC (mostly active with PtdCho); an enzyme clearing glycerylphosphatidylinositol (GPI)-anchored proteins (see Section 5.3.2) and PtdIns-specific PLC. As in mammals, the latter seem the most important and mainly hydrolyse PtdIns$(4,5)P_2$ (see Sections 8.1.1–8.1.3).

Traditionally, PLD (Fig. 4.20) has been thought of as a plant enzyme and it has been purified from several tissues including cabbage and carrots. In *Arabidopsis* there are five genes which produce enzymes with different characteristics. They are mostly active with PtdCho, PtdEtn and PtdGro but not with phosphoinositides.

In addition to catalysing hydrolysis of phospholipids, PLD will cause PtdOH exchange. This allows the formation of new phospholipids in the presence of an appropriate alcohol. For example:

$$\text{Phosphatidylcholine} + \text{glycerol} \rightleftharpoons \text{phosphatidylglycerol} + \text{choline}$$

Such transphosphatidylation reactions were first noticed because phosphatidylmethanol was formed when plant tissues were extracted with methanol. However, transphosphatidylation does *not* seem to be a physiological function for plant PLD because the PtdGro produced in the above reaction is a racemic mixture rather than having the two glycerols in the opposite

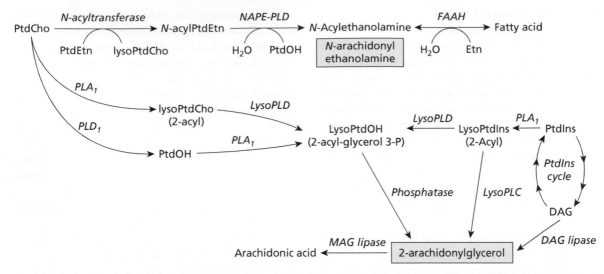

Fig. 4.24 Synthesis and degradation of *N*-acylethanolamines and other endocannabinoids. Abbreviations: FAAH, fatty acid amide hydrolase; NAPE–PLD, *N*-acylPtdEtn phospholipase D.

configuration, as occurs naturally. Nevertheless, the activity can be used in the laboratory for the preparation of phospholipids.

Plant PLD is also involved in the generation of the signal molecule *N*-acylethanolamine (see Section 4.6.6 & Fig. 4.24) from *N*-acyl-PtdEtn. The product plays a role in seed germination and plant stress (see Section 4.6.6).

Mammalian PLDs have been found in many tissues and it is clear that, like the plant enzymes, they play a role in lipid signalling reactions and the control of cellular activity (see Section 8.4). Interestingly, the product of PLD activity, PtdOH, binds to some of the plant PtdIns $(4,5)P_2$-hydrolysing PLC enzymes, causing their activation. This is an example of 'cross-talk' between signalling pathways where products or intermediates in one pathway affect the activity of a second pathway. There is also considerable recent interest in the separate activities of PLC (and other lipid enzymes) in the nucleus as opposed to the plasma membrane. Such dual distribution allows separate pools of signalling lipids to influence cellular regulation.

Bacterial PLDs are often toxins that lead to severe tissue damage of the infected tissues. For the bacteria, together with PLC, they help to provide essential nutrients like phosphate. The first crystal structure of a PLD is that from *Streptomyces sp*.

4.6.5 Phospholipids may also be catabolized by nonspecific enzymes

Not all phospholipid hydrolysis is catalysed by dedicated PLs. The action of two TAG lipases in this regard has already been mentioned. In addition, acyl hydrolases are present, particularly in plant leaves. These enzymes have activity towards a variety of lipids, including partial glycerides, glycosylglycerides and phosphoglycerides. The leaf acyl hydrolases have very high activities and they are rather resistant to denaturation. Failure to inactivate the enzymes leads to spoilage of vegetables during freezing.

4.6.6 Endocannabinoid metabolism

The endocannabinoid system in animals consists of neuromodulatory lipids based on arachidonate which are physiological ligands for cannabinoid receptors. Their metabolism and, hence, regulation is an interesting example of the use of multiple lipid sources as well as several different enzymes involved in their formation and inactivation. In animals the main endocannabinoids are anandamide (*N*-arachidonylethanolamide) and 2-arachidonylglycerol but other *N*-acylethanolamides are present, often in much higher amounts. The roles of endogenous cannabinoids and their receptors are described further in Chapter 7, see Table 7.5.

Several N-acylethanolamides are present in a variety of plant tissues (they were first discovered in the 1950s in soy lecithin and peanut meal). Their metabolism in plants is broadly similar to that in animals (see Fig. 4.24) and they have important actions in defence signalling, as inhibitors of PLDα activity and in mediating seedling development.

N-Acylethanolamines are formed using a Ca^{2+}-dependent N-acyltransferase and PLD action and PtdCho and PtdEtn substrates. Alternatively, the reverse reaction of FA amide hydrolase (FAAH) can be used, although this enzyme is usually employed to inactivate anandamide. The other main endocannabinoid is sn-2-arachidonyl-glycerol and this can be formed from phospholipids containing arachidonate at their sn-2 position either via lysoPtdOH or from PtdIns. In the latter case, there are again two routes. First, PtdIns can be utilized via lysoPtdIns or, second, it is converted into DAG (see Section 8.1.3) in the PtdIns cycle and DAG lipase forms MAG with arachidonate at the sn-2 position (Fig. 4.24).

4.7 Metabolism of glycosylglycerides

The addition of sugar molecules to DAGs to form glycosylglycerides involves the UDP-sugars familiar in other carbohydrate biosynthetic processes. By analogy with PLs, a battery of enzymes exists to remove selectively the acyl groups and the sugar moieties of glycosylglycerides.

4.7.1 Biosynthesis of galactosylglycerides takes place in chloroplast envelopes

Since the galactosylglycerides are confined (almost exclusively) to chloroplasts (see Section 2.3.2.3), it would seem natural to seek an active enzyme preparation from these organelles. First, it was shown that carefully prepared isolated chloroplasts could incorporate ^{14}C-galactose into the lipids. The nature of the substrates involved was not clear until Ongun and Mudd in California showed that an acetone powder of spinach chloroplasts would catalyse the incorporation of galactose from UDP-galactose into monogalactosyldiacylglycerol (MGDG) if the acetone-extracted lipids were added back. The acceptor proved to be DAG. For the biosynthesis of digalactosyldiacylglycerol (DGDG), a MGDG acceptor was needed. Further work on the enzymes involved was carried out by Joyard and Douce in Grenoble who showed that the

reactions were confined to the chloroplast envelope – hence the need for carefully prepared chloroplasts.

The reactions are therefore:

1, 2-diacylglycerol + UDP-galactose $\rightleftharpoons$
$$\text{monogalactosyldiacylglycerol} + \text{UDP}$$

Monogalactosylglycerol + UDP-galactose $\rightleftharpoons$
$$\text{digalactosyldiacylglycerol} + \text{UDP}$$

The UDP-galactose is produced in the cytosol. In contrast, the DAG is synthesized by a PAP localized in the envelope. The two galactosyltransferases have slight differences in their enzymic characteristics and, of course, result in the formation of β- and α-glycosidic bonds, respectively (Section 2.3.2.3).

More recently, in Wintermanns' laboratory in The Netherlands, it was noticed that isolated chloroplast envelopes were capable of forming DGDG from labelled MGDG by a reaction that did not require UDP-galactose. Further examination revealed that inter-lipid galactosyl-transfer was involved thus:

monogalactosyldiacylglycerol
$$+ \text{monogalactosyldiacylglycerol}$$
$$\rightleftharpoons \text{digalactosyldiacylglycerol} + \text{diacylglycerol}$$

This enzyme is referred to as a 'processive' galactose transferase because it is also capable of forming tri- or tetragalactosyldiacylglycerols. Because such galactosylglycerides are formed in significant amounts only in vitro, the physiological significance of galactolipid: galactolipid galactosyltransferase is unproven.

In *Arabidopsis* there are three genes coding for MGDG synthase. MGDG synthase 1 is highly expressed in all developmental stages and in all green tissues and is thought to form the bulk of the MGDG needed. MGDG synthases 2 and 3 have similar sequences, distinct from MGDG synthase 1 and tend to be expressed in particular tissues including nonphotosynthetic ones. Two DGDG synthase genes are found in *Arabidopsis* and account for all the DGDG made, as revealed in double deletion mutants. Whereas the MGDG synthases take the α-galactose in UDP-galactose and change it to a β-configuration in MGDG, the DGDG synthases preserve the α-configuration in DGDG.

One difference in the fatty acyl compositions of galactosylglycerides, which has been noticed repeatedly in the analysis of plant lipids, is the presence of hexadecatrienoate (16:3 n-3) in MGDG (but little in DGDG) from certain plants. The presence of 16:3 in some plants seems

Table 4.3 Fatty acid composition of galactosyldiacylglycerols from *Euglena gracilis* cells or spinach chloroplasts.

Lipid		Fatty acid (% total)						
E.gracilis		16:0	16:3	16:4	18:1	18:2	18:3	Other
	MGDG	6	–	32	9	6	41	6
	DGDG	17	–	7	19	12	26	19
Spinach								
	MGDG	trace	25	–	1	2	72	trace
	DGDG	3	5	–	2	2	87	1

MGDG, monogalactosyldiacylglycerol; DGDG, digalactosyldiacylglycerol.

to be related to the provision of palmitate at the *sn*-2 position of MGDG, where it acts as a substrate for FA desaturases (see Section 3.1.8.2 & Fig. 4.4). In such plants, the DAG for galactolipid biosynthesis is generated by PAP within the chloroplast. In contrast, the source of DAG for galactolipid biosynthesis in other plants comes from outside the chloroplast and it does not contain 16-carbon acids at the *sn*-2 position. A glance at Table 4.3 will also reveal that even in plants like spinach, which do contain hexadecatrienoate in their MGDG, little is present in the digalactosyl derivative. It is presumed that this is due to substrate selectivity of the DGDG synthases, as well as to the desaturation of palmitate to hexadecatrienoate on MGDG referred to above.

An interesting aspect of galactosylglyceride function has emerged recently when plants were grown under low phosphate conditions: DGDG biosynthesis is increased and this lipid replaces much of the phospholipid in extra-plastidial membranes.

4.7.2 Catabolism of galactosylglycerides

Enzymes are present in higher plant tissues that rapidly degrade glycosylglycerides. The initial attack is by an acyl hydrolase (see Section 4.6.5), which removes acyl groups from both positions and has very high activity in some tissues, such as runner bean leaves or potato tubers. Indeed, homogenization of some potato cultivars at a suitable pH in aqueous media results in the complete breakdown of all membrane lipids within a minute! The activity of acyl hydrolases (and other lipid degradative enzymes) in many vegetables even at low temperatures makes it necessary to blanch (boil) such products before storage in a deep-freeze.

Several acyl hydrolases, with slightly different specificities, have been purified from various plant tissues. The enzymes from runner bean (*Phaseolus vulgaris*) leaves are

remarkably stable to heating (only 10% activity is lost after 30 min at 70 °C) and to solvents and can be conveniently purified using hydrophobic chromatography. The patatin-like acyl hydrolases (see Section 4.6.2) that deacylate phospholipids are also active with galactolipids and, indeed, at least one patatin-like enzyme hydrolyses galactosylglycerides but not phosphoglycerides. Further breakdown of the galactosylglycerides occurs by the action of α- and β-galactosidases.

4.7.3 Metabolism of the plant sulpholipid

The plant sulpholipid, (sulphoquinovosyldiacylglycerol, SQDG), is an interesting molecule that was first studied in detail by Andy Benson. After helping to elucidate the 'Calvin cycle' (best called the Calvin-Benson-Bassham cycle) Benson moved from Berkeley to the Scripps Institute near San Diego in California, where he discovered all sorts of fascinating lipids in marine organisms. But because of the unusual structure of its sugar residue, the biosynthesis of sulpholipid remained an enigma for many years.

Eventually Harwood's group in Cardiff (UK) proposed a mechanism based on biochemical experiments. This was later confirmed by the isolation of the relevant genes by Benning's laboratory in Michigan. The UDP-sulphoquinovose synthase uses UDP-glucose and inorganic sulphite. A glucoseenide intermediate is involved (see Fig. 4.25), which permits reaction with sulphite. The enzyme is referred to as SQDG1. It appears to require some additional factors for full activity (see Shimojima (2011) in **Further reading**) and is present in the chloroplast stroma. Final production of the sulpholipid uses the SQDG synthase (SQDG2; Fig. 4.23) to transfer the sugar from UDP-sulphoquinovose to DAG. SQDG2 is localized to the chloroplast envelope.

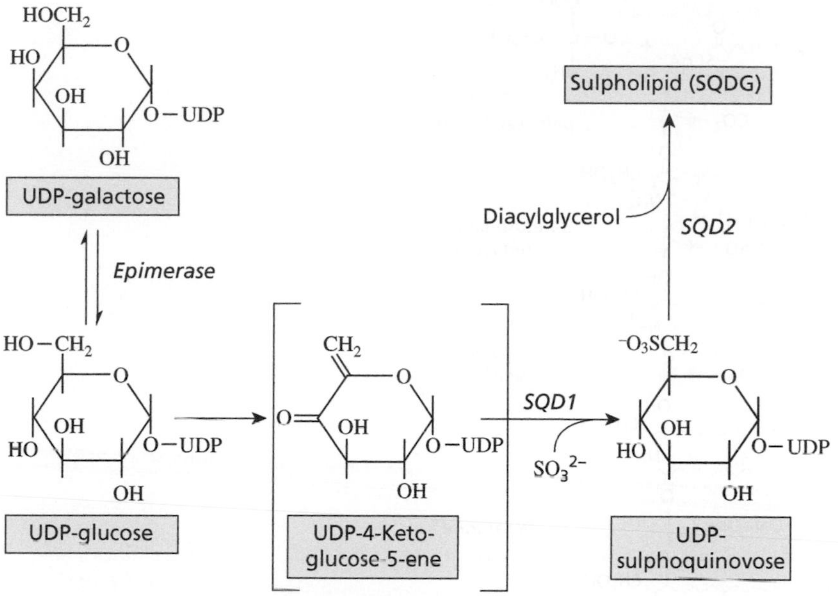

Fig. 4.25 Biosynthesis of the plant sulpholipid (SQDG) *SQD1* and *SQD2* are genes 1 and 2 involved in its synthesis. Brackets indicate the proposed intermediate that accepts sulphite to form sulphoquinovose.

Breakdown of SQDG is also intriguing because of the stable nature of the sulphonic acid residue on sulphoquinovose. As with galactosylglycerides, active acyl hydrolases release the fatty acyl residues. An α-glycosidase then hydrolyses the sulphoquinovose. Benson suggested that the latter could be catabolized through a 'pseudo-glycolytic' pathway with sulphonate derivatives substituting for phosphate intermediates. This is carried out by soil bacteria and the first few reactions have now been demonstrated by Harwood's group.

More SQDG is produced during phosphate deficiency, where it substitutes for the other negatively charged lipid in chloroplasts, PtdGro. Increases in SQDG during phosphate limitation are also found in photosynthetic bacteria as well as other species of bacteria that synthesize it. Such an adaptation is taken to extremes in the picocyanobacterium, *Prochlorococcus*, which uses less than 1% of the total phosphate taken up for phospholipid biosynthesis – 94–99% of its total lipids being MGDG, DGDG and SQDG.

4.8 Metabolism of sphingolipids

The sphingosine backbone of sphingolipids arises from a condensation of palmitoyl-CoA with serine. Acyl-CoAs donate acyl groups to the NH_2 moiety of sphingosine. In cerebroside biosynthesis, the sugar moieties are supplied from UDP-derivatives whereas in ganglioside biosynthesis, cytidine phosphates are used. A range of hydrolytic enzymes is available to degrade sphingolipids.

4.8.1 Biosynthesis of the sphingosine base and ceramide

The pathway by which sphingosine is formed was elucidated by Stoffel in Cologne, who demonstrated the pathway by isolating intermediates (Fig. 4.26). The first intermediate is 3-ketodihydrosphingosine (3-ketosphinganine), which is formed by the condensation of palmitoyl-CoA with serine using a pyridoxal phosphate-requiring enzyme. The selectivity of the serine palmitoyltransferase accounts for the prevalence of 18C bases in most sphingolipids. Some very potent inhibitors of the reaction have been recently developed, including a myriocin that is at least ten times better as an immunosuppressant than cyclosporin A. Sphingoid base biosynthesis is stimulated by a number of factors including endotoxin and cytokines. In both yeast and mammals, two gene products (which appear to associate) are necessary for serine palmitoyltransferase activity. 3-Ketosphinganine is short-lived and reduced rapidly to sphinganine, using NADPH.

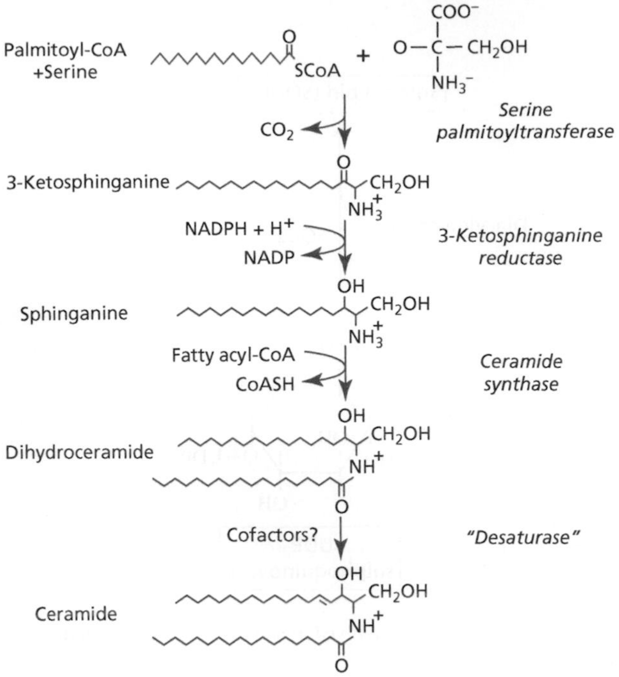

Fig. 4.26 Biosynthesis of ceramide.

The saturated long-chain base, sphinganine, is then acylated using acyl-CoA and a family of six ceramide synthases. The latter have different chain-length specificities, especially synthase 1, which is selective for stearoyl-CoA and is prevalent in the nervous system. Ceramide, which anchors all sphingolipids in the membrane, has several important functions per se, including signalling (see Section 8.5). Ceramide can also be produced by sphingomyelinase activity (see Section 4.8.7) either from sphingomyelin or more complex sphingolipids.

The last steps in ceramide biosynthesis or in that of their 4-hydroxy derivatives (phytoceramides) were characterized first in plants by Sperling and Heinz in Hamburg. The desaturase works with an intact lipid (see Section 3.1.8.2), dihydroceramide, to introduce a 4,5-*trans*-double bond. Genes for this enzyme have been cloned from a variety of organisms. In mice the enzyme is bifunctional and can produce 4-hydroxyceramides. Two mammalian genes are found; one can carry out only desaturation, while the second can perform both desaturation and hydroxylation. Potent inhibitors of ceramide synthase (fumonisins) are produced by some strains of *Fusarium verticilloides*, a very common

contaminant of corn. Similar mycotoxins are produced by fungi that grow on other plants. Fumonisins cause important veterinary diseases (e.g. horse leukoencephalomalacia, pig pulmonary oedema) and have been implicated in human cancer. Whereas all the above reactions are located on the ER, further conversions of ceramide to complex sphingolipids or to sphingomyelin are in the Golgi apparatus.

4.8.2 Cerebroside biosynthesis

Cerebrosides, such as galactosylceramide (GalCer) are made by uridine diphosphate (UDP)-gal:galactosylceramide synthases:

ceramide + UDP-gal ⇌ galactosylceramide + UDP

The reaction was first demonstrated by Morell and Radin and the UDP-galactose:ceramide galactosyltransferase has now been purified from rat brain. It is a membrane-bound enzyme that has to be solubilized with detergent and contains tightly bound phospholipid, even in its purified form. Interestingly, in view of the concentration of α-hydroxy FAs in galactosylceramides (Gal-Cer, Table 2.14) the enzyme was active only with ceramides that contain hydroxy FAs.

Table 4.4 Substrates used for the formation of sphingolipids.

Group to be added	Substrate used
Glucose	UDP-glucose
Galactose	UDP-galactose
N-Acetylgalactosamine	UDP-N-acetylgalactosamine
Fucose	GDP-fucose
Sialic acid (N-acetylgalactosamine)	CMP-NeuAc
Sulphate	PAPS

CMP-NeuAc, CMP-N-acetyl neuraminic acid; PAPS, phosphoadenosine-5′-phosphosulphate.

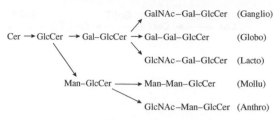

Fig. 4.27 Formation of different core glycosphingolipids from ceramide. Abbreviations: Glc, glucose; Gal, galactose; Man, mannose; GalNac, N-acetylgalactosamine; GlcNAc, N-acetylglucosamine; Cer, ceramide.

Glucosylceramide (Glu-Cer), which is a precursor for complex (neutral) glycosphingolipids and for gangliosides, is formed in an analogous reaction with ceramides that contain nonhydroxy FAs.

4.8.3 Formation of complex glycosphingolipids

Many higher homologues of ceramide are present in different tissues (Table 2.14). These are produced by direct transfer of a single sugar usually from its UDP derivative (Table 4.4). Each enzyme is specific for its individual glycosphingolipid substrate and, therefore, sugar addition is in a defined order. These reactions give rise to the different core sphingolipid structures illustrated in Fig. 4.27.

Whereas formation of the cerebrosides (Glu-Cer, Gal-Cer) occurs on the cytoplasmic face of the endoplasmic reticulum, the more complex glycosphingolipids are made in the lumen of the Golgi apparatus.

4.8.4 Ganglioside biosynthesis

Gangliosides contain sialic acid (N-acetylneuraminic acid, NANA) residues that are added using a cytidine derivative. This compound, CMP-neuraminic acid (Fig. 4.28) is formed by an enzymic reaction from CTP and is analogous to the cytidine derivatives used in phosphoglyceride biosynthesis.

The gangliosides are made by transfer of the neuraminic acid residue to the growing sugar chain at various stages using sialyltransferase enzymes. In mammals there are five of these (Fig. 4.29).

The picture of ganglioside biosynthesis was first built up largely by the work of Roseman, Brady and their colleagues in the USA. The way in which they worked out the sequence of sugar additions was by testing the specificity of each individual transferase for a specific acceptor. Each enzyme requires a specific substrate as a donor and the end-product of the previous step has much the greatest activity as acceptor. Pathways for the formation of major gangliosides are shown in Fig. 4.29 and Fig. 4.30.

The pattern of gangliosides within a particular tissue depends on the relative activities of the various enzymes involved in their biosynthesis. The overall regulation of ganglioside biosynthesis is due to both transcriptional and post transcriptional factors. These include large

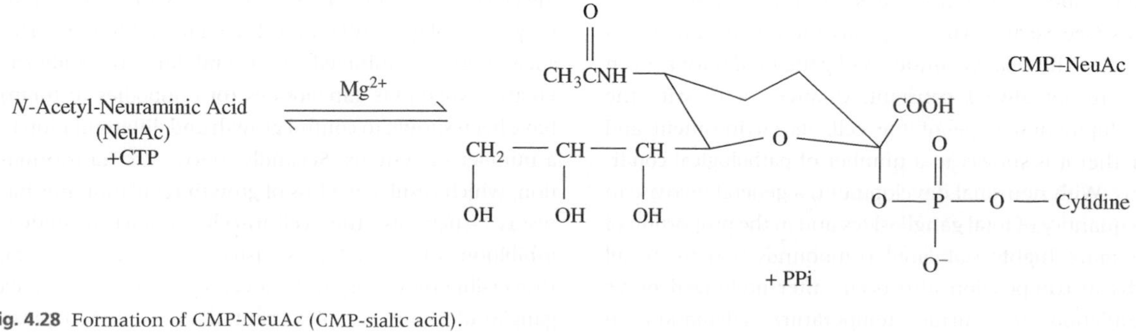

Fig. 4.28 Formation of CMP-NeuAc (CMP-sialic acid).

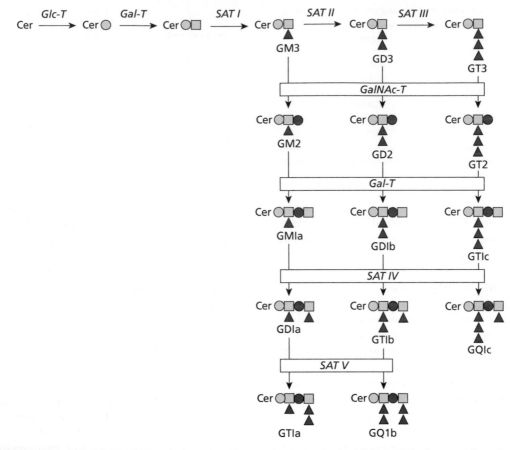

Fig. 4.29 Biosynthetic pathways for ganglioside formation. The transferases involved in adding carbohydrate or sialic acid moieties are indicated in italics. Enzyme abbreviations: Glc-T, glucosyltransferase; Gal-T, galactosyltransferase; SAT, sialyltransferase. The gangliosides are shown as containing one (GM), two (GD), three (GT) or four (GQ) sialic acids. Symbols: ○, glucose; ☐, galactose; ●, N-acetylgalactosamine; ▲, sialic acid.

changes upon oncogenic transformation (see Section 10.2.2 for more detail of cellular lipid changes in cancer).

In recent years considerable information has accumulated to show that gangliosides, like other glycosphingolipids (see Section 10.2.2.1), have important properties for cells. The relative amounts of gangliosides for a given cell are not always constant. Changes occur with the developmental stage of the cell, its environment and whether it is subject to a number of pathological conditions. With neuronal development, a general increase in the quantity of total gangliosides and in the proportion of the more highly sialylated compounds is seen. Small shifts in composition also occur after prolonged nerve stimulation or during temperature adaptation in poikilothermic animals, or during hibernation in mammals. Several drug-induced changes (e.g. with opiates) have also been noted.

The changes in ganglioside patterns, which occur during cell development and growth, imply that these lipids may also play a role in cell-contact inhibition. This suggestion is reinforced by several lines of evidence. Firstly, exogenous gangliosides (or antibodies to them) have been shown to control growth and differentiation in a number of systems. Secondly, oncogenic transformation, which results in a loss of growth regulation mechanisms (such as the cell-to-cell contact-dependent inhibition of growth), is usually paralleled by an irreversible reduction in the levels of the more complex gangliosides (and complex 'neutral' glycosphingolipids).

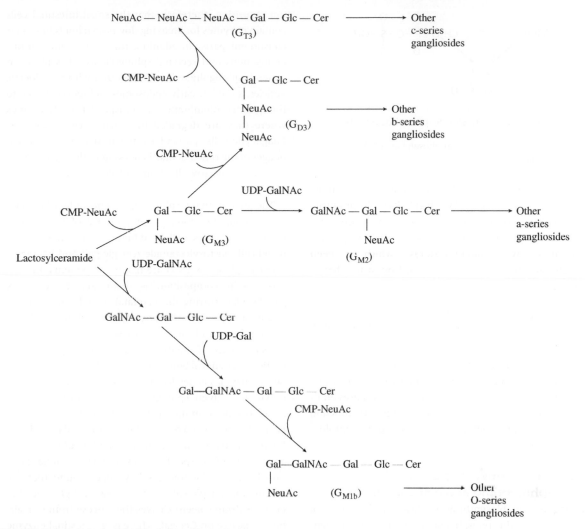

Fig. 4.30 Key steps in the initial formation of gangliosides.

Thirdly, gangliosides make an important contribution to the characteristic immunoexpression of individual cell types. Fourthly, they act as binding sites for a number of important compounds. Originally, several toxins, such as tetanus, botulinum and cholera toxins, were shown to interact tightly with gangliosides, but there is now some indication that they may also function as receptors for interferon and for certain cell-growth and differentiation factors. Further discussion of the role of glycosphingolipids in the immune system, especially in relation to cancer, can be found in Section 10.2.2.1.

4.8.5 Sulphated sphingolipids

The major sulphated glycolipid in mammalian tissues is Gal-Cer with a sulphate attached to the 3-position of the sugar (Table 2.14). It is commonly known as 'sulphatide'. However, various other sulphated glycolipids (e.g. lacto-sylceramide sulphate) have also been detected in small amounts in different animal tissues.

Sulphatide is an important constituent of the myelin sheath of nervous tissue and the incorporation of radioactive sulphate into the sulphatide molecule has been used to study myelination. The donor of the sulphate group is the complex nucleotide 3'-phosphoadenosine

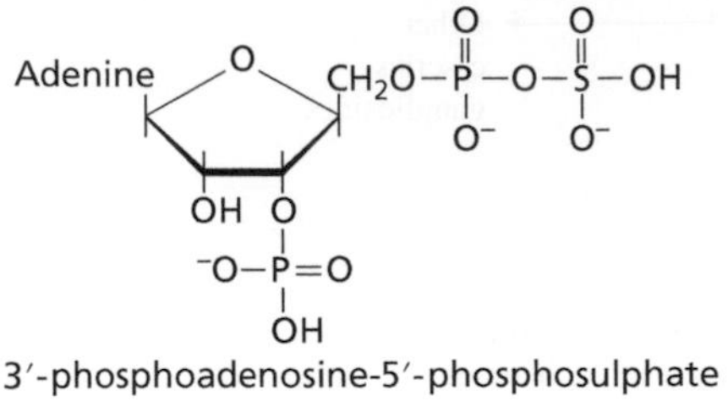

Fig. 4.31 Phosphoadenosine phosphosulphate (PAPS): the donor of sulphate groups.

5′-phosphosulphate, usually abbreviated in scientific papers to PAPS (Fig. 4.31). PAPS itself is produced from ATP in two steps via an adenosine 5′-phosphosulphate (APS) intermediate. The biosynthesis of sulphatide is catalysed by a sulphotransferase, which has been detected in the microsomal fraction from a number of tissues:

Galactosylceramide + PAPS ⇌

3-sulpho-galactosylceramide + PAP

The myelin sheaths of nerves in knock-out mice lacking the sulphotransferase are thinner than normal and the demyelination of their lower spinal cord causes hindlimb paralysis. This has confirmed the important function for sulphatide in myelin where its loss leads to neurological disorders.

4.8.6 Sphingomyelin is both a sphingolipid and a phospholipid

Sphingomyelin biosynthesis is closely tied to that of PtdCho, since the phosphocholine moiety of sphingomyelin is donated directly from PtdCho to ceramide using sphingomyelin synthase (SMS):

Phosphatidylcholine + ceramide ⇌ sphingomyelin
+ diacylglycerol

In mammals there are two SMSs. SMS1 is localized in the Golgi, while SMS2 is in the plasma membrane. For SMS1 an additional protein (ceramide transport protein) is involved in delivery of substrate. Sphingomyelin biosynthesis is regulated by many factors including aging and development.

4.8.7 Catabolism of the sphingolipids

Sphingolipids are widely distributed in animal tissues and constitute significant components of the human diet. It is

not surprising, therefore, that the small intestinal cells contain enzymes for breaking down such lipids into their constituent parts. In addition, turnover of the various endogenous and secreted sphingolipids takes place. In general, sphingolipids are internalized within endocytic vesicles, sorted in early endosomes and recycled back to the plasma membrane or transported to lysosomes where they are degraded by specific acid hydrolases. Phagocytic cells, particularly the histiocytes or macrophages of the reticuloendothelial system (located primarily in bone marrow, liver and spleen) play a prominent role here.

The predominant sphingolipid type differs between tissues and, therefore, the importance of the various catabolic enzymes will be different. For white or red blood cells, lactosylceramide (cer-glc-gal) and haematoside (cer-glc-gal-NANA) are major components. In contrast, brain composition is dominated by complex gangliosides. During the neonatal period, the turnover of gangliosides is particularly rapid as sphingolipids are broken down and then resynthesized.

Usually each of the catabolic enzymes is specific for a particular chemical bond. Thus, a combination of a number of enzymes is needed to ensure the complete breakdown of a given sphingolipid. Catabolism begins by attack on the terminal hydrophilic portions of the molecules by specific enzymes: glucosidases, galactosidases, hexosaminidases, neuraminidases and a sulphatase. As an example of the specificity of such enzymes it has been noted that β-galactosidases have been found that are specific for Cer-glc-gal and for Cer-glc-gal-gal. Another enzyme (from spleen) cleaves the glucose from Cer-glc, but is inactive on Cer-gal, whereas an intestinal enzyme is active with both substrates. Brain contains an enzyme that cleaves ceramide (but not cerebroside) to yield a NEFA and sphingosine. The sulphatases are responsible for the cleavage of the sulphate ester from sulphatides while NANA is hydrolysed from gangliosides by neuraminidases.

The breakdown of the various sphingolipids usually proceeds smoothly and it has been suggested that the various catabolic enzymes are aligned in an ordered fashion on the lysosomal membrane – thus ensuring more efficient hydrolysis than would be the case for different substrates allowed random access to enzymes freely admixed within the organelle. Sometimes one of the breakdown enzymes is missing or has very low activity. When this happens there is a build-up of one

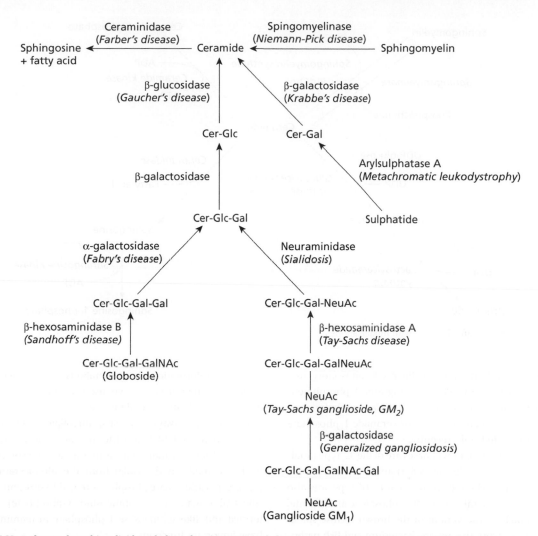

Fig. 4.32 Pathways for sphingolipid catabolism showing enzyme deficiencies in lipid storage disease (Section 10.1.1).

of the intermediate lipids. Such accumulation can impair tissue function and gives rise to sphingolipid storage diseases called '(sphingo)lipidoses'. These diseases are discussed in Section 10.1.1 and general pathways for sphingolipid breakdown are shown in Fig. 4.32 together with the enzyme deficiencies in the different storage diseases. An important laboratory that has contributed greatly to our understanding of the metabolic enzymes, as well as the lipidoses, is that of Sandhoff in Germany. In fact, one of the lipidoses is called Sandhoff's Disease after his discovery.

Because of their role in generating lipid-derived second messengers (see Sections 8.3 and 8.5), the catabolism of

sphingomyelin and ceramide is especially important (Figs. 4.32 & 4.33). Several sphingomyelinases with neutral to alkaline pH optima have been described. They are found in different cellular compartments including the plasma membrane, cytosol and nuclear membrane. In fact, five distinctly different sphingomyelinases have been identified in mammalian cells – based on their pH optima, cellular distribution and cation requirement.

When ceramide is generated, it can be acted upon by acidic or neutral ceramidases. The neutral ceramidases moderate release of sphingosine from ceramides produced in the plasma membrane for participation in signalling (see Section 8.5).

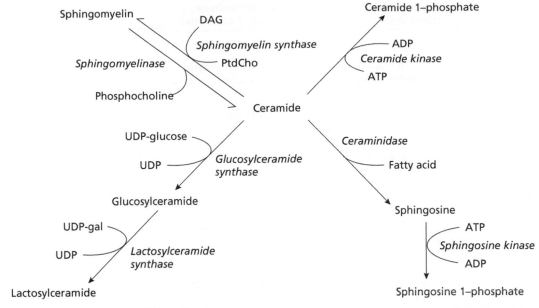

Fig. 4.33 The sphingomyelin cycle with modifications.

Both ceramide and sphingosine can be phosphorylated by kinases (Fig. 4.33). While sphingosine 1-phosphate is established to be important for cellular regulation (see Section 8.3), similar evidence for ceramide 1-phosphate has accumulated only recently.

While most sphingomyelinases are of the 'C' type (analogy with PLC – see Section 4.6.4) and liberate ceramide and phosphorylcholine, those of the 'D' type are also known. These liberate ceramide 1-phosphate and choline. Sources include the venom of the brown recluse spider (*Loxosceles reclusa*), the aquatic bacterium and fish pathogen (*Vibro damsela*) that can infect wounds, and the human pathogen *Arcanobacterium haemolyticum*. The sphingomyelinase in the venom causes most of the tissue damage that follows a bite by activating prolonged inflammation.

Sphingomyelinase can also catabolize lysoPtdCho to yield lysoPtdOH, a powerful inflammatory mediator (see Table 7.5 & Section 8.2 for further information).

4.8.8 Sphingolipid metabolism in plants and yeast

The yeast, *Saccharomyces cerevisiae*, has always been a useful experimental organism which, in many cases, can be used as a surrogate for mammalian experiments. Although yeast shares only the simpler sphingolipids with mammals, experiments with it have revealed important information about metabolism and function. Genes identified in yeast have also enabled their mammalian equivalents to be identified.

The main pathways of yeast sphingolipid metabolism are shown in Fig. 4.34, where the enzymes concerned and their substrates are indicated. As in animals, the complex lipids are made in the Golgi from the phytoceramide precursor. Various base phosphates (e.g. dihydrosphingosine-1-phosphate, phytosphingosine 1-phosphate) are formed and, like sphingosine 1-phosphate in mammals, have important functions.

In yeast, the sphingosine long-chain bases are signalling molecules that control growth, responses to heat stress, cell wall biosynthesis and repair and endocytosis. Again, as in mammals, yeast sphingolipids are important for lipid rafts or microdomains (see also Section 5.3.3), cellular aging, and cross-talk with other signalling lipids.

The first plant sphingolipid to be recognized as a significant component was the discovery of Glc-Cer in plasma membranes. Since that time, the metabolic products of its catabolism, ceramide and long-chain sphingoid bases (and their phosphorylated derivatives, ceramide 1-phosphate, sphingosine 1-phosphate) have been detected in small, but significant amounts. Another compound is inositol-phosphorylceramide, which may contain additional

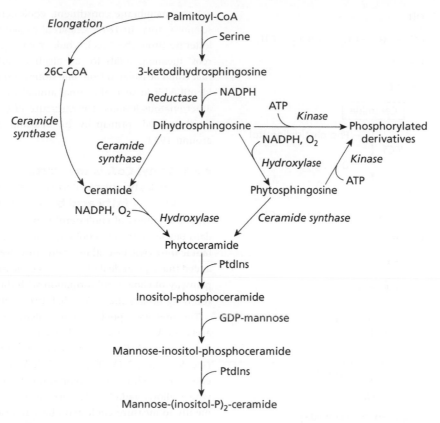

Fig. 4.34 Yeast sphingolipid metabolism.

sugars linked to the inosital ring. In leaves, the inositol-phosphorylceramides and their derivatives are the dominant sphingolipids.

A whole host of different bases is used by plants (see Section 2.3.3 & Fig. 2.10). In seeds, enrichment with dihydroxy bases is found, while leaves have trihydroxy long chain bases. The particular long chain bases (and their esterified FAs in ceramides) appear to have important roles in plant sphingolipid metabolism and function.

A detailed discussion of plant sphingolipid metabolism is beyond to scope of this book and readers are referred to the **Further reading** section. Production of bases and their incorporation into ceramides and glucosylceramide is broadly similar to mammals (Figs. 4.26 & 4.27). A difference applies to the production of inositolphosphorylceramide which is synthesized as shown in Fig. 4.35.

Sphingolipids serve three basic functions in plants:

1 They are membrane structural components, where they control permeability and are altered significantly

in drought and following temperature stress. They are enriched in membrane rafts (see Section 5.3.3).

2 As in animals, sphingolipids are involved in signalling and cell regulation. The bioactive molecules include free long-chain bases, ceramide and sphingosine 1-phosphate. Recent evidence provides information on ceramide 1-phosphate also.

3 Sphingolipids are involved in cell-cell interactions. This is a new research area but Glc-Cer may have important actions such as acting as an elicitor in hypersensitivity reactions to pathogen invasion.

4.9 Cholesterol biosynthesis

Cholesterol biosynthesis, in common with FA biosynthesis, starts from the 2-carbon compound acetyl-CoA. However, the condensation follows a quite different pathway. Hydroxymethylglutaryl (HMG)-CoA is formed

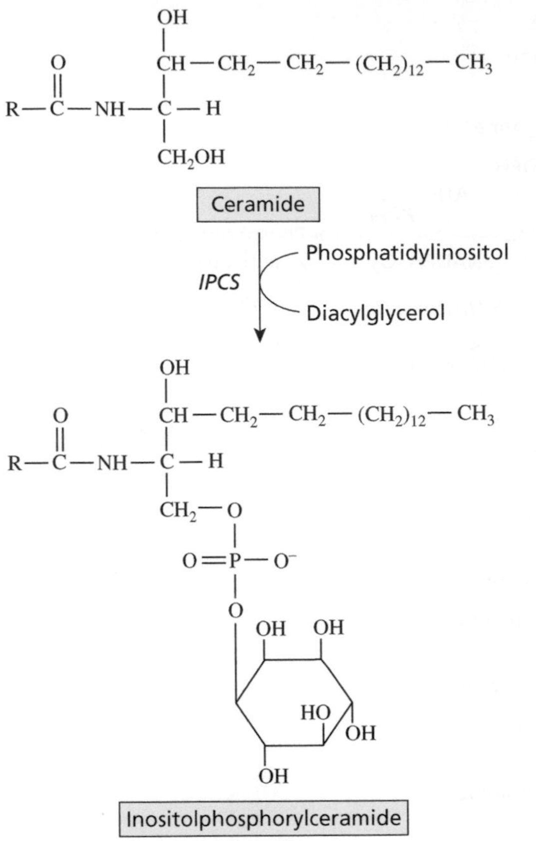

Fig. 4.35 Inositolphosphorylceramide synthase (IPCS) catalyses the transfer of inositol phosphate from phosphatidylinositol to ceramide.

and is a key intermediate in the biosynthesis of cholesterol and a host of other polyisoprenoids. After the stage at which squalene (a 30C straight-chain polyisoprenoid) is formed, a cyclization reaction gives rise to the multi-ring sterol structure and, after several more steps, cholesterol.

As mentioned in Section 2.3.4, cholesterol is a major and important constituent of animal membranes. Different sterols may be important in other organisms. Sterols are, however, a relatively small group of a very large class of biogenetically related substances – the polyisoprenoids (or terpenoids). These compounds are all derived from a common precursor, isopentenyl diphosphate. Polyisoprenoids can be open-chain, partly cyclized, or fully cyclized substances. They contain or derive from the basic structure of a branched-chain 5C (isoprenoid) unit.

For a membrane constituent, cholesterol has a long scientific story. In 1816, Chevreul coined the term cholesterine (from the Greek 'chole' meaning bile and 'stereos' meaning solid) for an alcohol-soluble substance that could be isolated from bile stones and in 1843 Vogel identified it in several normal animal tissues as well as atheromatous lesions. The structure of cholesterol was finally solved, principally by Wieland and Windaus around 1932.

4.9.1 Acetyl-CoA is the starting material for polyisoprenoid (terpenoid) as well as fatty acid biosynthesis

The biosynthesis of cholesterol is well covered in standard biochemistry textbooks and only the more important features will be outlined here. However, it should be added that a great deal of work has gone into solving the pathway of cholesterol formation with those concerned receiving no less than 13 Nobel prizes between them!

The precursor pool in mammalian cells is cytosolic acetyl-CoA. This may be derived, for example, from β-oxidation of FAs by mitochondria or microbodies (see Section 3.2.1). The acetyl-CoA pool is in rapid equilibrium with intracellular and extracellular acetate, which allows radiolabelled acetate to be used conveniently to measure cholesterol biosynthesis in tissues.

The first two steps involve condensation reactions catalysed by a thiolase and HMG-CoA synthase. Both enzymes are soluble and the first reaction is driven to completion by rapid removal of acetoacetyl-CoA in the second step (Fig. 4.36).

Cytosolic HMG-CoA synthase has been studied in considerable detail and the reaction mechanism defined. It shows a very high degree of specificity with regard to the stereochemistry of the acetoacetyl-CoA substrate and the condensation proceeds by inversion of the configuration of the hydrogen atoms of acetyl-CoA. In addition to cytosolic HMG-CoA synthase, a second synthase is found in mitochondria. Not only has this been shown to be a different protein, the HMG-CoA it forms has a different function. HMG-CoA in the cytosol is destined for mevalonate formation, whereas in mitochondria it is broken down by HMG-CoA lyase to yield acetyl-CoA and acetoacetate (see Section 3.2.1.6). Intriguingly, both isoforms of HMG-CoA synthase contain peroxisomal targeting sequences, so they may reside in multiple cellular compartments. The cytosolic HMG-CoA synthase seems to be one of the pathway enzymes important for

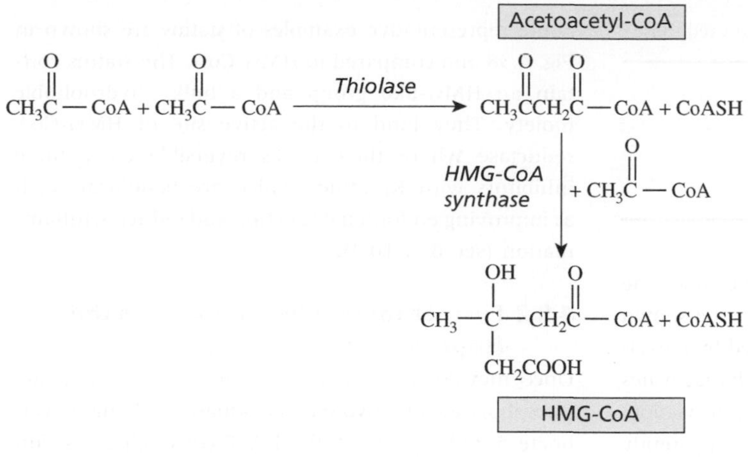

$$\underset{\text{O}}{\overset{\text{O}}{\underset{\|}{CH_3C}}} - CoA + \underset{\text{O}}{\overset{\text{O}}{\underset{\|}{CH_3C}}} - CoA \xrightarrow{\textit{Thiolase}} \boxed{\text{Acetoacetyl-CoA}}$$

Acetoacetyl-CoA

$$CH_3C - CoA + CH_3C - CoA \xrightarrow{\textit{Thiolase}} CH_3CCH_2C - CoA + CoASH$$

HMG-CoA synthase $+ CH_3\overset{O}{\overset{\|}{C}} - CoA$

$$CH_3 - \underset{CH_2COOH}{\overset{OH}{\underset{|}{\overset{|}{C}}}} - CH_2\overset{O}{\overset{\|}{C}} - CoA + CoASH$$

HMG-CoA

Fig. 4.36 Formation of hydroxymethylglutaryl-CoA (HMG-CoA).

controlling cholesterol biosynthesis and its activity is changed by transcriptional modulation.

The next reaction, which results in mevalonate production, is catalysed by the ER-localized HMG-CoA reductase. It is a highly regulated enzyme with a short half-life ($T^1/_2$ about 3 h) and it is usually considered to catalyse the reaction that exerts the most control over the rate of sterol biosynthesis. Even in a normal diurnal cycle, its activity will vary about tenfold and under extreme conditions changes of 200-fold can be seen!

HMG-CoA reductase activity is regulated at the transcriptional level and by post-transcriptional methods. When sterols are added to animal diets there is a decline in the mRNA levels for HMG-CoA reductase (together with HMG-CoA synthase, farnesyl diphosphate synthase and the low density lipoprotein (LDL)-receptor (see Section 7.3.1). These four mRNAs increase when cells are deprived of sterols. Brown and Goldstein showed that the 3′-flanking region of the gene for HMG-CoA reductase (and the LDL-receptor) contained one to three copies of a nucleotide sequence, sterol regulatory element-1 (SRE-1). Transcription factors SREBP-1a, -1c and -2 bind to SRE and induce transcription of genes involved in lipid biosynthesis. SREBP-1a is a strong activator of both cholesterol and FA formation. SREBP-1c activates FA biosynthesis, while SREBP-2 preferentially activates cholesterol biosynthetic genes (and the LDL receptor). Thus SREBP-2 specifically binds to SRE-1, which is within the promoter for HMG-CoA reductase (and the LDL-receptor). The transcription factor, SREBP-2, is synthesized as a 125 kDa precursor bound to the ER. In cells deprived of cholesterol, the factor is cleaved and a 68 kDa N-terminal fragment is released, which is targeted to the nucleus, where it binds to SRE-I and promotes expression of HMG-CoA reductase. The SREBP system, which also regulates FA biosynthesis, is described further in Section 7.3.1.

HMG-CoA reductase expression is also regulated by changes in mRNA translation and stability and by protein turnover. The degradation of HMG-CoA reductase protein in the ER is regulated through the eight trans-membrane domains, probably by the pathway intermediate lanosterol or via oxysterols. A newly discovered ER protein called Insig ('insulin induced gene') also plays a key role in regulating HMG-CoA reductase both transcriptionally and post-translationally (including proteolytic degradation).

HMG-CoA reductase is also regulated by a reversible phosphorylation/dephosphorylation cycle (Fig. 4.37).

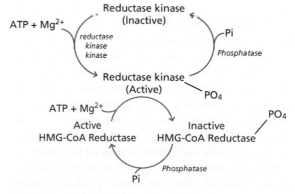

Fig. 4.37 Regulation of HMG-CoA reductase activity by phosphorylation and dephosphorylation.

Table 4.5 Methods to regulate activity of HMG-CoA reductase.

Transcription
mRNA stability
Translation
Protein stability
Protein phosphorylation/dephosphorylation

The phosphorylated reductase is inactive and the amounts of the phosphorylated enzyme can be shown to be increased when its activity is decreased by mevalonate or glucagon. The protein kinase, which inactivates HMG-CoA reductase, is itself subject to phosphorylation. Interestingly, both ATP and ADP are needed. Apparently ADP binds to a different site on the reductase kinase and acts as an allosteric effector. The phosphatases, which activate HMG-CoA reductase, are highly sensitive to NaF. The kinase and phosphatases are each present in the cytosol as well as ER. The protein kinase, which phosphorylates HMG-CoA reductase, has been identified as an AMP-activated protein kinase. Interestingly, this kinase also phosphorylates (and thus regulates) ACC (see Section 3.1.11.2). Under conditions when ATP concentrations decrease, those of AMP increase. The resultant activation of the kinase reduces both cholesterol and FA biosynthesis thus preserving cellular ATP and allowing its generation. (See Table 4.5 for factors regulation HMG-CoA reductase activity.)

Because of the involvement of cholesterol in the aetiology of arterio-vascular disease (see Section 10.5) considerable efforts have been made to develop suitable pharmaceutical agents able to reduce its formation. Two interesting compounds that inhibit HMG-CoA reductase are natural antibiotics isolated from the moulds *Penicillium spp.* and *Aspergillus terreus* and named compactin and mevinolin, respectively. These compounds are competitive inhibitors of HMG-CoA reductase with a K_i for the enzyme of about 10^{-9} M compared to 10^{-6} M for HMG-CoA. Surprisingly, they increase mRNA and protein levels for the enzyme, but this is consistent with a downstream metabolite of mevalonate inhibiting transcriptions and HMG-CoA reductase turnover (as discussed above). Nevertheless, these fungal metabolites rapidly enter cells and inhibit cholesterol biosynthesis.

HMG-CoA reductase is a therapeutic target for 'statins' (drugs developed from the above mentioned fungal products) which are widely used for the treatment of hypercholesterolaemia (see Sections 7.3.1 & 10.5.2.2).

Some representative examples of statins are shown in Fig. 4.38 and compared to HMG-CoA. The statins contain an HMG-like group and a bulky hydrophobic moiety. They bind to the active site of HMG-CoA reductase where they act as reversible competitive inhibitors with K_i values which are beneficial, such as improving endothelial function and reducing inflammation (see Box 10.3).

4.9.2 Further metabolism generates the isoprene unit

Once mevalonate has been formed it is sequentially phosphorylated by two separate kinases yielding mevalonate 5-diphosphate. A third ATP-consuming reaction involving a decarboxylase then generates the universal isoprene unit, isopentenyl diphosphate (Fig. 4.39). The function of the ATP in this reaction appears to be to act as an acceptor for the leaving OH group in the dehydration part of the reaction.

4.9.3 More complex terpenoids are formed by a series of condensations

Isopentenyl diphosphate (IPP) is potentially a bifunctional molecule. Its terminal vinyl group gives it a nucleophilic character, whereas when it is enzymically isomerized to 3,3-dimethylallyl diphosphate, the latter is electrophilic. Thus, longer-chain isoprenoids are formed by a favourable condensation of IPP, first with dimethylallyl diphosphate and, later, with other allylic diphosphates. The initial interconversion of IPP and dimethylallyl diphosphate is promoted by an isomerase. The successive condensations yield the 10C compound geranyl diphosphate and then the 15C farnesyl diphosphate. Farnesyl diphosphate synthase is one of a family of prenyltransferases that synthesize the backbone of the various isoprenoids. The latter include not only cholesterol but other steroids, prenylated proteins, dolichol, carotenoids, retinoids, chlorophyll, chical (in chewing gum) and rubber. The two molecules of farnesyl diphosphate condense to form presqualene diphosphate, which is reduced by NADPH to give the 30C open-chain isoprenoid squalene. Squalene synthase catalyses both the condensation and reduction reactions and is the first committed step for cholesterol (sterol) biosynthesis. It is strongly regulated by the cell's cholesterol content. Each condensation reaction with IPP represents a novel method of C—C bond formation since, in the formation

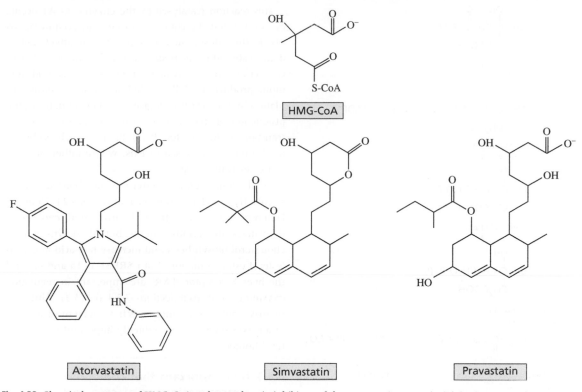

Fig. 4.38 Chemical structures of HMG-CoA and several statin inhibitors of the enzyme. Atorvastatin (Lipitor), pravastatin (Pravachol) and simvastatin (Zocor) are commonly prescribed cholesterol-lowering drugs.

of other types of natural products (peptides, sugars, FAs, etc.), the reactions involve Claisen- or aldol-type condensations.

4.9.4 A separate way of forming the isoprene unit occurs in plants

In plant chloroplasts a mevalonate-independent pathway for IPP formation is present. This pathway was discovered first by Rohmer when he was studying the biosynthesis of hopanoids (pentacyclic steroidlike molecules, see Section 2.3.4.3) in bacteria, but is used generally by algae and higher plants for plastid isoprenoid production. This pathway does not use mevalonate but, instead, uses central metabolites like pyruvate and glyceraldehyde 3-phosphate to make 1-deoxy-D-xylulose-5-phosphate which then, after a series of steps, yields IPP. Because IPP can be transported from the plastid, it may contribute to sterol biosynthesis in the cytosolic compartment. However, the portion of sterol precursor carbon originating from this route is not yet known.

4.9.5 Sterol biosynthesis requires cyclization

Formation of sterols from squalene involves cyclization. First, a microsomal mixed-function oxidase (squalene epoxidase) forms squalene-2,3-oxide in the presence of NADPH, FAD and O_2 (there is no requirement for cytochrome P450 in this reaction). The cyclization of the oxide to lanosterol then takes place by a concerted reaction without the formation of any stable intermediates. This conversion, which has been described as the most complex known enzyme-catalysed reaction, depends on a cyclase with a molecular mass of only 90 kDa. In plants and algae squalene-2,3-epoxide is cyclized to cycloartenol, which is the precursor of stigmasterol whereas lanosterol is the precursor of cholesterol and ergosterol (Fig. 4.40).

The conversion of lanosterol into cholesterol involves a 19-step reaction sequence catalysed by microsomal enzymes. The exact order of the reactions varies according to the tissue (see Fig. 4.41). The main features of the transformation are the removal of three methyl groups,

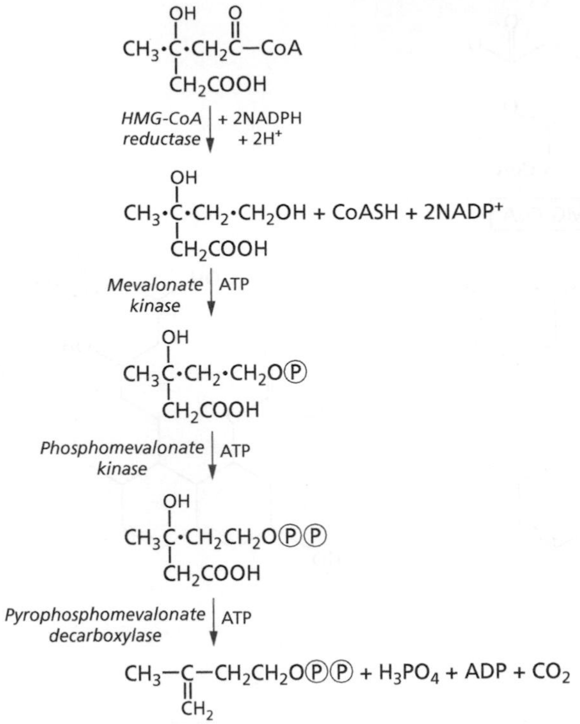

$$CH_3 \cdot \underset{\underset{CH_2COOH}{|}}{\overset{\overset{OH}{|}}{C}} \cdot CH_2 \overset{\overset{O}{\|}}{C} - CoA$$

HMG-CoA | + 2NADPH
reductase ↓ + 2H$^+$

$$CH_3 \cdot \underset{\underset{CH_2COOH}{|}}{\overset{\overset{OH}{|}}{C}} \cdot CH_2 \cdot CH_2OH + CoASH + 2NADP^+$$

Mevalonate | ATP
kinase ↓

$$CH_3 \underset{\underset{CH_2COOH}{|}}{\overset{\overset{OH}{|}}{C}} \cdot CH_2 \cdot CH_2O\textcircled{P}$$

Phosphomevalonate | ATP
kinase ↓

$$CH_3 \underset{\underset{CH_2COOH}{|}}{\overset{\overset{OH}{|}}{C}} \cdot CH_2 CH_2O\textcircled{P}\textcircled{P}$$

Pyrophosphomevalonate | ATP
decarboxylase ↓

$$CH_3 - \underset{\underset{CH_2}{\|}}{C} - CH_2CH_2O\textcircled{P}\textcircled{P} + H_3PO_4 + ADP + CO_2$$

Fig. 4.39 Formation of isopentenyl diphosphate from HMG-CoA.

reduction of the 24-double bond and isomerization of the 8-double bond to position 5 in cholesterol.

4.9.6 Cholesterol is an important metabolic intermediate

Cholesterol is not only an important membrane constituent in animals but also plays a vital role as a metabolic intermediate (Fig. 4.42). It is the precursor for the steroid hormones (glucocorticoids, aldosterone, oestrogens, progesterones, androgens), for the bile acids (and their salts – see Box 7.1) and can also be esterified with FAs. Even if cholesterol was not required continuously (because of membrane turnover), a considerable amount would be needed for bile production (see Box 7.1) – even though greater than 80% of the bile salts are reabsorbed from the large intestine and reutilized.

Because high levels of unesterified cholesterol are thought to be deleterious to cells, excess of this sterol is converted into cholesteryl ester:

acyl-CoA + cholesterol ⇌ cholesterylester + CoA

This reaction (catalysed by the enzyme ACAT occurs on the ER and the cholesteryl ester can accumulate as LDs in the cytosol. Such LDs are often relatively abundant in steroidogenic tissues where the cholesteryl ester can act as a readily available precursor for steroid hormone production. Cellular cholesterol homeostasis regulation by the acyl transferase is also dependent on the production of free cholesterol in the lysosome and its transport to the ER. Hereditary diseases (such as cholesterol ester storage disease) exist where either of these processes is impaired.

In addition, excess cholesterol is metabolized to oxysterols. Hydroxylation can occur at carbons 24, 25 or 27. Different tissues use these hydroxylated cholesterol derivatives to various extents but all down-regulate cholesterol biosynthesis and increase bile acid and cholesterol ester formation. Both SREBP (−1a and -2) and the liver X receptor (LXR) are important in regulating oxysterol action (see Sections 7.3.1 & 7.3.3). Although hydroxycholesterol compounds have been most studied, other oxysterols are undoubtedly important for cholesterol homeostasis.

4.9.7 It is important that cholesterol concentrations in plasma and tissues are regulated within certain limits and complex regulatory mechanisms have evolved

Cholesterol balance in cells is maintained by a number of factors (Table 4.6). In fact, it has been found that uptake of lipoproteins may influence cholesterol synthesis itself via the LDL-receptor mechanism described in Section 7.2.4.

Early experiments showed that the biosynthesis of cholesterol was reduced when animals were fed cholesterol and this led eventually to the identification of the SREBP system. The molecular basis for the regulation of the activities of the enzymes of cholesterol biosynthesis, involving the SREBP system, is described above in Section 4.9.1 and will be described in more detail in Section 7.3.1. Table 4.5 further summarizes the main mechanisms for the regulation of the key enzyme in the pathway, HMG-CoA reductase (see Section 4.9.1).

Although sterols are present in most mammalian body tissues, the proportion of sterol ester to free sterol varies markedly. For example, blood plasma, especially that of humans, is rich in sterols and like most plasma lipids they are almost entirely found as components of the

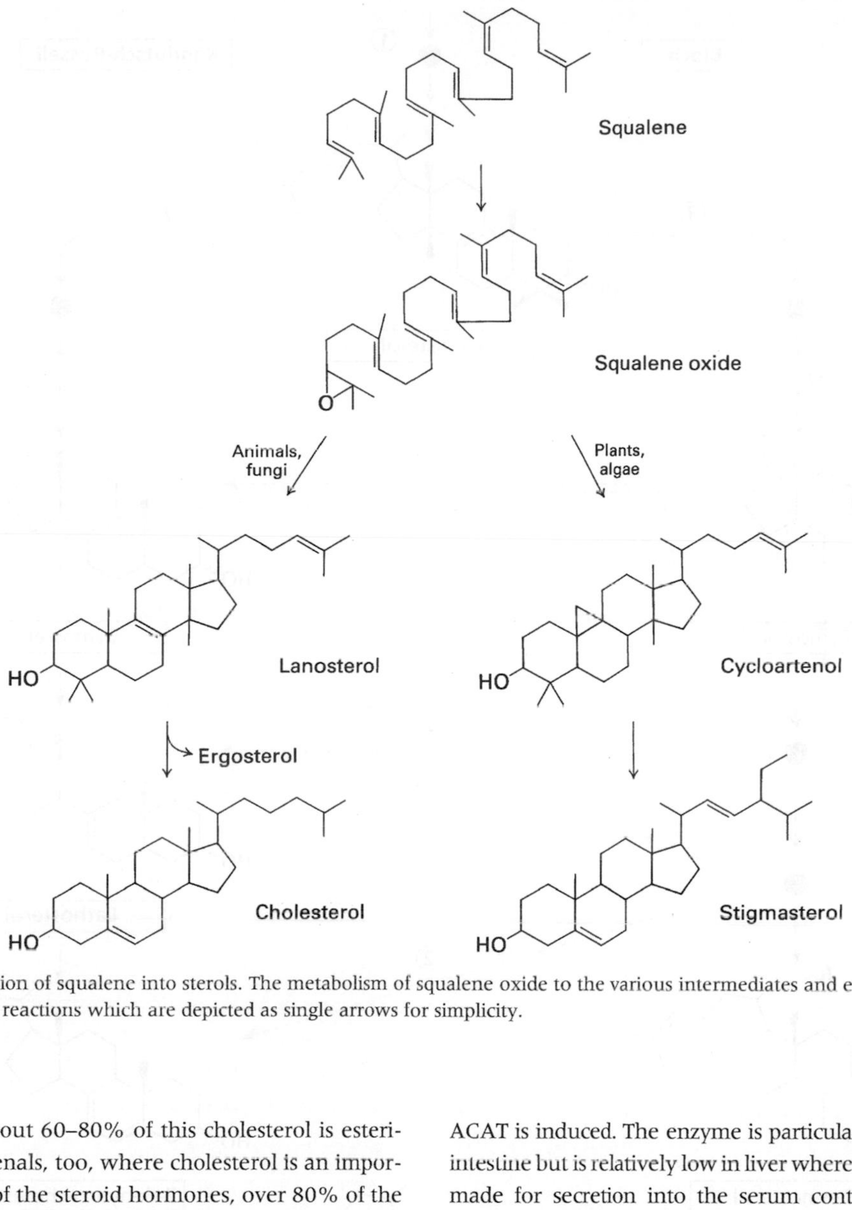

Fig. 4.40 Conversion of squalene into sterols. The metabolism of squalene oxide to the various intermediates and end-products involves multiple reactions which are depicted as single arrows for simplicity.

lipoproteins; about 60–80% of this cholesterol is esterified. In the adrenals, too, where cholesterol is an important precursor of the steroid hormones, over 80% of the sterol is esterified. However, in brain and other nervous tissues, where cholesterol is a major component of myelin, virtually no cholesteryl esters are present. Cholesteryl esters are formed by the action of a microsomal ACAT, which is present in most cells. Under normal conditions the enzyme is considered to tightly control the rate of cholesterol esterification. It is regulated by progesterone and may be modulated by phosphorylation/dephosphorylation like HMG-CoA reductase. Under conditions where cells take up a large amount of cholesterol, such as via LDL-receptors,

ACAT is induced. The enzyme is particularly important in intestine but is relatively low in liver where the lipoproteins made for secretion into the serum contain little if any cholesteryl ester.

In contrast, cholesteryl esters are formed in blood by another acyltransferase, the lecithin:cholesterol acyltransferase (LCAT – see Section 7.2.5 & Fig. 7.13). Most of the cholesterol that accumulates in arterial plaques during the development of arteriosclerosis is in the esterified form (see Section 10.5.1). An understanding of cholesterol transport and cholesteryl ester metabolism is crucial for the understanding of this disease, as discussed in Sections 7.2, 7.3 and 10.5.

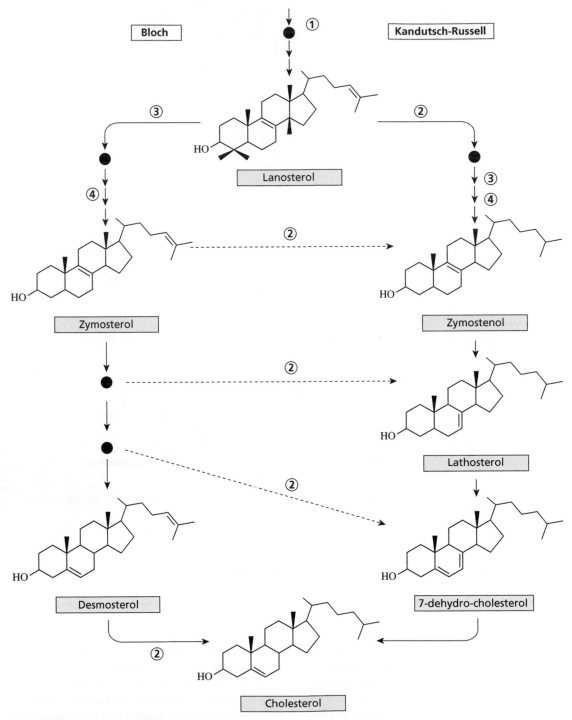

Fig. 4.41 Comparison of the production of cholesterol in astrocytes and neurons, showing how the variants (Bloch, Kandutsch-Russell) of the conversion of lanosterol into cholesterol are used differently.

Abbreviations: (1), squalene synthase; (2), 24-dehydrocholesterol reductase; (3), lanosterol 14-methylase; (4), part of the C4-demethylation enzyme complex. Adapted from FW Pfreiger & N Ungerer (2011) Cholesterol metabolism in neurons and astrocytes. *Prog Lipid Res* **50**:357–71. Reproduced with permission of Elsevier.

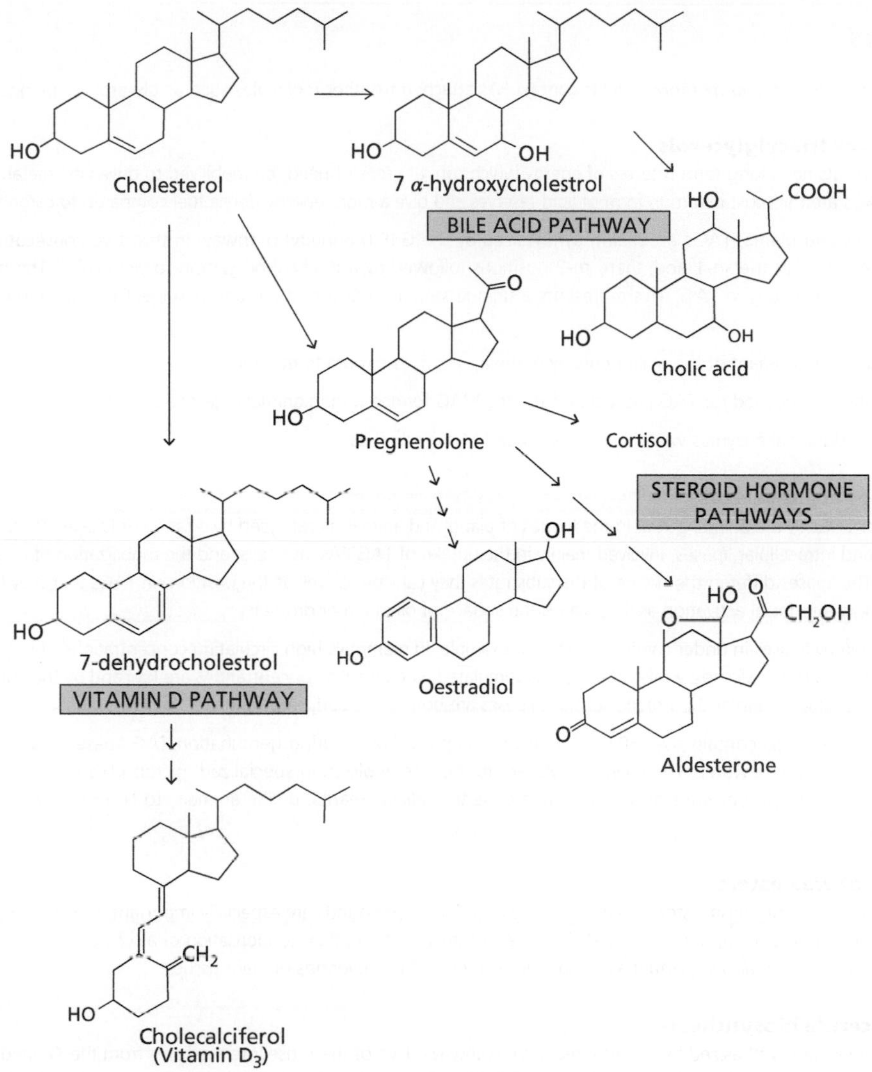

Fig. 4.42 Metabolism of cholesterol to bile acids, calciferols and steroid hormones.

Table 4.6 Factors influencing cellular cholesterol concentrations.

1. Uptake of intact lipoproteins via receptors.
2. Uptake of free cholesterol from lipoproteins by lipid transfer.
3. Cholesterol synthesis.
4. Cholesterol metabolism (e.g. to hormones).
5. Efflux of cholesterol.
6. Esterification of cholesterol by acyl-CoA:cholesterol acyltransferase.
7. Breakdown of cholesterol esters by neutral cholesterol esterase.

KEY POINTS

The most common storage and membrane lipids contain FAs attached to other molecules such as glycerol or sphingosine.

Biosynthesis of triacylglycerols

- Animals and plants need long-term reserves of energy which can, at times of need, be mobilized to drive vital metabolic processes. TAGs are the most common form of lipid reserves and give a more energy-dense fuel compared to carbohydrate.

- In both animals and plants, TAGs are mainly synthesized by the G3P (Kennedy) pathway. In this, two consecutive acylations take place at the *sn*-1 and, then, *sn*-2 positions followed by a dephosphorylation to yield DAG. The final step is the acylation of DAG to give TAG. This represents a unique step in TAG formation and is subject to regulation in animals and plants.

- The DHAP pathway, a variant of the main Kennedy pathway, provides a route to ether lipids.

- The MAG pathway is needed for TAG production from the MAG formed during animal digestion of fatty foods.

- Plants can use additional enzymes when they accumulate TAGs in oil seeds.

Catabolism of acylglycerols

- Breakdown of TAGs in the animal gut or in the tissues of plants and animals is catalysed by a range of lipases. There are extracellular and intracellular lipases, involved mainly in the uptake of TAG FAs into cells, and the mobilization of lipid stores, respectively. The lipases differ in the nature of the substrates they can use as well as the position on TAGs that they hydrolyse. They often show interfacial activation as they act on the surface of oil-in-water droplets.

- TAG breakdown by lipases in under careful control. For example, in mammals high circulating concentrations of insulin favour TAG biosynthesis and fuel storage while catabolism is stimulated when insulin concentrations are low and by the release of stress hormones. Lipoprotein and hormone-sensitive lipases are both particularly important.

- In plants, some oil seeds contain over 60% of their dry weight as lipid. During germination TAG lipase activity rapidly increases as seeds imbibe water. The liberated FAs are further metabolized in specialized microbodies called glyoxysomes. Glyoxysomes also contain enzymes of the glyoxylate cycle that allows plants, unlike animals, to convert lipid carbon into carbohydrate.

Metabolism of wax esters

- Instead of TAGs, some organisms store energy as wax esters. Such compounds are especially important in marine organisms and zooplankton are vital components of food chains. Wax esters are formed by condensation of alcohols with FAs (as acyl-CoAs). The wax esters are also important as components of the surface coverings of plant tissues.

Phosphoglyceride biosynthesis

- Phosphoglycerides are synthesized by one of three basic pathways. Two of these use intermediates from the Kennedy pathway (used for TAG formation). Either a CDP-base reacts with DAG to form a phosphoglyceride or, for anionic phosphoglycerides, CDP-DAG is generated from PtdOH and used to form phosphoglycerides like PtdIns or PtdGro. A third type of pathway is to form one phosphoglyceride and then convert it into another – an example being the methylation of PtdEtn in three steps to make PtdCho.

- Different organisms – mammals, plants, yeast, bacteria – make their phosphoglycerides in different ways. For example, *E. coli* makes all its phosphoglycerides from CDP-DAG. Moreover, whereas mammals make PtdSer by an exchange reaction, using another phosphoglyceride, microorganisms make it from CDP-DAG and then use PtdSer to form other phosphoglycerides.

- Plasmalogen (1-vinylether) derivatives of phosphoglycerides are formed by the same basic Kennedy pathway but with a 'plasmalogenic DAG' (1-alkenyl-2-acyl-*sn*-glycerol). An important alkyl ether lipid is PAF, which is the first well-documented example of a biologically active phosphoglyceride. While it has an ether link at *sn*-1, it has an acetyl at position-2.

Degradation of phosphoglycerides

- Catabolism of phosphoglycerides is mainly initiated by a variety of PLs. These usually operate best at the surface of immiscible solvents such as with lipid micelles. The micellar (or membrane) nature of the substrates makes classic enzyme kinetics difficult. Many PLs (and lipases) are extremely stable proteins which exhibit activity in organic solvents and at high temperatures. This has allowed their use in various industrial processes.

- Dependent on the position of attack, PLs are classified as A, B, C or D. PLs A remove a FA from a phosphoglyceride and are subdivided into A_1 or A_2, depending upon which FA they release. PLB removes the remaining fatty acid from a monoacylphospholipid or hydrolyses both positions. PLC gives rise to DAG and a phospho-base while PLD removes the base moiety to yield phosphatidate. In general, PLs have functions in digestion, in venoms, in remodelling reactions to modify phospholipids and for generating lipid signalling molecules.

Metabolism of glycosylglycerides

- In the photosynthetic membranes of cyanobacteria, algae and plants, the main lipids are glycosylglycerides – MGDG, DGDG and SQDG. MGDG is formed by transfer of a galactose from UDP-galactose to DAG. A second such transfer converts MGDG into DGDG. SQDG is formed a novel pathway beginning with UDP-glucose which accepts a sulphite through an intermediate to give a UDP-sulphoquinovose. Finally, this reacts with DAG to give SQDG.

- All three glycosylglycerides are catabolized by various acyl hydrolases to give water-soluble products and NEFAs. Some of the acyl hydrolases can be extremely active. Because the sulphoquinovose moiety is chemically stable, its further catabolism poses particular problems but various soil bacteria can carry out the process.

Metabolism of sphingolipids

- Sphingolipids are based on sphingosine bases. Condensation of palmitoyl-CoA with serine yields 3-ketosphingosine which can be modified to produce other bases. Acylation of the amino group yields ceramide.

- Once ceramides have been produced, the terminal alcohol moiety can be glycosylated to various degrees to yield cerebrosides, sulphatides and the more complex glyco-sphingolipids. Usually UDP-sugars are the source of the sugar moiety but the sialic acid moiety of gangliosides is supplied by a CMP-derivative. By contrast, sphingomyelin is produced by reaction of a ceramide with PtdCho.

- Sphingolipids are broken down by substrate-specific enzymes. Sialic acid residues are removed by neuraminidases, galactose by galactosidases, glucose by glucosidases etc. Almost all of the breakdown enzymes are found in lysosomes and their genetic loss gives rise to the respective substrate sphingolipid building up in tissues. This causes the various sphingolipidoses. Some catabolic enzymes are also important in the turnover of sphingolipid signalling molecules.

Sterol metabolism

- Cholesterol (and other sterols) is derived from acetyl-CoA. By a series of reactions, the 5C-isoprene unit IPP is formed and this can self-condense to give a series of 10C, 15C, 30C – etc. – isoprenoid molecules. Reduction of HMG-CoA is an important regulatory step in the overall process. HMG-CoA reductase is regulated at the transcriptional level and by post-transcriptional methods. To form sterols from the open-chain isoprenes requires cyclization followed by various other modifications.

- A mevalonate-independent pathway for IPP formation has been found in algae and plants. It uses central metabolites like pyruvate or glyceraldehyde 3-phosphate to make 1-deoxy-D-xylulose-5-phosphate rather than mevalonate as precursor of the isoprene unit.

- Cholesterol itself is an important metabolic intermediate – being converted into cholesteryl esters, bile acids, to cholecalciferol (and vitamin D) or to various steroid hormones in appropriate tissues. The biosynthesis of cholesterol and regulation of its plasma levels or conversion to other compounds is normally carefully regulated. Several enzymes of the cholesterol biosynthetic pathway are controlled through a specific transcription factor, SREBP. Furthermore, the ratio of cholesterol to cholesteryl esters is carefully regulated in different tissues by the activity of various acyltransferases.

Further reading

General

Fell D (1997) *Understanding the Control of Metabolism*. Portland Press, London.

Gunstone FD, Harwood JL, Dijkstra AJ, eds. (2007) *The Lipid Handbook*, 3rd edn. CRC Press, Boca Raton.

Lipid Library (http://lipidlibrary.aocs.org).

Murphy DJ, ed. (2005) *Plant Lipids: Biology, Utilisation and Manipulation*. Blackwell Science, Oxford.

Ratledge C & Wilkinson SG, eds. (1988). *Microbial Lipids*, **2** vols. Academic Press, London.

Vance DE & Vance JE, eds. (2008). *Biochemistry of Lipids, Lipoproteins and Membranes*, 5th edn. Elsevier, Amsterdam.

Triacylglycerol metabolism

Coleman RA & Lee DP (2004) Enzymes of triacylglycerol synthesis and their regulation, *Prog Lipid Res* **43**:134–76.

Coleman RA, Lewin TM & Muoio DM (2000) Physiological and nutritional regulation of enzymes of triacylglycerol synthesis, *Ann Rev Nutr* **20**:77–103.

Frayn KN, Coppack SW, Fielding BA & Humphreys SM (1995) Coordinated regulation of hormone-sensitive lipase and lipoprotein lipase in human adipose tissue *in vivo*; implications for the control of fat storage and fat mobilization, *Adv Enzyme Regul* **35**:163–78.

Gibbons GF, Islam K & Pease RJ (2000) Mobilisation of triacylglycerol stores, *Biochim Biophys Acta* **1483**:37–57.

Harwood JL, Ramli US, Tang M. *et al.* (2013) Regulation and enhancement of lipid accumulation in oil crops: the use of metabolic control analysis for informed genetic manipulation, *Eur J Lipid Sci Tech* **115**:1239–46.

Lehner R & Kuksis A (1996) Biosynthesis of triacyglycerols, *Prog Lipid Res* **35**:169–201.

McIntyre TM, Snyder F & Marathe GK (2008) Ether-linked lipids and their bioactive species. In DE Vance & JE Vance, eds., *Biochemistry of Lipids, Lipoproteins and Membranes*, 5th edn. Elsevier, Amsterdam, pp. 245–76.

Vance JE (1998) Eukaryotic lipid biosynthetic enzymes: the same but not the same, *Trends Biochem Sci* **23**:423–8.

Yamashita A, Hayashi Y, Nemoto-Sasaki Y *et al.* (2014) Acyltransferases and transacylases that determine the fatty acid composition of glycerolipids and the metabolism of bioactive mediators in mammalian cells and model organisms, *Prog Lipid Res* **53**:18–81.

Phosphoglyceride biosynthesis

Carman GM & Henry SA (1999) Phospholipid biosynthesis in the yeast *Saccharomyces cerevisiae* and interrelationship with other metabolic processes, *Prog Lipid Res* **38**:361–99.

Chen M, Hancock LC & Lopes JM (2007) Transcriptional regulation of yeast phospholipid biosynthetic genes, *Biochim Biophys Acta* **1771**:310–21.

Dormann P. (2005) Membrane lipids. In DJ Murphy, ed., *Plant Lipids: Biology, Utilisation and Manipulation*. Blackwell Scientific, Oxford, pp. 123–61.

McIntyre TM, Snyder F & Marathe GK (2008) Ether-linked lipids and their bioactive species. In DE Vance & JE Vance, eds., *Biochemistry of Lipids, Lipoproteins and Membranes*, 5th edn. Elsevier, Amsterdam, pp. 245–76.

Nagen N & Zoeller RA (2001) Plasmalogens: biosynthesis and functions, *Prog Lipid Res* **40**:199–229.

Parsons JB & Rock CO (2013) Bacterial lipids: metabolism and membrane homeostasis, *Prog Lipid Res* **52**:249–76.

Vance DE & Vance JE (2008) Phospholipid biosynthesis in eukaryotes. In DE Vance & JE Vance, eds., *Biochemistry of Lipids, Lipoproteins and Membranes*, 5th edn. Elsevier, Amsterdam, pp. 213–44.

Vance JE (1998) Eukaryotic lipid biosynthetic enzymes: the same but not the same, *Trends Biochem Sci* **23**:423–8.

Yamashita A, Hayashi Y, Nemoto-Sasaki Y *et al.* (2014) see Triacylglycerol synthesis above.

Degradation of phosphoglycerides

Gelb MH, Jain MK, Hanel AM & Berg OG (1995) Interfacial enzymology of glycerolipid hydrolases: lessons from secreted phospholipases A_2, *Annu Rev Biochem* **64**:653–88.

Liscovitch M, Czarny M, Fiucci G & Tang X (2000) Phospholipase D: molecular and cell biology of a novel gene family, *Biochem J* **345**:401–15.

Murakami M, Taketomi Y, Miki Y *et al.* (2011) Recent progress in phospholipase A_2 research: from cells to animals to humans, *Prog Lipid Res* **50**:152–92.

Ohara O, Ishizaki J & Arita H (1995) Structure and function of the phospholipase A_2 receptor, *Prog Lipid Res* **34**: 117–38.

Wilton DC (2008) Phospholipases. In DE Vance & JE Vance, eds., *Biochemistry of Lipids, Lipoproteins and Membranes*, 5th edn. Elsevier, Amsterdam, pp. 305–29.

Metabolism of glycosylglycerides

Dormann P (2005) Membrane lipids. In DJ Murphy, ed., *Plant Lipids: Biology, Utilisation and Manipulation*. Blackwell Scientific, Oxford, pp. 123–61.

Harwood JL & Okanenko AA (2003) Sulphoquinovosyldiacylglycerol (SQDG – the sulpholipid of higher plants) In YP Abrol and A Ahmad, eds., *Sulphur in Plants*. Kluwer, Dordrecht, pp. 189–219.

Holzl G & Dormann P (2007) Structure and function of glyceroglycolipids in plants and bacteria, *Prog Lipid Res* **46**: 225–43.

Kates M, ed. (1990) *Glycolipids, Phosphoglycolipids and Sulphoglycolipids*. Plenum Press, New York.

Shimojima M (2011) Biosynthesis and functions of the plant sulpholipid. *Prog Lipid Res* **50**:234–9.

Metabolism of sphingolipids

Hirabayashi Y, Igarashi Y & Merrill AH Jr, eds. (2006) *Sphingolipid Biology*. Springer, Tokyo.

Kanfer JN & Hakomori S, eds. (1983) *Sphingolipid Biochemistry*. Plenum, New York.

Lynch DV & Dunn TM (2004) An introduction to plant sphingolipids and a review of recent advances in understanding their metabolism and function, *New Phytol* **161**: 677–702.

Merrill AH Jr (2008) Sphingolipids. In DE Vance & JE Vance, eds., *Biochemistry of Lipids, Lipoproteins and Membranes*, 5th edn. Elsevier, Amsterdam, pp. 363–97.

Merrill AH Jr, Wang MD, Park M & Sullards MC (2007) (Glyco)sphingolipidology: an amazing challenge and opportunity for systems biology, *Trends Biochem Sci* **32**:457–68.

Sandhoff K (2013) Metabolic and cellular bases of sphingolipidoses, *Biochem Soc Trans* **41**:1562–8.

Tidhar R & Futerman AH (2013) The complexity of sphingolipid biosynthesis in the endoplasmic reticulum, *Biochim Biophys Acta* **1833**:2511–18.

Sterol metabolism

Brown MS & Goldstein JL (1986) A receptor-mediated pathway for cholesterol homeostasis, *Science* **232**:32–47.

Burg JS & Espenshade PJ (2011) Regulation of HMG-CoA reductase in mammals and yeast, *Prog Lipid Res* **50**:403–10.

Hammerlin A, Harwood JL & Bach TJ (2012) A *raison d'etre* for two distinct pathways in the early steps of plant isoprenoid biosynthesis?, *Prog. Lipid Res.* **51**:95–148.

Hong C & Tontonoz P (2014) Liver X receptors in lipid metabolism: opportunities for drug discovery, *Nat Rev Drug Discov* **13**:433–44.

Liscum L (2008) Cholesterol biosynthesis. In DE Vance & JE Vance, eds., *Biochemistry of Lipids, Lipoproteins and Membranes*, 5th edn. Elsevier, Amsterdam, pp. 399–421.

Supplementary reading
Triacylglycerol metabolism

Csaki LS, Dwyer JR, Fong LG *et al.* (2013) Lipids, lipinopathies and the modulation of cellular lipid storage and signalling, *Prog Lipid Res* **52**:305–16.

Dodds PF (1995) Xenobiotic lipids: the inclusion of xenobiotic compounds in pathways of lipid biosynthesis, *Prog Lipid Res* **34**:219–47.

Hajra AK (1995) Glycerolipid biosynthesis in peroxisomes (microbodies), *Prog Lipid Res* **34**:343–64.

Rincon E, Gharbi SI, Santos-Mendoza T & Mérida I (2012) Diacylglycerol kinase ζ: at the crossroads of lipid signaling and protein complex organization, *Prog Lipid Res* **51**:1–10.

Sul HS & Wang D (1998) Nutritional and hormonal regulation of enzymes in fat synthesis. Studies of fatty acid synthase and mitochondrial glycerol-3-phosphate acyltransferase gene transcription, *Ann Rev Nutr* **18**:331–51.

Zammit VA (1999) The malonyl-CoA-long-chain acyl-CoA axis in the maintenance of mammalian cell function, *Biochem J* **343**:505–15.

Phosphoglyceride synthesis

Aoyama C, Liao H & Ishidate K (2004) Structure and function of choline kinase isoforms in mammalian cells, *Prog Lipid Res* **43**:266–81.

Brites P, Waterham HR & Wanders RJA (2004) Functions and biosynthesis of plasmalogens in health and disease, *Biochim Biophys Acta* **1636**:219–31.

Chang, Y-F & Carman GM (2008) CTP synthetase and its role in phospholipid synthesis in the yeast *Saccharomyces cerevisiae*, *Prog Lipid Res* **47**:333–9.

Cornell RB & Northwood IC (2000) Regulation of CTP: phosphocholine cytidylyltransferase by amphitrophism and relocalization, *Trends Biochem Sci* **25**:441–7.

Cronan JE (2003) Bacterial membrane lipids: where do we stand?, *Ann Rev Microbiol* **57**:203–224.

de Kroon AIPM, Rijken PJ & De Smet CH (2013) Checks and balances in membrane phospholipid class and acyl chain homeostasis, the yeast perspective, *Prog Lipid Res* **52**:374–94.

Ishii S & Shimizu T (2000) Platelet-activating factor (PAF) receptor and genetically engineered PAF receptor in mutant mice, *Prog Lipid Res* **39**:41–82.

Karasawa K, Harada A, Satoh N, Inoue K & Setaka M (2003) Plasma platelet activating factor-acetylhydrolase, *Prog Lipid Res* **42**:93–114.

Liu Y & Bankaitis VA (2010) Phosphoinositide phosphatases in cell biology and disease, *Prog Lipid Res* **49**:201–17.

Prescott SM, Zimmerman GA, Stafforini DM & McIntyre TM (2000) Platelet activating factor and related lipid mediators, *Annu Rev Biochem* **69**:419–45.

Smith JD (1993) Phospholipid biosynthesis in protozoa, *Prog Lipid Res* **32**:47–60.

Sohlenkamp C, López-Lara IM & Geiger O (2003) Biosynthesis of phosphatidylcholine in bacteria, *Prog Lipid Res* **42**:115–62.

Vance JE & Steenbergen R (2005) Metabolism and functions of phosphatidylserine, *Prog Lipid Res* **44**:207–34.

Yamashita A, Hayashi Y & Nemoto-Sasaki Y (2014) Acyltransferases and transacylases that determine the fatty acid composition of glycerolipids and the metabolism of bioactive lipid mediators in mammalian cells and model organisms, *Prog Lipid Res* **53**:18–81.

Degradation of phosphoglycerides

Schaloske RH & Dennis EA (2006) The phospholipase A(2) superfamily and its group numbering, *Biochim Biophys Acta* **1761**:1246–59.

Yamashita A, Hayashi Y & Nemoto-Sasaki Y (2014) Acyltransferases and transacylases that determine the fatty acid composition of glycerolipids and the metabolism of bioactive lipid mediators in mammalian cells and model organisms, *Prog Lipid Res* **53**:18–81.

Metabolism of sphingolipids

Clarke CJ, Snook CF & Tani M (2006) The extended family of neutral sphingomyelinases, *Biochemistry* **45**:11247–56.

D'Angelo G, Capasso S, Sticco L & Russo D (2013) Glycosphingolipids: synthesis and functions, *FEBS J* **280**:6338–53.

Dickson RC, Sumanasekera C & Lester RL (2006) Functions and metabolism of sphingolipids in *Saccharomyces cerevisiae*, *Prog Lipid Res* **45**:447–65.

Jennemann R & Gröne H-J (2013) Cell-specific in vivo functions of glycosphingolipids: lessons from genetic deletions of enzymes involved in glycosphingolipid synthesis, *Prog Lipid Res* **52**:231–48.

Kolter T & Sandhoff K (2006) Sphingolipid metabolism diseases. *Biochim Biophys Acta* **1758**:2057–79.

Markham JE, Lynch DV, Napier JA, Dunn TM & Cahoon EB (2013) Plant sphingolipids: function follows form, *Curr Opin Plant Biol* **16**:350–7.

Merrill AH Jr (2002) *De novo* sphingolipid biosynthesis: a necessary, but dangerous, pathway, *J Biol Chem* **277**:25843–6.

Tidhar R & Futerman AH (2013) The complexity of sphingolipid biosynthesis in the endoplasmic reticulum, *Biochim Biophys Acta* **1833**:2511–18.

Sterol metabolism

Gotto AM Jr. & LaRosa JC (2005) The benefits of statin therapy – what questions remain?, *Clin Cardiol* **28**:499–503.

Hong C & Tontonoz P (2014) Liver X receptors in lipid metabolism: opportunities for drug discovery, *Nat Rev Drug Discov* **13**:433–44.

Oikkonen VM (2004) Oxysterol binding proteins and its homologues: new regulatory factors involved in lipid metabolism, *Curr Opin Lipid* **15**:321–7.

Pfreiger FW & Ungerer N (2011) Cholesterol metabolism in neurons and astrocytes, *Prog Lipid Res* **50**, 357–71.

CHAPTER 5
Roles of lipids in cellular structures

One of the most important roles of lipids in biological organisms is to provide the fundamental building blocks of cell membranes. Semipermeable plasma membranes based on lipid bilayer structures enable cells to separate themselves from their external environment and to selectively import, retain and export different types of molecules. Hence desirable substances such as nutrients may be imported into a cell, while undesirable molecules such as toxins can be excluded from entry. Equally, the plasma membrane enables cells selectively to export certain products of their metabolism, such as waste products or actively secreted compounds. Lipid-based membranes also surround the various specialized subcompartments, or organelles within eukaryotic cells, that enable each cell to carry out a wide range of metabolic functions in an efficient and well regulated manner. Over the past decade new imaging techniques have been combined with advances in biochemistry and molecular genetics greatly to improve our understanding of the structure and function of biological membranes (see Box 5.1 & Fig. 5.1).

In addition to forming membrane bilayers, some lipids can also form other types of useful macromolecular arrangements. Examples include various types of micelle as well as a wide range of hydrophobic structures of different shapes and sizes. The key property that enables many lipids to form such a diversity of structures is their amphipathic nature. This attribute enables lipids to come together spontaneously with other molecules such as proteins to form relatively large and stable assemblies. In this chapter we will examine the roles of lipids in forming such assemblies and in facilitating their biological functions. We will also look at recent evidence that is shedding light on how membrane lipids may have played vital roles in the origin of cellular life and how they continue to underpin the structure and metabolism of all biological organisms from the simplest bacteria to the most complex animals, including humans.

5.1 Lipid assemblies

By far the most common type of lipid assembly in cells is a mixed-composition sheetlike bilayer structure with two polar surfaces sandwiching a hydrophobic core. Most lipid bilayers in biological systems are made up of a mixture of glycerolipid and sterol molecules that form the matrix of cellular membranes. Other lipid assemblies commonly found in cells include spherical droplets, semicrystalline rods, polymer granules, and various forms of micelle. All these complex multimolecular structures can self-assemble spontaneously thanks to the amphipathic nature of lipid molecules, although their formation and turnover are normally closely regulated.

5.1.1 Lipids can spontaneously form macromolecular assemblies

Experiments in vitro have demonstrated that lipids do not require a biological environment to form higher order structures such as membranes. For example, many classes of lipids, including most phosphoglycerides, sterols, sphingolipids and hopanoids, will spontaneously self-assemble to form bilayer structures in an aqueous medium (Fig. 5.2). This ability to self-assemble is due to the amphipathic nature of these lipid molecules. An amphipathic molecule is made up of a relatively hydrophilic and water-soluble polar domain plus a more hydrophobic nonpolar domain that tends to be excluded from aqueous environments. The hydrophobic lipid domain is excluded because water molecules preferentially associate either with one another or with other relatively polar, i.e. soluble, molecules via hydrogen

Lipids: Biochemistry, Biotechnology and Health, Sixth Edition. Michael I. Gurr, John L. Harwood, Keith N. Frayn, Denis J. Murphy and Robert H. Michell.
© 2016 John Wiley & Sons, Ltd. Published 2016 by John Wiley & Sons, Ltd.

Box 5.1 The use of imaging to study membranes and other lipidic structures

Some of the most important advances in lipid biology over recent years have come from the use of new imaging technologies. These have contributed greatly to our understanding of the real-time dynamics of the behaviour of lipid assemblies in cells. Since 2005, there have been particular advances in the use of live imaging systems including live microscopy; vital staining in combination with fluorescence-activated cell sorting; and time-lapse adaptive harmonic generation microscopy.

Greatly improved analytical tools have also made key contributions to the study of what can be highly complex and rapidly changing lipid mixtures in various cellular compartments. These methods have been especially powerful when used in conjunction with genomic-based approaches such as gene insertions, gene knockouts, and site-directed mutagenesis. Studies employing these and other methods have revealed much detail about the dynamic nature of all of the lipid assemblies in cells and their contribution to fundamental cellular processes such as the trafficking of lipids, proteins, and entire membrane systems.

bonds. The strong tendency of water molecules to form hydrogen bonds is the driving force behind the formation of the various classes of lipid macromolecular assembly, the most important of which is the biological membrane.

Why are hydrophobic domains excluded from aqueous environments? According to the Second Law of Thermodynamics, physical systems that are in equilibrium tend to assume the lowest available energy state. An aqueous solution is at its lowest energy state when the maximum number of hydrogen bonds can be formed between adjacent water and other solute molecules. This occurs readily in liquid water or in an aqueous solution containing polar substances such as salts, e.g. sodium chloride, NaCl. In such solutions, each water molecule can fully satisfy its requirement for hydrogen bond formation by interacting either with neighbouring water molecules or with polar solutes like NaCl.

In contrast, nonpolar compounds such as hydrocarbons cannot form hydrogen bonds with water molecules. If we try to introduce such compounds into an aqueous phase, any water molecules adjacent to a hydrocarbon cannot satisfy their requirement for hydrogen bonds and the net free energy of the system increases. Because physical systems tend to minimize their free energy, the nonpolar hydrocarbons in our example will tend to be excluded from the aqueous phase. If the hydrocarbon is a liquid it will tend to form an oily phase separate from the water, and if it is solid it will remain undissolved. In each case, the aqueous phase minimizes its area of interaction with the hydrocarbon phase.

Polar materials such as NaCl that are readily soluble in water are termed hydrophilic, or 'water loving'. In contrast, nonpolar molecules like hydrocarbons that are unable to dissolve completely in water are called hydrophobic, or 'water fearing'. The crucial property of most

biological lipids is that they are amphipathic, i.e. they contain both polar and nonpolar regions and therefore have both hydrophilic and hydrophobic domains. When large numbers of such amphipathic molecules are dispersed in water, the lowest energy conformation will involve maximizing the interaction of their polar regions with the aqueous phase while minimizing the surface area of their nonpolar regions that are in direct contact with water. The most efficient way to achieve such a state is for the lipid molecules to aggregate so that their nonpolar regions form a separate hydrophobic phase that has little or no contact with the aqueous phase. This can be best achieved if the polar regions of the lipid molecules face the aqueous phase and maximize their capacity to form hydrogen bonds while their nonpolar, hydrophobic regions face inwards away from the water molecules. Examples of such lipid aggregates include bilayers, micelles, inverted micelles, and dispersed droplets.

5.1.2 The shapes of lipid molecules affect their macromolecular organization

Different lipid classes will form different kinds of macromolecular assembly depending on their headgroup and acyl composition and on other conditions such as temperature, salinity and pH. One crucial factor in determining the type of structure formed by a particular lipid is the overall shape of the lipid molecule. As shown in Fig. 5.3, most lipid molecules can be broadly described as either cylindrical, conical, or wedge shaped. Cylindrical lipids such as phosphatidylcholine (PtdCho) or cholesterol will preferentially form relatively flat bilayer structures. Wedge shaped lipids such as lysophosphoglycerides or nonesterified fatty acids (NEFAs) tend to induce convex curvature in a bilayer while at higher

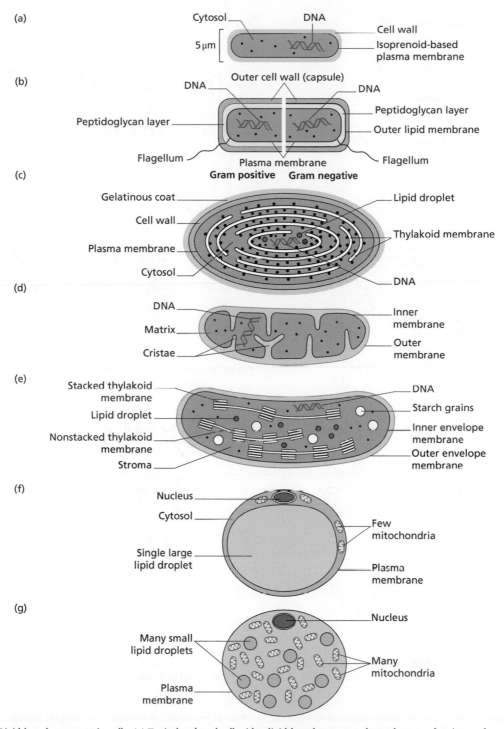

Fig. 5.1 Lipid-based structures in cells. (a) Typical archaeal cell with a lipid-based outer membrane but very few internal membranes. (b) Simple bacterial cell with a lipid-based outer membrane and several invaginated internal membranes. (c) Cyanobacterial cell with separate intracellular thylakoid network. (d) Mitochondrion from a eukaryotic cell. (e) Chloroplast from a photosynthetic plant cell. (f) Mammalian white adipocyte with a single large lipid droplet. (g) Mammalian brown adipocyte with many small lipid droplets.

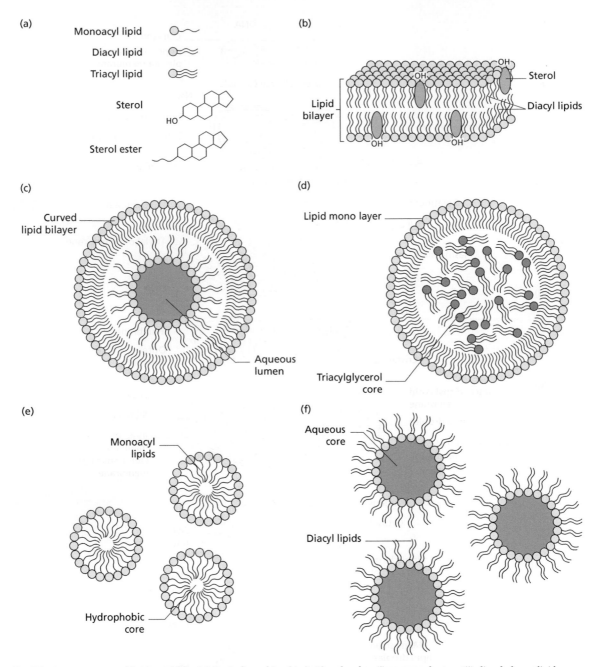

Fig. 5.2 Common types of lipid assembly. (a) Typical amphipathic lipid molecules: (i) monoacyl ester, (ii) diacyl glycerolipid, (iii) sterol. (b) Lipid bilayer. (c) Vesicle with aqueous lumen. (d) Nonpolar lipid droplet with lipidic core. (e) Micelle. (f) Reverse micelle.

concentrations they tend to form spherical or tubular micelles called H_I phases. Conical lipids such as phosphatidylethanolamine (PtdEtn) tend to induce a concave bilayer curvature and at higher concentrations they

may form inverted spherical or tubular micelles called H_{II} phases.

In any biological membrane there tends to be a mixture of relatively flat and curved regions. For this

(a)

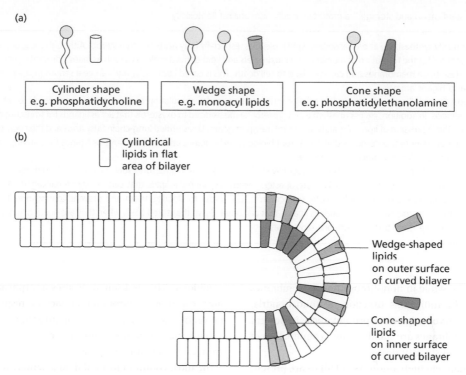

(b)

Fig. 5.3 Lipid shapes and membrane curvature. (a) Cylindrical, conical and wedge-shaped glycerolipids. (b) Membrane curvature caused by the presence of differently shaped lipids.

reason, all membranes tend to have evolved to consist of a mixture of lipid classes. It is this mixture of lipids that enables different membranes to assume a variety of shapes in response to varying physical or biological requirements. In a relatively flat or slightly curved membrane such as the plasmalemma, cylindrical lipids like PtdCho or cholesterol will predominate but conical or wedge shaped lipids may also be present to accommodate localized regions of higher curvature. Perhaps because it is so effective at forming flat membrane bilayers, PtdCho is the most common membrane lipid in most organisms. In organisms that cannot synthesize PtdCho, such as some bacteria and algae, other cylindrical molecules such as betaine lipids are present instead (see Section 2.3.2.4).

During processes such as budding and invagination very highly curved membrane domains can form temporarily but these tend to be mainly stabilized by proteins and are relatively short lived as described below. Because wedge-shaped lipids such as lysophosphoglycerides or NEFAs tend to form micelles rather than bilayers, such lipids will tend to destabilize a bilayer if present above a

certain concentration. For this reason, most cells have protective mechanisms to minimize the accumulation of lysophosphoglycerides and NEFAs. One of the most common ways to achieve this is to esterify such NEFAs to form triacylglycerols (TAG; see Section 4.1), which can then be safely stored away from the bilayer membrane in the form of cytosolic lipid droplets (LDs – see Section 5.5). One of the problems associated with obesity is that the ability of cells to store fatty acids (FAs) safely in this way is exceeded by the amount of circulating FAs, which then accumulate in cells other than adipocytes and cause a series of damaging effects known as lipotoxicity (see Box 5.2 & Section 10.4.1.3).

In many cases, localized accumulations of nonbilayer lipids are highly beneficial to membrane structure and function. For example, in mitochondrial cristae, endoplasmic reticulum (ER), or chloroplast thylakoids, highly curved regions are permanently present as part of the normal membrane structure. Such highly curved regions are greatly enriched in conical or wedge shaped lipids. The most abundant conical lipid in chloroplast thylakoids is the glycolipid, monogalactosyldiacylglycerol (MGDG – see

Box 5.2 Lipid esterification and storage – a protective mechanism against lipotoxicity

Cells are continually synthesizing and/or absorbing lipid molecules, especially fatty acids and free sterols. Although these acyl lipids and steroids are essential for the formation of membranes, lipid droplets and lipid-derived molecules such as many hormones, their presence in excess can destabilize membranes and can even lead to cell death. This is called lipotoxicity and can be a serious problem for all cells from bacteria to higher animals.

Free cholesterol crystallizes even at very low cellular concentration, and accumulation of cholesterol crystals leads to cell death by apoptosis or necrosis. In addition, excess unesterified cholesterol can be oxidized to oxysterols that act as ligands for transcription factors that upregulate the expression of lipogenic and cholesterol transport genes. Unesterified long-chain fatty acids and their acyl-coenzyme A (CoA) derivatives are potent detergents that can disrupt biological membranes and can also serve as ligands for transcription factors that regulate many cellular and metabolic processes.

Cells protect themselves from lipotoxicity by temporarily storing excess lipids in structures such as cytosolic lipid droplets. Firstly the fatty acids are converted in the ER into TAGs or sterol esters, which are nonpolar lipids that can readily be packaged into spherical droplets. These droplets are coated with specific surface proteins in order to stabilize them and prevent coalescence. The surface proteins also act as receptors for enzymes such as lipases that can break down the stored lipids when they are needed, e.g. for energy production, membrane formation, or hormone biosynthesis.

Section 2.3.2.3), while in nonphotosynthetic membranes PtdEtn normally fulfils this function. In mitochondria, where tightly curved cristae are present, these important respiratory membranes are stabilized by a combination of specific lipids and structural proteins.

In *E. coli*, relatively high amounts of PtdEtn are present in cell membranes to enable them to accommodate highly curved regions. The importance of lipid shape, rather than chemical structure is shown by experiments with *E. coli* mutants that lack PtdEtn. The mutant bacterial cells are severely defective in many membrane processes but can be restored to normal function if alternative conical lipids, such as MGDG, are provided via gene transfer from plant sources. Normal *E. coli* cells cannot synthesize MGDG but this wedge-shaped glycolipid can still substitute completely for PtdEtn in terms of overall membrane function.

5.1.3 The polymorphic behaviour of lipids

Polymorphism is the occurrence of different forms within a given population. As applied to lipids, it refers to the various types of structure into which assemblies of similar lipid molecules can self-organize under different conditions. We saw previously that different lipid molecules can form various types of flat or curved lamellar (bilayer) structures. Other types of lipid tend to form spherical structures such as various forms of micelle, cubic structures, and less common structures such as rhombic phases. A single class of lipid molecule may form different aggregate structures under different

conditions. This is known as mesomorphism and the most common environmental factors responsible for mesomorphic changes in lipid organization in biological systems include temperature, hydration, pH, and ionic strength of the aqueous phase.

The most common form of lipid assembly is the lamellar or bilayer phase, but this can exist in one of two forms, gel or liquid-crystalline (see Fig. 5.4). In the liquid-crystalline state the FA chains are freely mobile and tend to occupy a significantly larger volume than those of gel phase lipids, which are more static and rigid. Most biological membrane lipids tend to be in a liquid-crystalline state but in some membranes both liquid-crystalline and gel phases can coexist for short periods. In some cases localized regions or domains of mainly gel phase lipids can be useful structures as they serve to restrict the movement of proteins within such domains. Examples include some of the so-called 'lipid raft' structures that can be highly enriched in cholesterol, sphingolipids and relatively saturated glycerolipids. This mixture of lipids forms a rigid matrix that traps specific populations of proteins required for processes such as membrane budding as discussed in Section 5.4.3.

The phase behaviour of a particular lipid class, such as PtdCho, is determined mainly by its FA composition and the temperature. Other factors that affect lipid phase behaviour in a membrane include hydration and pressure as well as the presence of other lipid classes and protein content. Shorter and more unsaturated acyl chains will adopt a fluid liquid-crystalline state at relatively low temperatures while longer chain and more

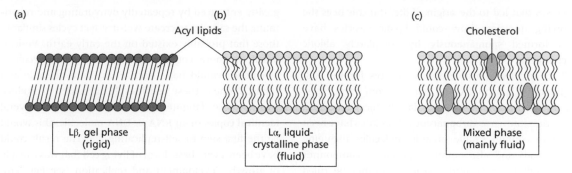

Fig. 5.4 Lipid polymorphism. (a) Rigid gel phase lipid bilayer – Lβ structure. (b) Fluid liquid-crystalline lipid bilayer – Lα structure. (c) Mixed lipid phases in a typical biological membrane.

saturated fatty acids (SFA) require higher temperatures before they undergo the phase transition from the gel to liquid-crystalline state. This phase transition from a gel to a liquid-crystalline state is analogous to the melting of a solid to form a liquid and is sometimes referred to as the melting transition. In a particular membrane, the temperature at which the melting transition occurs can also be influenced by the nature of the polar headgroup, the presence of other lipid classes, and by the protein composition.

5.2 Role of lipids in cellular evolution

During recent decades there has been a revolution in our understanding of the origin and evolution of living organisms. The discovery of previously unrecognized life forms, such as the archaea, and the advent of cheap and rapid DNA sequencing have changed many perceptions about how and when living organisms might have arisen. This research has also suggested that lipids might have played a pivotal role in the origin and evolution of cellular life.

5.2.1 Lipids and the origin of life

All living organisms on Earth are based on discrete cellular structures that reproduce via DNA/RNA-mediated processes. It is now widely believed that life originated in a so-called 'RNA world' in which very simple life forms used RNA both as a genetic molecule, able to copy itself, and as enzymes that catalysed a rather limited range of biochemical reactions. However, the Darwinian selection of any better adapted versions of these life forms would not have been possible unless each entity could be

physically separated both from other life forms and from the external environment. In other words, genetics-based evolution would not have been possible without some sort of a limiting membrane or other type of barrier between each replicating unit.

Another way of putting it is that selection of improved RNA or DNA forms cannot occur if they are all mixed together in a biochemical 'soup'. However, if the different RNA or DNA variants were packaged separately, these discrete 'cells' could compete with one another and the better adapted or 'fitter' cells would be more likely to survive and give rise to genetically similar offspring. This process of prebiological evolution by natural selection eventually gave rise to the relatively complex DNA/protein-based organisms bounded by lipidic cell membranes that now populate the biosphere.

The question of whether lipid-encapsulated protocells and RNA-based genetic material evolved at the same time, or if one arose before the other, has yet to be resolved. One suggestion is that RNA-based life first arose at submarine hydrothermal vents around which inorganic mineral-based compartments of 1–100 μm diameter can still be found. These small compartments might have enabled RNAs and other complex organic molecules to become sufficiently concentrated to evolve new types of biochemistry that eventually led to the formation of membrane-enclosed cells. However, an alternative theory suggests that life originally arose in lipidic vesicles that were formed spontaneously via abiotic processes at the same time as the RNA-based replicators. Once some of these replicators became enclosed within lipidic vesicles, the combined structures would effectively have become the first 'living' cells. In other words, lipids may have played a vital role in the earliest processes that led to the origin of life. But this begs the following question: how could lipidic vesicles have been formed spontaneously by completely abiotic mechanisms?

There are two possible abiotic sources for the first amphipathic lipid molecules, namely from extraterrestrial meteorites or from terrestrial chemical reactions. Meteorites known as carbonaceous chondrites have been found to contain simple organic molecules including amino acids and also amphipathic lipidic compounds able to self-assemble into vesicles. It cannot be ruled out that these compounds played a role in the origin of life. However, a terrestrial abiotic source of lipids is a more parsimonious, and hence more scientifically satisfactory, explanation because it does not require us to invoke the presence of rare and uncertain extraterrestrial agents such as carbon-rich meteorites.

It has been shown experimentally that a range of hydrocarbons can be formed from carbon monoxide and hydrogen in the presence of naturally occurring iron catalysts under conditions that still commonly occur in oceanic hydrothermal vents. Such conditions were more common at the dawn of life about 4 billion years ago. As shown in Fig. 5.5, some of the hydrocarbons produced in these reactions could readily have become oxidized to generate long-chain FA and fatty alcohol mixtures that are capable of forming micelles. In the presence of glycerol, a 3C polyalcohol, a range of glycerolipids could then be produced. Initially these would be mostly simple glycerolipids (see Sections 2.2 and 2.3) such as monoacylglycerols (MAG) and diacylglycerols (DAG), but in time it is possible that more complex phospholipids and glycolipids could have been produced.

Other experiments in vitro have shown that mixtures of even simpler abiotic molecules, such as 10C-acids, 10C-alcohols and 10C-glycerol monoesters, can spontaneously form semipermeable vesicles. Moreover, these vesicles are able to take up oligonucleotides that can then become trapped inside them. In these experiments the uptake of oligonucleotides by simple lipid vesicles was greatly enhanced by repeatedly dehydrating and rehydrating the samples to create wet-dry-wet cycles similar to those that probably existed on the early earth. Vesicles based on more complex phospholipid and glycolipid assemblies would have been even more versatile and robust. If one of these vesicles split or budded to produce two or more daughter vesicles that contained several identical copies of an RNA or DNA molecule, this would be the first step to self-replicating life. The result could have been a very basic form of living cell that was capable of growth, development and replication (see Fig. 5.6). The encapsulation of genetic material inside a lipidic membrane could therefore have been a decisive step in the evolution of cellular life, and remains a defining feature of all living organisms to this day.

5.2.2 Lipids and the evolution of prokaryotes and eukaryotes

If all life is descended from simple cells bounded by glycerolipid membranes, we would expect that the membrane lipids of the major groups of present-day organisms to be fundamentally similar. However, one of the most puzzling features of the membrane lipids of prokaryotes and eukaryotes is the presence of two quite distinctive stereochemical (chiral) forms of glycerolipids. Where two chiral forms of a molecule are possible, it is normally found that biological systems produce only one of the two possible versions. For example, biologically synthesized amino acids are normally found as L (*laevo*- or left-handed) forms while naturally occurring sugars are almost always found as D (*dextro*- or right-handed) chiral forms. It was therefore puzzling when two alternative chiral forms of glycerol phosphate were found in the lipids of living organisms. One chiral form was found in only bacteria and eukaryotes, while the glycerolipids of the archaea were all made up of the opposite chiral form.

The glycerolipids of bacteria and eukaryotic organisms are invariably based on *sn*-glycerol 3-phosphate (G3P), which is produced from dihydroxyacetone phosphate

Fig. 5.5 Theoretical mechanism for the abiotic synthesis of membrane lipids. (a) It is possible for hydrocarbons and glycerol to be produced abiotically from carbon monoxide and hydrogen in the presence of iron catalysts via the Fischer-Tropsch process under conditions found in modern hydrothermal vents. (b) Similar conditions would have been common in early oceans leading to formation of long-chain hydrocarbons that would readily oxidize into mixtures of long-chain carboxylic acids and alcohols. (c) Glycerol can condense with long-chain carboxylic acids and alcohols to form phospholipids. (d) As the concentrations of these lipids increased they would spontaneously assemble into mixed micelles, and eventually into lipid bilayers to form vesicles, some of which might encapsulate abiotically produced RNA or DNA molecules.

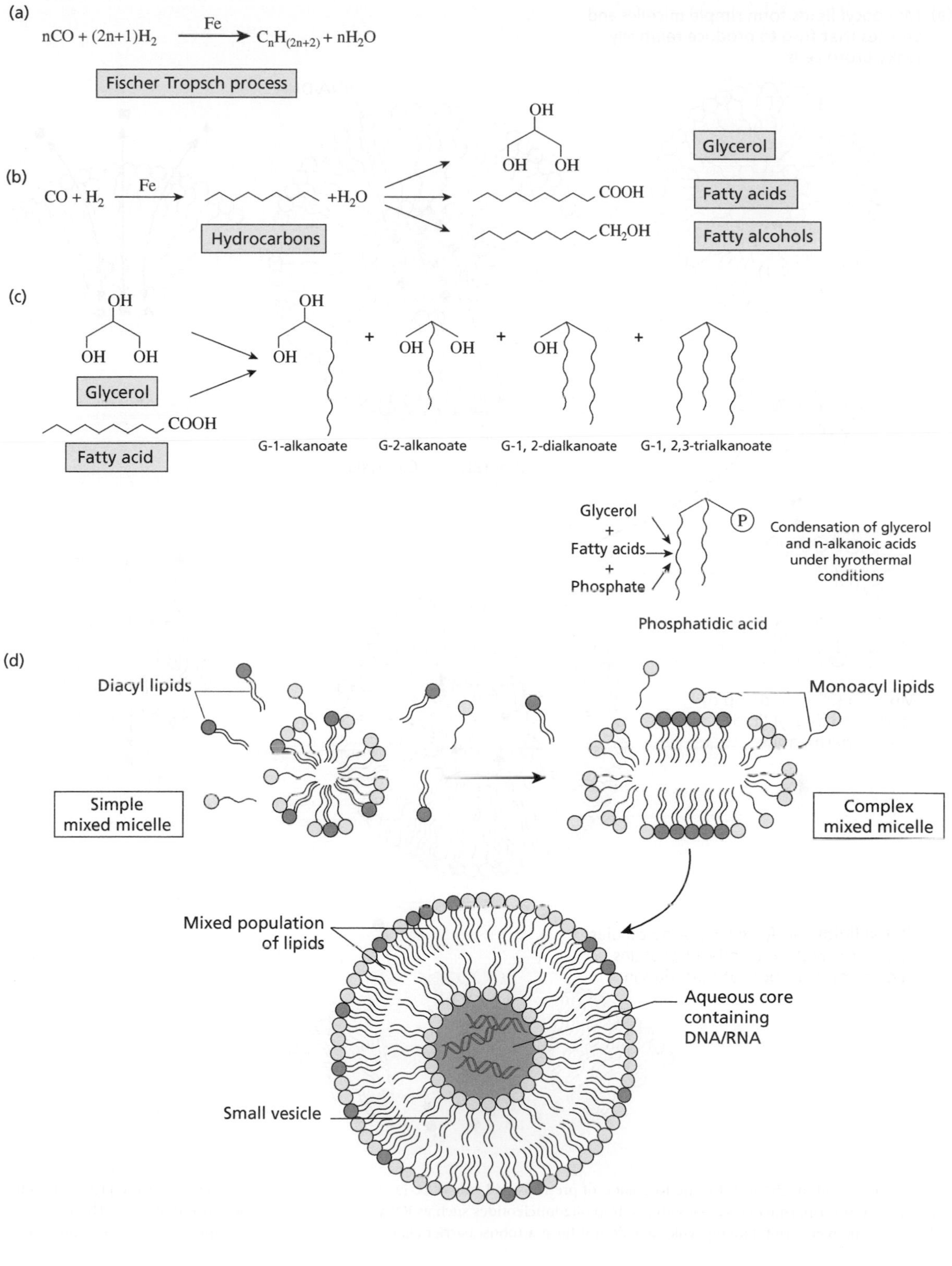
(a)
nCO + (2n+1)H2 —Fe→ CnH(2n+2) + nH2O
Fischer Tropsch process
(b)
CO + H2 —Fe→ +H2O
Hydrocarbons
OH
OH OH
Glycerol
COOH
Fatty acids
CH2OH
Fatty alcohols
(c)
Glycerol
COOH
Fatty acid
G-1-alkanoate
G-2-alkanoate
G-1, 2-dialkanoate
G-1, 2,3-trialkanoate
Glycerol
+
Fatty acids
+
Phosphate
P
Condensation of glycerol and n-alkanoic acids under hyrothermal conditions
Phosphatidic acid
(d)
Diacyl lipids
Monoacyl lipids
Simple mixed micelle
Complex mixed micelle
Mixed population of lipids
Aqueous core containing DNA/RNA
Small vesicle

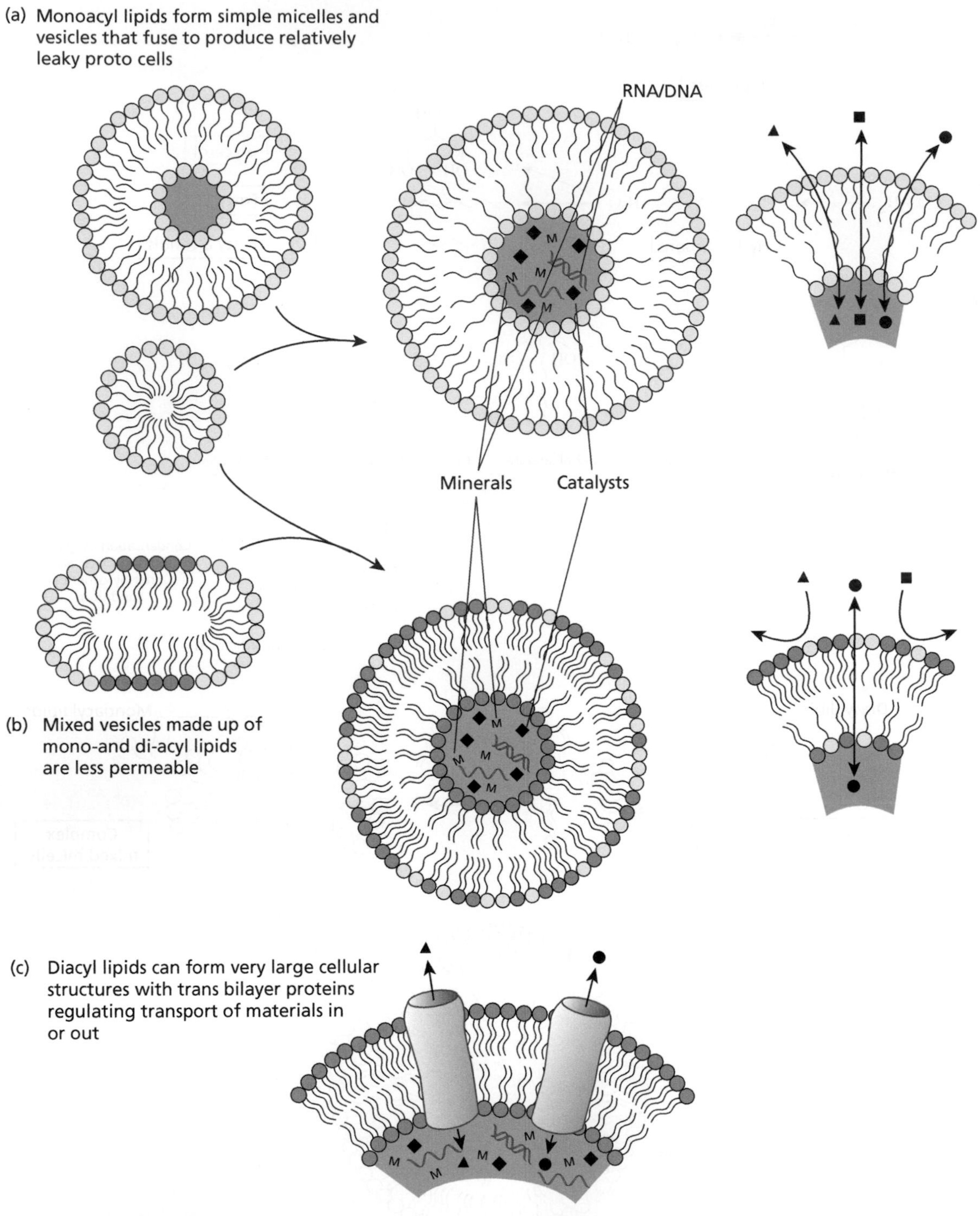
(a) Monoacyl lipids form simple micelles and vesicles that fuse to produce relatively leaky proto cells
RNA/DNA
M
Minerals
Catalysts
(b) Mixed vesicles made up of mono-and di-acyl lipids are less permeable
(c) Diacyl lipids can form very large cellular structures with trans bilayer proteins regulating transport of materials in or out

Fig. 5.6 Theoretical mechanism for the formation of pre-cells and true cells. (a) Simple hydrocarbons such as decane (10C) can self-assemble into semi-permeable vesicles able to trap oligonucleotides such as RNA to form a very basic form of pre-cell. These hydrocarbon bilayers are relatively leaky and do not form a robust barrier against the external environment. (b) Vesicles based on

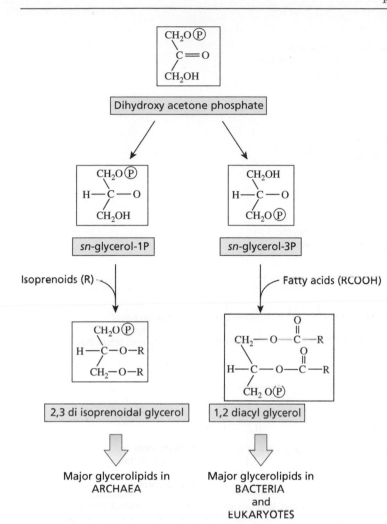

Fig. 5.7 Glycerolipid biosynthesis in archaea, bacteria and eukaryotes. The major archaeal glycerolipids are based on G1P-based instead of G3P. This difference in the chiral form of glycerol phosphate is due to the altered stereochemical specificity of glycerol phosphate dehydrogenase in archaea compared to the same enzyme in bacteria and eukaryotes. Three other important differences are that in archaea the lipid side chains are ether-linked instead of ester-linked, isoprene-based instead of fatty acid-based, and are highly branched instead of linear

(DHAP) by a distinctive form of glycerol phosphate dehydrogenase (see Fig. 5.7). This means that in bacteria and eukaryotes the glycerolipid polar headgroup is always attached to the C3 position of the glycerol molecule. However, archaeal cells have a completely different type of glycerol phosphate dehydrogenase compared to all other organisms. This archaeal enzyme produces *sn*-glycerol 1-phosphate (G1P) from DHAP. For this reason archaeal cell membranes contain glycerolipids with their polar headgroups at only the C1 position rather than the more familiar C3 position found in bacteria and eukaryotes (see Fig. 5.8).

This situation is rather puzzling because most evolutionary models assume that archaeal and bacterial cells fused to produce a composite cell that gave rise to all eukaryotic organisms including protists, animals, plants

mixtures of hydrocarbons with more complex fatty acids and diacylglycerols are more robust and may have been the next stage of cellular evolution. These membranes are less permeable and form better barriers against the external environment. (c) Finally, the modern bilayer membrane is formed from glycerolipids plus additional stabilizing lipids such as sterols or hopanoids. The enclosed RNA and DNA begin to direct synthesis of enzymatic proteins and cellular metabolism becomes possible. These glycerolipid-based membranes are able to incorporate proteins and thereby act as a highly effective selectively permeable barrier to the external environment.

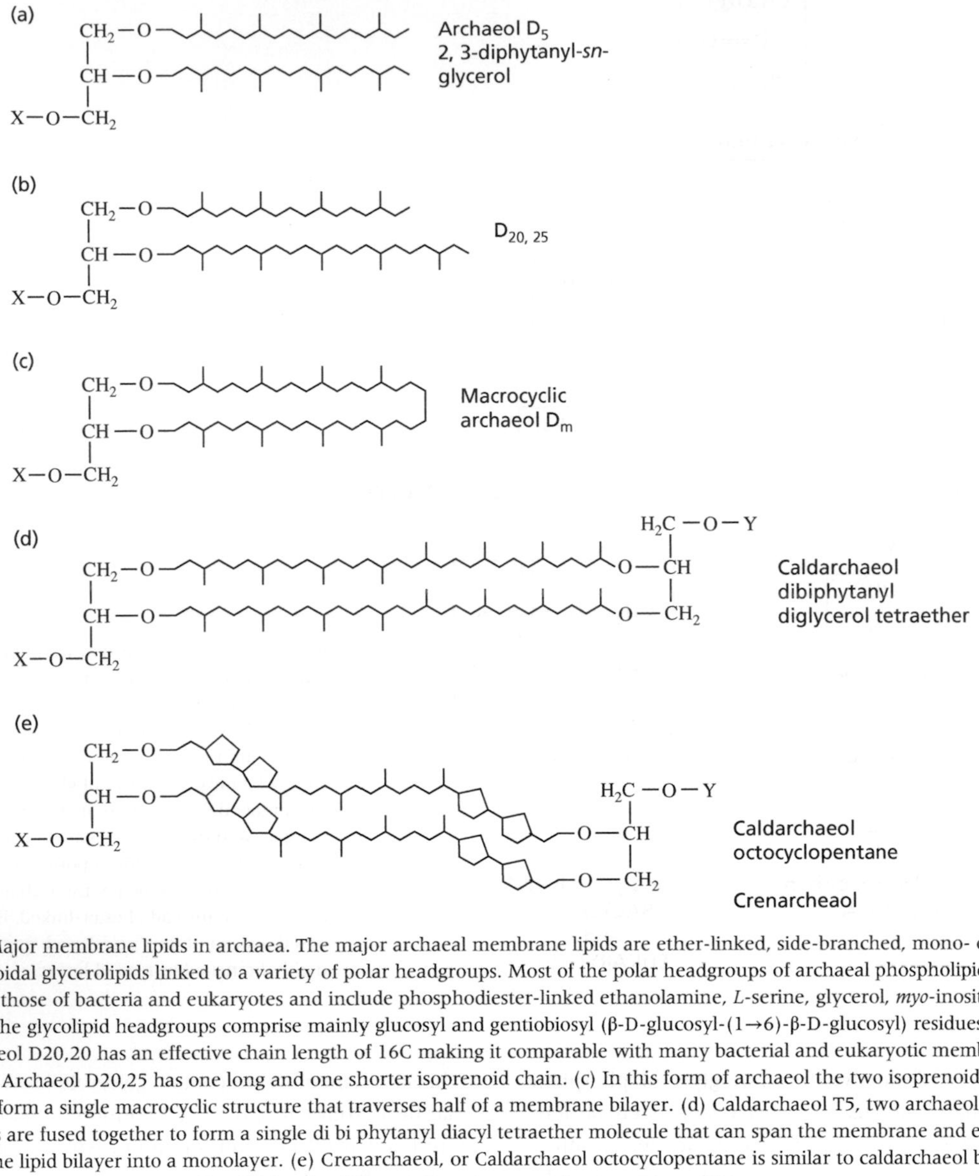

Fig. 5.8 Major membrane lipids in archaea. The major archaeal membrane lipids are ether-linked, side-branched, mono- or diisoprenoidal glycerolipids linked to a variety of polar headgroups. Most of the polar headgroups of archaeal phospholipids are similar to those of bacteria and eukaryotes and include phosphodiester-linked ethanolamine, L-serine, glycerol, myo-inositol, and choline. The glycolipid headgroups comprise mainly glucosyl and gentiobiosyl (β-D-glucosyl-(1→6)-β-D-glucosyl) residues. (a) Archaeol D20,20 has an effective chain length of 16C making it comparable with many bacterial and eukaryotic membrane lipids. (b) Archaeol D20,25 has one long and one shorter isoprenoid chain. (c) In this form of archaeol the two isoprenoid chains are linked to form a single macrocyclic structure that traverses half of a membrane bilayer. (d) Caldarchaeol T5, two archaeol-like molecules are fused together to form a single di bi phytanyl diacyl tetraether molecule that can span the membrane and effectively convert the lipid bilayer into a monolayer. (e) Crenarchaeol, or Caldarchaeol octocyclopentane is similar to caldarchaeol but with eight cyclopentane rings to provide additional stability under extreme environmental conditions.

and fungi. Eukaryotic cells should therefore contain a mixture of glycerolipids with polar headgroups at both the C1 or C3 positions. Indeed, in the earliest protocells that were theoretically derived from the abiotic processes described in Fig. 5.6 there would probably have been an approximately equal (racemic) mixture of G1P and G3P glycerolipids. Because both enzymic forms of glycerol phosphate dehydrogenase were present in the earliest

eukaryotes, these cells probably produced both chiral forms of glycerolipids. This leads to two questions:

1 Why does every modern organism produce only one chiral form of glycerolipids rather than a chirally mixed population of glycerolipids?
2 Why do archaea use one chiral form of glycerolipids while bacteria and eukaryotes use the opposite chiral form?

We can answer these two questions as follows. As we saw above, the original abiotically formed membranes could have contained equimolar mixtures of G1P and G3P glycerolipids. However, once these lipids were produced biologically via a dehydrogenase, only one of the chiral forms would have been produced, depending on the stereochemical specificity of the original enzyme. The question of which chiral form was produced would depend which type of dehydrogenase evolved in a particular lineage. It appears that G1P-producing dehydrogenases evolved in the archaea, whereas an unrelated family of dehydrogenases that produced G3P evolved in the bacteria.

When an archaeal and bacterial cell first combined to produce the ancestor of all extant eukaryotes, it seems that only the lipid synthesizing enzymes of the dominant bacterial partner were retained and all traces of G1P-based archaeal lipids eventually disappeared from the eukaryotic lineage. However, eukaryotes did retain the isoprenoid pathway needed from their archaean ancestors (the pathway is not present in bacteria) although these compounds were no longer used to synthesize membrane lipids (see Section 5.3.2) but assumed a range of alternative functions instead.

Another evolutionary puzzle is the origin of sterol biosynthesis. We know that sterols are present in almost all eukaryotes and that they are also found in three bacterial lineages (but not in any of the archaea). Those bacteria that can synthesize sterols contain closely related homologues of eukaryotic sterol biosynthetic genes but such genes are absent from the vast majority of present-day bacteria. One possible explanation for this situation is that a common ancestor of bacteria and eukaryotes was capable of sterol biosynthesis and that most bacteria subsequently lost these genes. However, as a result of more extensive genome sequencing, this possibility has now been questioned. Instead, it is proposed that sterol biosynthesis was not universal in bacteria and that the earliest eukaryotes did not synthesize sterols either. However, a gene cluster, or operon, containing all the bacterial sterol biosynthetic genes may have been acquired by early eukaryotes soon after their divergence from bacteria. These genes could have been acquired via horizontal gene transfer, for which there is good evidence in the case of many other genes during the course of evolution. Such a bacterial origin for sterol biosynthesis in eukaryotes remains strictly an unproven working hypothesis.

5.2.3 Archaeal lipids are unusual but are well adapted for their lifestyle

As we saw above, the archaeal membranes contain different chiral forms of glycerolipids compared with all other organisms. But that is far from the complete story of the lipid components of these highly versatile cells. Although many archaea are able to synthesize fatty acyl ester-based glycerophospholipids, by far their most abundant membrane glycerolipids are based instead on ether-linked 20C methyl-branched isoprenoids. An even stranger lipid that is relatively abundant in some archaeal membranes is a tail-to-tail fusion of two di-20C isoprenoids to produce a single 80C tetraether molecule that can span the entire width of the membrane (see Fig. 5.9). This creates a doubled headed monolayer membrane structure instead of a conventional bilayer. Physical studies have shown that such membranes maintain a relatively fluid liquid crystalline organization over a much greater range of physiological conditions than membranes made up of FA-based lipids. Another feature of membranes made from archaeal ether lipids is that they are extremely impermeable, particularly to protons, which enables them to maintain proton gradients at temperatures well in excess of 100 °C.

It is assumed that the major reason for the presence of isoprenoid ether lipids rather than fatty acyl ester lipids in archaeal membranes is to allow them to grow and reproduce in a variety of diverse and often harsh environmental conditions, for example: extremely acid (pH <2), alkaline (pH >9), salty (>2 M NaCl), cold (−15 °C), or hot (123 °C) conditions. Archaea thus thrive in a far greater range of environments than any other forms of extant life. This is due, at least to some extent, to the uniquely stable and versatile nature of their isoprenoid-based membrane lipids. In another adaptation to changing environmental conditions, some archaea are able radically to alter their membrane lipids to form even more stable and flexible structures. Thus hyperthermophiles (e.g. *Thermococcus* and *Archaeoglobus*) can combine two di-20C tetraethers to form a single 80C membrane-spanning tetraether molecule (namely caldarcheol), while also generating numerous cyclopentane rings in their isoprenoid chains. The result is a bulky but extremely stable lipid monolayer structure that is capable of withstanding environmental conditions that would almost immediately destroy any other type of biological membrane.

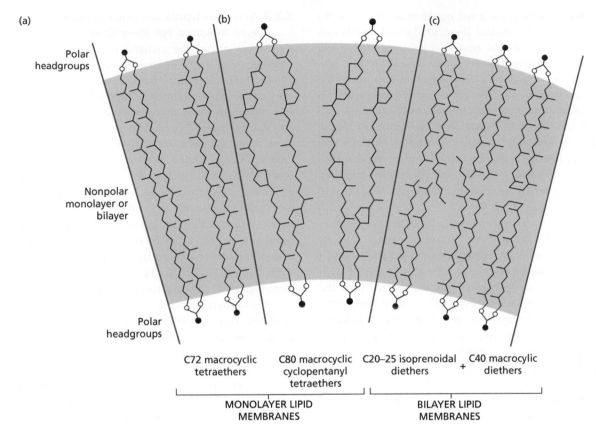

Fig. 5.9 Lipid organization in archaeal membranes. The lipid organization in archaeal membranes is more varied that than in bacteria and eukaryotes. (a) In many extremophile archaea, the membrane is in the form of a monolayer typically made up of 36C macrocyclic tetraethers. (b) Under the most extreme conditions, such as hydrothermal vents where the temperature can be at or above the boiling point of water, some archaea are able to survive by forming exceptionally rigid membranes made up of 40C macrocyclic tetraethers containing numerous cyclopentane rings. (c) In some archaea, 20–25C isoprenoid diethyl glycerolipids and 40C macrocyclic diethyl glycerolipids form a bilayer that is not very different in overall structure from those found in 16C/18C fatty acyl diester glycerolipids of bacteria and eukaryotes.

It is now appreciated that the isoprenoid ether lipids of archaea are one of the major factors in their amazing success in withstanding extremes of temperature, salinity or dehydration that would be fatal to organisms with 'conventional' fatty acyl ester-based membrane lipids. However, there is a price to be paid for such stable membranes and archaeal cells tend to grow much more slowly than their bacterial or eukaryotic counterparts. In general it is found that fatty acyl ester-based lipids perform better in the more 'normal' environmental conditions that exist in most present-day ecosystems. Under such conditions, fatty acyl ester-based cells are able to grow and divide much more rapidly than cells containing isoprenoid ether-based lipids. As a result fatty acyl ester lipids are now by far the most widespread lipids and the bacterial and eukaryotic species that contain them are present in most terrestrial habitats. In contrast, many of the slower growing but highly versatile archaea are more frequently found in unusual and extreme niches, such as superheated volcanic vents, where environmental conditions are far too harsh for organisms with conventional fatty acyl ester-based membrane lipids.

5.3 Membrane structure

Biological membranes tend to look very similar when viewed in electron micrographs but their structure at the molecular level differs considerably in various cell types and even between different membrane systems in the same cell. We have already seen how cell membranes are based on relatively straightforward lipid bilayer structures that can adopt different shapes depending on their lipid composition. However, cell membranes also contain a wide range of proteins that are either embedded within or lie immediately adjacent to the lipid bilayer. These proteins can have a considerable influence on the structure and function of a given membrane.

The most widely used way of depicting the main structural features of a biomembrane is termed the fluid-mosaic model. This model figuratively describes the membrane as a bilayer consisting of a relatively fluid 'sea' of small lipid molecules in which much larger proteins are embedded to various extents (see Fig. 5.10). Proteins and lipids can diffuse more or less freely in the lateral plane of the membrane. As we will see below, the real situation is rather more complicated with many membrane systems having additional features that are still being discovered. Examples of such features include the transverse and lateral heterogeneity of lipids and proteins; trans-bilayer flip-flop of lipids; relatively rigid lipid rafts with highly selective protein populations; and the tethering of certain membrane proteins to various intra- and extracellular components.

5.3.1 The fluid-mosaic model of membrane structure

The fluid mosaic model of membrane organization was originally proposed by Singer and Nicholson in 1972. In this model, which applies to most nonarchaeal cells, the fundamental matrix of the biological membrane is a bilayer that is mainly composed of relatively mobile glycerolipids. In a typical membrane these lipids usually contain two fatty acyl residues of 16C, 18C or 20C chain length, with each chain having between zero and three double bonds. Shorter and longer fatty acyl chains are also found in membranes, but these are usually minor components. Under the most common normal physiological conditions, namely between 0 °C and 50 °C, the acyl chains will be relatively fluid, enabling the lipid molecules to flex, to rotate freely and to diffuse laterally in the plane of the membrane. SFAs are also present in most membranes and in some specialized membrane regions, such as lipid rafts, they can be major components.

A membrane bilayer made up of glycerolipids in a fluid liquid crystalline phase can provide a two-dimensional structure into which integral membrane proteins can be fully embedded while still able to diffuse freely within the plane of the bilayer. Studies with fluid artificial lipid bilayers have shown that integral membrane proteins

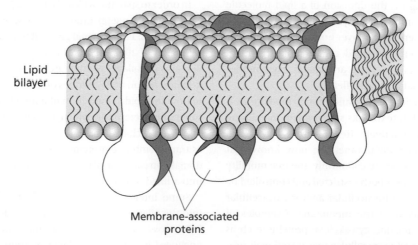

Fig. 5.10 Classic fluid-mosaic model of membrane organization. According to this model, a biological membrane can be considered as a relatively fluid two-dimensional liquid-like structure. This allows most lipid and protein molecules to diffuse freely in the lateral plane of the membrane while transverse movement of lipids across the bilayer is much more constrained.

are capable of rapid lateral diffusion rates – with coefficients typically of the order of 10^{-8} to $10^{-7}\,cm^2\,sec^{-1}$. Because the relatively bulky protein molecules tend to impede one another, these diffusion rates decrease as the bilayer protein concentration increases. However, lateral diffusion rates of proteins in biological membranes tend to be considerably lower than in model systems – typically only about $10^{-10}\,cm^2\,sec^{-1}$ for intact plasma membranes. This implies that additional factors constrain protein movement in biological membranes. These factors can include specific interactions between membrane proteins, or between proteins and lipids, that impede their movement.

Most of the lipid molecules in a membrane bilayer are free to diffuse laterally within the bilayer leaflet in which they are located. As discussed below, there are some exceptions to this rule – as in microdomains such as membrane 'lipid rafts', in which particular lipids may transiently associate with proteins involved in processes such as signal transduction and import/export. Another exception is the boundary lipid layer, typically an annulus one molecule thick that surrounds some membrane proteins. Some of these boundary lipids associate with specific proteins so do not diffuse freely throughout the bilayer. Experiments in model membranes containing lipid molecules tagged with chromophores that can be detected using 'fluorescence recovery after photobleaching' (FRAP) show that most of the common membrane lipids can laterally diffuse within along the bilayer rapidly – exchanging places with a neighbour on a timescale of $\sim 10^{-7}$ sec. In contrast, the diffusion of a lipid molecule between the two leaflets of a bilayer, which is termed 'flip-flop', occurs only very slowly, over hours. This means that transverse asymmetries of lipid composition are much more easily maintained than are lateral asymmetries within an individual bilayer leaflet.

The original version of the fluid mosaic model that is still often depicted in textbooks shows a few membrane proteins diffusing relatively freely within the lateral plane of a lipid-rich membrane. We now know that this is rarely the case. More commonly, the free mobility of membrane proteins is both restricted and controlled by their interactions with intracellular and/or extracellular components adjacent to the membrane. Examples of such components include cytoskeletal proteins such as actin and dynein and extracellular matrix/cell wall proteins or polysaccharides. In addition, many membrane proteins are parts of bulky macromolecular polypeptide

complexes that have very limited mobility. This mobility can be further impeded by the presence of one or more layers of quite tightly bound lipid molecules – the so-called 'boundary lipids' that were mentioned above. A further restriction on protein mobility in many membranes is the relative abundance of intrinsic proteins. For example it is estimated that the inner and outer surfaces of chloroplast thylakoid membranes in plants and synaptic vesicles in mammals are almost entirely covered in proteins, with the lipids simply filling in the gaps between what is mostly a protein-dominated structure.

5.3.2 Extrinsic and intrinsic membrane proteins

Membrane-associated proteins can be broadly divided into two groups, intrinsic and extrinsic (see Fig. 5.11). Intrinsic proteins are embedded partially or completely within the bilayer and normally cannot be removed without disrupting the membrane, for example by detergent solubilization. Because intrinsic proteins are partially or completely embedded in the membrane, a substantial part of the protein will be in contact with the hydrophobic core of the bilayer. The most common conformation for the bilayer-embedded part of an intrinsic protein is in the form of one or more α-helical regions made up of nonpolar or hydrophobic amino acid residues. Examples of the relatively hydrophobic amino acids that commonly make up such α-helical regions include valine, leucine, isoleucine, alanine, phenylalanine and methionine.

In order to span the width of a typical bilayer membrane each α-helical region must contain about 20–25 amino acid residues. For a normal glycerolipid bilayer, an α-helical region of 20–23 residues may suffice but in lipid raft domains the bilayer width is greater and might require an α-helical region of 25 residues or more. This is an example of how the lipid composition of a membrane domain can sometimes determine which proteins are admitted and which are excluded. Bilayer-spanning proteins are known as transmembrane proteins. They usually expose a polar domain on each side of the membrane, and this locks them into a fixed orientation.

Some integral proteins are only partially embedded into the bilayer and their hydrophobic domain(s) do not fully cross the membrane, so they are not as firmly anchored into the lipid bilayer as transmembrane proteins. The polar regions of these nontransmembrane integral proteins are present only on one side of the

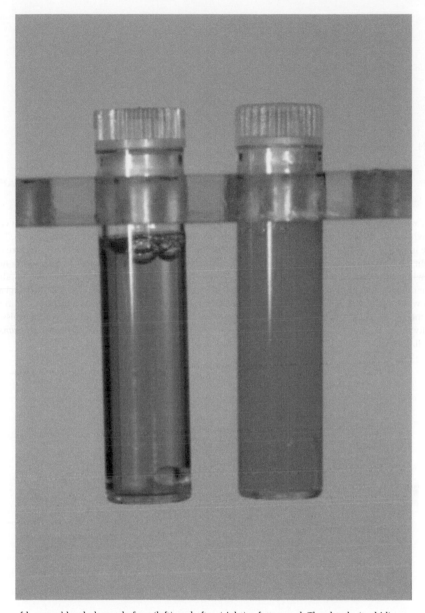

Fig. 7.6 Appearance of human blood plasma before (left) and after (right) a fatty meal. The cloudy (turbid) appearance after a fatty meal is due to the presence of TAG-rich chylomicrons, which are large enough to scatter visible light.

Lipids: Biochemistry, Biotechnology and Health, Sixth Edition. John L. Harwood, Keith N. Frayn, Denis Murphy, Robert Michell and Michael I. Gurr.
© 2016 John Wiley & Sons, Ltd. Published 2016 by John Wiley & Sons, Ltd.

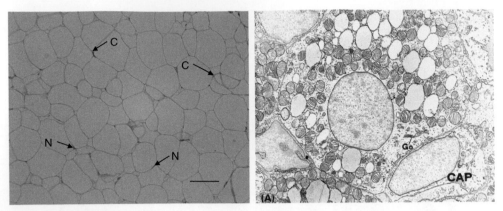

Fig. 9.1 Histology of white and brown adipose tissues. Left, white adipose tissue under the light microscope. Each cell consists of a large lipid droplet (white) surrounded by a narrow layer of cytoplasm. The nucleus (N) can be seen in some cells. There are capillaries (C) at the intersections of the cells: some are marked. The scale bar represents 100 μm (0.1 mm). Picture courtesy of Rachel Roberts. Right, an electron micrograph of brown adipose tissue. In this high-powered view, one adipocyte nearly fills the picture. Unlike the white adipocytes shown above, it has multiple lipid droplets (white areas) and many mitochondria (white adipocytes also have mitochondria, but not so densely packed). CAP is a capillary adjacent to the cell, Go the Golgi apparatus. The picture represents a width of about 14 μm (i.e. it is about 14 times more enlarged than the left hand picture). From S Cinti (2001). The adipose organ: morphological perspectives of adipose tissues. *Proc Nutr Soc* **60**:319–28, reproduced with permission from Cambridge University Press. The combined picture is reproduced from KN Frayn (2010) *Metabolic Regulation: A Human Perspective*, 3rd edn. Wiley-Blackwell, Oxford.

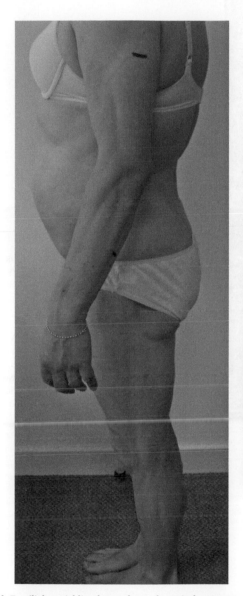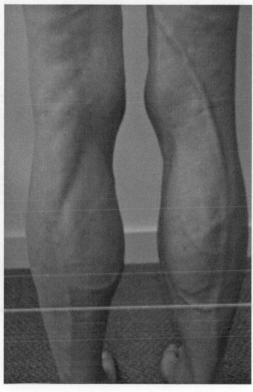

Fig. 10.11 Familial partial lipodystrophy. Left, typical appearance of a patient with familial partial lipodystrophy. Loss of subcutaneous adipose tissue is particularly apparent on the arms and legs, so that the muscles show through clearly. Right, phlebectasia – pronounced veins – see on the legs. Photographs courtesy of Dr Sara Suliman, with consent of the patient.

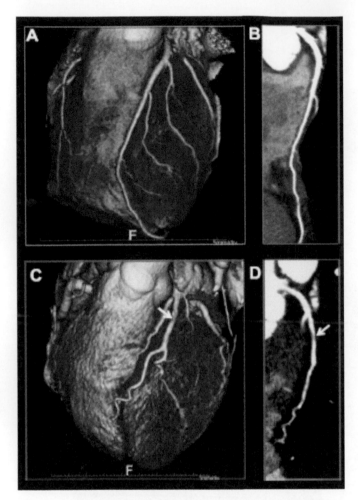

Fig. 10.12 Atherosclerosis demonstrated using the technique of computed tomography coronary angiography. This technique allows construction of a 3-dimensional image of the heart and the coronary tree (a, c) and a planar image (b, d). (a), (b) show a healthy heart; (c), (d) show a diseased (atherosclerotic) left anterior descending (LAD) artery. The white arrow denotes an atheromatous plaque resulting in a significant LAD stenosis (narrowing). Note also the diffuse vessel disease in the distal LAD (D) (mottling along the vessel). Picture courtesy of Drs Charalambos Antoniades and Alexios Antonopoulos.

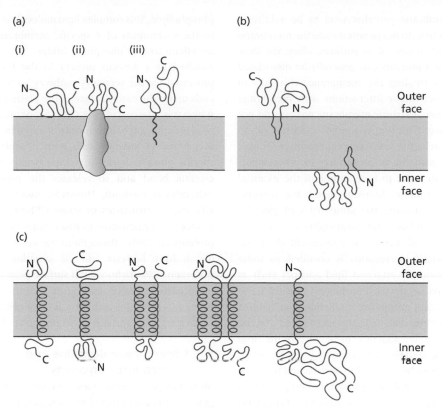

Fig. 5.11 Some common conformations of membrane proteins. (a) Extrinsic proteins do not penetrate into the membrane but may be attached to the membrane surface via electrostatic interactions with (i) lipid headgroups, (ii) with other membrane proteins, or (iii) with other extrinsic proteins bind to the membrane via fatty acyl chains that partially penetrate into the bilayer. In some cases the acyl chains shown in (iii) may be removable, leading to dissociation of the protein from the membrane, while in other cases they are permanently attached and the protein is always bound to the membrane. (b) Some intrinsic proteins partially penetrate into the bilayer via a hydrophobic domain, but the C- and N-terminals of the protein remain on the same side of the membrane. (c) The most securely bound intrinsic proteins extend right across the lipid bilayer via membrane-spanning stretches of relatively hydrophobic α–helical amino acids. These proteins have domains on both sides of the membrane and, according to the number of membrane-spanning α–helices, their N- and C- terminals may be either on the same side or on different sides of the membrane. In some cases, the main bulk of a large protein may be located externally from the membrane and attached to it only by a single trans-bilayer α-helical domain.

membrane. These are not as common as transmembrane proteins, but there are numerous examples of partially embedded proteins that extend about halfway across the bilayer – on one side or the other.

The main coat components of the lipid droplets (LDs) that are now believed to be virtually ubiquitous in cells (see Section 5.5) are of this type. Cytosolic LDs are made up of a large core of nonpolar lipids surrounded by a monolayer of phospholipid into which various proteins are inserted. The surface of a LD is bounded by a lipid monolayer rather than a bilayer. Therefore, most transmembrane proteins with polar groups on either side of a bilayer membrane cannot be accommodated on the surface unless they are able to change their conformation. However, there are numerous examples of soluble cytosolic proteins that can also bind to LDs under certain circumstances. This could be achieved by a modification, such as phosphorylation, which causes the soluble protein to assume a different conformation where a patch of hydrophobic residues becomes exposed and promotes its tight binding to the hydrophobic interior of the LD. If the modification is reversed, e.g. by dephosphorylation, the protein will revert to its original conformation and dissociate from the LD.

Extrinsic membrane proteins tend to be relatively polar molecules that do not penetrate into the membrane but are attached to one of its surfaces, often via ionic interactions. Such proteins can generally be dissociated quite readily by treating the membrane with salts to disrupt their electrostatic interactions with membrane lipids and proteins. Extrinsic membrane proteins do not come into contact with the hydrophobic interior of the lipid bilayer. Instead they associate either with polar lipid headgroups or with the polar surfaces of other membrane proteins in order to bind to one of the external faces of the membrane. Extrinsic membrane proteins therefore tend to assume the same types of globular conformations found as polar nonmembrane proteins.

A stronger form of membrane attachment than that due to electrostatic interactions is provided to some proteins by covalently attached lipid anchors such as fatty acyl, prenyl or glycosylphosphatidylinositol (GPI) groups. Lipid anchor groups are frequently attached to a nascent protein immediately after translation so that the protein immediately attaches to a membrane and remains attached for its lifetime. In some cases, however, the lipid anchor can be attached to or detached from the protein as part of a regulatory process, bringing about its reversible binding to a membrane. Common FA anchor groups include myristate (14:0) or palmitate (16:0). Myristate is normally covalently attached via an amide bond to an N-terminal glycine residue while palmitate is attached via a thioester bond to a cysteine residue. The FA group, which is on the outside surface of the protein, inserts into the lipid bilayer and enables an otherwise soluble protein to become a tightly membrane-bound protein. Examples of acylated membrane proteins include the lipoprotein component of Gram-negative bacterial membranes, and some protein kinases and guanosine nucleotide-binding proteins (G-proteins) that act as molecular switches. Other lipid-anchored proteins interact with the membrane bilayer via prenyl chains such as farnesol, dolichol or geranylgeraniol. A common group of prenylated proteins is the Rab (Ras-related in brain) family of G-proteins that are involved in many aspects of membrane trafficking including vesicle budding and fusion and the movement of vesicles along the cytoskeletal network (see Section 5.4).

A GPI anchor is attached to the C terminus of a membrane protein as a posttranslational modification. The core structure of this anchor consists of ethanolamine phosphate, trimannoside and glucosamine bound to an inositol phospholipid. This complex lipid molecule is in turn linked to the C-terminus of a specific membrane protein via an ethanolamine phosphate bridge. The GPI-anchor is attached to a nascent protein in the ER and is then processed via the secretory pathway so that the protein ends up on the external face of the plasma membrane to which it is attached via the GPI-anchor. Some GPI-anchored proteins can be specifically released from the plasma membrane by treating cells with phosphatidylinositol-specific phospholipases C, which cleaves the phosphoglycerol bond and so releases the protein into the extracellular medium. However, modifications to the GPI anchor structures of some GPI-anchored proteins render them insensitive to this treatment. GPI-anchored proteins are found throughout the eukaryotes, they are absent from bacteria and the situation in archaea is uncertain. GPI-anchored cell surface proteins tend to be localized in lipid raft domains (see Section 5.3.3) and have numerous functions including as enzymes, receptors, cell surface antigens and cell adhesion molecules.

5.3.3 Membrane domains and micro-heterogeneity

Most biological membranes are considerably less homogeneous than is depicted in most simplified representations of the fluid mosaic model. These differences are often highly dynamic with considerable changes in lipid composition in part of a membrane occurring over a timescale of milliseconds, which can make them difficult to investigate. There are two major types of lipid asymmetry in membranes, namely transverse and lateral. Both types of asymmetry play important roles in membrane function and both are often found in the same membrane domain.

Transverse asymmetry relates to differences in composition between the two leaflets of a bilayer membrane. As discussed above, the rate of lipid diffusion between the leaflet of a typical bilayer, also known as 'flip-flop', is very low. Therefore in all membranes examined to date, the two bilayer leaflets have significantly different lipid compositions. However, in some cases experimentally determined rates of flip-flop in biological membranes can be much higher than those measured in model membranes of similar lipid composition. This is because biological membranes often contain specialized transmembrane proteins such as flippases and phospholipid translocases that accelerate the movement of certain lipids across the bilayer. This form of assisted movement of lipids across a bilayer is sometimes essential for membrane biogenesis. For example, unless they are trafficked via vesicles, newly synthesized phospholipids are inserted into only the cytoplasmic face of the plasma membrane and require flippases for transfer to the exoplasmic face.

In other cases a lipid such as phosphatidylserine (PtdSer) is normally retained exclusively on cytoplasmic face during cell development and its flippase-mediated appearance on the exoplasmic face is an early indicator of programmed cell death, or apoptosis. This effect is also seen in neurons exposed to the amyloid-β peptide and it may be an important stage in the sort of neurodegeneration found in Alzheimer's disease. In many cases, flippases are made up of several proteins that assemble transiently to form an enzymically active complex. Some of the most pronounced examples of transverse bilayer asymmetry are found in the plasma membranes of eukaryotes where 80–98% of the five major lipid classes are found on only one or other side of the bilayer.

Lateral asymmetry or heterogeneity of both lipid and protein distributions in membranes is now recognized as being widespread in all organisms. It is also an essential mechanism in cellular processes such as endocytosis and vesicle formation. In eukaryotic plasma membranes and chloroplast thylakoid membranes, the transverse and/or lateral heterogeneity is relatively long lasting and occurs on a large scale (see Fig. 5.12). More commonly, however, membranes form tiny domains of micro-heterogeneity that may only exist for a fraction of a second, which is nevertheless long enough to carry out important biological functions. Examples of such short-lived micro-heterogeneity include the specialized lipid-synthesizing domains in the ER and the lipid rafts involved in secretion and endocytosis in the trans-Golgi network.

Membrane lipid rafts are among the best-studied examples of membrane micro-heterogeneity (see Fig. 5.13). The exact nature of lipid rafts remains the subject of much debate but we can regard them as being transient nano-assemblies consisting of specific proteins and lipids. Estimates of their size vary from 10 nM to 200 nM in diameter, although the lower end of this range is rather small as it corresponds to the size of a single large protein complex plus its associated shell of boundary lipids. The presence of lipid rafts has been demonstrated by the punctate pattern of some plasma membrane proteins, such as sugar transporters, when labelled by fluorescent tags. Lipid rafts are highly dynamic, sterol- and sphingolipid-enriched, membrane domains that

compartmentalize a wide range of cellular processes including transport and signalling. Small rafts can sometimes be stabilized to form larger platforms through protein-protein interactions. Lipid rafts are typically short-lived (10–20 msec) with highly asymmetric lipid and protein concentrations. In the planar lipid raft depicted in Fig. 5.13a, the outer (exoplasmic) bilayer leaflet contains almost exclusively sphingolipids and cholesterol. The inner bilayer leaflet is mostly made up of phospholipids but the mobility of these lipids is highly constrained by the presence of sphingolipids and cholesterol and the lipids collectively form a relatively rigid gel phase (see Section 5.1.3).

Because sphingolipids often contain very long chain (<26C) acyl groups, they form a relatively immobile gel phase raft domain in which the bilayer is significantly thicker than elsewhere in the membrane. The highly unusual bilayer composition within lipid raft domains enables specific groups of proteins to partition preferentially into these membrane regions. Examples include proteins with relatively long transmembrane α-helical segments that are better accommodated in the thicker raft bilayer and proteins containing saturated acyl groups, such as palmitate (16:0) or myristate (14:0), which favourably partition into the rigid gel phase lipids of the raft. The presence of these proteins further stabilizes the raft lipids and enables these highly organized asymmetric domains to persist long enough to carry out complex functions such as the formation, processing and transport of vesicles as described in Section 5.5.

There are two major types of lipid raft in eukaryotic cells: planar rafts and caveolae. Planar rafts tend to be relatively flat regions that are continuous with the major plane of the plasma membrane. In contrast, caveolae are flask-shaped invaginations of the plasma membrane that are enriched in caveolin proteins. Although caveolins are found in all eukaryotic cell types, they are especially abundant in cells involved in signalling processes. Examples include many neuronal cells, especially in brain cell types such as astrocytes, oligodendrocytes and Schwann cells.

Planar rafts and caveolae have a similar lipid composition that is highly enriched in sterols and sphingolipids but have different protein populations. In particular, planar rafts are often enriched in flotillins, whereas caveolae are enriched in caveolins. Both flotillins and caveolins are able to recruit signalling molecules into their respective raft domains. Lipid rafts play important roles in processes as

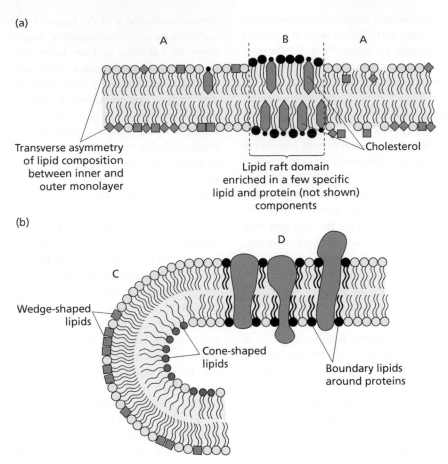

(a)

Transverse asymmetry of lipid composition between inner and outer monolayer

Cholesterol

Lipid raft domain enriched in a few specific lipid and protein (not shown) components

(b)

Wedge-shaped lipids

Cone-shaped lipids

Boundary lipids around proteins

Fig. 5.12 Membrane domains. (a) In many flat (planar) membranes, the lipids are distributed randomly in the lateral plane of each bilayer but are asymmetrically distributed in the transverse plane, i.e. between the top and bottom bilayers. (b) In other relatively flat membranes, such as the ER, lipids may be asymmetrically distributed in the lateral plane and may form specialized regions such as lipid rafts (see Fig. 5.13) or metabolons such as sites of TAG formation and lipid droplet budding (see text). (c) Highly curved membrane regions may be formed due to the asymmetric distribution of nonbilayer cone- or wedge-shaped lipids and/or due to the action of proteins such as caveolins or SNAP (see Fig. 5.13 and text). (d) Some proteins are surrounded by a layer of closely associated boundary lipids that are relatively motion-restricted and tend to move with the protein.

diverse as membrane transport, stress responses, polarized cell growth, neurotransmission in animals and the cell-cell communication processes required to establish the symbiosis between certain soil-dwelling, nitrogen-fixing bacteria and their plant hosts.

5.4 Membrane function

In order to grow, develop, respond to the environment, and divide, cells must constantly form new bilayer membranes and other lipidic structures such as LDs. Cells also turn over the components of existing membranes, sometimes to remove damaged components or to insert new proteins or lipids that may be required. In eukaryotic cells, many routine metabolic processes also require a considerable amount of membrane trafficking and turnover. Examples include the complex vesicle-based import and secretion mechanisms that may involve the plasmalemma, Golgi, and ER membrane systems. The growth and trafficking of membranes and other lipidic structures involves processes, such as the swelling, budding and fusion of lipid assemblies, that until recently have been rather poorly understood. While there is still

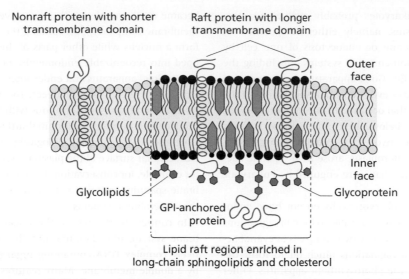

Nonraft protein with shorter transmembrane domain

Raft protein with longer transmembrane domain

Outer face

Inner face

Glycolipids

GPI-anchored protein

Glycoprotein

Lipid raft region enriched in long-chain sphingolipids and cholesterol

Fig. 5.13 Lipid raft structures. Lipid rafts are relatively short-lived (10–20 msec) microdomains ranging from 10 nm to 200 nm in diameter that are a feature of the plasma membranes of eukaryotic cells. The lipid components of the rafts are mainly long-chain sphingolipids and cholesterol that form a rigid gel phase that is thicker than a normal lipid bilayer. This provides a specific environment into which certain specialized proteins are able to partition. Examples of such proteins include glycoproteins involved in signalling and transport and GPI-anchored proteins.

much to be learned about the detailed molecular mechanisms of these vital processes, there have been considerable advances in recent years in our knowledge of these forms of dynamic membrane behaviour. In particular, it has become clear that proteins such as SNAP and SNARE (Soluble N-Ethylmaleimide-sensitive Fusion Attachment Receptor Protein) play key roles in eukaryotic membrane trafficking functions, as discussed below (see also Box 5.3).

5.4.1 Evolution of endomembranes and organelles in eukaryotes

Eukaryotic cells are characterized by an extensive network of organelles and endomembranes, many of which have unique lipid compositions. The evolution of this network of intracellular membranes is largely responsible for the ability of eukaryotic cells to achieve sizes that typically range from 1000 to 10,000 times greater than those of bacteria or archaea. The intracellular

Box 5.3 Proteins mediating endomembrane dynamics

Clathrin: a large protein complex involved in formation of coated vesicles especially in import to and export from cells.

COPI: *Coat protein I*, a protein involved in retrograde transport from Golgi to ER membranes.

COPII, *Coat protein II*, a protein involved in anterograde transport from ER to Golgi membranes.

NSF: *N-ethylmaleimide-sensitive factor*, a protein that acts with SNARE proteins in membrane fusion.

Rab: *Ras-related in brain*, a large protein family involved in many intercellular processes including lipid attachment to membrane proteins and membrane trafficking.

Sec: *secretory*, a group of proteins involved in vesicle budding.

SNAP: *Soluble NSF Attachment Protein*, a component of the SNARE complex involved in vesicle budding and fusion.

SNARE: *Soluble NSF Attachment Protein Receptor*, a protein family mainly made up of elongated α-helical domains that plays a key role in membrane budding and fusion.

Syntaxin: a specialized group of SNARE proteins involved in exocytosis.

membranes in eukaryotes probably evolved via two separate mechanisms, namely either invagination of the plasma membrane or endocytosis of one cell by another. Most endomembrane systems – including the nuclear envelope, ER, Golgi apparatus, endosomes and lysosomes – may have evolved from invaginations of the plasma membrane that occurred in ancestral prokaryotic cells, as described below. In contrast, there is good evidence that the two DNA-containing organelles, mitochondria and chloroplasts, arose from separate bacterial endosymbionts that were engulfed by eukaryotic host cells.

Although most prokaryotic cells do not have endomembranes, a few bacteria are able to form invaginations of their plasma membrane. One example is in cyanobacteria where such invaginations pinch off to form the vesicles that house the photosynthetic apparatus. Other bacteria in the planctomycetes have even more complex endomembrane systems including a nuclear region partially enclosed by a double membrane that has some similarity to the nuclear envelope of eukaryotes. However, unlike eukaryotes, the endomembrane system of planctomycetes remains connected with the plasma membrane and can be regarded as a more complex form of the limited invaginations seen in Gram-negative bacteria such as *E. coli*. Planctomycetes can also take up proteins by endocytosis, which is another characteristic that was previously thought to be confined to eukaryotes. Therefore the beginnings of a eukaryote-like endomembrane system are already present in several groups of bacteria. This has enabled researchers to propose a mechanism for the evolution of eukaryotic endomembrane systems, as shown in Fig. 5.14(a).

Initially there would have been a relatively basic network of internalized membranes to which ribosomes and chromosomes were attached. Gradually this network became more complex with a two-layered envelope of membrane completely surrounding the chromosomes to form a nucleus while other parts of the network developed into recognizable endomembrane structures such as ER, Golgi apparatus and endosomes. This mechanism explains why the space between the inner and outer nuclear membranes is continuous with the lumen of the ER. It also accounts for the luminal surfaces of the ER and Golgi membranes being topologically equivalent to the extracellular surface of the plasma membrane, allowing, for example, for conservation of the orientation of membrane-spanning proteins as they are trafficked between these membrane systems.

In contrast to the rest of the eukaryotic endomembrane system, mitochondria and chloroplasts are relatively complex DNA-containing organelles surrounded by a double membrane. Many features of these organelles, such as their DNA organization, ribosomal structure and protein biosynthesis are very similar to bacteria and there is convincing evidence that they arose from the uptake of an α-proteobacterium in the case of mitochondria and a cyanobacterium in the case of chloroplasts. In each case, the endocytosis of the ancestral bacterial cell gave rise to an outer organelle membrane that was derived from the plasma membrane of the host cell that enclosed the bacterial endosymbiont, as depicted in Fig. 5.14(b).

The signature of this endosymbiotic process can still be traced in the lipid composition of the different membranes of chloroplasts and mitochondria. This is particularly evident for chloroplasts where the inner envelope and thylakoid membranes, which are derived from the original cyanobacterial cell, are overwhelmingly dominated by galactolipids and sulpholipids and contain no PtdCho. Galactolipids and sulpholipids are major components of cyanobacterial membranes. In contrast, the

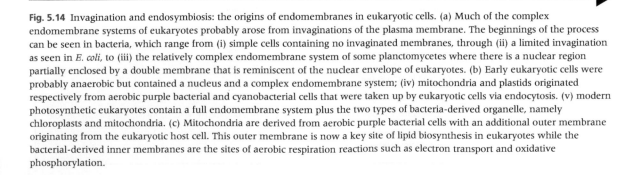

Fig. 5.14 Invagination and endosymbiosis: the origins of endomembranes in eukaryotic cells. (a) Much of the complex endomembrane systems of eukaryotes probably arose from invaginations of the plasma membrane. The beginnings of the process can be seen in bacteria, which range from (i) simple cells containing no invaginated membranes, through (ii) a limited invagination as seen in *E. coli*, to (iii) the relatively complex endomembrane system of some planctomycetes where there is a nuclear region partially enclosed by a double membrane that is reminiscent of the nuclear envelope of eukaryotes. (b) Early eukaryotic cells were probably anaerobic but contained a nucleus and a complex endomembrane system; (iv) mitochondria and plastids originated respectively from aerobic purple bacterial and cyanobacterial cells that were taken up by eukaryotic cells via endocytosis. (v) modern photosynthetic eukaryotes contain a full endomembrane system plus the two types of bacteria-derived organelle, namely chloroplasts and mitochondria. (c) Mitochondria are derived from aerobic purple bacterial cells with an additional outer membrane originating from the eukaryotic host cell. This outer membrane is now a key site of lipid biosynthesis in eukaryotes while the bacterial-derived inner membranes are the sites of aerobic respiration reactions such as electron transport and oxidative phosphorylation.

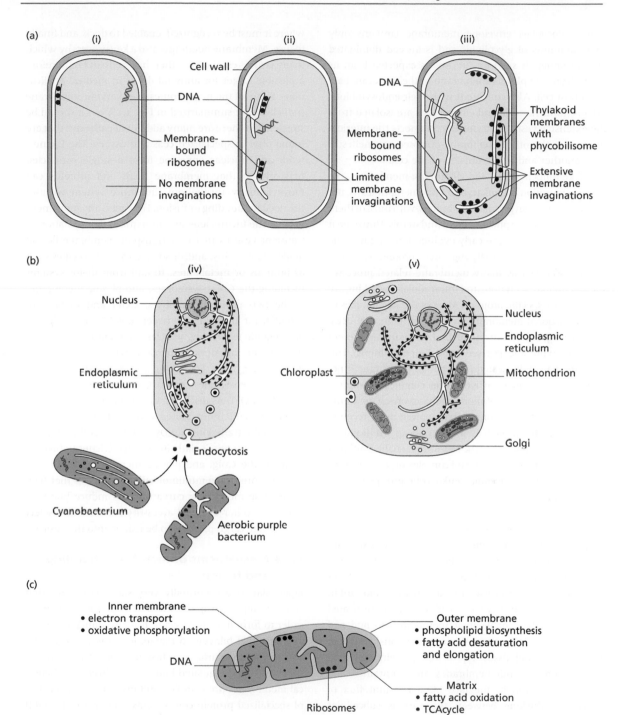

outer chloroplast envelope membrane contains only small amounts of glycolipids and is instead dominated by phospholipids such as PtdCho, as expected from its origin from the plasma membrane of the noncyanobacterial host cell. Also consistent with their endosymbiotic origins, mitochondria and chloroplasts are isolated from the extensive vesicular traffic that connects the interiors of most of the other membrane-enclosed organelles to each another and to the outside of the cell.

Unlike mitochondrial and chloroplast membranes, the lipid composition of eukaryotic endomembranes is not very different from plasma membranes, with most of them being dominated by phospholipids and sterols. However, it is possible to deduce the early evolutionary origin of the endomembranes by analysing their protein contents. As discussed below, many membrane-related processes such as uptake, secretion, and intracellular trafficking are regulated by specific proteins. Most of the key families of these functional membrane proteins, including SNAREs, COPI, COPII, NSF, clathrin, rab and syntaxin, are present in all eukaryotic lineages (see Box 5.3 for definitions). This suggests that sophisticated intracellular membrane functions had already evolved in the more complex prokaryote that was the ancestor of modern eukaryotes. One of the few endomembrane proteins that evolved more recently is caveolin. This protein is found only in multicellular animals and may be one of the factors that enabled animals to become relatively mobile and complex organisms compared to other multicellular eukaryotic groups such as plants and fungi.

5.4.2 Membrane trafficking

Membrane fission (budding) and membrane fusion are crucial processes in cells without which intracellular trafficking, import, export and cell division would not be possible. These processes are particularly important in eukaryotes, which have evolved highly sophisticated mechanisms that enable their membranes to bud and fuse in a controlled manner that can vary considerably in different cell and tissue types. The protein and lipid compositions of endomembranes are maintained and regulated by, and in the face of, a constant flux of vesicular trafficking between the various subcellular compartments, and with the external environment via the plasma membrane.

These processes normally involve the budding of a membrane vesicle from a donor membrane followed by its regulated transport to an acceptor membrane where it must be recognized, enabled to dock, and finally to fuse. Membrane budding is also a key process by which many viruses escape from their host cells and is therefore a possible target for antiviral drugs in medical applications. Some of the major routes of eukaryotic membrane trafficking are summarized in Fig. 5.15, but it should be stressed that there are many additional pathways that are found only in specialized cells and tissues. The ER network and nuclear envelope form a single connected network, so that membrane lipids and proteins can move laterally throughout the entire system without the need for budding or fusion. However, any movement beyond the ER/nucleus system requires the formation of transient vesicles that may transport membrane-bound lipids and proteins, and/or soluble cargo molecules such as proteins or metabolites, to and from major systems including the Golgi, lysosomes, and plasma membrane.

The two protein complexes, COPI and COPII, are essential for trafficking between the ER and Golgi. In anterograde trafficking, vesicles coated in COPII bud off from the ER and dock at the Golgi where they fuse and deliver their cargo. Trafficking in the opposite direction, known as retrograde transport, is mediated by vesicles coated in COPI. As discussed in Section 5.4.3, the role of COPI and COPII is not limited to vesicle transport but also includes the trafficking of LDs in the cell. A third type of vesicle coat protein is clathrin, which mediates trafficking between the Golgi and plasma membrane. These coat proteins form large multimolecular complexes that help to increase membrane curvature and induce budding. They, and other essential accessory proteins, also select appropriate cargo proteins to be loaded into the vesicles.

5.4.3 Mechanisms of membrane budding and fusion

Lipid bilayers are normally very stable structures that resist attempts to rupture them or to pull them apart in order to form buds or vesicles. It is also difficult to force adjacent lipid bilayers to merge, for example to enable two membrane surfaces to fuse together. The processes involved in the regulated budding and fusion of biological membranes are complex and involve several groups of specialized protein components such as COPI, COPII and clathrin as described above. However, although these and other proteins play key roles in membrane trafficking, there is growing evidence of the importance of lipid behaviour and lipid metabolism in membrane dynamics.

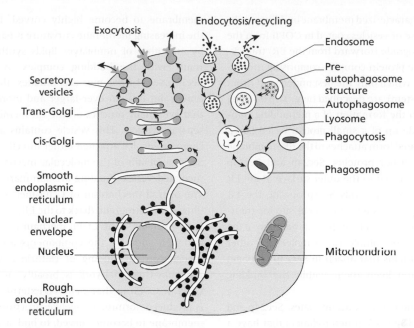

Exocytosis

Endocytosis/recycling

Secretory
vesicles

Trans-Golgi

Cis-Golgi

Smooth
endoplasmic
reticulum

Nuclear
envelope

Nucleus

Rough
endoplasmic
reticulum

Endosome

Pre-
autophagosome
structure

Autophagosome

Lyosome

Phagocytosis

Phagosome

Mitochondrion

Fig. 5.15 Membrane trafficking in eukaryotic cells. Eukaryotic cells exhibit multiple modes of membrane trafficking leading to exocytosis, endocytosis/recycling, or phagocytosis. Molecules can also be trafficked between these pathways. Exocytosis typically begins with the accumulation of molecules destined for export in specialized ER domains that then bud off as vesicles that move to and fuse with the *cis* face of the Golgi network. As the Golgi becomes loaded with such fused vesicles they gradually move towards it *trans* face and are released to form the *trans*-Golgi network of vesicles that are trafficked to the plasma membrane where a final fusion event results in their release from the cell. Endocytosis/recycling involves the uptake of extracellular molecules that initially accumulate in pits that form in the plasma membrane. These pits bud off as endosome vesicles into the cytosol that deliver their cargoes to lysosomes before being recycled to the plasma membrane. Phagocytosis involves the ingestion of larger structures, including entire cells, via an invagination of the plasma membrane that buds off to form a phagosome, which delivers its contents to lysosomes from where they may be either broken down and excreted or recycled within the cell.

For example, lipases or acyltransferases can create localized accumulations of nonbilayer forming lipids that help to destabilize membranes prior to their budding or fusion. A specific example of this is described for the budding of LDs from the ER membrane (see Section 5.5.5). In some cases, the accumulation of specific lipids and proteins at a particular place helps to destabilize the bilayer organization transiently, but for long enough to enable it to break apart and then reseal as part of budding or fusion. Other lipids may have signalling functions or may interact with specific proteins. One example is phosphatidylinositol-4-phosphate (PtdIns4P), which was previously regarded as merely an intermediate in the biosynthesis of the important signalling lipid PtdIns(4,5)P_2 (see Section 8.1.3). More recently, PtdIns4P has been found to have a much wider role as a key regulator of important cell functions, particularly in

Golgi trafficking. One of the roles of PtdIns4P is to recruit proteins to specific locations in a membrane.

Trafficking proteins have several functions such as enabling selected membranes to approach one another in a highly organized manner prior to fusion. Other proteins assist in the 'pinching off' of the narrow neck of a vesicle that is about to bud from a membrane. Although lipids play important roles in membrane budding and fusion, it is the proteins that ensure that these processes occur at the correct time and place during cellular function. In most cases of budding, such as the formation of caveolae, the process requires both specific proteins and the formation of lipid micro-domains. However, there are also several examples of budding that is more or less exclusively driven by proteins, such as clathrin and COPI/COPII-mediated vesicle formation.

One of the best characterized membrane budding mechanisms is the release of vesicles coated in COPII from the ER as part of anterograde (forward from the ER) trafficking. COPII is a large protein complex comprising numerous components of which only the best understood (from the yeast model system) are described here. The budding process begins with the formation of a prebudding complex, which is made up of a small protein called Sar1p whose phosphorylated form attaches to the ER membrane and recruits two further proteins, Sec24p and Sec23p that also become attached to the outer surface of the ER membrane. While Sec23p binds Sar1p outside the ER membrane, Sec24p acts as an attachment site for transmembrane proteins that will form part of the cargo of the vesicle. In the case of soluble cargo molecules, other transmembrane proteins that attach to Sec23p will bind these molecules and keep them within the budding vesicle.

As the prebudding complex accumulates, Sec24p and Sec23p recruit Sec13p.Sec31p heterodimers that have a positively charged concave surface. When the Sec13p.Sec31p heterodimers polymerize they form a highly curved coat on the outer surface of the ER membrane. This curved protein coat in turn forces the attached membrane to become highly curved. It is likely that the increasing membrane curvature is partially stabilized by an influx of nonbilayer lipids synthesized concurrently with the budding complex. As more of the Sec13p.Sec31p polymer accumulates, the membrane is drawn out into an ever-larger and more spherical bud that eventually pinches off the ER membrane to form a separate vesicle. This vesicle contains a specific set of cargo molecules attached either directly or indirectly to Sec24p. Details of the molecular mechanisms of COPII-mediated vesicle budding are summarized in Fig. 5.16. The overall mechanism of COPI-mediated vesicle budding in the retrograde (back to the ER) transport pathway is very similar to the COPII mechanism described above, although some of the components are different. The mechanism of budding of clathrin-coated vesicles in the *trans*-Golgi network is broadly similar to that of COPI or COPII in that a curved external coat of protein is produced, forming a cagelike structure that forces the membrane to become curved, to bud, and eventually to pinch off to form a vesicle. However, because the formation and composition of clathrin coats are much more complex than either COPI or COPII they will not be described in detail.

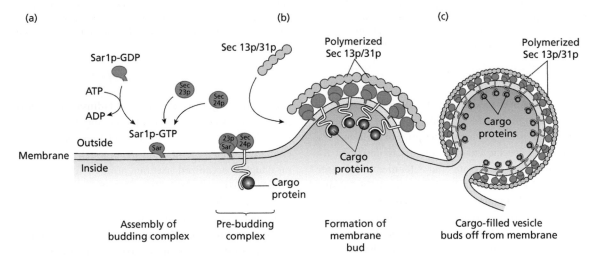

Fig. 5.16 Budding mechanisms in eukaryotic cells. This scheme depicts the mechanism of vesicle budding from the ER as mediated by the COPII complex. (a) As described in the text, the phosphorylated form of Sar1p becomes attached to the cytosolic face of the ER membrane and recruits a series of additional proteins such as Sec23p, Sec24p that aggregate to form a curved complex that deforms the lipid bilayer to form a bud. Sec24p has a domain that extends into cytosol and acts as a binding site for various Cargo proteins that are destined for secretion from the cell. (b) Other cytosolic proteins such as Sec13p and Sec31p then bind to the external face of this aggregate to form a continuous coat that forms a curved bud. (c) This bud eventually grows into a spherical vesicle that buds off from the ER membrane. The overall mechanisms of COPI-mediated budding in retrograde transport and clathrin-mediated budding in endocytosis are similar to that shown here for COPII although the components are different.

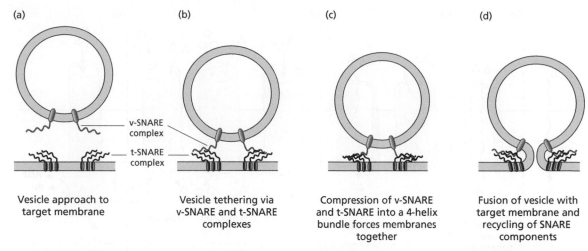

(a)   (b)   (c)   (d)

v-SNARE complex
t-SNARE complex

Vesicle approach to target membrane

Vesicle tethering via v-SNARE and t-SNARE complexes

Compression of v-SNARE and t-SNARE into a 4-helix bundle forces membranes together

Fusion of vesicle with target membrane and recycling of SNARE components

Fig. 5.17 Fusion mechanisms in eukaryotic cells. Most mechanisms of membrane fusion in eukaryotes involve various SNARE proteins. (a) Vesicles are initially tagged for fusion by attachment of v-SNARE proteins that contain large coiled helices that are able to interact with similar helices attached to t-SNAREs on other membranes. (b) As the v- and t-SNARE helices approach each other they form a *trans*-SNARE complex that tethers the vesicle to the target membrane. (c) The *trans*-SNARE complex becomes highly compressed into a tight 4-helix bundle that draws the two lipid bilayers ever closer together. (d) Eventually the two bilayers are forced to fuse so that the vesicle becomes part of the target membrane and its contents are released into the compartment bound by the target membrane. The v-SNARE proteins are then recycled onto another vesicle so that another fusion event can begin.

Mechanisms of membrane fusion can vary to some extent in different trafficking pathways but in most cases SNARE proteins are involved (Box 5.3). These are members of a large membrane-associated protein family found in organisms as diverse as yeast and mammals. According to the SNARE hypothesis, vesicles are tagged for fusion by the attachment of v-SNARE proteins (also called synaptobrevin) to their membranes. These proteins contain a single transbilayer segment attached to a large polar domain that can interact with similar proteins to form a tightly coiled helix. The receptor membrane with which the vesicle will fuse contains t-SNARE proteins including SNAP-25 and syntaxin. Like the v-SNAREs, t-SNAREs are attached to their membrane via a single transbilayer segment attached to large polar domains that can form helices. As the vesicle approaches the receptor membrane, the polar domains of the v-SNAREs and t-SNAREs come together to form a very stable four-helix bundle that pulls the vesicle and receptor membranes into very close contact. This is termed a *trans*-SNARE complex.

As the helix bundle becomes ever more tightly coiled, the two bilayers are forced together until they finally fuse into a single membrane. This fusion means that the transmembrane domains of the two proteins are now next to each other, forming a *cis*-SNARE complex on the receptor membrane with the helical domains facing outwards. The final step in membrane fusion is to recycle the v-SNARE components back to their membranes of origin to enable further rounds of vesicle fusion to occur. These processes are summarized in Fig. 5.17.

5.4.4 Transport mechanisms in membranes

The lipid bilayer of a typical cell surface membrane is impermeable to most polar molecules. As long as the lipids remain in a relatively fluid conformation and do not contain significant packing defects, they form an effective seal that protects the cell or organelle from its external environment. Among the few molecules that can cross a typical lipid bilayer are water and gases such as CO_2, although the rates of such transport are very low. However, a highly impermeable membrane would be of no use to a living cell. All cells need to import and export a wide range of molecules with different shapes, sizes and physical properties. Examples include sugars, ions such as Ca^{2+}, Na^+ and Cl^-, and large and bulky proteins to name only a few. Moreover the transport of these diverse

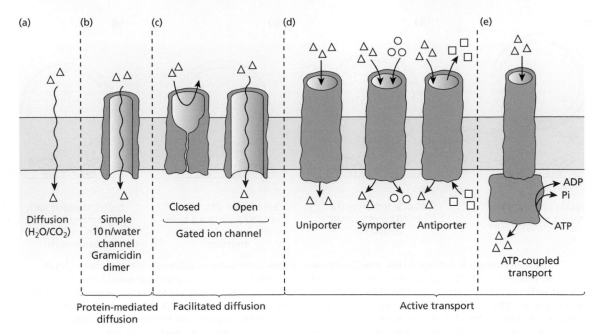

Fig. 5.18 Membrane transport mechanisms. (a) Simple diffusion across a typical lipid bilayer is a slow process that is only possible for a few small molecules such as liquid water and gases such as CO_2. (b) Much more rapid transport for ions and water can be achieved by small proteins such as the gramicidins, which dimerize to form hydrophilic pores through the lipid bilayer. (c) A more regulated transport of ions can be mediated by gated channels such as aquaporin which can be either closed or open via an active energy-requiring mechanism, depending on cell requirements. (d) More complex transporter proteins can mediate the simultaneous actively regulated influx and/or efflux of several molecules (symport) or the exchange of an imported molecule for one that is exported (antiport). (e) In other cases, molecules such as hydrogen ions are actively transported across a membrane via a direct ATP-coupled mechanism as in the ATPases involved in establishing proton gradients in mitochondrial and plastidial membranes.

molecules must be carried out in a highly regulated fashion that can change with time – sometimes even from minute to minute. In order to move these various molecules across their membranes, cells contain a wide range of membrane transporter systems, all of which are based on membrane-spanning proteins, as shown in Fig. 5.18.

There are three main kinds of transport mechanism for small molecules across membranes, namely simple diffusion, facilitated diffusion and active transport. Simple diffusion is the direct movement of molecules across the membrane bilayer and is mainly restricted to a slow movement of water molecules. Facilitated diffusion is much more common than simple diffusion and involves the use of protein channels to enable movement of charged molecules, which are otherwise unable to cross lipid bilayers, to diffuse freely in and out of the cell. Protein channels are most commonly used to move small ions like Na^+, K^+ and Cl^- and the rate of facilitated

transport is proportional to the number of available protein channels. There are also many uniport (one-way) transporters that enable hydrophilic – but often nonionic – nutrients and metabolites enter and leave cells. For example, glucose is absorbed into enterocytes by an electrogenic symport (two-way) in the brush border but leaves them into the circulation by a facilitated diffusion uniport through the basolateral membrane.

One of the simplest forms of protein channel is present in the family of 15-mer peptides – named gramicidin. This molecule forms a single helix in a lipid bilayer with a hydrophobic outer face of the helix enabling the peptide to span the entire the bilayer. In contrast, the inner face of the helix is lined with polar amino acid residues that form a hydrophilic channel that readily allows water and ions to cross the membrane. One of the drawbacks of this mechanism is that the channel is always 'open' meaning that water and ions can continue to leak out of a cell until their external concentration equals that inside the cell.

Unsurprisingly, gramicidin is an antibiotic that is produced as a toxin by *Bacillus brevis* to kill other bacteria.

A rather more complex facilitated diffusion mechanism is found in gated ion channels such as the protein aquaporin. This large protein contains six membrane-spanning domains that form a hydrophilic channel at its core. Depending on the conformation of the protein the channel can be either open or closed, which enables the movement of molecules to be regulated. The main function of aquaporins is to transport water selectively into or out of a cell, although a few other small noncharged solutes such as glycerol, CO_2, ammonia, and urea can also be transported. Facilitated diffusion is limited by only being able to move molecules down a concentration gradient, i.e. from where they are more concentrated to where they are more dilute. The only way to move molecules against a concentration gradient is by active transport where the cell expends energy, often in the form of ATP hydrolysis.

In primary active transport, chemical energy from a process such as ATP hydrolysis is used directly to move molecules in or out of a cell against their concentration gradient. In secondary active transport, an electrochemical gradient is first established by pumping ions across the membrane. The energy stored in this gradient can then be used to move other molecules such as glucose across the membrane against their concentration gradient. Where two or more molecules are actively transported in the same direction it is known as symport, whereas the coupled simultaneous transport of molecules in opposite directions is called antiport.

An example of primary active antiport is the calcium-sodium exchanger that is found in many cell types. This ATP-dependent pump will allow 3 sodium ions into the cell in exchange for one calcium ion that is exported. Another common ATP-dependent antiporter is the Na^+/K^+ ATPase that simultaneously transports K^+ into and Na^+ out of the cell.

An example of secondary active symport is found in *E. coli*, which initially pumps out protons (H^+) via an ATPase to create a gradient across the membrane. These protons are then able to move back into the cell down their concentration gradient via any one of several transmembrane proteins, such as lactose permease. As the proton moves across the membrane there is a coupled movement of lactose into the cell. The lactose is able to move up its own concentration gradient by being coupled with protons moving down their concentration

gradient. *E. coli* uses a similar symport mechanism to transport ribose, arabinose, and several amino acids into the cell.

5.5 Intracellular lipid droplets

Intracellular lipid droplets (LDs) are typically spherical bodies with a core of nonpolar lipid surrounded by a phospholipid monolayer and a specific population of proteins. Until recently intracellular LDs were regarded as relatively inert storage deposits mostly found in specialized lipid-storing cells such as adipocytes in mammals or seeds in plants. However, recent advances in cell biology and genetics have shown that LDs are present in most cells and have a broad range of dynamic roles. For example, in multicellular organisms, LDs are involved in processes such as cell- and body-level lipid homeostasis, protein sequestration, membrane trafficking and various types of signalling. LDs are also implicated in some important diseases in which pathogens of plants and animals subvert the LD metabolism of their host as part of their infection process. Finally, LD malfunctions are implicated in several degenerative diseases in humans, among the most serious of which is the increasingly prevalent group of illnesses, such as obesity and insulin resistance, that are associated with metabolic syndrome (see Section 10.4).

5.5.1 Prokaryotes

Unlike the dynamic lipid bodies of eukaryotes, prokaryotic lipid bodies appear to act exclusively as energy stores often laid down in response to nutrient limitation such as a low carbon/nitrogen ratio. The most common storage lipids in prokaryotes are the polyhydroxyalkanoates (PHAs), which are found in wide range of archaeal and bacterial cells. The monomers that make up PHAs are synthesized from acetyl-CoA via a short pathway. The most important enzyme in this pathway is PHA synthase which assembles monomers such as hydroxybutyrate or hydroxyvalerate into either homopolymers, such as polyhydroxybutryate, or copolymers, such as poly(3-hydroxybutyrate-co-3-hydroxyvalerate). The manipulation and biotechnological uses of PHAs and other storage lipids are discussed in Sections 11.5 and 11.6.

Although the majority of bacteria, and many archaea, store PHAs, a subset of bacteria, primarily nocardioform

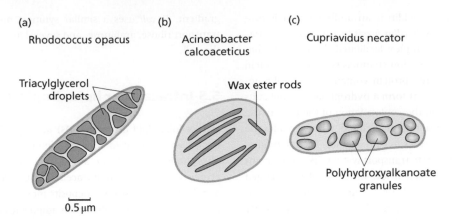

Fig. 5.19 Storage lipid droplets, rods and granules in prokaryotes. (a) During conditions of nutrient limitation, spheroidal TAG droplets of 0.2–0.4 µm diameter can make up 80% of cell volume in *Rhodococcus opacus*. (b) Semisolid wax ester rods of 1 µm in length are the preferred method of carbon storage in *Acinetobacter calcoaceticus*. (c) Semi-solid 0.2–0.5 µm diameter granules of biopolymers formed from polyhydroxyalkanoates, are found in *Cupriavidus necator*.

actinomycetes, streptomyces and some Gram-negative species, accumulate LD enriched in TAGs (see Fig. 5.19).

The biosynthesis of wax esters and/or TAGs is catalysed by a plasma membrane-associated multifunctional wax ester synthetase/diacylglycerol acyltransferase (WS/DGAT) that is found in many bacteria. The amino acid sequence of this bacterial enzyme is unrelated to that of any previously identified WS or DGAT in animals, fungi or plants. Evidence suggests that each WS/DGAT enzyme might give rise to a single microdroplet of about 60 nM diameter. These droplets may then coalesce into mature droplets of 300 nM in a process regulated by the protein, TAG accumulation deficient (TAD-A).

The highest amounts of TAG accumulation have been reported mainly in nocardioform bacteria and in some streptomycetes. For example, the *Rhodococcus opacus* PD630 strain can accumulate LD of 50–400 nM diameter that form more than 75% of cellular dry weight, with a potential daily TAG production from organic wastes of almost 60 mg l^{-1}. Spherical wax ester-rich droplets of about 200 nM in diameter have been reported in some *Acinetobacter* spp., although other species accumulated rectangular or rod-shaped wax ester structures.

5.5.2 Plants and algae
All major groups of plants, from unicellular algae to the most complex angiosperms, can form cytosolic LDs in at least some of their cells/tissues. In most plants, their cytosolic LDs contain TAGs, although the desert shrub,

jojoba (*Simmondsia chinensis*), accumulates fluid wax esters instead (see Section 2.2.3). As with many other unicellular organisms, some unicellular algal species accumulate cytosolic LDs as storage reserves in response to nutrient limitation or abiotic stress. In some cases, these LDs can make up as much as 86% cell dry weight. Although there have been fewer studies into plant LDs compared with their nonphotosynthetic counterparts, a similar picture is emerging regarding the mechanism of LD formation on specific domains of the ER or plastidial thylakoid membranes. LDs also have roles in several aspects of plant development and function. Some of the major classes of plant LD are depicted in Fig. 5.20.

The physical mechanism of cytosolic LD formation in plant cells is similar to that of animals. There is probably a dynamic two-way flow of lipids between the ER and LDs in most plant cells whereby small numbers of LDs are constantly being formed and recycled back to the ER. In order to increase the accumulation of LDs, their recycling must be reduced or prevented, and this may be one of the roles of the highly abundant plant LD-binding proteins such as oleosin and caleosin. The accumulation of LDs in plant cells is regulated by a hierarchy of transcription factor proteins of which one of the most important is WRINKLED1 (WRI1). Several recent studies have shown that WRI1 is one of the key master switches leading to TAG accumulation in higher plants. Ectopic expression of WRI1 in tissues that do not normally accumulate large amounts of LDs, such as leaves, leads to the formation of

(a)   (b)   (c)

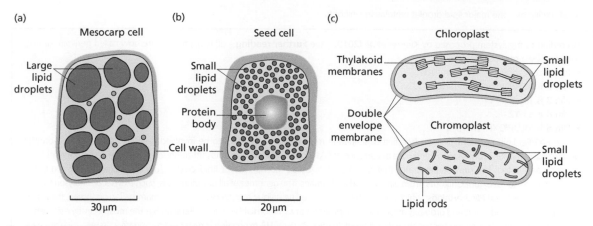

Fig. 5.20 Lipid droplets in plants. (a) Large irregular TAG droplets are found in the mesocarp tissue of lipid-rich fruits such as olive and oil palm. (b) Smaller and more regular TAG droplets of 0.5–1.0 μm diameter can make up >50% of cell volume in seed tissues of many higher plants. (c) Lipid droplets in plastids can be spherical or rod shaped depending on their lipid and protein compositions.

numerous TAG-rich oleosin-bound LDs in a manner that is normally seen only in seeds.

5.5.3 Protists and fungi

Many protists and fungi act as parasites or pathogens, and the ability to accumulate cytosolic LDs is often a key part of their success as infectious agents. Several of these organisms can stimulate their host cells to form large numbers of LDs that are then mobilized by the infectious agents as energy sources. In the case of the malarial parasite, *Plasmodium falciparum*, an essential factor for the proliferation of the parasite within infected human erythrocytes, is its ability to induce the accumulation and subsequent mobilization of large amounts of TAGs in LDs. Whilst some of these lipids are transferred to the parasite and accumulate in its cytosol as TAGs, many are released into the infected erythrocyte as the parasite reaches the merozoite stage of its life cycle. A sudden release of FAs may cause the membrane lysis that leads to cell rupture and the release of merozoites into the blood vessel where they can infect new cells. In this case, therefore, host LDs can be used not just as a nutrient source for the parasite but also as a cellulolytic mechanism to enable *P. falciparum* cells to escape from their original host cells and infect new ones.

There have been relatively few recent studies on the occurrence and function of cytosolic LDs in free-living protists. However, it now appears likely that protists share a common mode of LD regulation with all other Unikonts. The Unikonts form a phylogenetic supergroup

that includes Amoebozoa (e.g. slime moulds), Metazoa (multicellular animals) and Fungi. Evidence of a common method of LD regulation comes mainly from comparative genomics, which has revealed the occurrence of 'PAT' or 'perilipin'-like genes in each of these diverse groups of Unikont organisms. In much of the literature after the year 2000, the term 'PAT proteins' was commonly used, but this has now been superseded by Perilipin (Plin), as described in Box 5.4.

Many pathogenic fungi use specialized structures, termed appressoria, to break through the tough surfaces of their hosts. LDs appear to be crucial for appressorium function and especially in the production of the high turgor pressure that is required for virulence. In the rice blast fungus *Magnaporthe grisea* and the insect pathogen *Metarhizium anisopliae*, LDs originate in fungal spores and redistribute to the incipient appressorium. Whilst the underlying mechanism is unknown, several kinases are implicated. LDs also play roles in colonization and sexual development in other fungi, including *Fusarium graminearum*, a wheat pathogen, and *Aspergillus nidulans*, a soil-dwelling fungus that is an opportunistic human pathogen.

The brewers' yeast, *Saccharomyces cerevisiae*, shares many features of LD organization and function found in much more complex animals such as insects and mammals. Yeast cells take up NEFAs, which are activated to acyl-CoAs and either catabolized via ß-oxidation or converted into complex lipids such as TAGs and sterol esters (SE). The biosynthesis of TAGs and the formation

Box 5.4 Perilipins – the major lipid droplet proteins in Unikonts

According to the system proposed by Kimmel *et al.* (2010) – see **Further reading** – all PAT proteins are now called perilipin, or Plin, followed by a number, as follows:

- Plin 1 corresponds to Perilipin/LSD1
- Plin 2 is ADRP/adipophilin/fatvg/LSD2
- Plin 3 is TIP 47/PP17
- Plin 4 is S3-12
- Plin 5 is MLDP/OXPAT/LDSP5

The PAT/Perilipin group of proteins was first described in mammals and is known to be involved in many aspects of lipid droplet organization and function. More recently, the occurrence of PAT-like genes in all Unikonts (although some of these genes have yet to be correctly annotated) has provided evidence that a PAT protein-based mechanism for lipid droplet function originally evolved in the common ancestor of extant Unikonts, from slime moulds to humans. The genomes of all nonmammals, such as insects, analysed to date only contain *Plin1* and *Plin2* orthologs, which implies that *Plin3-5* may be unique to mammals. In some mammalian genomes, the *Plin 3, 4,* and *5* genes are adjacent and probably arose due to duplication of a primordial *Plin3* gene. Perilipin 5 is the most recently established family member and its function has yet to be determined. In mice *Plin5* mRNA expression is restricted to oxidative tissues, such as heart, slow-twitch fibres of skeletal muscle, brown adipose tissue, and liver and the gene is strongly induced by fasting.

of LDs are regulated in part by acyl-CoA oxidases. If LD formation is blocked, excess exogenous FAs become toxic when they accumulate in the yeast cells. This indicates that one of the functions of LDs is to minimize the risk of FA toxicity by serving as a reservoir for the sequestration of excess acyl groups. Further parallels between yeast and mammals are shown by the presence in yeast of homologues of LD-associated proteins involved in human lipodystrophy, such as seipin and lipin (see Section 10.4.2 & Table 10.5). As discussed below, mutations in these genes in humans can result in defective adipogenesis with often severe clinical outcomes. In yeast, a functional homologue of human seipin, Fld1p, regulates the LD size, which is consistent with seipin's proposed role in humans in the assembly and maintenance of LDs. In yeasts and other fungi, cells can contain a mixture of TAG-rich and sterol ester (SE – see Section 2.3.4.1)-rich LDs.

5.5.4 Animals

5.5.4.1 Invertebrates

Most research on invertebrate LDs has used the fruit fly, *Drosophila melanogaster*. About 1.5% of the expressed *Drosophila* genome, or 370 genes, is involved in LD function, which shows their importance for the insects. Early-stage *Drosophila* embryos were one of the first systems where it was shown that cytosolic LDs are able to move bidirectionally within cells in association with dynein, a motor protein that can move along microtubules whilst carrying a cargo. The proteome of *Drosophila* LDs is similar to those of other animals in containing PAT/Plin family proteins, although only Plin1 and Plin2 are present. Plin3–5 are apparently restricted to vertebrates. Plin1-deficient mutants have much larger LDs than wild-type insects, and in some cases, fat-storage cells were dominated by a single giant LD, reminiscent of mammalian white adipocytes. These mutant flies also had an obesity phenotype that was clearly related to LD dysfunction and suggests that LD status can determine feeding behaviour, possibly via an adipokine or other signals from fat-storing organs.

An example of the utility of *Drosophila* as a model for studying LD function and regulation comes from studies that have elucidated the role of the COPI complex as a regulator of lipid homeostasis. As discussed in Section 5.4.2, COPI and COPII vesicles are essential components of the trafficking system between the ER and Golgi. If either COPI or COPII complexes are disrupted, the Golgi function is abolished, but only COPI is required for LD mobilization. In the absence of COPI function, LDs accumulate to an abnormal degree. COPI components changed the LD protein coat composition, particularly by removing Plin3 and hence promoting lipase binding and the initiation of TAG hydrolysis. COPI-mediated regulation of LD droplet turnover is now believed to be a general mechanism common to all animals.

5.5.4.2 Mammals

Mammalian cells have a LD system similar to other animals. However, in mammals, these basic LD mechanisms are supplemented by additional layers of complexity and functional redundancy that have been adaptive during the evolution of this group of relatively large and long-lived eukaryotes. For example, mammalian cells appear to have several distinct global mechanisms for the formation and turnover of LD that are regulated by different effectors and/or are active in different tissues. Even at the metabolic level, mammals tend to have several alternative and redundant forms of key enzymes, such as DAG acyltransferase (DGAT – see Section 4.1) or lipases (see Section 4.2).

The morphology and function of mammalian LDs tend to be more variable than those of other organisms. For example, mammals have two broad types of storage cells that are highly enriched in LDs, namely white and brown adipocytes, which have very different structures and physiological functions (see Sections 9.1 & 9.2). In mammals, cytosolic LDs are also prominent constituents of steroidogenic cells in the adrenal gland and reproductive organs (for steroid hormone biosynthesis), mammary gland epithelial cells (for milk fat synthesis), hepatocytes and enterocytes (for lipid metabolism and lipoprotein formation), and leukocytes (for synthesis of lipidic mediators such as eicosanoids). More recently, the list of processes involving LDs in mammalian cells has expanded considerably, as has the number of proteins that are associated with these organelles (Fig. 5.21).

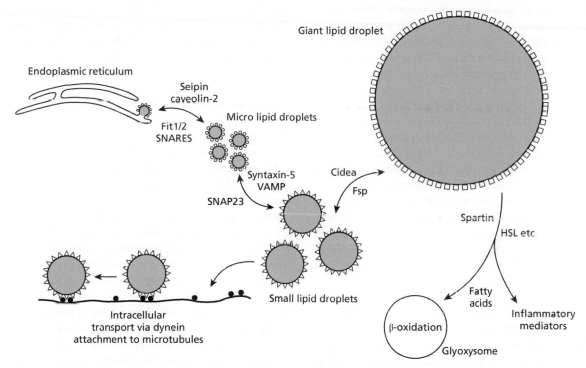

Fig. 5.21 Regulation of cytosolic lipid droplets in mammals. This is a brief summary of some of the major processes, components and organelles involved in the regulation of lipid droplet (LD) formation, maturation and turnover in mammalian cells. Regulation of lipid droplets in mammalian cells is a highly complex process that plays a key role in cellular homeostasis, and its dysfunction is implicated in many pathological processes ranging from some forms of cancer to obesity and coronary heart disease. Nascent neutral lipid esters are initially released from the ER as micro lipid droplets in a process regulated by various proteins such as caveolin2 and seipin. In most cells, the micro lipid droplets fuse to produce mature 0.5–2 μm small lipid droplets via a SNARE-mediated process that includes proteins such as SNAP23, VAMP and Syntaxin-5. In mammalian adipocytes, the proteins, Cidea and Fsp1 stimulate small lipid droplets to fuse further to form giant large lipid droplets of 50–200 μm diameter. In most cells lipid droplets are able to move around by means of the cytoskeletal network via proteins such as Dynein and ERK2. Mobilization of lipid droplets to form fatty acids, inflammatory mediators, and other breakdown products is regulated via several hormones and protein factors such as Spartin. Scheme adapted from DJ Murphy (2012) The dynamic roles of intracellular lipid droplets: from archaea to mammals, *Protoplasma* **249**:541–85, fig. 4.

Multifunctional LDs are required in diverse cell types to store and release acyl groups that have been tailored to accommodate cell-specific requirements and functions. Such functions include rapid FA storage/release into or from blood by white adipocytes, mitochondrial FA oxidation for thermal regulation by brown adipocytes, FA oxidation for long-term mobility demands by slow-twitch fibres in skeletal muscles, lipidation during very low-density lipoprotein (VLDL) production and mitochondrial ß-oxidation, milk production in mammary epithelial cells and surfactant production in type II alveolar pneumatocytes in the lung (see Section 10.3.7.3). Mammalian LDs have been implicated in acyl lipid turnover, cholesterol homeostasis, in hepatitis C infection, in proteasomal and lysosomal degradation of apolipoprotein B, in and protein phosphorylation and in the regulation of protein trafficking by small GTPases.

As in other animals, the major group of LD-associated proteins in mammals is the PAT/Plin family, although the remainder of the mammalian LD proteome is more variable and much larger than in nonmammals. The abundant LD-binding phosphoprotein, Plin1, has a major role in regulating LD size and access by lipases, whilst Plin2/Adipophilin acts as the major coat protein that is especially important in the formation of LDs *de novo* and their maturation after release from the ER. In addition to these two basic members of the PAT/Plin family that are common to all animals, mammals have three additional members, Plin3-5, which act as regulators of LD formation and turnover and may be capable of functionally substituting for Plin 1. A basic overview of the regulation of LD formation/turnover in mammals is shown in Fig. 5.21.

5.5.5 Cytosolic lipid droplet formation/ maturation

Cytosolic LDs in eukaryotes are produced on specialized domains of the ER that are enriched in enzymes of TAG and SE biosynthesis and may be related to lipid rafts (see Section 5.3.3). Some lipid raft proteins, such as Flotillin-1 and -2 and Stomatin, colocalize on various cellular membranes and on LDs. In particular, Flotillin-1 appears to associate with nascent LDs and could induce vesiculation at the ER or plasma membrane following loading of the bilayer with nonpolar lipids. Flotillin is a determinant of a clathrin-independent endocytic pathway in mammalian cells, which implicates this protein in vesicular trafficking, while its expression in insect cells can induce

the formation of caveolae-like vesicles. Newly formed TAGs and SE molecules may be prevented from lateral diffusion within the ER bilayer by transmembrane proteins that can effectively pen in these nonpolar lipids so that they form a bulge in the bilayer (see Fig. 5.22).

The increased membrane curvature around growing TAG/SE bulges can be stabilized by phospholipid demixing. The demixing of bilayer phospholipids can be generated by the selective biosynthesis of cone-shaped lipids in TAG/SE-rich regions of the ER. Such lipids have headgroups with relatively small cross-sectional areas and longer acyl chains. Two of the most common cone-shaped lipids that can stabilize regions of high negative membrane curvature are PtdEtn and DAG. In addition, signalling events can lead to the localized accumulation of highly charged lipids such as phosphorylated phosphoinositides or phosphatidic acid (PtdOH) and lyso-phosphatidic acid (lysoPtdOH) on one bilayer leaflet. Such lipids can themselves induce curvature and/or recruit specific proteins to such localized domains that then influence curvature.

As shown in Fig. 5.22, the directionality of LD release could be promoted by the selective accumulation of PtdEtn on the cytosolic monolayer of TAG/SE-producing ER domains. Providing the TAG/SE and PtdEtn molecules in such domains were constrained from lateral diffusion within the bilayer, as discussed above, the presence of localized PtdEtn clusters could facilitate an outward, cytosol-facing bulge in the ER that could then be stabilized by recruitment of structural LD-binding proteins such as Plin2 (in Unikonts) or oleosin (in plants). This process would bring about the formation of a large cytosol-facing bulge enriched in nonpolar lipids and enclosed/stabilized by a phospholipid/protein coat. It is possible that in some cells these LD bulges stay attached to the ER via a narrow neck stabilized by lipids such as PtdEtn and specialized proteins such as SNAREs.

LD biogenesis requires the presence of coat proteins such as Plin2, oleosin, or phasins, but additional ER-resident proteins are also essential for LD formation and release. For example, the lateral escape of nonpolar lipids from the incipient droplet would be further constrained by the presence of transmembrane proteins, possibly including seipin, which could form an ever-diminishing collar around the neck of the droplet (see Fig. 5.22). Some LDs remain physically and/or functionally connected with the ER, but most are released into the cytosol as microdroplets about 50–60 nM in diameter.

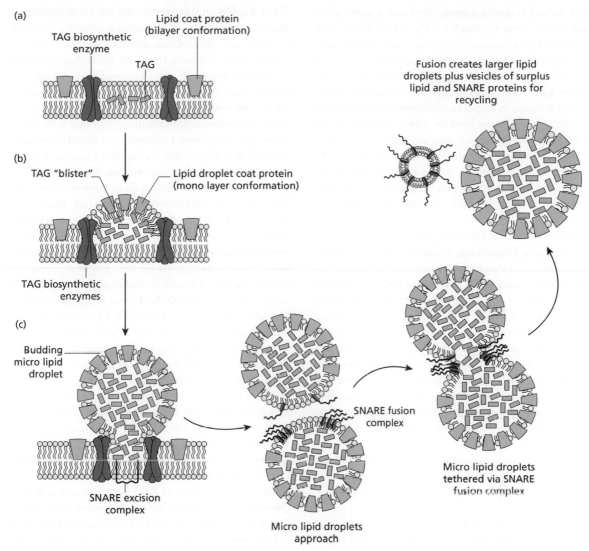

Fig. 5.22 Formation and maturation of cytosolic lipid droplets. (a) LD formation occurs in discrete ER microdomains that are enriched in enzymes of nonpolar lipid ester (e.g. TAG) and nonplanar bilayer lipid (e.g. PE) biosynthesis, plus LD coat proteins such as perilipin or oleosin. Lateral diffusion of the TAG and PE products is physically restricted by a ring of transbilayer proteins around their site of formation. (b) The accumulating TAG (grey rectangles) within the ER bilayer is forced into a cytosol-facing bulge or 'blister'. This structure is stabilized at its concave neck by localized domains enriched in lipids such as PtdEtn, and by specialized LD coat proteins that are recruited into the bulge, possibly thanks to the favourable free energy transition conferred by moving from a bilayer (larger blue wedges) to a monolayer (smaller blue wedges) environment. (c) As more TAG enters the bulge and its neck is increasingly constrained, it begins to assume a spherical shape. The final excision of the neck to release a nascent micro lipid droplet may be facilitated by a SNARE excision complex. Micro lipid droplets can then fuse to form larger structures in a SNARE-mediated process similar to that depicted in Figs. 5.17 and 5.21.

Microdroplets grow via fusion and/or the accretion of nonpolar lipid from the ER to produce the mature lipid droplets of about 500–2000 nM diameter as found in most cell types in most organisms.

In most mammalian cells, the accumulation of small LDs appears to be the default mechanism. However, in a very small number of cell types, most notably white adipocytes, small LDs continue to fuse with each other until the cell is almost entirely filled with a single, giant unilocular LD (see Section 9.1 & Fig. 9.1). This process is mediated by LD maturation genes, such as *fsp27* (fat-specific protein 27); lesions in such genes can result in the failure of white adipocytes to accumulate giant unilocular LDs. In addition to regulatory proteins such as fsp27, LD fusion in adipocytes involves microtubule and SNARE proteins. These proteins, which also play roles in vesicle budding and fusion, may facilitate fusion of LDs. In mammals, LD fusion may also be promoted by the activation of phospholipase D, which cleaves headgroups from membrane phospholipids resulting in the generation of localized accumulations of the nonplanar bilayer lipid, PtdOH.

The importance of LD fusion can be appreciated from the fact that it requires the fusion of about 1000 micro-LDs to produce a unilocular LD in a white adipocyte. LD fusion probably occurs via a simplified version of SNARE-mediated bilayer vesicle fusion described previously in Section 5.4.3. According to this model, two small LDs are tethered together via a SNARE fusion complex consisting of a four-helix bundle between Syntaxin-5, SNAP23, and VAMP4. As shown in Fig. 5.22, the SNARE fusion complex draws the two droplets together until they fuse into a single larger droplet. It is likely that additional proteins, such as members of the Plin family stabilize the new enlarged droplet or enable further fusion to create even larger droplets.

5.6 Extracellular lipid assemblies

All lipids are synthesized within cells but certain lipid molecules are then exported to extracellular sites where they can have a variety of roles such as the protection of external layers and in intercellular transport. For example, lipids are key components of bacterial cell walls, the exoskeleton of arthropods, and the sebum layer in the epidermis of mammals. Lipid-derived coatings such as cutin, suberin and waxes also form the external surfaces of most land plants. In many animals, from insects to mammals, nonpolar lipids as part of lipoprotein assemblies provide one of the major ways in which carbon and energy are transported around the body via their circulatory systems. Recent evidence also suggests that even relatively simple organisms such as some fungi can use analogous lipid assemblies for the transport of carbon and energy to their reproductive structures.

5.6.1 Lipids in extracellular surface layers

Bacteria and *archaea* have several different types of external cell wall. Gram-positive bacteria and the archaea have relatively thick cell walls that are based on various forms of polysaccharide and polypeptide matrix. The cell walls and membranes of most Gram-positive bacteria contain a series of highly anionic polymers. Most important among these is teichoic acid, which is a polymer of glucose 1-phosphate or ribitol phosphate. Membrane teichoic acids are based on glucose 1-phosphate and a proportion may be linked to glycolipids to give lipoteichoic acids. The details of these structures and the amount of substitution of the teichoic acids differ between bacterial species. Apart from the glycolipids linked to teichoic acid, the only significant lipid in most Gram-positive bacteria is that in the plasma membrane. Notable exceptions are the mycobacteria, where 25–30% of the cell wall is made up of unusual lipids such as mycosides, sulpholipids, and trehalose mycolate. The latter lipid is also called the cord factor and enables virulent strains of tubercle and related bacteria to form long strings or cords as part of their pathogenic activity.

In contrast, Gram-negative bacteria have relatively thin cell walls that consist of a few layers of peptidoglycan surrounded by a second outer lipid bilayer membrane enriched in various lipoproteins and lipopolysaccharides. Although it is based on a glycerolipid bilayer structure, the lipidic phase of this outer membrane is mostly made up of FAs esterified to internally facing proteins (lipoproteins) and externally facing lipopolysaccharides such as lipid A, which is a key contributor to pathogenicity in Gram-negative bacteria (see Fig. 5.23 & Section 10.3.7.1). Phospholipid molecules are missing from the outer membrane in many Gram-negative bacteria but are present in some species. In those bacteria containing phospholipids in the outer membrane, PtdEtn is, by far, the most common constituent ($<85\%$ $^w/_w$).

The lipopolysaccharide is involved in several aspects of pathogenicity. It is a complex polymer in four parts. Outermost is a carbohydrate chain of variable length (called the O-antigen), which is attached to a core polysaccharide. The core polysaccharide is divided into the outer core and the backbone. These two structures differ between bacteria. Finally, the backbone is attached to a glycolipid called lipid A. The link between lipid A and the rest of the molecule is usually via a number of 3-deoxy-D-manno-octulosonic acid (KDO) molecules. The presence of KDO is often

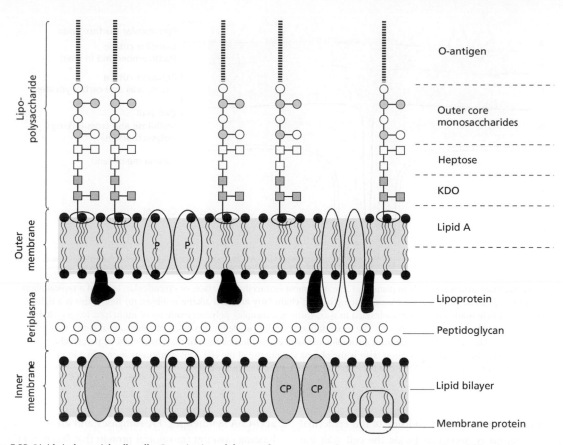

Fig. 5.23 Lipids in bacterial cell walls. Organization of the membrane system of a Gram-negative bacterium. Abbreviations: P, porin; A, transmembrane peptidoglycan-associated protein; CP, carrier protein. Based on Fig. 6.25 from *Lipid Biochemistry*, 5th Edn.

used as a marker for lipopolysaccharide (or outer membrane) even though it is not present in all bacterial lipopolysaccharides. Lipid A consists of two phosphorylated glucosamine groups esterified to several acyl chains as shown in Fig. 5.23. Phosphate and KDO groups are also substituted. Unsaturated and cyclopropane FAs, which are common in other lipid types, are absent from the lipopolysaccharide.

Lipid A, which anchors lipopolysaccharide in the membrane, is the first component to be synthesized. Hydroxy acids are added to the disaccharide, followed by KDO and then SFA. The hydroxy FAs come from acyl-CoA substrates whereas CMP-KDO is the source of the second addition units. After the addition of SFA, sugars are added from nucleotide diphosphate derivatives. These reactions build one half of the molecule. Another lipid, PtdEtn, has been suggested to be intimately involved in the binding of the transferase enzymes to

the lipopolysaccharide acceptor. Lipopolysaccharide biosynthesis occurs on the inner (plasma) membrane from which the molecule must be transferred to the outer membrane. This occurs at specific membrane sites generally where the two membranes adhere to each other.

Plants have a relatively impermeable lipidic protective layer around all their external surfaces that is commonly termed the cuticle. The cuticle limits water loss and blocks the entry of microbial pathogens: it is one of the key adaptations that enabled plants to colonize the land surface. In all aerial parts of the plant body, the cuticle is primarily made up of a FA-derived polymer called cutin plus a complex mixture of long-chain waxes. In the underground parts of most plants, such as roots and tubers, and in the aerial parts of certain drought-tolerant plants a different lipid derived polymer called suberin is present. While cutin is deposited on the

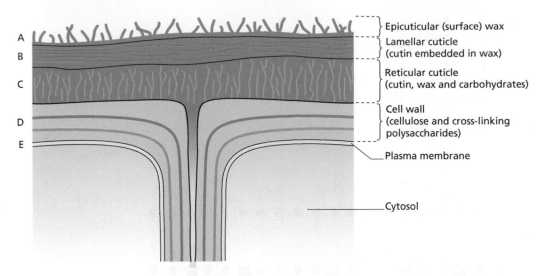

Fig. 5.24 Extracellular structural lipids in plants. (a) The outermost extracellular surface, or epicuticular, layer of a typical plant consists of semi-crystalline rods of a solid wax made up of long-chain fatty acyl and alkane residues. (b) Below this is a tightly packed lamellar cuticle made up of cutin embedded in wax. Cutin is a complex polymer made up of numerous longer-chain ω-hydroxy fatty acid and epoxide residues. (c) The more diffuse reticular cuticle layer consists of a mixture of cutin, wax and polysaccharides. (d) The cell wall is made up of a complex network of parallel cellulose fibrils cross-linked by polysaccharides such as hemicellulose and pectin. (e) The innermost layer is the plasma membrane of the cell. This membrane is enriched in enzymes responsible for the biosynthesis of all the components that make up the four outer layers.

external facing surface on the outside of the epidermal cell wall, suberin is deposited inside the cell wall but outside the plasma membrane (see Fig. 5.24).

Cutin consists of interesterified ω-hydroxy FAs, cross-linked by epoxy and ester bonds, and makes up the major matrix (40–80% $^w/_w$) of the cuticle. Another nonsaponifiable hydrocarbon-based polymer called cutan is sometimes present in the cuticle of drought-adapted plants. Embedded in the cutin matrix are cuticular waxes, which are complex mixtures of long- and very-long-chain FA derivatives (16C–36C) that can also include triterpenoids and minor components, such as sterols, alkaloids, phenylpropanoids and flavonoids. Cuticular waxes form an intracuticular layer in close association with the cutin matrix. They also form a separate epicuticular film on the outside of the cuticle, which may include epicuticular wax crystals.

Cutin and suberin are both polyesters consisting mainly of 16C–18C hydroxyacids, diacids and epoxyacids esterified to each other and to glycerol. Analysis of *Arabidopsis* mutants has shown that the biosynthesis of these extracellular lipidic polymers is regulated by acyl-transferases with unique specificities, FA hydroxylases,

acyl-CoA synthetases, FA elongases and an ABC (ATP-binding cassette) transporter protein that facilitates their export across the plasma membrane.

Animals have numerous types of lipid-based external layers such as the waxes that coat the chitinous exoskeleton of arthropods and the sebum coating of mammalian skin. Many of the most detailed studies on animal waxes have been done with insects. Their waxes consist of a complex mixture that includes straight-chain saturated and unsaturated hydrocarbons, wax esters, sterol esters, ketones, alcohols, aldehydes and acids. The composition may be influenced by the plants on which the insect feeds and differ widely between species. For example, the amount of hydrocarbon can vary from 100% in the surface lipids of the field cricket (*Nemobius fasciatus*) to less than 1% in tobacco budworm pupae. Not only do insect waxes prevent desiccation, they may also have other diverse functions such as acting as pheromones or kairomones, helping in recognition and in thermoregulation.

Mammalian skin contains a mixture of lipids that forms a key protective layer around the body. Deficiencies in the formation or function of skin lipids are associated with several diseases in humans. The epidermis, which is separated from a predominantly acellular dermis by a basement membrane, is composed of four layers. The outermost of these four layers, the stratum corneum, consists of impermeable cellular layers composed of flattened, densely packed, anucleate cells embedded in a waxy lamellar matrix of complex intercellular lipids. The latter are derived from the lipid-filled membrane-coating granules of the cells of the next layer, the stratum granulosum. In addition, the sebaceous glands secrete sebum, a lipid mixture.

The composition of human skin surface lipids differs over various parts of the body, but in sebaceous gland-enriched areas (such as the face) the secretion, sebum, may represent 95% of the total. When sebum is freshly formed it contains mainly TAGs, wax esters and squalene. However, NEFAs are rapidly released by bacterial action. These acids are both an irritant and comedogenic (give rise to 'blackheads') and have been implicated in acne vulgaris – an extremely common complaint of puberty. It is often stated that the severity of acne correlates with excessive secretion of sebum (seborrhoea), and many treatments are designed primarily to remove such lipids from the skin surface.

More serious than acne, but less prevalent, are the skin diseases atopic eczema and psoriasis. Lipids have been suggested as being involved in both of these complaints, although the evidence at present is equivocal. It has been noted in patients with atopic eczema that there is a disturbance in the body's normal complement of polyunsaturated fatty acids (PUFA). For convenience, blood samples have usually been analysed and these usually exhibit reduced amounts of γ-linolenic (all-c6,9,12-octadecatrienoic, n-6) acid. γ-Linolenate is an intermediate in the normal conversion of the essential fatty acid (EFA), linoleate (18:2, n-6), to the eicosanoid precursor arachidonate (20:4,n-6). Psoriasis, which affects 2% of the global population, and ranges from relatively benign to highly debilitating forms. It is responsible for more days spent in hospital than any other skin complaint and present treatments are rudimentary and rather unsatisfactory. It is now believed that psoriasis is an autoimmune disorder that results in abnormalities in various cellular differentiation and developmental processes. These symptoms can include excessively high amounts of some arachidonate metabolites and deficiencies in phosphoinositide metabolism.

In humans, EFA deficiency can result in pathological alterations of skin lipids that result in dryness, flaking, scales, loss of pigmentation and poor healing of wounds. The most important EFA, linoleic acid, is the precursor of several classes of eicosanoids, leukotrienes, prostaglandins, thromboxanes and related compounds. Until recently it was thought that the classic skin symptoms resulting from EFA deficiency (see Section 6.2.2) were due to a lack of eicosanoids. However, the roles of n-6 acids (such as linoleate) and n-3 acids (such as α-linolenate) in membrane function and prostaglandin formation cannot explain why they are essential for skin function. For example, systemic treatment with aspirin, which effectively prevents prostaglandin formation, does not result in the appearance of EFA-deficiency symptoms. It is also generally accepted that at least some cell lines do not have an EFA requirement in culture, suggesting that EFAs are not essential for the formation and function of cellular membranes in general. Several linoleic acid-rich lipids such as acylglucosylceramide, acylceramide and a specific wax ester have been identified in human, pig and rat epidermis. These lipids form part of the intercellular lipid-rich matrix serving as a water permeability barrier. In EFA deficiency the linoleate of these lipids is replaced by oleate, a change that diminishes the barrier function.

5.6.2 Lipids in extracellular transport

The extracellular transport of lipids in animals, particularly in the context of diet and assimilation, is described in Chapter 7. However, the formation of such lipid assemblies, called lipoproteins, is also linked with LD function as described in Section 5.5.4 of this chapter. The term 'lipoprotein' as used here refers to extracellular assemblies of nonpolar and polar lipids stabilized by specific proteins termed apolipoproteins. However, it should be noted that as mentioned in Section 5.3.2 the term 'lipoprotein' is also sometimes used to refer to a completely different class of proteins that are covalently bound to fatty acyl residues or other lipid moieties.

Some mammalian cells, such as hepatocytes and enterocytes, produce LDs in the ER lumen as well as in the cytosol. There are three main types of LDs in the ER lumen. The first types are luminal LDs, similar in size and composition to cytosolic LDs, the second are the VLDLs produced in hepatocytes, and the third are the chylomicrons produced in enterocytes (see Chapter 7). In each case the formation and maturation of these LDs have similarities as well as differences to the processes described above for cytosolic LDs, as shown in Fig. 5.21.

The ER lumen LD that are compositionally similar to those produced in the cytosol are probably also made by a process regulated by members of the fat-specific protein (Fsp) family. The major difference is that the droplets bud off from the internal face of the ER membrane. In contrast, VLDLs and chylomicrons are assembled via a more complex multistage mechanism that involves the initial lipidation of a newly synthesized full-length apolipoprotein B (apoB) molecule produced on the rough ER. In hepatocytes, VLDLs are formed when nascent apoB100 is lipidated with TAG in a process that requires a soluble protein called microsomal TAG-transfer protein (MTP). This results in the formation of small pre-VLDL droplets in the smooth ER that then fuse with luminal LDs in a process that is regulated by a protein called cell death-inducing DFFA-like effector b (CIDE-B) to produce full sized VLDL droplets that are incorporated into specialized transport vesicles for export from the cell via the Golgi. If the *cide-b* gene function is disrupted, the result in an abnormal accumulation of LDs, resulting in obesity and fatty liver syndrome (hepatic steatosis).

In enterocytes, a truncated version of apoB100, called apoB48, is partially lipidated with cholesterol, phospholipids and TAG via MTP to produce small relatively dense droplets called prechylomicrons. These prechylomicron droplets acquire more nonpolar lipids and a coating of apolipoprotein AIV before they are moved into transport vesicles for trafficking to the Golgi where they undergo several additional process steps, including further MTP-dependent lipidation and a coating of apolipoprotein AI, before they are finally ready for export from the cell.

The lipoproteins involved in extracellular transport in mammals are discussed in Chapter 7 (Table 7.1). The two classes of lipoprotein that most closely resemble cytosolic LDs are VLDLs and chylomicrons, both of which are formed in the ER lumen in cells of the liver and intestinal wall. After their synthesis, these two lipoproteins are processed for export from the cell via the Golgi. VLDLs are encapsulated into transport vesicles that bud off from the hepatocyte ER via a COPII-mediated mechanism that is similar to that used for other forms of anterograde transport from the ER to the Golgi (see Section 5.4.2). However, chylomicrons are believed to bud off from the enterocyte ER via a COPII-independent process although their further trafficking through the Golgi involves the COPII mechanism. The trafficking of both lipoproteins through the Golgi and their eventual release from the cell into the circulation (in the case of the chylomicrons via the lymphatic system) involves several rounds of vesicle budding from and fusion with different membranes. These processes are mediated by the SNARE protein complex as previously described for bilayer vesicles (Fig. 5.17) and cytosolic LDs (Fig. 5.22).

The physiological roles of lipid transport in chylomicrons and VLDL will be considered in more detail in Section 7.2.3.

KEY POINTS

- Lipids are amphipathic molecules that spontaneously form a range of macromolecular assemblies including bilayer membranes, micelles and spheroidal droplets.

- Lipid-bound membranes are the defining feature of all cellular life forms.

- Biological cells are enclosed by a plasma membrane made up of a glycerolipid bilayer in which proteins are fully or partially embedded.

- Eukaryotic cells also contain a complex network of endomembranes that are involved in various lipid and protein trafficking, import, secretion and signalling processes.

- Lipids can also form nonbilayer assemblies the most common of which are the nonpolar lipid droplets involved in energy storage and other metabolic processes.

- Extracellular lipid such as waxes and polyesters play important roles in the protection of many organisms from the external environment.

Further reading

Major cellular lipids

Galea AM & Brown AJ (2008) Special relationship between sterols and oxygen: Were sterols an adaptation to aerobic life? *Free Radical Bio Med* **47**:880–9.

Chalfant C & Del Poeta M, eds. (2010) *Sphingolipids as Signaling and Regulatory Molecules*. Landes Bioscience, Austin, TX & Springer Science, New York.

Poralla K. & Kannenberg E (1987) Hopanoids: sterol equivalents in bacteria. In *Ecology and Metabolism of Plant Lipids*, ACS *Symposium Series*, **325**, American Chemical Society.

Saenz JP, Sezgin E, Schwille P & Simons K (2012) Functional convergence of hopanoids & sterols in membrane ordering, *Proc Nat Acad Sci USA* **109**:14236–40.

Weete JD, Abril M & Blackwell M (2010) Phylogenetic distribution of fungal sterols, *PLoS ONE* **5**,e10899.

Welander PV, Hunter RC, Zhang L, *et al.* (2009) Hopanoids play a role in membrane integrity and pH homeostasis in *Rhodopseudomonas palustris* TIE-1, *J Bacteriol* **191**:6145–56.

Zakelj-Mavric M, Kastelic-Suhadolc T, Plemenitas A, Rizner TL & Belic I. (1995) Steroid hormone signalling system and fungi, *Comp Biochem Physiol Part B* **112**:637–42.

Lipid assemblies

Cooke IR & Deserno M (2006) Coupling between lipid shape and membrane curvature, *Biophys J* **91**:487–95.

Dowhan W & Bogdanov M (2002) Functional roles of lipids in membranes. In DE Vance & JE Vance, eds, *Biochemistry of Lipids, Lipoproteins & Membranes*, 4th edn. Elsevier, New York, pp. 1–35.

Hesse D, Jaschke A, Chung B & Schürmann A (2013) *Trans*-Golgi proteins participate in the control of lipid droplet and chylomicron formation, *Biosci Rep* **33**:1–9.

Khandelia H, Duelund L, Pakkanen KI & Ipsen JH (2010) Triglyceride blisters in lipid bilayers: implications for lipid droplet biogenesis and the mobile lipid signal in cancer cell membranes, *PLoS ONE* **5** (9):e12811.

Vadim A, Shnyrova AV & Zimmerberg J (2011) Lipid polymorphisms and membrane shape, *Cold Spring Harb Perspect Biol* 2011;**3**:a004747.

Roles of lipids in biological evolution

Boucher Y, Kamekura M & Doolittle WF (2004) Origins and evolution of isoprenoid lipid biosynthesis in archaea, *Mol Microbiol* **52**:515–27.

Chen IA & Walde P (2010) From self-assembled vesicles to protocells, *Cold Spring Harb Perspect Biol* 2010;**2**:002170.

Desmond E. & Gribaldo S (2009) Phylogenomics of sterol synthesis: insights into the origin, evolution, and diversity of a key eukaryotic feature, *Genome Biol Evol* **1**:364–81.

Koga Y & Morii H (2005) Recent advances in structural research on ether lipids from archaea including comparative and physiological aspects, *Biosci Biotech Bioch* **69**:2019–34.

Lombard J, López-García P & Moreira D (2012) The early evolution of lipid membranes and the three domains of life, *Nat Rev Microbiol* **10**:507–15.

Mansy SS, Schrum JP & Krishnamurthy M (2008) Template-directed synthesis of a genetic polymer in a model protocell, *Nature* **454**:122–6.

Matsumi R, Atomi H, Driessen AJ & van der Oost J (2011) Isoprenoid biosynthesis in archaea – biochemical and evolutionary implications, *Res Microbiol* **162**:39–52.

Pearson A, Budin M & Brocks JJ (2004) Phylogenetic and biochemical evidence for sterol synthesis in the bacterium, *Gemmata obscuriglobus*, *Proc Nat Acad Sci USA* **100**:15352–7.

Pereto J, López-García P & Moreira D (2004) Ancestral lipid biosynthesis and early membrane evolution, *Trends in Biochem Sci* **29**:469–77.

Pohorille A & Deamer D (2009) Self-assembly and function of primitive cell membranes, *Res Microbiol* **160**:449–56.

Schrum JP, Zhu TF & Szostak JW (2010) The origins of cellular life, *Cold Spring Harb Perspect Biol* 2010; **2**:a002212.

Simoneit BRT, Rushdi AL & Deamer DW (2007) Abiotic formation of acylglycerols under simulated hydrothermal conditions and self-assembly properties of such lipid products, *Advances in Space Research* **40**:1649–56.

Wächtershäuser G. (2003) From pre-cells to Eukarya – a tale of two lipids, *Mol Microbiol* **47**:13–22.

Wikström M, Kelly AA & Georgiev A (2009) Lipid-engineered *E. coli* membranes reveal critical lipid headgroup size for protein function, *J Biol Chem* **284**:954–65.

Membrane organization

Fuerst JA & Sagulenko E (2012) Keys to eukaryality: planctomycetes and ancestral evolution of cellular complexity, *Front Microbiol* **3**:art. 167.

Lingwood D & Simons K (2010) Lipid rafts as a membrane-organizing principle, *Science* **327**:46–50.

Malinsky J, Opekarová M, Grossmann G & Tanner W (2013) Membrane microdomains, rafts, and detergent-resistant membranes in plants and fungi, *Annu Rev Plant Biol* **64**:1–25.

Maxfield FR (2002) Plasma membrane microdomains, *Curr Opin Cell Biol* **14**:483–7.

Owen DM, Williamson D, Rentero C & Gaus K (2009) Quantitative microscopy: protein dynamics and membrane organisation, *Traffic* **10**:962–71.

Rowland AA & Voeltz GK (2012) Endoplasmic reticulum-mitochondria contacts: function of the junction, *Natu Rev Mol Cell Biol* **13**:607–15.

Santarella-Mellwig R, Pruggnaller S, Roos N, *et al.* (2013) Three-dimensional reconstruction of bacteria with a complex endomembrane system, *PLOS Biol* **11**(5):e1001565.

Simon-Plas F, Perraki A & Bayer E (2011) An update on plant membrane rafts, *Curr Opin Plant Biol* **14**:1–8.

Simons K & Gerl MJ (2010) Revitalizing membrane rafts: new tools and insights, *Nat Rev Mol Cell Biol* **11**:688–99.

Simons K & Sampaio JL (2011) Membrane organization and lipid rafts, *Cold Spring Harb Perspect Biol* 2011;**3**:a004697.

Singer SJ & Nicholson GL (1972) The fluid mosaic model of the structure of cell membranes, *Science* **175**:720–31.

Voelker DR (2002) Lipid assembly into cell membranes. In DE Vance & JE Vance, eds. *Biochemistry of Lipids, Lipoproteins and Membranes*, 4th edn. Elsevier, New York, pp. 449–82.

Zick M, Rabl R & Reichert AS (2009) Cristae formation – linking ultrastructure and function of mitochondria, *Biochim Biophys Acta* **1793**:5–19.

Membrane function

Bankaitis VA, Garcia-Mata R & Mousley CJ (2012) Golgi membrane dynamics and lipid metabolism, *Curr Biol* **22**:R414–424.

Bonifacino JS & Glick BS (2004) The mechanisms of vesicle budding and fusion, *Cell* **116**:153–66.

Dacks JB & Field MC (2007) Evolution of the eukaryotic membrane-trafficking system: origin, tempo and mode, *J Cell Sci* **120**:2977–85.

Glebov OO, Bright NA & Nichols BJ (2006) Flotillin-1 defines a clathrin-independent endocytic pathway in mammalian cells, *Nat Cell Biol* **8**:46–54.

Hurley JH, Boura E, Carlson LA & Różycki B (2010) Membrane budding, *Cell* **143**:875–87.

Jacobson K, Mouritsen OG & Anderson RGW (2007) Lipid rafts: at a crossroad between cell biology and physics, *Nat Cell Biol* **9**:7–14.

Manford AG, Stefan CJ, Yuan HL, Macgurn JA & Emr S.D (2012) ER-to-plasma membrane tethering proteins regulate cell signaling and ER morphology, *Dev Cell* **23**:1129–40.

Südhof TC & Rothman JE (2009) Membrane fusion: grappling with SNARE and SM proteins, *Science* **323**:474–7.

Tamm LK, Crane J & Kiessling V (2003) Membrane fusion: a structural perspective on the interplay of lipids and proteins, *Curr Opin Struc Biol* **13**:453–66.

Intracellular lipid droplets

Bozza PT & Viola JPB (2010) Lipid droplets in inflammation and cancer, *Prostag Leukotr Ess* **82**:243–50.

Greenberg AS & Coleman RA (2011) Expanding roles for lipid droplets, *Trends Endocrin Met* **22**:195–6.

Kimmel AR, Brasaemle DL, McAndrews-Hill M, Sztalryd C & Londos C (2010) Adoption of PERILIPIN as a unifying nomenclature for the mammalian PAT-family of intracellular lipid storage droplet proteins. *J Lipid Res* **51**:468–71.

Martin S & Parton RG (2006) Lipid droplets: a unified view of a dynamic organelle, *Nat Rev Mol Cell Biol* **7**:373–8.

Murphy DJ (2012) The dynamic roles of intracellular lipid droplets: from archaea to mammals, *Protoplasma*, **249**:541–85.

Wältermann M & Steinbüchel A (2005) Neutral lipid bodies in prokaryotes: recent insights into structure, formation, and relationship to eukaryotic lipid depots, *J Bacteriol* **187**:3607–19.

Walther TC & Farese RV (2008) The life of lipid droplets, *Biochim Biophys Acta* **1791**:459–66.

Extracellular lipid assemblies

Hesse D, Jaschke A, Chung B & Schürmann A (2013) Trans-Golgi proteins participate in the control of lipid droplet and chylomicron formation, *Biosci Rep* **33**:1–9.

Jonas A (2002) Lipoprotein structure. In DE Vance & JE Vance, eds., *Biochemistry of Lipids, Lipoproteins and Membranes*, 4th edn. Elsevier, New York, pp. 483–504.

Palm W, Sampaio JL, Brankatschk M, *et al.* (2012) Lipoproteins in *Drosophila melanogaster* – assembly, function, and influence on tissue lipid composition, *PLoS Genet* **8**(7):e1002828.

Vance JE (2002) Assembly and secretion of lipoproteins. In DE Vance & JE Vance, eds., *Biochemistry of Lipids, Lipoproteins and Membranes*, 4th edn. Elsevier, New York, pp. 505–26.

Supplementary reading

Glansdorff N, Xu Y & Labedan B (2008) The last universal common ancestor: emergence, constitution and genetic legacy of an elusive forerunner, *Biol Dir* **3**:29.

Kandler O (1994) The early diversification of life. In S Bengston, ed., *Early Life on Earth*. Columbia University Press, New York, pp. 152–60.

Koonin EV & Martin W (2005) On the origin of genomes and cells within inorganic compartments, *Trends Genet* **21**:647–54.

Russell MJ & Hall AJ (1997) The emergence of life from iron monosulphide bubbles at a submarine hydrothermal redox and pH front, *J Geol Soc* **154**:377–402.

CHAPTER 6
Dietary lipids and their biological roles

The main emphasis of this chapter is on lipids in the diet, and their functions in the body. The emphasis will be on lipids in the diet of mammals, especially humans. We eat food lipids that originate from plants and animals, and these lipids may be modified by processing before they are eaten. Some dietary lipids are essential to life, but others may be toxic.

6.1 Lipids in food

Lipids form an integral part of the diet of humans and other animals. Dietary lipids have many functions in the body other than simply provision of energy. There are many connections between dietary lipids and health and disease.

There are two important aspects of lipids in the human diet, which are often confused. One is the total amount of fat eaten, expressed either in g/day or as a percentage of total energy intake. The other aspect is the content of different types of fatty acids (FAs): saturated (SFA), monounsaturated (MUFA) and polyunsaturated (PUFA) (see Section 2.1). These two aspects of dietary fat are often referred to as 'quantity' and 'quality'.

More than 90% of dietary lipids are triacylglycerols (TAGs), which originate from the adipose tissue or milk of animals, or from plant seed oils, mainly in the form of manufactured products. Some 35–45% of dietary energy in industrialized countries comes from TAGs. Total dietary fat content ('quantity') is therefore of clear relevance to body weight regulation. Worldwide, however, the contribution that fat makes to the diet can vary enormously.

Dietary fat quality, on the other hand, is not related directly to body weight regulation, but is closely related to health and disease particularly via regulation of the blood cholesterol concentration. This will be described later (see Section 10.5.5.2).

6.1.1 The fats in foods are derived from the membrane and storage fats of animals and plants

Lipids in our diet originate from living matter, are therefore the membrane and storage lipids of animals and plants. Meat provides animal storage fats (mainly TAGs with some cholesterol and fat-soluble vitamins), especially so if the meat contains adipose tissue. Suet, used as a hard fat in cooking, is bovine or ovine internal adipose tissue. Lard is a cooking fat derived from porcine adipose tissue. Milk fat occurs in almost all dairy products. Plant storage fats (again mainly TAGs with some fat-soluble vitamins) are present in their original form in nuts, cereal grains and some fruits such as the avocado. They are also familiar in cooking oils, salad oils and mayonnaise. Both animal and plant storage fats are used in the manufacture of spreading fats. Egg yolks are rich sources of fat. The lipid is present as lipoproteins in which TAGs, cholesterol and phospholipids are associated together with proteins. The lipoproteins are either in a high-density form ('lipovitellin', 78% protein, 12% phospholipid, 9% TAG) or a low-density form (18% protein, 22% phospholipid, 58% TAG, or richer still in TAG). They are evolutionarily related to the plasma lipoproteins described in Section 7.2.1.

The predominance of fat in foods like cooking oils, margarines, butter and meat fat is obvious: these are called 'visible fats'. Where fat is incorporated into the structure of the food, either naturally (e.g. in peanuts or cheese), or when it is added in the cooking or manufacturing process (e.g. in cakes, pastries, meats), it is less obvious to the consumer and is referred to as 'hidden fat'.

Lipids: Biochemistry, Biotechnology and Health, Sixth Edition. Michael I. Gurr, John L. Harwood, Keith N. Frayn, Denis J. Murphy and Robert H. Michell.
© 2016 John Wiley & Sons, Ltd. Published 2016 by John Wiley & Sons, Ltd.

Membrane fats in foods are predominantly phospholipids and glycolipids with, in animal tissues, cholesterol, and in plant tissues, plant sterols. Thus, the lean part of meat contains the membrane lipids, phospholipids and cholesterol of the muscle membranes and minor quantities of glycolipids, along with storage TAGs that are present in the muscle cells. Dairy products contribute small amounts of membrane lipids because of the presence of the milk fat globule membrane. The membrane lipids of plants that are important in the diet are those in green leafy vegetables, which are predominantly galactolipids (Table 2.11, Sections 2.3.2.3 & 5.4.1) and the membrane lipids of cereal grains, vegetables and fruits.

6.1.2 The fatty acid composition of dietary lipids and how it may be altered

6.1.2.1 Determinants of dietary lipid composition

When we describe a fat as saturated, monounsaturated or polyunsaturated, we are referring to the constituent FAs. All natural fats contain mixtures of all three types of FAs. Fats that contain a large proportion (not necessarily the greatest proportion) of SFA, and which tend to be solid at room temperature, are usually categorized as 'saturated fats' to contrast them with fats in which PUFAs predominate, and which tend to be more liquid. For example, lard is frequently categorized as a 'saturated fat', although over half (about 55%) of the FAs are unsaturated (see Table 2.7). Similarly, among vegetable oils, palm oil is usually categorized as 'saturated' even though 50% of its FAs are unsaturated.

Mammalian storage fats are characterized by a predominance of SFAs and MUFAs (Table 2.7). Storage fats from ruminant animals tend to have a greater proportion of SFAs because of the extensive biohydrogenation of the animal's dietary fats in the rumen (see Section 3.1.9). Nevertheless, ruminant fats contain a substantial proportion of MUFAs because of the presence in mammary gland and in adipose tissue of desaturases (see Section 3.1.11.5). Milk fat is unusual among animal fats in containing significant amounts of SFAs with chain lengths of 12 carbon atoms or less. The proportions of 4:0, 6:0, 8:0, 10:0 and 12:0 are highly dependent on species (Table 2.7 & Section 6.2 below). These short- and medium-chain length FAs are synthesized in the mammary gland itself (see Section 3.1.3.3). The FA composition of simple-stomached animals is more dependent on their diet since they do not extensively hydrogenate the dietary unsaturated fatty acids (UFAs), but incorporate them directly into adipose tissue TAGs.

Seed oils contain a wide variety of FAs, the composition of which is characteristic of the family to which the plant belongs. Generally one of the more common FAs predominates (Table 2.6): for instance, palmitic, oleic, or linoleic, as exemplified by palm oil, olive oil and sunflower seed oils. Coconut oil and palm kernel oil are unusual among seed oils in having a large proportion of medium-chain SFAs. It is therefore an oversimplification to characterize all vegetable oils as unsaturated.

The FA composition of the membrane lipids of the animal tissues used as foods, mainly phospholipids, is remarkably uniform irrespective of the animal's diet or whether it is simple-stomached or a ruminant (Table 6.1), although diet can influence membrane composition in subtle ways (see Sections 6.1.2.2, 6.2.4.1 & 6.2.4.2). An important PUFA in animal membrane lipids is arachidonic acid (20:4n-6, Table 6.1). Meat provides most of our dietary supply of this FA, although it can also be produced in mammals from linoleic acid (see Section 3.1.8.3).

In quantitative terms, the major source of PUFAs in the human diet is plant lipids. Of these, the major proportion is usually n-6 PUFA. Some plant seed oils, notably flax seed and linseed oils, are enriched in n-3 PUFAs, mainly α-linolenic acid (18:3n-3). Fish and shellfish provide the major source of the longer chain PUFAs of the n-3 family, especially eicosapentaenoic acid (EPA) and docosahexaenoic acid (DHA) (Table 2.7).

6.1.2.2 Manipulation of fatty acid composition at source

Because the FA composition of simple-stomached animals is determined in part by the diet, porcine adipose tissue, for instance, can be significantly enriched in linoleic acid by supplementing the pigs' diets with soybean oil. Ruminant fats can be enriched with PUFAs, but only if the dietary PUFAs are protected from hydrogenation in the rumen (see Section 3.1.9). This can be achieved by treating the oilseeds used to supplement the feed with formaldehyde, which cross-links the protein and renders the seed oil unavailable to the rumen microorganisms. When the food reaches the acid environment of the true stomach, the cross-links are digested and the PUFAs are available for absorption just as in simple-stomached animals. Table 2.7 illustrates the wide range of FA compositions that can occur in animal

Table 6.1 The fatty acid composition of some structural (membrane) lipids important in foods.

	Fatty acid (g/100 g total fatty acids*)								
	16:0	**16:1**	**18:0**	**18:1**	**18:2**	**18:3**	**20:4**	**LCPUFA**[a]	**total**
Foodstuff									
Beef: muscle	16	2	11	20	26	1	13		89
Lamb: muscle	22	2	13	30	18	4	7		96
Lamb: brain	22	1	18	28	1		4	14	88
Chicken: muscle	23	6	12	33	18	1	6		99
Chicken: liver	25	3	17	26	15	1	6		99
Chicken: egg yolk	29	4	9	43	11				96
Pork: muscle	19	2	12	19	26	8			86
Cod: flesh	22	2	4	11	1	4		52	96
Green leaves	13	3		7	16	56			95

[a] Longer chain polyunsaturates, mainly 20:5n-3 and 22:6n-3

* The figures are rounded to the nearest whole number; therefore the content of an individual fatty acid of less than 0.5% is not recorded. The figures across a row rarely sum to 100% because each sample contains a large number of components each contributing less than 1%. Thus samples from ruminants contain a wide variety of odd and branched chain fatty acids.

storage fats and the changes in PUFA content that can be achieved by dietary manipulation.

However, although such techniques work experimentally, polyunsaturated ruminant products are not generally available: they are expensive and subject to oxidative deterioration (see Section 3.3), which may result in poor taste and appearance. The technique of genetic modification may provide a way to alter the composition of dietary fats. A clear example comes from the provision of EPA and DHA in the human food chain. (These FAs have particular roles in human development (see Section 6.2.4.1) and in protection against cardiovascular disease (see 10.5.4.3).) At present these are essentially available only from marine fish and shellfish, or products derived from them, such as fish oils. Fish do not produce these FAs themselves: they incorporate them from plankton, algae and other marine food sources (see also 11.4.3). Fish raised by farming techniques must be fed fish meal if they are to retain these health-giving longer-chain PUFA of the n-3 family. Ultimately there is a limit to the amount of fish meal that can go into the food chain. To overcome this limitation, there is extensive experimental work at present to introduce biosynthetic pathways for the longer chain PUFA of the n-3 family into plants, and even into fish such as salmon (discussed further in Section 11.6.1).

The FA composition of plant fats can also be modified by conventional breeding techniques. The best example of this is the change that was brought about by breeding the erucic acid out of rapeseed in exchange for its metabolic precursor, oleic acid (see Section 6.1.3.2).

6.1.2.3 Processing may influence the chemical and physical properties of dietary fats either beneficially or adversely

Although food FA composition can be manipulated at source, as described above, by far the biggest influence is industrial processing, and domestic processing to a lesser extent. Processing can alter the nature of the constituent FAs considerably, mainly converting UFAs into SFAs but also potentially generating toxic products. The topic is discussed further in Sections 11.3.2, 11.4.2 & Table 11.1).

6.1.2.4 Structured fats and other fat substitutes

'Structured fats' describes semisynthetic mixed-acid TAGs containing both very long- and very short-chain FAs. They may find applications in reduced-energy foods. There are two main principles. First, the short-chain acids have an intrinsically lower energy value than the longer-chain FAs and are, in any case, metabolized in the liver in such a way that they are not deposited in adipose tissue (see Section 7.1.3). Second, the very long-chain acids are less efficiently absorbed. The combination of reduced gross energy and poorer absorption reduces the overall metabolizable energy of the fat. There is as yet no convincing evidence that such products contribute effectively to weight-reducing diets in the long term. Another type of structured TAG, which contains palmitic acid at position *sn*-2 to mimic a major molecular species in human milk TAGs, has been introduced commercially for use in infant feeds. Molecules with this structure are digested and absorbed more efficiently by the human infant (see Section 6.2.4.2).

Other products that are aimed at reducing dietary energy, including sucrose polyesters (see Section 11.4.1 for further detail), are also now approved in the USA for limited food use. These are not TAGs, but have fatlike texture. They are not attacked by pancreatic lipase (see Section 7.1.1) and so are not absorbed and contribute no energy value to the diet. Much discussion has taken place about potentially harmful effects of ingesting such compounds. These include significant reductions in the absorption of fat-soluble vitamins (which can be overcome by fortification), gastrointestinal discomfort and fatty stools.

6.1.3 A few dietary lipids may be toxic

Much of the discussion in this chapter is concerned with the positive contributions made by dietary lipids, especially FAs, to nutrition and health. It is also important to recognize that some dietary lipids can have adverse or toxic effects.

6.1.3.1 Cyclopropenes

FAs containing a cyclopropene ring (see Section 2.1.4) exhibit toxicity as a result of their ability to inhibit the 9-desaturase. One result of this is to alter membrane permeability as seen in 'pink-white disease'. If cyclopropene FAs are present in the diet of laying hens, the permeability of the membrane surrounding the yolk is increased, allowing release of pigments into the white of the egg. Rats given diets containing 5% of dietary energy as sterculic acid (see Table 2.4) died within a few weeks and at the 2% level, the reproductive performance of females was completely inhibited. Cottonseed oil is the only important oil in the human diet that contains cyclopropene FAs. However, their concentration in the natural oil is low (0.6–1.2%) and is reduced still further to harmless levels (0.1–0.5%) by processing. There has been no evidence that consumption of cottonseed oil in manufactured products has had any adverse nutritional effects.

6.1.3.2 Very long-chain monounsaturated fatty acids

The very long-chain MUFA, erucic acid (*c*13-22:1; see Table 2.2), was present in high concentration (up to 45% of total FAs) in the seed oil of older varieties of rape (*Brassica napus*). When young rats were given diets containing more than 5% of dietary energy as high erucic rapeseed oil, their heart muscle became infiltrated with TAGs. After about a week on the diet, the hearts contained three to four times as much lipid as normal hearts. With continued feeding, the fat deposits decreased but other pathological changes were noticeable. These included the formation of fibrous tissue in the heart muscle. The hydrolysis of TAGs that contain erucic acid is somewhat slower than with FAs of shorter chain length and this may have contributed to the accumulation of lipid deposits. Despite lack of evidence for harmful effects in man, an extensive breeding programme was undertaken to replace older varieties of rape with new 'zero-erucic' varieties (see Table 2.6). The use of low erucic varieties in manufactured products is now mandatory in most industrialized countries.

High concentrations of very long-chain MUFAs also occur in some fish oils and therefore contribute to the diets of people consuming these fish, as well as to diets containing certain fat spreads that incorporate hardened fish oils (see Table 2.7). The nutritional and toxicological consequences of long-term consumption of marine long-chain MUFAs have been less extensively studied than have those of high erucic acid-containing rapeseed oil.

6.1.3.3 *Trans*-unsaturated fatty acids

Dietary *trans*-unsaturated FAs originating either naturally in ruminants or from catalytic hydrogenation may have adverse effects on cardiovascular disease risk: these are considered later (see Section 10.5.4.2 & Box 10.5; Section 11.3.2 & Box 11.4).

6.1.3.4 Lipid peroxides

Lipid peroxides may be formed from PUFAs when they are in contact with oxygen (Chapter 11, Table 11.1). As described in Section 3.3 and Table 11.1, the reaction is accelerated in the presence of catalysts such as transition metal ions or haem (US: heme) compounds and by heating. When lipid hydroperoxides are ingested, they are rapidly metabolized in the mucosal cells of the small intestine to various oxyacids that are then rapidly oxidized to carbon dioxide. High concentrations of lipid peroxides in the gut may damage the mucosa and potentiate the growth of tumours. However, there is little evidence for the absorption of unchanged hydroperoxides, or for their incorporation into tissue lipids. Hydroperoxyalkenals, lower molecular weight breakdown products of lipid hydroperoxides, are absorbed and may be toxic. Rats given diets enriched in these compounds had enlarged livers, increased concentrations of malondialdehyde, peroxides and other carbonyl compounds and decreased concentrations of α-tocopherol and linoleic acid in tissues. Lipid peroxidation in the walls of arteries is a process that probably contributes to the development of atherosclerosis (see Section 10.5.1) but there is no evidence that these oxidized lipids arise directly from the diet.

6.2 Roles of dietary lipids

6.2.1 Triacylglycerols provide a major source of metabolic energy especially in affluent countries

The presence of fat contributes substantially to the palatability of the diet. However, the main nutritional contribution of TAGs is to supply metabolic energy. The useful energy available to the body (metabolizable energy, ME) depends on the digestibility of the fat (see Section 7.1.1) and the chain length of the constituent FAs. Most people efficiently digest common food fats, with insignificant differences in their digestible energies. As discussed in Section 6.1.2.4, some practical use has been made of differences in chain length to limit the metabolism of certain TAGs. Metabolic energy is derived from TAGs is through β-oxidation of the constituent FAs (see Section 3.2.1).

6.2.2 Dietary lipids supply essential fatty acids that are needed for good health but cannot be made in the animal body

Many lipids that are needed for tissue growth and development can be synthesized by mammalian cells. However, the FAs linoleic ($18:2n$-6) and α-linolenic ($18:3n$-3) cannot be made by animal cells and these so-called 'essential fatty acids' must be supplied by the diet. Once incorporated into cells they may be further elongated and desaturated to longer-chain polyunsaturated products. These structural and metabolic relationships

were described in Section 3.5, together with a discussion of some the difficulties of defining 'essentiality'.

6.2.2.1 Historical background: discovery of essential fatty acid deficiency

Until the early part of the 20th century it was thought that dietary fat was important only as a source of energy. In 1929 the Americans Burr and Burr described how acute deficiency states could be produced in rats by feeding fat-free diets and that these deficiencies could be eliminated by adding specific FAs to the diet. It was shown that linoleic and arachidonic acid were responsible for this effect and the term vitamin F was coined for them. They are now known universally as 'essential fatty acids' (EFAs). As described below, we now consider that linoleic acid (n-6 family) and α-linolenic acid (n-3 family) are the true EFAs in humans.

EFA deficiency can be produced in a variety of animals, but data for the rat are the best documented. The disease is characterized by skin symptoms, such as dermatosis; the skin becomes more 'leaky' to water. Growth is retarded, reproduction is impaired and there is degeneration or impairment of function in many organs of the body. Biochemically, EFA deficiency is characterized by changes in the FA composition of tissues and especially of the membranes in cells, whose function is impaired. In the mitochondria, the efficiency of oxidative phosphorylation is much reduced.

Well-documented EFA deficiency in man is rare. It was first seen in children fed virtually fat-free diets. Four hundred infants were fed milk formulas containing different amounts of linoleic acid. (This study was published in 1963. It would almost certainly be considered as unethical today.) When the formulas contained less than 0.1% of the dietary energy as linoleic acid, clinical and biochemical signs of EFA deficiency ensued. The skin abnormalities were very similar to those described in rats and these and other signs of EFA deficiency disappeared when linoleic acid was added to the diet. EFA deficiency in adults has been noted in several patients receiving all their nutrition through an artificial route, either intravenously or by gastric tube, because of disease of the small or large bowel. In the early days of intravenous feeding, this consisted of just glucose and amino acid solutions. Once symptoms of EFA deficiency were recognized, it was found that daily rubbing of sunflower oil into the skin would cure the problem, demonstrating that EFA need not necessarily be absorbed through the conventional route to be effective. Nowadays intravenous nutrition given over a prolonged period always contains a source of essential fat.

6.2.2.2 Biochemical index of essential fatty acid deficiency

In the tissues of animals with EFA deficiency, a PUFA that is normally either undetectable or present in extremely small concentration accumulates: this is all-c5,8,11-eicosatrienoic acid (20:3n-9), known as 'Mead acid' after the American biochemist, James Mead. The ratio of the concentration of the Mead acid to arachidonic acid in tissues or plasma phospholipids (the 'triene/tetraene ratio') is used as an index of EFA deficiency. In health, the ratio is about 0.1 or less, rising to 1.0 in severe EFA deficiency.

The relationship between the proportion of linoleic acid in the diet and the magnitude of the triene/tetraene ratio in plasma phospholipids is similar in all species and all tissues examined (Fig. 6.1). It is a sensitive indicator

since it increases significantly before overt EFA deficiency signs are apparent.

The explanation for the elevation of the concentration of 20:3n-9 in EFA deficiency can be readily appreciated from Fig. 6.2. During the course of evolution, animals lost the ability (retained by plants) to insert double bonds in positions 12 and 15. Each 'family' of UFAs (n-3, n-6, n-9) is biochemically distinct and its members cannot be interconverted in animal tissues. The 'parent' or 'precursor' FA of the n-9 family, oleic acid, can originate from the diet or from biosynthesis in the body, whereas linoleic and α-linolenic acids, the precursors of the n-6 and n-3 families, respectively, are essential nutrients that must be obtained from the diet. All three precursors can compete for desaturation by the same 6-desaturase enzyme as illustrated in Fig. 6.2. The affinity of the substrates for the 6-desaturase is in the order 18:3 > 18:2 > 18:1. In humans and many other mammals, quantitatively the most important role of the 6-desaturase is the conversion of linoleic acid into arachidonic acid. Normally, the diet contains sufficient

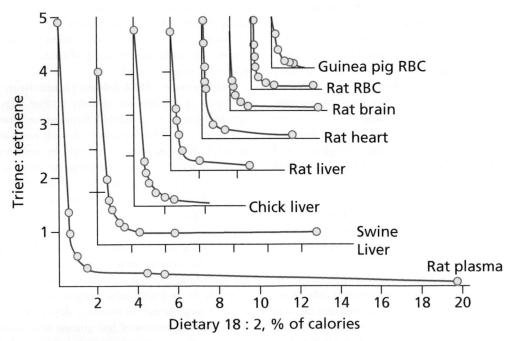

Fig. 6.1 Influence of dietary linoleic acid on the ratio of 20:3n-9 to 20:4n-6 (the triene/tetraene ratio) in different species. The relative scales are similar in all plots; the inset plots indicate the similarity in the relationship over a wide range of tissues and species. From RT Holman (1970) Biological activities of and requirements for polyunsaturated acids. *Prog Chem Fats Other Lipids* **9**: 607–82. Reproduced with kind permission of Professor RT Holman and Elsevier.

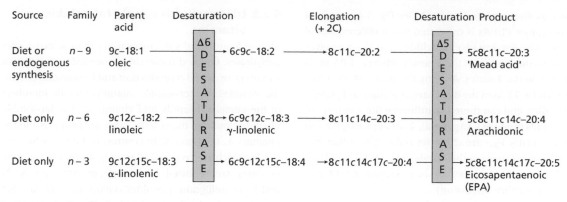

Source	Family	Parent acid	Desaturation		Elongation (+ 2C)	Desaturation	Product
Diet or endogenous synthesis	$n-9$	9c–18:1 oleic	Δ6 DESATURASE	→ 6c9c–18:2	→8c11c–20:2	Δ5 DESATURASE	→ 5c8c11c–20:3 'Mead acid'
Diet only	$n-6$	9c12c–18:2 linoleic		→ 6c9c12c–18:3 γ-linolenic	→8c11c14c–20:3		→ 5c8c11c14c–20:4 Arachidonic
Diet only	$n-3$	9c12c15c–18:3 α-linolenic		→ 6c9c12c15c–18:4	→8c11c14c17c–20:4		→ 5c8c11c14c17c–20:5 Eicosapentaenoic (EPA)

Fig. 6.2 Principal metabolic pathways for desaturation and elongation of 'parent acids' to long-chain polyunsaturated fatty acids in the n-9, n-6 and n-3 families. A minor family, n-7 (whose 'parent' acid is palmitoleic acid, 9c-16:1, or 16:1n-7) has been omitted for simplicity. The figure aims mainly to illustrate the sequence of alternate desaturations and elongations of the parent acids in each family and the competition between families for the Δ5 and Δ6-desaturases. For example, it should be readily apparent how the Mead acid accumulates in preference to arachidonic acid when there is a dietary deficiency of 18:2n-6 and an excess of 18:1n-9. An important continuation of the n-3 pathway via further elongation to C22 and further metabolism to form docosahexaenoic acid (DHA, 22:6n-3) has also been omitted (see Fig. 3.25). No Δ4-desaturase exists in mammals; the pathway to 22:6n-3 is discussed in Section 3.1.8.3.

linoleic acid for this pathway to supply the quantity of arachidonic acid needed by the membranes of body tissues. When the amount of linoleic acid in the diet is very low, however, the n-9 pathway takes precedence and there is a relative accumulation of the Mead acid and a relative depletion in arachidonic acid. Although a relatively large proportional increase in 20:3n-9 may be observed early in the progression to EFA deficiency, noticeable depletion of arachidonic acid stores requires long-term consumption of diets containing little or no linoleic acid.

It will be clear from Fig. 6.2 that there is also potential for competition for the 6-desaturase between 18:2n-6 and 18:3n-3 as well as the competition between the n-6 and n-9 substrates discussed above. Furthermore, there may also be competition at the level of the 5-desaturase. Notwithstanding the high affinity of α-linolenic acid for the 6-desaturase, the flux through the n-3 pathway is generally low because of the small proportion of α-linolenic acid in most diets compared with linoleic acid.

6.2.2.3 Functions of essential fatty acids

EFA are essential for two main roles: as membrane components (but see also Section 5.6.1) and as precursors for biologically active metabolites, the eicosanoids (see Section 3.5).

Among the results of EFA deficiency are changes in the properties of biological membranes of which PUFA are

major constituents of the membrane lipids. These changes in properties, for example the permeability of the membrane to water and small molecules such as sugars and metal ions, can be correlated with changes in the FA composition of the membrane. The permeability of the skin to water is particularly sensitive to the amount of linoleic acid in the epidermis, probably because of its role in ceramide synthesis (see Sections 2.3.3 & 4.8.1). In nervous tissue, reproductive organs and the retina of the eye, there is a greater proportion of the longer-chain FAs with 5 or 6 double bonds, predominantly of the n-3 family, compared with other tissues. This suggests that these FAs have specific roles in vision, reproduction and nerve function. For instance, in the retina, 50–60% of total phospholipid FA content is DHA. The presence of these highly unsaturated FAs in membranes has traditionally been considered to increase membrane fluidity, in this case in the photoreceptor outer membrane, so as to accommodate rapid conformational changes in the retinal-protein complex rhodopsin (see Section 6.2.3.1). However, this simple interpretation has been questioned more recently: there is evidence that fluidity as measured in artificial membranes does not increase much with additional double bonds beyond the first (e.g. oleate), but phospholipids containing highly unsaturated fatty acids have 'conformational plasticity' that allows the membrane proteins to operate in an appropriate environment.

As was discussed in Chapter 3 (see Fig. 3.43), each of the precursor PUFAs is converted into a different set of eicosanoids. For instance, arachidonic acid (n-6) gives rise to prostaglandins of the 2-series, whereas EPA (n-3) gives rise to the 3-series. Altering the amounts and types of n-6 and n-3 FAs in the diet changes plasma and platelet FA profiles, and can therefore influence the spectrum of eicosanoids produced. In general, 2-series prostaglandins (PG; e.g. PGE_2, $F_{2\alpha}$) are characterized as 'pro-inflammatory', whereas those of the 3-series (e.g. PGE_3, $F_{3\alpha}$) have anti-inflammatory properties (see Section 3.5.13 and 10.3.3 for further discussion).

6.2.2.4　Which fatty acids are essential?

Linoleic acid seems to be essential for almost every animal species at a dietary level of about 1% of energy, although there is evidence that some insects and protozoa can produce this FA. Other EFAs differ somewhat by species. In rodents, linoleic and arachidonic acids were initially described as the EFAs. In humans and other primates, the requirement of arachidonic acid can be met by biosynthesis from linoleic acid under most circumstances, but there is also a clear, but smaller, need for α-linolenic acid. Fish, whose lipid metabolism is geared to processing a high dietary intake of n-3 FAs, seem to have a definite and high requirement (3% of energy) for FAs of this family. Cats are unusual among mammals in that neither linoleic nor α-linolenic acid alone is sufficient to protect them against the effects of EFA deficiency: they require arachidonic acid in their diets because they lack both the 6- and the 12-desaturases.

6.2.2.5　What are the quantitative requirements for essential fatty acids in the human diet?

Individual countries make slightly different recommendations. In the UK, for example, recommended intakes to avoid deficiency are not less than 1% of dietary energy as 18:2n-6 (approximately 1–2.5 g/d) and 0.2% of dietary energy as 18:3n-3 (approximately 0.4–0.5 g/d). Some authorities may recommend intakes of up to 3% of dietary energy as n-6 PUFA during pregnancy and lactation to account for increased needs of the foetus (US: fetus) or to provide milk for the newborn. However, this assumes that there is no metabolic adaptation. For example adipose tissues store large amounts of EFA, which could be mobilized for this purpose.

6.2.3　Dietary lipids supply fat-soluble vitamins

Minute amounts of substances other than proteins, carbohydrates, fats and minerals are needed in the diet to sustain growth and reproduction and to maintain health: the vitamins. Water-soluble vitamins include members of the vitamin B family and vitamin C. The fat-soluble vitamins, found in many fatty foods, are now classified as vitamins A, D, E and K. In contrast to EFA (see Section 6.2.2), which are required in gram quantities, fat-soluble vitamins are required in microgram (vitamins A, D and K) or milligram quantities (vitamin E) per day. Fat is necessary in the diet for the absorption and utilization of fat-soluble vitamins. There is little evidence that the amount of dietary fat significantly affects the utilization of fat-soluble vitamins within the range of fat intakes on typical 'western' diets, but low fat intakes in developing countries may impair their absorption: this is true for diets with <20% energy from fat with regard to dietary carotenes. In addition, if fat absorption is impaired, insufficient fat-soluble vitamins may be absorbed, leading to a deficiency state. This can occur when the secretion of bile salts is restricted (as in biliary obstruction), when sections of the gut have been removed or damaged by surgery or in diseases, such as tropical sprue and cystic fibrosis, all of which are associated with poor intestinal absorption. It may also occur when fat absorption is inhibited pharmacologically (see Box 10.2). Deficiencies of fat-soluble vitamins are widespread (see individual descriptions below). Some fat-soluble vitamins, notably vitamin D, are present in a limited number of foods; so deficiency diseases can then become a significant problem when economic conditions limit food supplies.

6.2.3.1　Vitamin A

The term vitamin A was initially used to refer to a biological activity (prevention of visual defects). Now it refers to a group of related compounds with diverse biological functions (the vitamin A family). (Related compounds with similar vitamin activity are known as 'vitamers'.) All-*trans*-retinol (Fig. 6.3A) is the 'parent molecule' of the family. Dietary vitamin A (retinol, retinaldehyde and retinoic acid, known collectively as 'preformed vitamin A') is found in animal sources (notably dairy products, liver and fish liver oils), retinol largely in the form of long-chain FA esters. Plant sources such as dark green leaves, yellow and red fruits and vegetables,

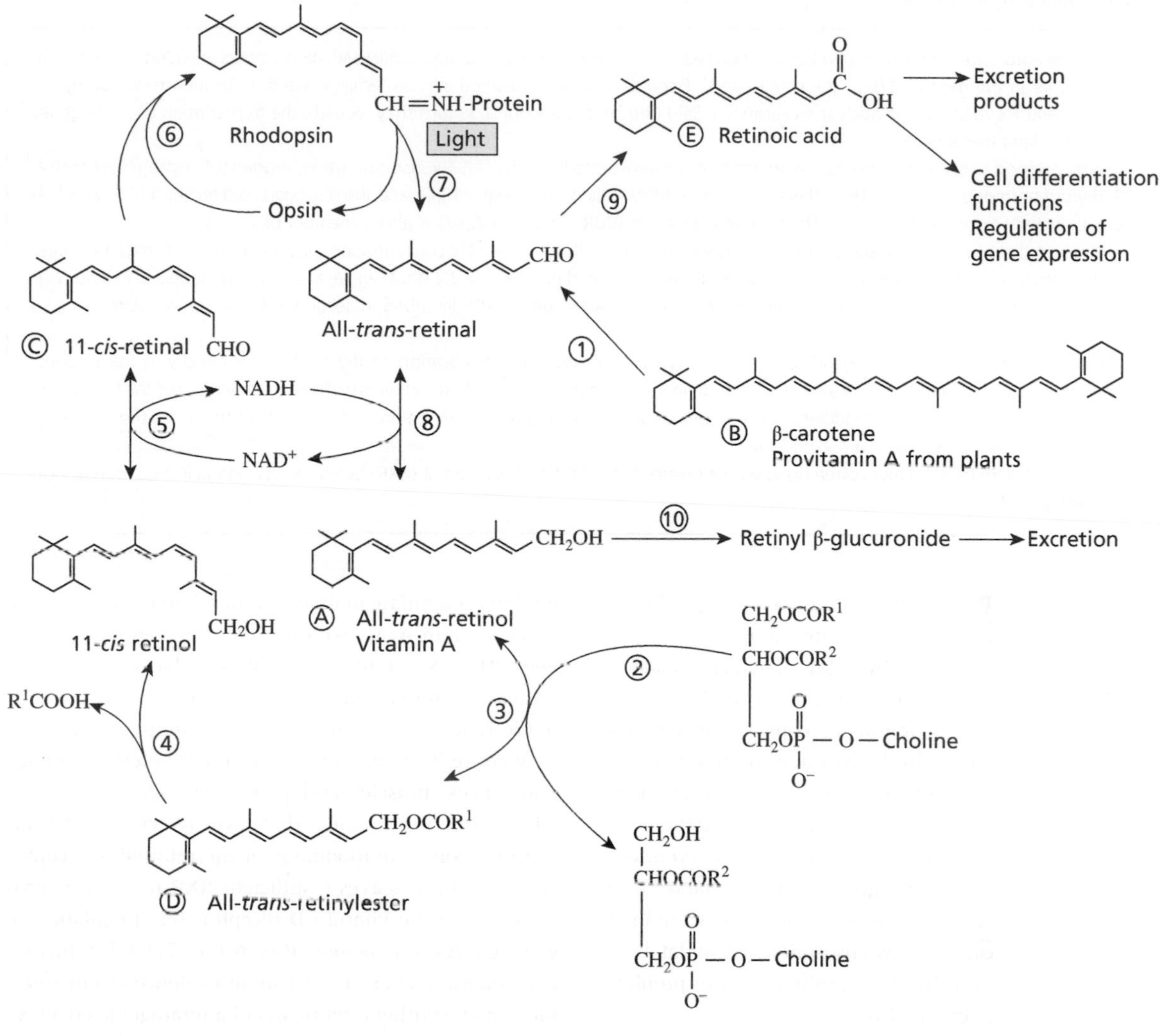

Fig. 6.3 Some important metabolic transformations of vitamin A. The parent compound of the vitamin A group is all-trans-retinol (A). It can be obtained directly from the diet or formed from the breakdown of dietary β-carotene (provitamin A) from plants (B). The active form of vitamin A in the visual process is 11-cis-retinal (C). Retinyl esters (mainly palmitate) provide a storage form of the vitamin in tissues, especially the visual pigment at the back of the eye (D). All-trans-retinal can be converted into retinoic acid (E), which has no visual function but is involved in processes of cellular differentiation. Vitamin A can be excreted via the formation of glucuronides from retinoic acid or from vitamin A itself. Reactions 1–10 are described in further detail in the text.

and vegetable oils such as palm oil, provide carotenoids (mainly α-, β- and γ-carotenes and cryptoxanthin) that can be cleaved to yield retinol; these are known as provitamin A carotenoids. β-Carotene (Fig. 6.3B) is the most abundant dietary carotenoid and the most potent in terms of vitamin A activity, but α-carotene is the predominant form in carrots and other yellow-orange vegetables

No more than 750 μg vitamin A is required by the average person daily; compare this with the gram quantities required for EFA intake (see Section 6.2.2.5). However, worldwide, vitamin A deficiency is widespread. The World Health Organization (WHO) and The United Nations Children's Fund (UNICEF) estimate that 280 million preschool children are affected by clinical vitamin A deficiency. Its most tragic manifestation is blindness in

Box 6.1 Nuclear receptors

Nuclear receptors are ligand-activated transcription factors. They are regarded as a superfamily with 48 members encoded in the human genome, 49 in the mouse (270 in the nematode *C. Elegans*). They are classified into six families based on sequence homology.

The ligand for many of the nuclear receptors is a lipid or related compound. In turn they regulate the transcription of many genes involved in lipid metabolism.

Some respond to circulating hormones, which must be transported into the cell: for instance, the oestrogen (US: estrogen) receptor, ER; thyroid hormone receptors, TRs – there are three isoforms called TRα1, TRβ1, TRβ2; and the glucocorticoid receptor (GR) which in humans mainly responds to cortisol. The vitamin D receptor (VDR: Section 6.2.3.2) is also a member of this family.

However, there are also nuclear receptors that respond to intracellular metabolite concentrations, including those of lipids. Examples are the peroxisome proliferator-activated receptors (PPARs) described in Section 7.3.2, that regulate fatty acid metabolism in a number of tissues, and the farnesoid-activated receptor (FXR) and liver X-receptor (LXR) described in Section 7.3.3, which regulate bile acid synthesis and other functions.

Nuclear receptors bind their ligand, either in the cytosol, followed by translocation to the nucleus, or directly in the nucleus, depending upon the receptor. They almost all form dimers, either homodimers or heterodimers with another member of the family. The activated dimer recruits further modulator proteins before binding directly to DNA in the promoter regions of the genes they regulate. They may increase or repress gene expression.

Those that form heterodimers often do so with a member of the retinoid X receptor (RXR) family, whose ligand is 9-*cis* retinoic acid (see Section 6.2.3.1).

young children. The first effects are seen as night blindness, followed by severe eye lesions, known as xerophthalmia, which are eventually followed by keratomalacia with dense scarring of the cornea and complete blindness. Xerophthalmia is one of the commonest preventable causes of blindness in the world. Although there are large-scale programmes for the supplementation of children's diets with vitamin A, these are difficult to implement successfully, and several interventions encouraging people to eat green leafy vegetables as a source of carotene have failed to improve vitamin A status, mainly because of low absorption on low-fat diets. Technological approaches to the solution of this problem are described in Section 11.6.1.

Epidemiological evidence shows that people with above-average blood retinol concentrations or above-average β-carotene intakes have a lower than average risk of developing cervical or breast cancer, and lower risk of death from lung cancer. However, attempts to prevent or treat cancers by supplementation have shown the opposite effect: an increase in mortality from lung cancer. In one large study, the so-called CARET trial, dietary supplementation with retinyl palmitate combined with β-carotene increased the risk of dying from lung cancer in smokers to the extent that the trial was terminated early, perhaps implying an interaction with oxidative stress.

In contrast to most water-soluble vitamins, excessive intakes of fat-soluble vitamins can be harmful, since they tend to accumulate in tissues rather than being excreted. Thus, vitamin A, if taken in excess, accumulates in the liver. The extent to which intracellular stores can be protein-bound is limited, and free retinol has a membrane-lytic action. Chronic overconsumption may not only cause liver necrosis but also permanently damage bone, eyes, muscles and joints. (One reason for the adverse effects on bone is that excessive vitamin A leads to formation of homodimers of the retinoid X receptor (RXR) and this leaves insufficient RXR to form heterodimers with the vitamin D receptor: see 'Regulation of gene expression' below, Box 6.1 & 7.3.2 for further explanation.) There is also some evidence (from treatment with synthetic retinoids) of a teratogenic effect, so pregnant women are advised to avoid foods with high vitamin A content such as liver. However, neonates have no store of vitamin A, and are reliant on vitamin A intake from milk, so it is important that the mother's vitamin A status is adequate. Vitamin A can be excreted via its conversion into glucuronides formed from retinoic acid (Fig. 6.3, reaction 9) or from all-*trans*-retinol itself (Fig. 6.3, reaction 10).

Absorption and metabolism

β-Carotene (Fig. 6.3B) is split by an enzyme in the small intestinal mucosa, β-carotene 15,15′-oxygenase-1 (or beta-carotene dioxygenase; gene name *BCMO1*), yielding two molecules of all-*trans*-retinal (also known as retinaldehyde) (Fig. 6.3, reaction 1). However, the yield of retinol from dietary β-carotene is much lower than this, for several reasons: poor absorption of dietary carotene; low activity of the dioxygenase; and the fact that another isoenzyme, beta-carotene oxygenase 2 (gene *BCO2*), does not cleave symmetrically and yields only one molecule of retinoic acid. β-Carotene 15,15′-oxygenase-1 is a PPAR-regulated enzyme and its expression appears to be determined both by PPARγ activation (PPARs are peroxisome proliferator-activated receptors: see Section 7.3.2 for fuller description), and by retinoid availability. Its expression is increased during retinoid deficiency. All-*trans*-retinal is reduced to all-*trans*-retinol (vitamin A) by an NADH-dependent reaction (Fig. 6.3, reaction 8). After absorption, or conversion from carotene, retinol is esterified mainly with palmitic acid (irrespective of the composition of the dietary esters) in the enterocytes in a reaction catalysed by one of two microsomal enzymes. Phosphatidylcholine:retinol acyltransferase (gene name *LRAT*), which transfers palmitic acid from position *sn*-1 of membrane phosphatidylcholine (PtdCho; Fig. 6.3, reaction 2), accounts for around 90% of retinol esterification. The alternative reaction uses acyl-CoA as a substrate and is catalysed by diacylglycerol (DAG) acyltransferase-1 (DGAT1, gene name *DGAT1*, an enzyme involved in TAG biosynthesis: see Section 4.1.1); when catalysing this reaction, the enzyme is known as acyl-CoA:retinol acyltransferase.

Retinoids, as lipids, are not soluble in the aqueous cytosol, and in all cells they are bound to members of a family of retinoid-binding proteins (RBP). The isoform expressed at high levels in the small intestine is also known as cellular retinol-binding protein II (CRBPII) or as retinol binding protein 2 (gene name *RBP2*).

Retinyl esters are incorporated into chylomicrons (see Section 7.1.3) and exported from the enterocyte into the circulation. They mostly remain with the lipoprotein particle after its TAG content has been hydrolysed, and are taken up by the liver in the chylomicron remnant (see Section 7.2.5). This property of retinyl esters has been used experimentally, to mark chylomicron particles in the circulation. A large dose of retinyl palmitate is usually given, along with a fatty test meal. The concentration of retinyl palmitate in the circulation rises and then falls, and is a marker for the rate at which chylomicron remnants can be cleared. However, it is now clear that around 25% of retinyl palmitate entering the circulation in chylomicrons is delivered to peripheral tissues rather than the liver. This may involve hydrolysis, along with the TAG, by lipoprotein lipase (LPL; see Section 7.2.5). This pathway probably accounts for the fact that people who lack RBP (see below) are quite healthy, provided that the diet contains adequate preformed retinol and fat. Moreover, RBP biosynthesis can become inadequate during protein-energy malnutrition, as a result of which functional vitamin A deficiency can occur even when there are adequate liver reserves.

Within the liver, some retinyl ester is retained in cytosolic lipid droplets of liver stellate cells. However, most dietary retinyl esters are hydrolysed by a retinyl palmitate esterase (Fig. 6.3, reaction 3), to release retinol. The activity of this enzyme can increase some 100-fold during vitamin A deficiency. This enzyme activity appears to belong to a more general carboxylesterase, carboxylesterase 1 (gene name *CES1*), and probably other hepatic lipases and esterases. Retinol may then be transported in the circulation to other tissues bound to specific RBPs: these belong to the family of proteins mentioned above, that bind retinoids within cells as well as in the circulation. The main RBP in human plasma is RBP4 (gene name *RPB4*). RBP4 is a small protein that circulates in a 1:1 complex with the larger protein transthyretin, which prevents rapid renal excretion of the retinol-RBP complex. Retinol is delivered to target tissues (e.g. visual cells) by interaction of the RBP with cell-surface receptors. The nature of this receptor, or receptors, is not entirely clear. One candidate receptor is known as STRA6 ('stimulated by retinoic acid family member 6'), but this is not expressed in all cells that metabolize retinoids.

Vitamin A has several distinct functions: in vision, in regulation of gene expression, in embryogenesis and in the immune response.

Vision

This is the best-defined function in molecular terms. The reactions involved in the visual process take place in the outer segment of rod cells and in the colour-sensing cone cells of the retina. All-*trans*-retinol, carried in the circulation on RBP, is taken up by the visual cells and converted into 11-*cis*-retinol. The reaction is complex, since it involves first the esterification of retinol to retinyl palmitate by transfer of an acyl group from PtdCho (Fig. 6.3, reaction 2) in the pigment epithelium at the back of the eye. Then retinyl palmitate breaks down in a concerted reaction that couples isomerization of the retinol and release of a FA (Fig. 6.3, reaction 4). 11-*cis*-Retinol is oxidized to the aldehyde, 11-*cis*-retinal (Fig. 6.3, reaction 5), which then reacts with a lysine residue in the protein opsin, through a Schiff base linkage, to form the complex known as rhodopsin (Fig. 6.3, reaction 6), also called 'visual purple'. There is a family of opsins, reacting to different wavelengths of light. This arrangement of 11-*cis*-retinal bound to an opsin is common to all visual organs from squid to man.

The bound 11-*cis*-retinal can absorb a photon of visible light (Fig. 6.3, reaction 7), converting it into all-*trans*-retinal which then dissociates. Opsins belong to the family of G protein-coupled receptors (GPCRs: see Section 7.3.4 for more detailed description). Unusually for a GPCR, (covalent) binding of 11-*cis*-retinal brings the receptor into its quiescent state. Conversion into all-*trans*-retinal activates it, producing a large conformational change, which opens up a G protein binding pocket. The G protein involved is the very specific 'transducin' (G_t), which produces a change in membrane potential, transmitted along the optic nerve to the brain, where it is perceived as sight. All-*trans*-retinol dissociates from opsin after the conformational change, and may be reconverted to the active 11-*cis* form by the reaction sequence 8, 3, 4, 5 shown in Fig. 6.3.

Regulation of gene expression

Whereas 11-*cis*-retinal is the analogue of vitamin A involved in vision, all-*trans*-retinoic acid and 9-*cis*-retinoic acid (Fig. 6.3E, reaction 9) are the vitamers mainly involved in regulation of gene expression. They act via nuclear receptors, which are described more fully in Box 6.1. There are two principal types of high-affinity receptors for retinoic acid, designated RAR (retinoic acid receptors) and RXR (retinoid X receptors). Each exists in three isoforms: α, β and γ (gene names *RARA, RARB, RARG* and *RXRA, RXRB, RXRG* respectively). These genes are expressed at different times and locations during differentiation. RARs bind either all-*trans*- or 9-*cis*-retinoic acids, whereas RXRs bind only the 9-*cis* isomer. Retinoic acid isomers, together with their nuclear receptors, can both activate and suppress gene expression, depending on the types of interactions described above. Ligand-activated RAR and RXR act as a heterodimer, but RXR also interacts (forms heterodimers) with a number of other nuclear receptors such as the vitamin D receptor, the thyroid hormone receptor and the PPARs (see Section 7.3.2). This raises the possibility of interesting interactions between vitamin A and other chemical messengers (see the later sections on vitamin D, below, and on PPARs, Section 7.3.2).

Retinoid signalling through these mechanisms plays a crucial role in differentiation and development. Site- and time-specific signalling is brought about not only by regulation of the expression of the RAR and RXR receptors, but also by the synthesis and degradation of retinoic acid. This pathway is a target for pharmacological intervention. Synthetic RAR and RXR agonists (the latter known as 'rexinoids') are being used in the treatment of a number of cancers. Patients with acute promyelocytic leukemia, a type of leukaemia in which chromosomal rearrangements disrupt RARα, have been very successfully treated with all-*trans* retinoic acid. The synthetic rexinoid bexarotene has increased survival significantly among some patients with lung cancer. Experimental work is ongoing in the treatment of metabolic diseases such as obesity, type 2 diabetes and the associated 'metabolic syndrome' (see Section 10.5.2.4 for more on this syndrome).

Retinoic acid can also form covalent bonds with certain proteins. These retinoylated proteins are similar in size to the nuclear retinoic acid receptors and may play roles similar to those of lipid-anchored proteins described in Section 5.3.2.

Immunity

Risk of infection is increased markedly in vitamin A deficiency and both humoral and cell-mediated immune responses are impaired. Several randomized controlled trials have shown that vitamin A supplementation in at-risk populations reduces mortality and morbidity, with fewer deaths from diarrhoeal diseases in children a major benefit. Important sites of vitamin A action in the immune response are the T lymphocytes, some subsets of which are especially involved in intestinal mucosal immunity. Retinoic acid drives the differentiation of T lymphocytes into different populations of T-helper cells and modulates their pattern of cytokine secretion. The reader is referred to **Further reading** for more information on this complex area.

6.2.3.2 Vitamin D

Vitamin D is the name originally given to the lipid factor with the property of preventing the bone disease rickets. It is not strictly a vitamin in the conventional sense, as it can be synthesized within the human body, and indeed synthesis in skin in response to sunlight is the major source under most conditions. Only when sunlight exposure is inadequate is there a need for a dietary supply. Vitamin D also differs from the other fat-soluble vitamins in that its most active metabolite, 1,25-dihydroxycholecalciferol (also called calcitriol), is a true hormone: it is mainly produced in one organ (the kidney) and signals through specific receptors in distant tissues.

It is difficult to decide upon a precise dietary requirement for vitamin D because much is derived from the action of sunlight on skin lipids. Two groups of people, however, may have a special need to obtain vitamin D from the diet. In the first group are children and pregnant and lactating women, whose requirements are particularly high. In the second group are people who are little exposed to sunlight, such as the housebound elderly, people in far northern latitudes and those who wear enveloping clothes. Dark-skinned immigrants to Northern Europe are especially vulnerable. Infants and children who obtain too little vitamin D develop rickets, with deformed bones that are too weak to support their weight. These degenerative changes soon become permanent if supplementation is not started early enough. This the reason why, in the UK and some other countries, vitamin D preparations are provided for children and pregnant women and margarine is fortified with it in most European countries and in Canada.

Like vitamin A, vitamin D is toxic in high doses. Even amounts five times the normal intake can be toxic. Too high an intake causes more calcium to be absorbed than can be excreted, resulting in excessive deposition in (calcification), and damage to, the kidneys, other soft tissue and blood vessels. Vascular calcification leads, in turn, to impaired blood vessel function and hypertension.

Metabolism of vitamin D

There are two parent forms of vitamin D, known as D_2 and D_3. (The terms D_2 and D_3 are historic and not related to chemical structure; D_1 was the term originally given to the first synthetic preparation, a crude extract of irradiated ergosterol from yeast.) Both D_2 and D_3 are biologically inactive without further metabolism. Cholecalciferol (vitamin D_3), also known as calciol, is produced in the skin by ultraviolet irradiation of 7-dehydrocholesterol, which is present in the skin surface lipids. It is the main source of the vitamin for most humans. Dietary sources are fish liver oils (e.g. cod), eggs, liver and some fat spreads. Ergocalciferol (vitamin D_2) is a synthetic form, produced by irradiation of the plant sterol ergosterol, found especially in fungi, and is used in fortification of foods. Naturally occurring ergocalciferol is also present in some foods, especially fungi. Ergocalciferol is metabolized, and acts physiologcally, in much the same way as cholecalciferol, whose metabolism is described below.

The active metabolites are formed in the liver and kidneys. After absorption from the diet and transport to the liver in chylomicrons (see Section 7.2.5), or formation in the skin, vitamin D is carried in the circulation bound to a specific transport protein of the α-globulin class, Vitamin D Binding Protein (DBP). There is a large molar excess of DBP so effectively all circulating vitamin D is bound to it. DBP also binds the hydroxylated forms of vitamin D, described below. Unlike vitamin A, there is no tissue storage of vitamin D, so most of it is in the plasma bound to DBP (mostly in the form of calcidiol; see below).

In the liver, cholecalciferol is hydroxylated to 25-hydroxycholecalciferol, also known as calcidiol (Fig. 6.4) by an enzyme that is a member of the cytochrome P450 (CYP) family. There are several possible candidates but CYP2R1 seems to play the major role in humans: one individual has been described who is homozygous for inactivating mutations in the gene, and displays the symptoms of rickets (functional vitamin D deficiency). The 25-hydroxylase requires NADPH, molecular oxygen and Mg^{2+}. Calcidiol is the main transport form of vitamin D and its concentration in plasma is used as an indicator of vitamin D status. Although this metabolite has little biological activity, it is carried to the kidney, bound to DBP, where it is further hydroxylated to 1,25-dihydroxycholecalciferol (also called calcitriol) by a 1-α-hydroxylase, again a CYP enzyme, known as CYP27B1. An alternative fate in the kidney is hydroxylation at position 24 to produce the inactive 24,25-dihydroxy derivative (Fig. 6.4) by another CYP enzyme, CYP24A1. This reaction may be seen as part of the regulation of calcitriol formation, since the two renal enzymes CYP27B1 and CYP24A1 are regulated reciprocally.

1,25-Dihydroxycholecalciferol, the active vitamin D metabolite, functions as a hormone. Its production in the kidney, and hence its circulating concentration, is tightly controlled as described below. It acts on target tissues by binding to a specific receptor, the vitamin D receptor (VDR), which is a member of the nuclear hormone receptor superfamily (see Section 6.2.3.1).

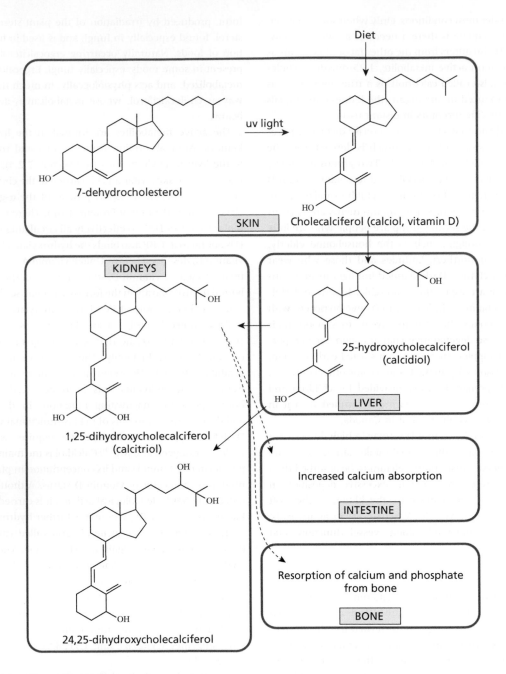

Fig. 6.4 Metabolism of vitamin D.

VDR heterodimerizes with RXR (see Section 6.2.3.1) and acts to regulate the expression of genes whose regulatory regions contain a vitamin D responsive element (VDRE).

The three main influences on renal calcitriol production are direct negative feedback by calcitriol itself, i.e. by vitamin D status; by parathyroid hormone (PTH) as a signal of calcium status; and by fibroblast growth factor

23 (FGF23) as a signal of phosphate homeostasis. The first of these is achieved by VDR-mediated suppression of expression of the renal 1-α-hydroxylase, CYP27B1. It represents a negative feedback mechanism for regulation of circulating calcitriol concentrations (when the concentration is high, synthesis is suppressed, and vice versa). The roles of PTH and FGF23 are described below.

The long-established functions of vitamin D relate to calcium and phosphate homeostasis. However, it is now recognized to have a much wider role in health and disease.

Calcium and phosphate homeostasis

Vitamin D was first identified as the dietary factor preventing the bone deficiency disease of rickets in children, and osteomalacia, the adult form. It is indeed essential for normal bone growth and maintenance, and its major role in this respect is to enhance bone mineralization through its promotion of dietary calcium and phosphate absorption. Calcitriol increases the efficiency of absorption of dietary calcium in the small intestine. There appear to be several coordinated mechanisms including increased calcium transport into the enterocyte via the calcium channel 'transient receptor potential vanilloid type 6' (TRPV6), up-regulation of expression of calbindin-D, an intracellular calcium-binding protein, and increased export of absorbed calcium through the basolateral membrane of the enterocyte.

Vitamin D also exerts direct effects on bone cells. In fact, all types of bone cells, including the osteoblasts that synthesize new bone and the osteoclasts that resorb (digest) bone, as well as the chondrocytes (cartilage cells), express both the VDR and the enzyme CYP27B1 that synthesizes calcitriol. This calcitriol that is synthesized outside the kidney probably has local actions. The role of calcitriol in bone is complex and it participates in regulation of the overall process of bone turnover. One effect is to activate bone osteoclasts to resorb calcium from bone, so as to maintain an appropriate plasma calcium concentration. A second is to stimulate bone formation by osteoblasts. PTH is the regulator of calcitriol biosynthesis in relation to calcium homeostasis. PTH is released from the parathyroid glands when the circulating Ca^{2+} concentration falls, through the action of an unusual GPCR known as the calcium-sensing receptor (CaSR). The CaSR acts as a homodimer, and is activated by binding several Ca^{2+} ions, suppressing PTH secretion.

Thus, when circulating Ca^{2+} concentrations fall, PTH is secreted and induces renal calcitriol synthesis by increasing CYP27B1 expression, and hence raises the circulating calcitriol concentration. This, in turn, activates osteoclasts, adding Ca^{2+} ions to the circulation.

Bone mineral comprises both calcium and phosphate. Vitamin D is also involved in phosphate homeostasis, with similar mechanisms to those described for calcium. In this case, however, feedback to the kidney to regulate calcitriol biosynthesis is via FGF23, which is secreted from bone. FGF23 exerts dual effects to reduce circulating calcitriol concentrations: it suppresses renal CYP27B1 expression and stimulates catabolism of calcitriol by CYP24A1 (forming the inactive 1,24,25-trihydroxycholecalciferol, also known as calcitetrol).

Other functions

The discovery of receptors for calcitriol in many tissues other than those involved in calcium and bone metabolism (e.g. pancreatic islet cells, skin keratinocytes, mammary epithelium and some neurons) suggests its wide involvement in tissue development and homeostasis. Indeed, calcitriol seems to be a general developmental hormone, inhibiting proliferation and promoting differentiation in many tissues. Present knowledge suggests functions that include regulation of gene products associated with mineral metabolism, energy-yielding metabolism, differentiation of cells in skin and in the immune system and regulation of DNA replication and cell proliferation. There is also an emerging literature suggesting rapid actions mediated by mechanisms other than the 'classic' regulation of gene expression through the VDR pathway. One suggestion is that these rapid effects, also known for other hormones acting via nuclear receptors, are mediated through a subpopulation of the same receptors acting outside the nucleus.

There are also widespread reports of the association between low circulating calcidiol concentrations (i.e. low vitamin D status) and various disease conditions, including some cancers, Alzheimer's disease, cerebrovascular disease (stroke), and insulin resistance and type 2 diabetes. Conversely, higher vitamin D status appears to be protective against development of insulin resistance and type 2 diabetes. Vitamin D supplementation has therefore been suggested to be useful in counteracting these and other conditions, including multiple sclerosis, but definitive studies are lacking. But it is important to note, as described above, that excess vitamin D intake is toxic.

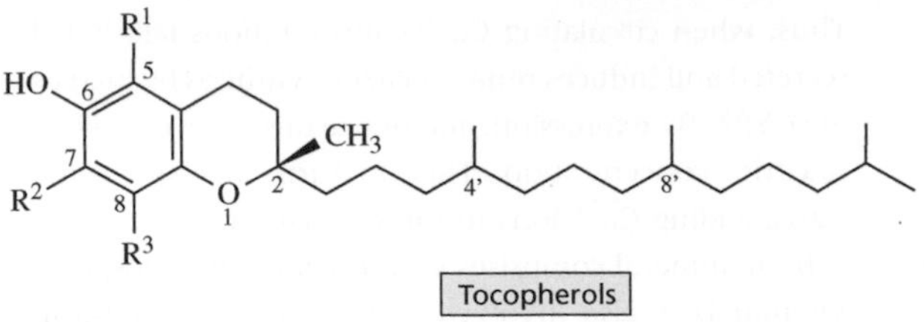

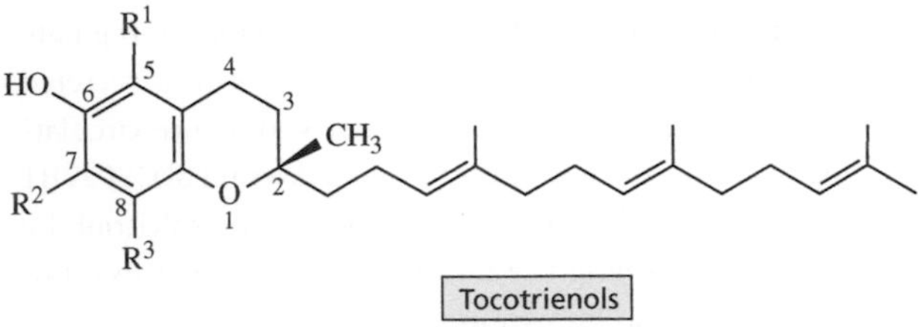

R^1

HO 6 5

7

R^2 8 O 2 CH_3

1

R^3

Tocopherols

R^1

HO 6 5 4

3

7

R^2 8 O 2 CH_3

1

R^3

Tocotrienols

Fig. 6.5 Structures of compounds with vitamin E activity. All molecules comprise a chromanol ring system and a 16C phytyl side chain. Tocopherols have a saturated side chain; tocotrienols contain 3 unsaturated bonds in the side chain. Analogues are named α, β, γ and δ-tocopherols or tocotrienols as follows:

$$\alpha : R^1 = CH_3; R^2 = CH_3; R^3 = CH_3$$
$$\beta : R^1 = CH_3; R^2 = H; R^3 = CH_3$$
$$\gamma : R^1 = H; R^2 = CH_3; R^3 = CH_3$$
$$\delta : R^1 = H; R^2 = H; R^3 = CH_3$$

6.2.3.3 Vitamin E

Vitamin E, originally called 'factor X', was identified in 1922 as a 'specific vitamin for reproduction'. Evans and Bishop discovered a factor present in vegetable oils that prevented sterility in rats reared on chemically pure protein, fat and carbohydrate diets. It was not isolated in pure form and characterized chemically until 1937.

Vitamin E is widespread in foods. In humans, dietary deficiency states were not identified until the 1980s. Essentially, vitamin E deficiency is seen only in premature infants with low fat stores and in pathological conditions such as severe malabsorption (e.g. in cystic fibrosis, chronic liver disease, intestinal resection). It is also found in the rare genetic condition abetalipoproteinaemia, which arises from lack of lipoproteins containing apolipoprotein B, causing deficient transport of lipids around the body. The main symptoms of vitamin E deficiency in humans are abnormal erythrocyte membrane morphology due to oxidative stress, and neurological abnormalities as a result of long-term deficiency. Patients with untreated abetalipoproteinaemia suffer from very severe neurological damage; they can be maintained in good health by providing very large supplements of vitamin E. Despite the original description of vitamin E, there is no evidence for a role in human fertility. The richest dietary sources are vegetable oils, cereal products and eggs.

Vitamin E activity is possessed by four tocopherols and four tocotrienols (Fig. 6.5). α-Tocopherol is the most potent of these, the other compounds having between 10% and 50% of its activity. α-Tocopherol is also the most abundant form of vitamin E in animal tissues, representing 90% of the mixture. γ-Tocopherol is the most abundant dietary form. The form used in commercial preparations is synthetic racemic α-tocopherol, often in the acetylated form as a protection from oxidation. Since the natural compounds comprise only one of the eight stereoisomeric forms, the synthetic vitamin E is less biologically active. The optical isomers have equal antioxidative capacity but some are not so well absorbed or retained in cells.

Vitamin E is absorbed from the small intestine with other lipids, and is carried in the blood by the plasma lipoproteins (see Section 7.2.3). Intestinal absorption is dependent upon lipid transport systems similar to those used by cholesterol (see Box 7.2). The scavenger receptor SR-BI (see Section 7.2.4.4) has been shown to play an important role in vitamin E absorption in human cellular models and in rodents. In the liver, however, a specific α-tocopherol transfer protein (gene name *TTPA*) is expressed, which mediates α-tocopherol transport between cellular compartments and its incorporation into very low density lipoproteins (VLDL). It is this protein that has a higher affinity for α-tocopherol than the other vitamers. *TTPA* expression is up-regulated by oxidative stress. Inactivating mutations in this gene result in low plasma vitamin E concentrations and progressive neurodegeneration.

Vitamin E as an antioxidant

The only well-characterized function of vitamin E at a molecular level is its role as a lipid-soluble antioxidant, preventing in particular the peroxidation of long-chain PUFAs. α-Tocopherol is present in the lipid bilayers of biological membranes and may play a structural role there. It has been suggested that it partitions into membrane domains enriched in polyunsaturated phospholipids and also that it partitions into lipid rafts (see Section 5.3.3). It is also present in all lipoprotein classes, including those rich in cholesterol, namely low density (LDL) and high density (HDL) lipoproteins (see Section 7.2.5). Its main function in membranes and in LDL is thought to be the prevention of the oxidation of unsaturated lipids (see Section 3.3). The products of lipid peroxidation can cause damage to cells if the oxidative process is not kept in check. Such damage appears to be exacerbated in animals fed diets deficient in vitamin E. α-Tocopherol primarily acts as a terminator of the lipid peroxidation chain reactions by donating a hydrogen atom to a lipid radical (see Section 3.3). The resulting tocopheryl radical is relatively unreactive and unable to attack adjacent UFAs because the unpaired electron becomes delocalized in the aromatic ring structure. There is approximately one molecule of vitamin E per 2000–3000 phospholipid molecules in membranes, so it would rapidly become depleted, but vitamin C (ascorbic acid), which is water-soluble, can regenerate the antioxidant form of vitamin E. This allows one vitamin E molecule to scavenge many lipid radicals. Oxidized ascorbic acid (dehydroascorbic acid) is in turn re-reduced by interaction with thiol-based reducing agents such as glutathione.

Because of this biological function, it has been suggested that dietary requirements for vitamin E should be considered in relation to the PUFA content of the diet, rather than in absolute amounts. A ratio of vitamin E to PUFA intake of about 0.4 mg/g is generally recommended. Vegetable oils that contain high concentrations of PUFAs are sufficiently rich in vitamin E to give adequate protection.

The 'oxidative theory' of cardiovascular disease (see Section 10.5.1) suggests that vitamin E might be protective against this condition. This was supported by several epidemiological studies that suggested lower cardiovascular disease risk in those with higher vitamin E intake. However, randomized controlled trials of dietary vitamin E supplementation have not in general borne out this idea. As with β-carotene supplementation, they sometimes showed more harm than protection.

Other functions of vitamin E

There is accumulating evidence that vitamin E may play other biological roles that do not involve its antioxidant function. A possible structural role in the maintenance of cell membrane integrity has been mentioned. Some observations imply diverse roles in immune function. Vitamin E has anti-inflammatory effects, possibly by interacting with the prostaglandin synthetase that produces inflammatory eicosanoids (Fig. 3.42), and also appears to stimulate the immune response. It has been suggested that such effects are mediated by forms other than α-tocopherol (e.g. γ- and δ-tocopherol). α-Tocopherol is involved in the regulation of intercellular signalling and cell proliferation through the modulation of protein kinase C activity. α-Tocopherol has been shown to exert effects on the expression of a number of genes in diverse pathways including blood coagulation, although the molecular mechanisms are not known. There is evidence that tocotrienols (but not tocopherols) can affect gene expression by interaction with the (o)estrogen receptor-β (ER-β, a member of the nuclear hormone receptor family) and this may underlie some putative anticancer effects. The details of many of these other roles of vitamin E are still far from clear.

6.2.3.4 Vitamin K

Vitamin K was identified in 1929 by Danish scientist Henrik Dam, who noted that chickens fed a low-fat, cholesterol-free diet developed haemorrhages. He identified a missing dietary factor which was called 'coagulation vitamin', and given the letter K for 'Koagulations-Vitamin' (when published in German).

Vitamin K is the generic name given to a group of compounds, having in common a naphthoquinone ring system (menadione) with different side chains (Fig. 6.6). Plants synthesize phylloquinone (vitamin K_1) with a phytyl side chain identical to that in chlorophyll. Bacteria synthesize menaquinones (vitamin K_2), the side chains of which comprise 4–13 isoprenyl units (see Section 4.9.2).

In vitamin K deficiency, the time taken for blood to clot is prolonged and the activities of factors II, VII, IX and X are reduced. It is rare to see vitamin K deficiency in an

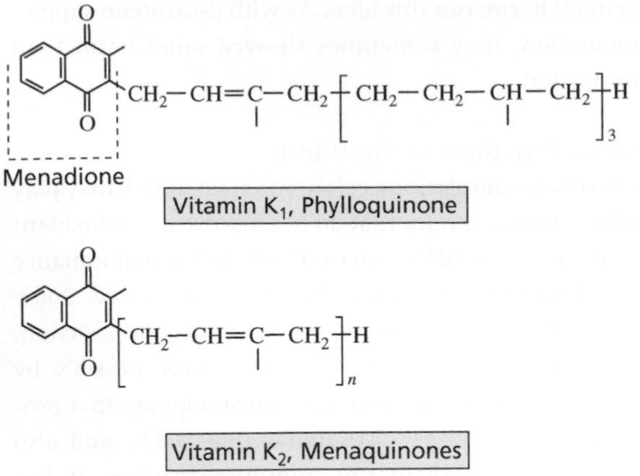

Menadione

Vitamin K$_1$, Phylloquinone

Vitamin K$_2$, Menaquinones

Fig. 6.6 Structures of compounds with vitamin K activity.

adult except when fat absorption is impaired. Deficiency signs occur more frequently in infants. This is because they have an immature liver that synthesizes prothrombin only slowly. They also have an undeveloped intestinal flora that might otherwise provide some menaquinones; and, finally, human milk contains only small amounts of vitamin K. Dietary requirements are thought to be about 10 µg/day for infants and between 40 and 80 µg/day for adults. Insufficient intakes of vitamin K in some new-born infants give rise to bleeding into tissues, known as haemorrhagic disease of the new-born. In some countries, including Australia, the US, UK and other European countries, vitamin K (as phylloquinone) is given routinely at birth. There is little evidence of harm from high intakes of natural vitamin K, although high intakes of the synthetic form, menadione, can lead to haemolysis and liver damage in infants.

Major dietary sources of vitamin K$_1$ are fresh green leafy vegetables, green beans and some seed oils. Cereals are poor sources and, of animal foods, only beef liver is a significant source. Vitamin K$_2$ is found in fermented foods, such as cheeses and yoghurt and in ruminant liver. Bacteria in the human gut synthesize menaquinones, but their bioavailability is low, partly because bile salts, necessary for their absorption, are not present in the colon; hence their significance as a source of vitamin K has been questioned.

Free phylloquinone is absorbed from the small intestine with about 80% efficiency, but only about a tenth of the vitamin K$_1$ present in dark green leafy vegetables is absorbed (although, as with other fat-soluble vitamins, absorption efficiency is affected by the amount of fat in the diet). Vitamin K is carried in chylomicrons (see Section 7.2.5) and delivered mainly to the liver, in chylomicron remnants. About two-thirds of the vitamin K intake is soon excreted from the body (into the faeces via the bile) and as there is no evidence of an enterohepatic circulation, it is presumed that a constant intake is essential.

Function in enzymic carboxylation reactions

Like many water-soluble vitamins, but in contrast to other fat-soluble vitamins, vitamin K has a specific function as an enzyme cofactor. The enzyme, an integral membrane protein present in the endoplasmic reticulum of liver cells and in some other tissues, is γ-glutamyl- (or vitamin K-dependent) carboxylase (gene name *GGCX*). This carboxylase has no known homology to other enzyme families. It catalyses the post-translational conversion of glutamate (Glu) residues in proteins to γ-carboxyglutamate (Gla). Clusters of several Glu residues are carboxylated within each of the specific calcium-binding proteins affected. The Gla residues are located at the Ca^{2+} binding site and are critical to the protein's action. These proteins are known collectively as the vitamin K-dependent (VKD) proteins. The carboxylation reaction requires both molecular oxygen and carbon dioxide. The active form of vitamin K in this reaction is the reduced quinol form, which donates hydrogen to glutamic acid, being transformed in the process to an epoxide. The 2,3-epoxide is then transformed to the quinone form by vitamin K oxidoreductase in a dithiol-dependent reaction, and the quinone is reduced to the quinol, thus completing the so-called vitamin K epoxide cycle.

Originally proteins involved in blood coagulation were recognized as VKD proteins, and these are synthesized mainly in the liver. Now, however, it is recognized that proteins functioning in many pathways are VKD proteins, and many of these are synthesized, and act, in tissues outside the liver. For more information on emerging roles of VKD proteins see **Further reading**.

In vitamin K deficiency, under-carboxylated proteins are produced. These have been used as the basis for a biochemical test of vitamin K status, by measurement of so-called 'protein(s) induced by vitamin K absence or antagonists (PIVKA)'.

Role in blood coagulation

Four of the procoagulant proteins of the blood coagulation cascade depend on the presence of vitamin K for their function (factor II, or prothrombin, and factors VII, IX and X). The formation of Gla residues in these proteins provides efficient chelating sites for Ca^{2+} ions, enabling them to form ion bridges with the surface phospholipids of platelets and endothelial cells. The anticoagulant (and rodent poison) warfarin inhibits the vitamin K oxidoreductase that generates the reduced quinol form of vitamin K required as a cofactor for carboxylation. Therefore there is decreased carboxylation of Glu residues, and increased dissociation of partially carboxylated, inactive VKD proteins. If vitamin K intake is high, then there can be more or less normal carboxylation; hence, vitamin K administration can be used to overcome warfarin-induced anticoagulation.

Role in bone metabolism

In 1975 it was discovered that bone tissue contains a Gla-containing protein, osteocalcin (formerly known as bone Gla-protein), which accounts for up to 15% of noncollagenous proteins in bone. Fully carboxylated osteocalcin contains three Gla residues and is able to form a strong complex with hydroxyapatite, the calcium phosphate mineral component of bone. Much circumstantial evidence suggests a key role for vitamin K in bone health. For example, treatment of pregnant mothers with vitamin K anticoagulants, such as warfarin, can lead to bone defects in their infants (known as foetal warfarin syndrome). Measurement of 'undercarboxylated' species of osteocalcin has been explored as an index of vitamin K deficiency. At least two other VKD proteins contribute to bone health. Gla-rich protein (GRP) contains 16 Gla residues in its 72 amino-acid sequence, the highest proportion of Gla reported among the VKD proteins. GRP appears to regulate extracellular calcium. Periostin is a VKD protein expressed in collagen-rich connective tissues, including bone, with a possible role in extracellular matrix mineralization.

The clearest evidence for a role of vitamin K deficiency in bone health relates to postmenopausal bone loss and associated fractures, particularly of the hip, which are more pronounced in women with low vitamin K status. Vitamin K antagonists (coumarin derivatives such as warfarin, used as oral anticoagulants) may have adverse effects on bone. Serious bone deformations can develop in human foetuses born to women having oral anticoagulant treatment but the evidence for adverse effects on bone in adults is less clear.

6.2.4 Dietary lipids in growth and development

6.2.4.1 Foetal growth

From conception, the cells of the growing foetus must incorporate lipids into their rapidly proliferating membranes. The foetus is totally dependent on the placental transfer of substrates from the mother's circulation, which are then elaborated into lipids. Glucose is a major substrate for the foetus, which expresses the enzymes for its conversion into FAs and for the production of the glycerol moiety of glycerides via G3P. A crucial question in developmental biology is how EFA are acquired and metabolized at this early stage.

Placental transfer of fatty acids

Fig. 6.7 illustrates several pathways by which the foetus may obtain the PUFAs that it requires for growth. The placenta of most mammals, including humans, is permeable to nonesterified fatty acids (NEFA) although there are large differences between animal species in rates of transfer. There are also large species differences in the lipid biosynthetic capacity of the placenta and of foetal tissues, so that generalizations cannot be made from research on any one species. Amounts of $18:2n-6$ and $18:3n-3$ and of their elongation/desaturation products in the mother's diet influences the FA composition of the developing foetus. Desaturation activity is present in developing foetal brain and liver (Fig. 6.7, reaction 7), so that the foetus may not be totally dependent on a supply of very long-chain PUFAs from the mother (Fig. 6.7, reaction 6). There is, however, little evidence for placental desaturase activity in simple-stomached animals, so that reaction 5 in Fig. 6.7 is unlikely to provide significant amounts of very long-chain PUFAs.

Concentrations of all lipoprotein classes increase in the maternal circulation from about the 12th week of human pregnancy, mediated by oestrogen and by the insulin resistance that accompanies pregnancy. This provides a second means of transport of maternal lipids to the developing foetus. Two lipases are expressed in human placenta: LPL (a TAG lipase) and endothelial lipase (a

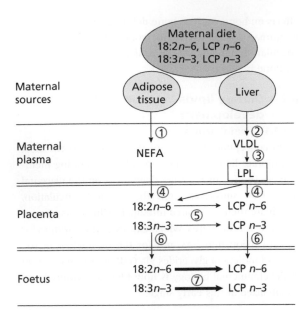

Fig. 6.7 Origins and metabolism of polyunsaturated fatty acids in the foetus. The maternal diet contains a mixture of *n*-6 and *n*-3 parent and long chain polyunsaturated fatty acids (LCP) which will vary widely among species and individuals. After digestion and absorption of the dietary lipids, the re-esterified fatty acids are transported in the maternal plasma as lipoproteins (Section 7.1.3) and incorporated into maternal tissues, e.g. adipose tissue and liver. Nonesterified fatty acids (NEFA) are released from adipose tissue stores (stage 1 in the figure; also see Chapter 4). The liver exports fatty acids as triacylglycerols in very low density lipoproteins (VLDL; stage 2 in figure) into the maternal plasma. The maternal face of the placenta contains lipases that release NEFA from VLDL and other lipoproteins (Stage 3: lipoprotein lipase (LPL) is a triacylglycerol lipase (Sections 4.2.2 & 7.2), and also endothelial lipase, which releases NEFA from lipoprotein phospholipids). The placenta is permeable to NEFA (4). The human placenta contains minimal desaturase activity and stage 5 is probably not a significant source of LCP. Parent polyunsaturates and LCP are transported to the foetal circulation (6) and the foetus can convert a proportion of the parent polyunsaturates into LCP (7).

phospholipase). These are members of the same lipase family (see Section 4.2.2). Release of FAs from maternal lipoproteins by placental lipases thus generates substrates for biosynthesis of lipids by the foetus. Transfer of intact or remnant lipoproteins by a receptor-mediated pathway is also thought to occur, as is transfer of lyso-phospholipids (formed by the action of endothelial lipase), which are likely to be enriched in very long-chain PUFAs.

In simple-stomached animals, arachidonic and some other very long-chain PUFAs, notably DHA, are present in higher concentration in foetal than in maternal plasma and in foetal tissues than in foetal plasma. These very long-chain PUFAs are selectively incorporated and trapped into placental phospholipids for export to the foetal circulation. The term 'biomagnification' has been coined for a process in which the proportion of very long-chain PUFAs in phospholipids increases progressively from maternal blood, to cord blood, to foetal liver and to foetal brain. The biomagnification process probably reflects the combined effects of placental FA uptake, selective protection of specific FAs against β-oxidation and selective direction of particular FAs into membrane or storage lipid biosynthesis.

In ruminants, biohydrogenation in the rumen severely limits the availability of EFAs to the mother (see Section 3.1.9). A ratio of 20:3*n*-9/20:4*n*-6 (the triene/tetraene ratio) above 0.4 is a biochemical marker of EFA deficiency (see Section 6.2.2.2) in simple-stomached animals but in foetal lamb tissues the ratio is 1.6. The ratio falls to 0.4 by 10 days after birth and by 30 days to 0.1. These values are well within the normal range despite the extremely low concentration of linoleic acid in ewe's milk (0.5% of energy). Ruminants are, therefore, able to conserve linoleic acid with supreme efficiency. Sheep placenta transfers linoleic acid at a relatively slow rate, but has a very high 6-desaturase activity by comparison with nonruminants, providing the major source of arachidonic acid for the foetus. This metabolite is concentrated into glycerophospholipids whereas the linoleic acid precursor is in higher concentration in the TAGs. This molecular compartmentation has the effect of conserving arachidonic acid and directing it into membranes.

Storage as triacylglycerols in adipose tissue
In some species, reserves of FAs (as TAGs) are built up in the foetus. For example, the development of human fat cells begins in the last trimester and a baby weighing 3.5 kg has, on average, 560 g of adipose tissue at birth (see 9.1.1 for more information on adipose tissue). Guinea pigs are also born with a large amount of adipose tissue but pigs, cats and rats have little or none. There is a remarkable parallelism between the placental permeability to NEFAs and the tendency to accumulate adipose tissue by the foetus, suggesting that an important source of foetal fat reserves is derived from circulating maternal

lipids. The adipose tissue composition of the foetal and new-born fat will, therefore, reflect the FA composition of the maternal diet.

Brain development

A large proportion of the very long-chain PUFAs synthesized or accumulated during the perinatal period is destined for the growth of the brain, of which 50% of the acyl groups may consist of 20:4n-6, 22:4n-6, 22:5n-6 and 22:6n-3. As with adipose tissue, there are large species differences in the extent of brain development at birth. The peak rate of brain development occurs in guinea-pigs in foetal life; in the rat, postnatally; while in man and pig it reaches a peak in late gestation and continues after birth. It has been suggested that transfer from the placenta is the major source of very long-chain PUFAs for the human foetal brain. This has been studied in detail using pigs as a model, because of their remarkable similarity to man in the timing of brain development and its lipid composition. Very long-chain derivatives of linoleic acid accumulate in brain from mid-gestation to term. Little linoleic acid, however, accumulates until birth, after which its concentration increases threefold while the concentrations of its elongation/desaturation products remain constant. Labelling experiments with [1^{14}C]linoleic acid in vivo have shown that linoleic acid is metabolized to very long-chain PUFAs by piglet brain and liver throughout the perinatal period. The contribution of the liver was many-fold greater than the brain at all stages. Whether foetal tissues can supply all the needs of the nervous system for very long-chain PUFAs without the need for the maternal transfer of intact very long-chain PUFAs is still to be resolved. In regard to maternal transfer, it has been proposed that an intracellular α-fetoprotein, derived from maternal plasma α-fetoprotein, may play a role in delivering very long-chain PUFAs to foetal brain. It binds FAs avidly, especially the very long-chain PUFAs such as 20:4n-6 and 22:6n-3. Indeed plasma nonesterified DHA seems to be almost exclusively transported on α-fetoprotein despite the quantitative predominance of albumin.

An outstanding feature of the composition of brain phospholipid is its remarkable consistency, irrespective of species or diet. The concentrations of the parent EFAs are extremely low (18:2n-6, 0.1–1.5% and 18:3n-3, 0.1–1.0%) while arachidonate (20:4n-6, 8–17%) and DHA (22:6n-3, 13–29%) predominate in all species.

In contrast to the conserved lipid composition of brain, liver lipids exhibit much greater variation. The parent EFAs are present in much greater concentrations than they are in brain and there are major differences in the elongation/desaturation products. For example, EPA is the major n-3 UFA in the liver lipids of ruminants and other herbivores while DHA predominates in the carnivores and omnivores. UFAs of the n-6 family usually predominate in liver phosphoglycerides, even when the n-3 UFAs dominate the dietary intake. Thus, zebra and dolphin, both species that have an overwhelming excess of n-3 UFAs in the diet, attain a preponderance of n-6 UFAs in the liver phosphoglycerides.

6.2.4.2 Postnatal growth

At birth, large changes in lipid metabolism occur. Whereas the foetus had synthesized (nonessential) FAs and TAGs from circulating glucose, the sole source of nutrition for the new-born is milk, from which about 50% of energy comes from fat. The enzymes of FA biosynthesis are suppressed and the baby's metabolism becomes geared to using fat directly from the diet.

Milk composition

The lipid composition of human milk is therefore the main factor that determines the availability of lipids for the breast-fed baby's development, especially in respect of the EFAs. TAGs comprise about 98% of the total lipids in milk and provide most of the required linoleic and α-linolenic acids. The glycerophospholipids represent only about 1% of total milk lipids but they provide about 50% of the very long-chain n-3 and n-6 PUFAs. The FA composition of human milk (and indeed that of any simple-stomached animal) is highly variable since it is strongly influenced by the FA composition of the mother's diet. However, the milk and dietary FA compositions are not related in a linear manner because the maternal adipose tissue stores also make a contribution to milk FA composition. Table 6.2 illustrates this variability and also compares the composition of human milk with that of cow's milk, which provides the base for most commercial infant formulas. Women eating a vegetarian diet tend to have more 18:2n-6 and 18:3n-3 in their milk and concentrations of 20:5n-3 and 22:6n-3 are increased many-fold when the diet is supplemented with fish oils. The increase in n-3 very long-chain PUFAs is not,

Table 6.2 The fatty acid composition of human and cow's milk fat (g/100 g total fatty acids).

	Cow's Milk	Human Milk	
		Vegans	Omnivores
Fatty acid			
4:0	3		
6:0	2		
8:0	1		
10:0	3		
12:0	4	4	3
14:0	12	7	8
16:0	26	16	28
16.1	3	1	4
18:0	11	5	11
18:1	28	31	35
18:2	2	32	7
18:3	1	1	1
20:4		1	1
Total	96	98	98

The data for human milk are adapted from the review by Jensen *et al.* (1990) and those from cow's milk from Gurr (1992). Figures are rounded to the nearest whole number. Note the high proportion of saturated fatty acids in cow's milk fat and especially those with short and medium chain lengths. The components not listed are mainly odd-chain and branched-chain fatty acids, individually in very small amounts. Note the high proportion of linoleic acid in the milk for of human vegans, which mainly replaces palmitic acid. The components not listed are mainly longer chain polyunsaturated fatty acids of the *n*-3 and *n*-6 families each in very small amounts.

References:

Jensen RG, Ferris AM, Lammi-Keefe CJ & Henderson RA. (1990) Lipids of bovine and human milks: a comparison. *J Dairy Sci* **73**, 223–40.
Gurr M (1992) *Role of Fats in Food and Nutrition*, 2nd edn, Elsevier Applied Science, London & New York, p. 28.

however, accompanied by a reduction in arachidonate (20:4*n*-6).

The FA composition of the milk feed given to newborn infants is quickly reflected in the composition of the adipose tissue.

Brain development

A dietary supply of 18:2*n*-6, 18:3*n*-3 and their very long-chain metabolites is important because, as discussed earlier, human brain development continues for quite a long time after birth. Data on the sequential development of human brain and nervous system are difficult to collect, but such data as exist, together with information from pigs, suggest a preferential

accumulation of *n*-6 UFAs in early foetal development, followed by an accumulation of *n*-3 UFAs later in gestation. The deposition of myelin (the lipid-rich material that forms an insulating layer around neurons) occurs late in the development of the central nervous system, predominantly after birth in human babies. The FA composition of immature myelin reflects the composition of the plasma membrane of the oligo-dendrocyte from which it is made. The high lipid content of the myelin sheath surrounding neurons, and the rapid rate of myelin synthesis in early postnatal life, put great demands on the supply of milk very long-chain PUFAs. Later, there is a marked decrease in the very long-chain PUFA content of myelin and a concomitant increase in MUFA and SFA, thus reducing the demands on the milk PUFA supply.

Infant formulas traditionally contained cow's milk fat having a low concentration of EFAs. A major development was replacement of the cow's milk fat by vegetable oils providing high intakes of 18:2*n*-6. More recent developments have been to supply a higher content of *n*-3 PUFAs, but the optimal ratio of 18:3*n*-3 to its longer chain metabolites is still debated. Limited data suggest that a dietary supply of preformed very long-chain *n*-3 PUFAs (EPA and DHA) results in significantly higher deposition of these acids in developing organs than when dietary 18:3*n*-3 alone is supplied, so the practice of adding fish oils to infant formulas is becoming more widespread. Reasons for the superiority of the very long-chain PUFAs may be the significantly higher β-oxidation of 18:3*n*-3 compared with very long-chain PUFAs and limited activity in the elongation/desaturation pathway.

An adequate supply of very long-chain PUFAs in milk is even more important in infants born prematurely, since these acids would normally have been supplied to the foetus from the mother's blood or metabolized from parent acids in foetal tissues. There is limited evidence that giving formula enriched with fish oil improves the 22:6*n*-3 status and visual function of premature babies but their 20:4*n*-6 status deteriorates.

Despite the above, however, randomized clinical trials of the supplementation of formula milk, for term infants, with very long-chain PUFAs have not produced clear evidence of benefit. Thus, the optimum amounts of parental EFAs and their elongation/desaturation products, and the optimal ratios between *n*-3 and *n*-6 families remain crucial unresolved topics in paediatric nutrition research.

KEY POINTS

The nature and amounts of dietary lipids

- Lipids in the diet of humans and other animals are derived from storage and membrane lipids of plants and animals. Both the quantity (total amount) and quality (nature of the constituent FAs) of dietary fat are relevant for human health.

- More than 90% of dietary lipids are TAGs, which originate from storage fats: the adipose tissue or milk of animals, or plant seed oils.

- Mammalian storage fats are characterized by a predominance of SFAs and MUFAs, whereas plant-derived lipids are a major source of PUFAs, especially *n*-6 PUFAs.

- Dietary lipids may be manipulated at source, by changing the feeding of the source animal or by altering the genetic makeup of the animal or plant. Extensive modification may also occur during industrial processing, especially saturation of UFAs (see also Chapter 11).

Dietary lipids also supply essential nutrients

- EFAs are those FAs that are required for health, but cannot be synthesized within the body. In humans, linoleic acid (18:2*n*-6) and α-linolenic acid (18:3*n*-3) are generally considered to be the true EFAs. They are required at low levels in the diet (1% of dietary energy for 18:2*n*-6, 0.2% of dietary energy for 18:3*n*-3).

- Fat-soluble vitamins are constituents of dietary lipids. They require adequate dietary fat for absorption and transport around the body.

- Vitamin A (retinol and related compounds) is important for sight, as a component of the photoreceptor rhodopsin. Retinol derivatives are also involved in regulation of gene expression through binding to particular nuclear receptors (ligand-activated transcription factors).

- The most active metabolite of vitamin D, 1,25-dihydroxycholecalciferol (also called calcitriol), is a true hormone, produced in one organ (the kidney) and signalling through specific receptors in distant tissues. Vitamin D also acts through a particular nuclear receptor and regulates calcium and phosphate homeostasis.

- Vitamin E (tocopherols and related compounds) is unusual in that it has no well-characterized molecular function other than its role as a lipid-soluble antioxidant, preventing the peroxidation of VLCPUFAs.

- Vitamin K refers to a group of compounds with a naphthoquinone ring system (menadione). It is essential for normal blood clotting and bone health. It acts as the cofactor of an enzyme that catalyses the post-translational conversion of glutamate (Glu) residues in proteins to γ-carboxyglutamate (Gla).

Dietary lipids and development

- Lipids are essential to the developing foetus. Essential PUFAs are supplied from the mother via the placenta and are needed especially for brain development.

- Large changes in lipid metabolism occur at birth, with a switch from mainly glucose as a fuel, to milk, in which about 50% of energy comes from fat.

- The newborn human infant is dependent upon a dietary supply of 18:2*n*-6, 18:3*n*-3 and their long-chain metabolites because brain development continues after birth.

Further reading

Dietary lipids

Garcia C & Innis S (2013) Structure of the human milk fat globule, *Lipid Tech* **25**:223–26.

Essential fatty acids

Heird WC & Lapillonne A (2005) The role of essential fatty acids in development, *Annu Rev Nutr* **25**:549–71.

Lands B (2012) Consequences of essential fatty acids, *Nutrients* **4**:1338–57.

Fat-soluble vitamins

Bender DA (2007) *Introduction to Nutrition and Metabolism*, 4th edn. CRC Press, Boca Raton, USA.

Booth SL (2009) Roles for vitamin K beyond coagulation, *Annu Rev Nutr* **29**:89–110.

D'Ambrosio DN, Clugston RD & Blaner WS (2011) Vitamin A metabolism: an update, *Nutrients* **3**:63–103.

Galli F & Azzi A (2010) Present trends in vitamin E research, *Biofactors* **36**:33–42.

Henry HL (2011) Regulation of vitamin D metabolism, *Best Pract Res Clin Endocrinol Metab* **25**:531–41.

Ross AC (2012) Vitamin A and retinoic acid in T cell-related immunity, *Am J Clin Nutr* **96**:1166S–1172S.

Nuclear receptors

Kurakula K, Hamers AA, de Waard V & de Vries CJ (2013) Nuclear receptors in atherosclerosis: a superfamily with many 'Goodfellas', *Mol Cell Endocrinol* **368**:71–84.

Sever R & Glass CK (2013) Signaling by nuclear receptors, *Cold Spring Harb Perspect Biol* doi: 10.1101/cshperspect.a016709.

Growth and development

Larqué E, Demmelmair H & Gil-Sánchez A (2011) Placental transfer of fatty acids and fetal implications, *Am J Clin Nutr* **94**:1908S–1913S.

Lindquist S & Hernell O (2010) Lipid digestion and absorption in early life: an update, *Curr Opin Clin Nutr Metab Care* **13**:314–20.

Muskiet FAJ, van Goor SA & Kuipers RS (2006) Long-chain polyunsaturated fatty acids in maternal and infant nutrition, *Prostag Leukotr Ess* **75**:135–44.

CHAPTER 7
Lipid assimilation and transport

This chapter will consider how lipids, which are by definition not water-soluble, are assimilated into the body and how they are transported around the body in the aqueous environment of the blood plasma. The emphasis will be on humans, but lipid assimilation and transport in other mammals and in other organisms will also be touched upon.

7.1 Lipid digestion and absorption

7.1.1 Intestinal digestion of dietary fats involves breakdown into their component parts by a variety of digestive enzymes

Typically 90–95% of fat in the human diet is provided by triacylglycerols (TAGs), with smaller contributions from phospholipids and cholesterol. It is a general rule that TAGs do not cross cell membranes intact, other than as components of lipoprotein 'particles'. The dietary TAGs must therefore be partially hydrolysed before the body can assimilate them. It is useful to divide the digestion and assimilation of dietary fat into several stages: emulsification, hydrolysis, solubilization and absorption into the enterocytes. This is followed by the secretion of the absorbed fat into the circulation, and its delivery to tissues, described in later sections.

The processes are somewhat different in infants and older individuals. At birth, the new-born animal has to adapt to the relatively high fat content of breast milk after relying mainly on glucose as an energy substrate in foetal life (see Section 6.2.4.2). At this stage, the pancreatic secretion of lipase is rather low and the immature liver is unable to provide sufficient bile salts to emulsify the digested lipids. These problems are even more acute in the premature infant. Yet the new-born baby can digest fat, albeit less efficiently than the older child or adult.

There are two mechanisms that allow this to occur. Breast milk contains a lipase that is stimulated by bile salts, known as bile salt-stimulated lipase (BSSL, gene name *CEL* for carboxyl ester lipase). Milk also contains lipoprotein lipase (LPL, gene *LPL*), although this is not thought to play a role in fat digestion. BSSL may assist the neonate in digesting the milk TAGs. BSSL is more active than is pancreatic lipase (the main enzyme involved in fat digestion in adults) against the ester bonds linking the very long-chain polyunsaturated fatty acids (PUFAs) such as arachidonic, docosahexaenoic (DHA) and docosapentaenoic to glycerol in TAGs. This is important in the ability of the new-born to accumulate these fatty acids (FAs) in the central nervous system (see Section 6.2.4.2). In addition, the neonate secretes a TAG lipase from glands in the stomach, named gastric lipase (gene *LIPF*), secreted from the chief cells in the fundic mucosa of the stomach. This is an acid lipase, with a pH optimum of around 4–6, although it is still active even at a pH of 1 and in the presence of pepsin (which digests other proteins). In other mammals a homologous lipase may be secreted higher up the gastrointestinal tract – from the serous glands of the tongue in rodents (lingual lipase) and from the pharyngeal region in ruminants (pharyngeal lipase). There may be some expression of human gastric lipase from the serous glands in the tongue, but its secretion in both neonates and adults is primarily from the stomach. These lipases that are involved in fat digestion before the entry of pancreatic lipase through the pancreatic duct into the duodenum are often called 'preduodenal lipases'. There is also evidence from mouse experiments for a role of another lipase in neonatal fat digestion. This is pancreatic lipase-related protein-2 (PLRP2, human gene name *PNLIPRP2*). PLRP2 and another protein, PLRP1, which appears not to have enzymic activity, are closely related to pancreatic lipase (described below) and their genes form a cluster. PLRP2

Lipids: Biochemistry, Biotechnology and Health, Sixth Edition. Michael I. Gurr, John L. Harwood, Keith N. Frayn, Denis J. Murphy and Robert H. Michell.
© 2016 John Wiley & Sons, Ltd. Published 2016 by John Wiley & Sons, Ltd.

is expressed at an earlier developmental stage than pancreatic lipase, and newborn PLRP2-deficient mice cannot digest dietary fat. Its role is currently unclear in human neonates, however.

The products of these lipases are mainly 2-monoacylglycerols (MAGs), diacylglycerols (DAGs) and non-esterified fatty acids (NEFAs), the latter being relatively richer in short- and medium-chain FAs than the original acylglycerols. The milk fat of most mammals is relatively rich in short- and medium-chain FAs (8C–12C) rather than the usual 16C and 18C FAs (see Table 6.2). The relative ease with which lipids containing short- and medium-chain FAs can be absorbed certainly helps lipid uptake in babies. Human milk TAGs (along with those of some other mammals including pigs) have a predominance of palmitic acid esterified at position sn-2. This will lead to the production of 2-palmitoylglycerol, which is readily absorbed (see below). Since the absorption of nonesterified saturated fatty acids (SFA) (which are solid at body temperature unless emulsified, see Section 2.1.1) may be less efficient than that of unsaturated fatty acids (UFAs), the structure of human milk TAGs seems to be a way of optimizing absorption of palmitic acid.

As the baby is weaned onto solid food, the major site of fat digestion shifts from the stomach to the duodenum, which is the pattern also in adults. Gastric lipase continues to play a role, however. It has been estimated that gastric lipase is responsible for 25% or maybe up to 40% of the partial TAG hydrolysis necessary for absorption to occur. In addition, the action of gastric lipase seems to produce lipid droplets (LD) that are better substrates for the later action of pancreatic lipase. The stomach also plays a role with its churning action, creating a coarse oil-in-water emulsion, stabilized by phospholipids. Also, proteolytic digestion in the stomach serves to release lipids from the food particles where they are generally associated with proteins.

The acidic fat emulsion that enters the duodenum from the stomach is neutralized and modified by mixing with bile and pancreatic juice. Bile supplies bile salts, which are powerful detergents that aid in emulsifying dietary fats (Box 7.1). In humans, the bile salts are mainly the glycine and taurine conjugates of tri- and di-hydroxy-cholanic acids (usually known as cholic and chenodeoxycholic acid), formed from cholesterol in the liver, and phospholipids. Much of the intestinal phospholipid in humans is of biliary origin and is estimated at

between 7 and 22 g/day, compared with a dietary contribution of 4–8 g/day. Biliary secretion is enhanced by the hormone cholecystokinin, secreted into the bloodstream from the I-cells of the duodenal mucosa, in response to entry of the acidic mixture from the stomach (chyme) into the duodenum. (I-cells are members of the family of enteroendocrine cells, i.e. cells of the gastrointestinal tract that sense chemical changes and secrete hormones.) Pancreatic juice, whose secretion is also stimulated by cholecystokinin, supplies bicarbonate to neutralize the acidic chyme, and enzymes that catalyse the hydrolysis of FAs from TAGs, phospholipids and cholesteryl esters (Fig. 7.1).

In healthy adults, the process of fat digestion is very efficient and the hydrolysis of TAGs is mainly accomplished in the small intestine by a TAG lipase secreted from the pancreas, pancreatic lipase (gene PNLIP). This enzyme is related to LPL and hepatic lipase (see Section 4.2.2); the preduodenal lipases are more distantly related. Pancreatic lipase attacks TAG molecules at the surface of the large emulsion particles (the oil–water interface), but before lipolysis can occur, the surface and the enzyme must be modified to allow interaction to take place. Firstly, bile salt molecules accumulate on the surface of the LD, displacing other surface-active constituents. As amphipathic molecules, they are adapted for this task since one side of the rigid planar structure of the steroid nucleus is hydrophobic and can essentially dissolve in the oil surface. The other face contains hydrophilic groups that interact with the aqueous phase (Fig. 7.2). The presence of the bile salts donates a negative charge to the oil droplets, which attracts to the surface the 10 kDa protein colipase. Colipase, also secreted from the pancreas, is essential for the activity of pancreatic lipase, which is otherwise strongly inhibited by bile salts. Thus, bile salts, colipase and pancreatic lipase interact in a ternary complex, which also contains calcium ions that are necessary for the full lipolytic activity. Pancreatic lipase catalyses the hydrolysis of FAs from positions sn-1 and -3 of TAGs to yield 2-MAGs. There is very little hydrolysis of the ester bond at position 2 and very little isomerization to the 1(3)-MAGs, presumably because of rapid uptake of the MAGs into epithelial cells.

Phospholipase A_2 (PLA$_2$ – see Section 4.6.2), also secreted in the pancreatic juice, hydrolyses the FA at position 2 of phospholipids, the most abundant being phosphatidylcholine (PtdCho). There is a large family of PLsA$_2$, and this particular one belongs to the subfamily of

Box 7.1 Bile acids and salts

These are derivatives of cholesterol (see Section 4.9.6), synthesized in the liver. A typical bile acid is shown below. This is cholic acid. Chenodeoxycholic acid lacks the hydroxyl group at carbon 12. They are secreted in the bile in the form of covalent conjugates, formed with a base: either glycine as shown here, or taurine ($^+H_3NCH_2CH_2SO_3^-$). The conjugate shown is sometimes known as glycocholate. They are amphipathic molecules, with a predominantly nonpolar ring structure but a highly polar acidic group (especially in the conjugated form).

The first committed step in bile acid biosynthesis from cholesterol is hydroxylation of carbon 7 (marked on figure). This is brought about by an enzyme formerly known as cholesterol 7α-hydroxylase. This enzyme, like many involved in hydroxylation reactions, is a member of a large family of haem (US: heme) proteins that has the characteristic of absorbing light at a wavelength of 450 nm (when it has bound CO), known generally as cytochrome P-450. The enzyme now has the 'family name' CYP7A1. CYP7A1 expression is controlled primarily at the transcriptional level by the relative levels of cholesterol and bile acids in the hepatocyte, through two nuclear receptor/transcription factors. These are LXR (liver X-receptor), which responds to levels of cholesterol-derived oxysterols, and activates CYP7A1 expression; and FXR (farnesoid-activated receptor), which responds to levels of bile acids and suppresses CYP7A1 expression. See Section 7.3.3 for further description.

There is a large family of bile acid transporters, expressed in a tissue-specific manner in the liver (for taking up bile salts from the plasma and exporting them to the gall bladder) and the small intestine. For more details see Azer (2004) in **Further reading**.

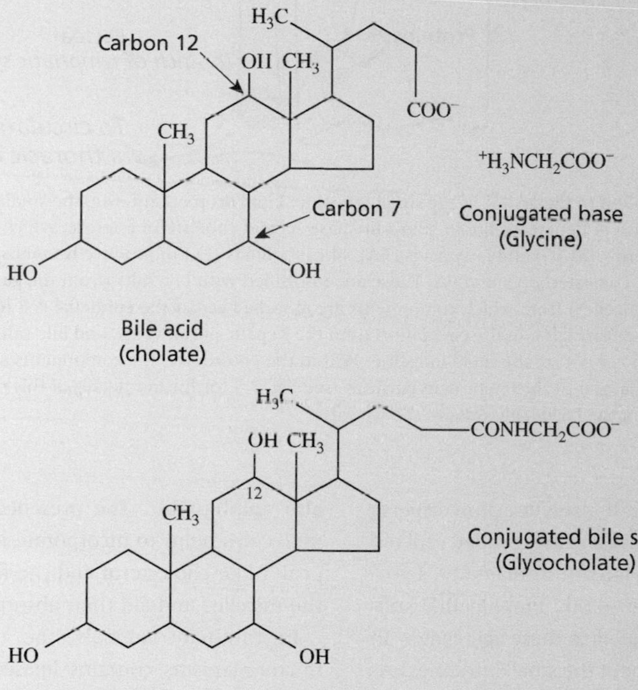

Carbon 12

OH CH$_3$

H$_3$C

COO$^-$

CH$_3$

$^+H_3NCH_2COO^-$

Carbon 7

Conjugated base
(Glycine)

HO

OH

Bile acid
(cholate)

H$_3$C

OH CH$_3$

—CONHCH$_2$COO$^-$

12

CH$_3$

Conjugated bile salt
(Glycocholate)

HO

OH

Text and figure based on KN Frayn (2010). *Metabolic Regulation: A Human Perspective*, 3rd edn. Wiley-Blackwell, Oxford. Reproduced with permission of John Wiley & Sons.

secreted lipases (gene *PLA2G1B*). The enzyme is present as an inactive proenzyme (or zymogen) in pancreatic juice and is activated by the tryptic hydrolysis of a heptapeptide from the N-terminus. The major digestion products that accumulate in intestinal contents are lyso-phospholipids. Hydrolysis of cholesteryl esters in the small intestine is carried out by a pancreatic cholesteryl ester hydrolase, usually called carboxyl ester lipase (gene *CEL*). This is actually a rather nonspecific enzyme, and is identical to BSSL mentioned earlier.

As these enzymes act upon the contents of the large emulsion particles, which may be around 1 μm in

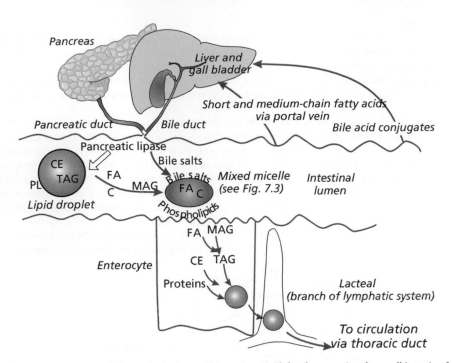

Fig. 7.1 The digestion and absorption of dietary fat in the small intestine. Lipid droplets entering the small intestine from the stomach are subjected to the action of pancreatic lipase, phospholipase A_2 and cholesterol esterase, which hydrolyse triacylglycerol (TAG) to produce monoacylglycerols (MAG) and fatty acids (FA), phospholipids (PL) to produce lysophosphatidic acids and FA; and cholesterol esters (CE) to liberate cholesterol (C) and FA. These are emulsified with bile salts (from the gall bladder) to produce a micellar suspension (the mixed micelles) from which components are absorbed across the epithelial cell (enterocyte) membranes. Short and medium-chain FAs pass through into the circulation (into the hepatic portal vein), and bile salts are reabsorbed, along with further cholesterol, in the lower part of the small intestine. Within the enterocyte, the components are reassembled, and packaged into chylomicrons, the largest of the lipoprotein particles (see Fig. 7.5 for further details of this process). The chylomicrons are secreted into small branches of the lymphatic system, the lacteals.

diameter, the products of their hydrolytic action disperse and form multimolecular aggregates called mixed micelles (Fig. 7.3), typically 4–6 nm in diameter. Lyso-phospholipids and MAGs are highly amphipathic substances and, with bile salts, stabilize these aggregates. In the more neutral environment of the small intestine, FAs are found largely in the ionized form and are therefore

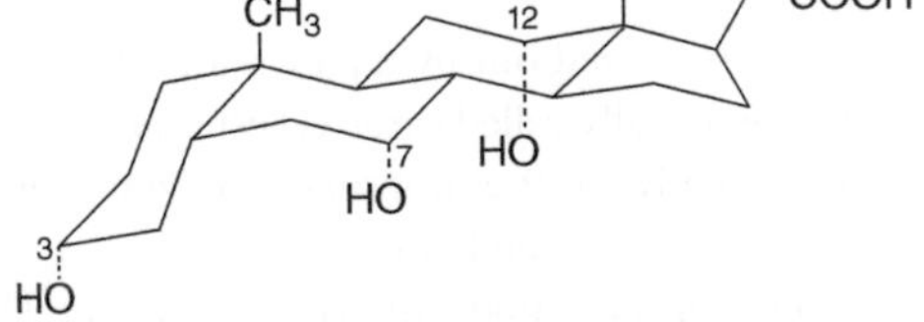

also amphipathic. The presence of these amphipathic molecules helps to incorporate insoluble nonpolar molecules like cholesterol and the fat-soluble vitamins into the micelles and aid their absorption.

In ruminant animals, the complex population of microorganisms contains lipases that split TAGs completely to glycerol and NEFAs. Some of these are fermented to acetic and propionic acids, which are absorbed directly from the rumen and carried to the liver, where they are substrates for gluconeogenesis. A proportion of the remaining long-chain UFAs undergoes several metabolic transformations catalysed by enzymes in the rumen microorganisms (Fig. 3.25) before passing into the small intestine where they are absorbed. Principal among these is biohydrogenation, in which double bonds are reduced by a process that is strictly anaerobic (see Section 3.1.9). During hydrogenation, some double bonds are

Fig. 7.2 The structure of a bile acid (cholic acid).

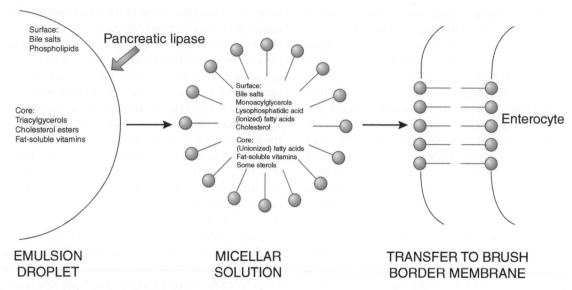

Surface:
Bile salts
Phospholipids

Pancreatic lipase

Core:
Triacylgycerols
Cholesterol esters
Fat-soluble vitamins

Surface:
Bile salts
Monoacylglycerols
Lysophosphatidic acid
(Ionized) fatty acids
Cholesterol

Core:
(Unionized) fatty acids
Fat-soluble vitamins
Some sterols

Enterocyte

EMULSION
DROPLET

MICELLAR
SOLUTION

TRANSFER TO BRUSH
BORDER MEMBRANE

Fig. 7.3 Role of the mixed micelle in fat absorption.

isomerized from the *cis* to the *trans* geometrical configuration. Positional isomerization also occurs, in which the double bonds, both *cis* and *trans*, migrate along the carbon chain. The result is a complex mixture of FAs, generally less unsaturated than those in the ruminant's diet and containing a wide spectrum of positional and geometrical isomers. Because there is very little MAG present in the digestion mixture, the mixed micelles in ruminants are composed largely of NEFAs, lysophospholipids and bile salts

7.1.2 The intraluminal phase of fat absorption involves passage of digestion products into the absorptive cells of the small intestine

Lipid absorption in humans begins in the distal duodenum and is completed in the jejunum. The principal molecular species passing across the brush-border membrane of the enterocyte are the MAGs and long-chain NEFAs. The bile salts themselves are not absorbed in the proximal small intestine, but pass on to the ileum where they are absorbed and recirculated in the portal blood to the liver, from whence they are resecreted in the bile. This is also true for some cholesterol that is a constituent of bile. The recirculation of bile constituents in this way is referred to as the entero-hepatic

circulation. It may be interrupted by resins (given by mouth) that bind cholesterol and bile salts and prevent their reabsorption, leading the liver to synthesize new bile salts from cholesterol. This is therefore a mechanism for depleting the body of cholesterol and may be useful in the treatment of high plasma cholesterol concentrations (see Section 10.5.2.2).

The digestion products encounter two main barriers to their absorption. At the surface of the microvillus membrane is a region known as the unstirred water layer. The mixed micelles are small enough to diffuse readily into this layer. It is a few hundred micrometres thick and is retained by mucopolysaccharides secreted by the epithelial cells (mucosa). It is relatively acidic. This will promote the protonation of NEFAs (i.e. they become uncharged) so that they can more easily leave the micelles and move into the epithelial cell membrane. This membrane is highly convoluted (hence the description 'brush-border') to increase its absorptive capacity. There is a suggestion that pancreatic lipase may be loosely bound to the brush-border membrane, which would bring the products of its action even more closely into contact with the absorptive surface.

The second barrier is the brush-border membrane itself. It has been speculated for many years that FAs in their protonated form might diffuse across this membrane (and other cell membranes) by a process known as 'flip-flop', down a concentration gradient which is maintained by intracellular sequestration of the FAs (described below). However, it is now clear that there are specific transport proteins to facilitate their movement into the cell, although passive diffusion may also contribute. Two candidate proteins have been identified in small intestinal enterocytes: fatty acid transport protein-4 (FATP4, gene *SLC27A4* as it is member 4 of family 27 of the solute-carrier proteins) and fatty acid translocase (FAT, gene *CD36*). FAT is identical to the cell-surface thrombospondin receptor CD36, which is a member of the family of scavenger receptors (see Section 7.2.4.4). The relative importance of these two proteins in FA uptake by human enterocytes is unknown, but this is an area of intense interest at present. MAGs must also cross the brush-border membrane. There is some evidence from cellular studies that this uptake is saturable (implying a carrier mechanism) but no candidate proteins have been identified.

The absorption of cholesterol is slower and less complete than that of the other lipids. About half of luminal cholesterol typically reaches the bloodstream. Dietary cholesterol intake on a typical Western diet is around 400 mg/day, but to this is added around 800–1400 mg/day excreted in bile. The absorption of cholesterol is largely carrier-mediated and is discussed in Box 7.2.

7.1.3 The intracellular phase of fat absorption involves recombination of absorbed products in the enterocytes and packing for export into the circulation

For absorption into the enterocytes to occur, an inward diffusion gradient of lipolysis products must be maintained. Two cellular events contribute. First, the long- and very long-chain FAs entering the cells bind to a cytosolic fatty acid binding protein (FABP). There are two members of the FABP family expressed in enterocytes. One is found only in the intestine and is called I-FABP (also known as FABP2, gene name *FABP2*); the other is liver FABP (L-FABP, gene *FABP1*) and is also expressed in liver and kidney. They are small proteins (molecular mass 14–15 kDa) and bind one or two FAs, respectively. The FABPs are believed to play a role in targeting FAs within the cell, and also in protecting the cell from the potentially cytotoxic effects of high FA concentrations. They also provide the concentration gradient that ensures efficient FA uptake from the intestinal

lumen. It is not clear whether the two FABP isoforms have different roles, although it has been suggested that I-FABP is involved in the uptake of dietary FAs for TAG biosynthesis, whilst L-FABP in enterocytes is involved in uptake of FAs delivered in the blood, which may be substrates for oxidation and for phospholipid biosynthesis. Both these FABPs bind long-chain UFAs with higher affinity than SFAs, and this may explain the more rapid absorption of oleic than stearic acid.

Up to this stage, the absorption process is not dependent on a source of energy. The next phase, which removes FAs, thereby maintaining a gradient, is the energy-dependent re-esterification of the absorbed FAs into TAGs and phospholipids. The first step is the ATP-dependent 'activation' of FAs to their acyl-CoA thioesters by an acyl-CoA synthase (ACS) (see Section 3.1.1). Again, the preferred substrates are the long- and very long-chain FAs. There is a family of ACS proteins, one of which, ACS5 (gene *ACSL5*, L for long- and very long-chain FAs), is prominent in the intestine. The ACSs are all membrane-associated and are situated on the endoplasmic reticulum (ER) as well as on mitochondrial and peroxisomal membranes. However, there is intriguing evidence that the long- and very long-chain-specific ACS members (acting preferentially on FAs with ≥ 18 carbons) are identical to the FATP proteins mentioned above (SLC27A family members). This would mean that at least some FAs are transported across the cell membrane and activated in one process, efficiently maintaining a concentration gradient for continued uptake. Long- and very long-chain acyl-CoAs are themselves bound by a specific, 10 kDa, cytosolic acyl-CoA binding protein (gene name is *DBI*, diazepam binding inhibitor, because that is how the protein was first identified) which, like the FABPs, may direct the acyl-CoA within the cell.

In humans and other simple-stomached animals, the major acceptors for esterification of acyl-CoA are the 2-MAGs that together with the NEFAs are the major forms of absorbed lipids. Resynthesis of TAGs, therefore, occurs mainly via the MAG pathway (see Section 4.1.5). In ruminant animals, the major absorbed products of lipid digestion are glycerol and NEFAs and resynthesis occurs via the glycerol 3-phosphate (G3P) pathway (see Section 4.1.1) after phosphorylation of glycerol catalysed by glycerol kinase. It should be noted, however, that lipids usually form a minor part of ruminant diets, so that glucose is probably the major precursor of G3P in the ruminant enterocyte.

Box 7.2 Cholesterol absorption in the small intestine and its specificity

Dietary cholesterol is accompanied by plant sterols (phytosterols) that are very similar in structure; and yet these phytosterols are present in plasma only in minute concentrations compared with cholesterol. Yet this specificity does not seem to be mediated at the uptake step.

Current understanding is that the protein Niemann-Pick C1-like protein-1 (NPC1L1, gene *NPC1L1*), expressed at the apical surface of the enterocyte, acts as a nonspecific sterol transporter. Cholesterol and plant sterols thus enter the enterocyte. After binding of sterols, the NPC1L1 protein is internalized and transports the bound sterols to the endoplasmic reticulum, where they may be esterified by the enzyme acyl-CoA:cholesterol acyltransferase (ACAT, discussed in Section 4.9.7). There are two isoforms of this enzyme, ACAT1 and ACAT2: ACAT2 is involved in the enterocyte. (The gene name for ACAT2 is *SOAT2*, for sterol O-acyltransferase 2; *ACAT2* is the gene for a specific acetyl-CoA:cholesterol acetyltransferase.)

Plant sterols are less effective substrates than cholesterol for ACAT2, and hence are directed back to the apical membrane for re-export to the intestinal lumen by the combined action of two proteins of the ABC (ATP-binding cassette) family: ABC-G5 and ABC-G8 (genes *ABCG5* and *ABCG8*). (Mutations in these genes give rise to the condition of sitosterolaemia discussed in Section 10.5.2.1.) There is also export (excretion) of a proportion of intracellular free cholesterol. This may be a mechanism for excretion of excess cholesterol from the body in a pathway known as trans-intestinal cholesterol efflux (TICE). This will be discussed again later (under HDL metabolism, Section 7.2.5). Esterified cholesterol is combined with triacylglycerols (TAG) and apoB48 to form chylomicrons as described in the text.

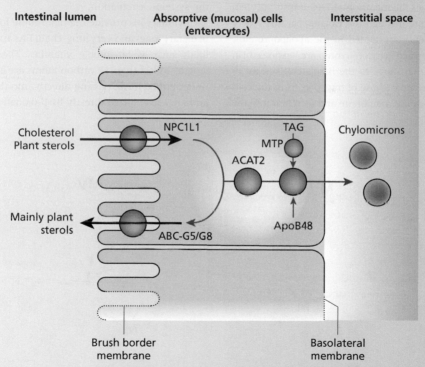

Diagram based on F Lammert & DQ-H Wang (2005) New insights into the genetic regulation of intestinal cholesterol absorption. *Gastroenterology* **129**: 718–34.

The main absorbed product of phospholipid digestion is 2-monoacylphosphatidylcholine (lysophosphatidylcholine, lysoPtdCho). A FA is re-esterified to position 1 to form PtdCho by an acyl transferase located in the villus tips of the intestinal brush-border. The function of this phospholipid is to stabilize the TAG-rich particles (chylomicrons) exported from the cell as described later. The PtdCho used for the biosynthesis and repair of membranes in the enterocytes (which turn over rapidly, with a typical lifetime of 5–6 days in humans) is synthesized by the classic

CDP-choline pathway (see Section 4.5.5) in cells at the villus crypts.

During fat absorption, the biosynthetic activity of the enterocyte is geared to packaging the resynthesized absorbed lipids in a form that is stabilized for transport in the aqueous environment of the blood. Within minutes of absorption products entering the enterocyte, fat droplets can be seen within the cysternae of the smooth ER, where the enzymes of the MAG pathway are located (Fig. 7.4). The rough ER is the main site for the biosynthesis of phosphoglycerides (see Section 4.5) and apolipoproteins (see Section 7.2.2), which provide the coat that stabilizes the LDs. One apolipoprotein in particular, apolipoprotein B48 (apoB48), provides a 'skeleton' which associates with lipid during its biosynthesis, forming the immature chylomicron particle. These immature particles gain further TAG in a process that involves the microsomal TAG transfer protein (MTP). They migrate through the Golgi apparatus where carbohydrate moieties are added to the apolipoproteins, and the fully formed chylomicrons are exported in secretory vesicles that move to the basolateral surface of the enterocyte. The final phase of transport from the cells involves fusion with the membrane and secretion into the intercellular space by exocytosis (Fig. 7.1). Chylomicron assembly in the enterocyte is similar to the assembly of very low density lipoprotein (VLDL) particles in hepatocytes (see Section 7.2.3). The process is illustrated in Fig. 7.5.

The chylomicrons do not enter the plasma directly. Instead they are secreted into the tiny lymph vessels that are found inside each of the intestinal villi, called 'lacteals' because of their milky appearance after the ingestion of fat. From here the chylomicrons pass via the thoracic duct (a main branch of the lymphatic system) and enter the circulation in the subclavian vein, from where they reach the heart for distribution around the body. Thus, dietary lipids, contained in the chylomicrons, are unique amongst the products of intestinal digestion and absorption in that they do not enter the hepatic portal vein and traverse the liver before entering the systemic circulation.

Chylomicrons provide the main route for the transport of dietary long- and very long-chain FAs and other dietary lipids such as fat-soluble vitamins. Those with chain lengths of less than 12 carbon atoms are absorbed in the nonesterified form, passing directly into the portal blood and are metabolized directly by β-oxidation in the liver.

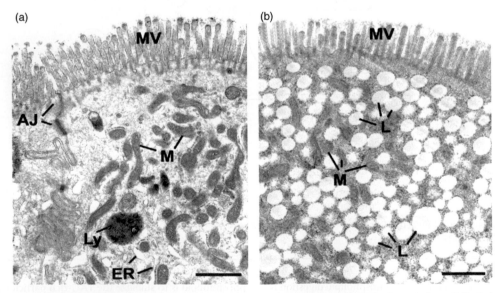

Fig. 7.4 (a) An enterocyte from the human jejunum showing the brush border (microvilli, MV). (b) A similar cell displaying multiple lipid droplets (L) a few hours after consuming a fatty meal. M shows mitochondria. AJ is an apical junctional complex between cells. Strands of rough endoplasmic reticulum (ER), and a lysosome (Ly) are visible. Scale bar is 1 μm. Reproduced, with permission from BMJ Publishing Group Ltd, from MD Robertson, M Parkes, BF Warren *et al.* (2003) Mobilization of enterocyte fat stores by oral glucose in man. *Gut* **6**: 833–8.

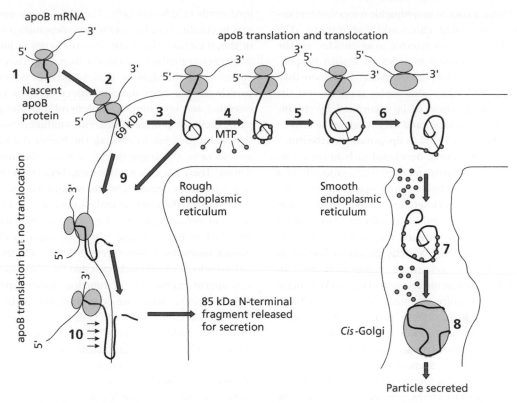

Fig. 7.5 Assembly of the triacylglycerol-rich lipoprotein particles, chylomicrons in the enterocyte, and VLDL in the hepatocyte. Apolipoprotein B-48 (apoB100 in hepatocytes) is synthesized on the rough endoplasmic reticulum (ER) (upper left). At step 4, the N-terminal region acquires a small amount of 'core lipid' (triacylglycerols and cholesteryl esters) in a process mediated by the microsomal triacylglycerol transfer protein (MTP). After further protein chain synthesis (step 5), a lipid-poor 'primordial' particle is produced. There is debate about whether this is released into the ER lumen as shown at 6, or whether the next stage, of bulk lipid addition (7), occurs whilst the primordial particle is still attached to the ER membrane. There is also debate about whether MTP is involved in this stage also (consensus is that it probably is). The left side of the diagram indicates that if lipid availability is low and lipid addition fails to occur, then the nascent apoB is directed into a degradative pathway (steps 9 and 10). Reproduced, with permission of the Nutrition Society, from White DA, Bennett AJ, Billett MA & Salter AM (1998) The assembly of triacylglycerol-rich lipoproteins: an essential role for the microsomal triacylglycerol transfer protein. *Br J Nutr* **80**: 219–29.

There are several reasons for this partition. Firstly, short- and medium-chain FAs are more readily hydrolysed from TAGs and since they occupy mainly position 3, are not retained in the 2-MAGs. Secondly, they are more likely to diffuse into the aqueous phase rather than the mixed micelles and for this reason are more rapidly absorbed. A few g per day short- and medium-chain FAs enter the diet from dairy products or foods that incorporate coconut or palm kernel oils. The short-chain FAs, acetate, propionate and butyrate, derived by microbial fermentation of nonstarch polysaccharides (dietary fibre) in the colon are also absorbed and contribute to lipid metabolism (see also Section 10.2.1 for relevance to colon cancer).

7.2 Transport of lipids in the blood: plasma lipoproteins

7.2.1 Lipoproteins can be conveniently divided into groups according to density

Lipids absorbed from the intestine, as well as some of the lipids synthesized in tissues, need to be transported in the bloodstream between organs. Since lipids are, by definition, not soluble in water (nor in blood plasma), this poses a problem, which has been solved during evolution by emulsification. Aggregates of hydrophobic molecules (particularly TAG and cholesteryl esters) are stabilized with a coat of amphipathic compounds: phospholipids, unesterified cholesterol and proteins. The resulting droplets, often referred to as particles, are the lipoproteins. The protein moieties are known as apolipoproteins and, as will be discussed later, have more than a stabilizing role. They also confer specificity on the particles, allowing them to be recognized by specific receptors on cell surfaces, and they regulate the activity of some enzymes involved in lipoprotein metabolism.

The lipoprotein system developed early in evolution. Birds, fishes, amphibians and even round-worms have a system for delivery of lipids from lipogenic organs to the developing egg that is closely related to the mammalian lipid transport system, and some of the proteins involved are homologous. For instance, the protein involved in lipid transport in these various groups is known as vitellogenin and is related to mammalian apolipoprotein B. The vitellogenin receptor of the chicken oocyte will, in fact, recognize mammalian apolipoprotein B, suggesting a common origin with mammalian lipoprotein receptors (see Section 7.2.4). The mammalian lipase gene family (pancreatic, lipoprotein, hepatic and endothelial lipases; see Section 4.2.2) is homologous to proteins in egg yolk of *Drosophila* that are presumably involved in related functions. See Section 7.2.6 for more information on species differences in lipid transport.

There are several types of lipoproteins with differing chemical compositions, physical properties and metabolic functions (Table 7.1), but their common role is to transport lipids from one tissue to another, to supply the lipid needs of different cells. The different lipoproteins may be classified in a number of ways depending on their origins, their major functions, their composition, physical properties or method of isolation (Box 7.3). They differ according to the ratio of lipid to protein within the particle as well as in the proportions of lipids (TAGs, esterified and nonesterified cholesterol and phospholipids). These compositional differences influence the density of the particles: in general, the higher the lipid to protein ratio, the larger the particle and the lower its density. There is a strong relationship between biological function and the density class into which a particle falls. It is convenient, therefore, to make use of density to separate and isolate lipoproteins by ultracentrifugation as described in Box 7.3, and it is now usual to classify plasma lipoproteins into different density classes. It is fortunate for the lipid biochemist that this classification is also approximately one of function. From lowest to highest density, these are: chylomicrons, VLDL, low density lipoproteins (LDL) and high density lipoproteins (HDL). (An additional class of intermediate density lipoproteins, IDL, is sometimes included between VLDL and LDL.) As density increases, so the ratio of TAGs to phospholipids and cholesterol decreases (Table 7.1): chylomicrons and VLDL particles are often grouped as the TAG-rich lipoproteins, whereas LDL and HDL are more important as carriers of cholesterol.

The lipoprotein classes are not homogeneous: there is a wide variety of particle sizes and chemical compositions within each, and there is overlap between them. It is also

Table 7.1 Composition and characteristics of the human plasma lipoproteins

	Chylomicrons	VLDL	LDL	HDL
Protein (% particle mass)	2	7	20	50
Triacylglycerols (% particle mass)	83	50	10	8
Cholesterol (% particle mass; free + esterified)	8	22	48	20
Phospholipids (% particle mass)	7	20	22	22
Particle mass (daltons)	$100–1000 \times 10^6$	$10–100 \times 10^6$	$2–3.5 \times 10^6$	$1.75–3.6 \times 10^5$
Density range (g/ml)	<0.95	0.95–1.006	1.019–1.063	1.063–1.210
Diameter (nm)	80–1000	30–90	18–22	5–12
Apolipoproteins	AI, AII, AIV, (AV), B-48, CI, CII, CIII, E	(AV), B-100, CI, CII, CIII, E	B-100	A1, A2, AIV, (AV), CI, CII, CIII, D, E

NB: Apolipoprotein AV (AV above) is only found in very low concentrations in the circulation: for instance, it has been estimated that only 1 in 24 VLDL particles will carry a molecule of AV.

Box 7.3 Methods for separation of lipoproteins

Separation of lipoproteins is based on physical properties. It must be remembered that physical properties and biological function are closely, but not absolutely, related. Thus, separation of a so-called chylomicron fraction on the basis of flotation of large, TAG-rich particles will produce a population of particles that contain some large particles secreted by the liver (which, on biological function, should be called VLDL). Similarly, the next population to be harvested, which will be mainly VLDL, will also contain small chylomicrons and chylomicron remnants. The main physical methods are flotation in the ultracentrifuge, electrophoresis, gel filtration and precipitation. They are compared in the table:

Method	Advantages	Disadvantages
Ultracentrifugal flotation	Regarded as the 'reference method'.	Relatively low capacity (usually performed in a swinging-bucket rotor with only 4 or 6 buckets).
	Can be used to prepare large amounts of material for further analytical work.	Alterations in particle composition are known to occur during isolation.
Electrophoresis (usually on agarose gel)	Quick to perform on large numbers of samples.	Semi-quantitative. Not preparative.
Gel filtration	Appears to cause least alteration to the particles.	Very limited capacity (usually performed in FPLC mode, running one sample at a time).
Precipitation	Can be applied to large numbers of samples.	Separation is often based upon empirical findings and the basis is not always understood. possibility for artefacts therefore exists.

The principles of sequential flotation were developed by the Californian pioneer in lipid research, Richard Havel, in the 1950s. Essentially, the plasma is laid in a tube under a salt solution of known density, which, on ultracentrifugation, allows one species of particle to be 'floated'. The particles are harvested and the density readjusted to prepare the next fraction. An alternative is to prepare a density gradient in which all types of particle will be separated in one centrifugation (usually faster but giving less clear separation).

important to remember that lipoprotein particles are not molecules – they are aggregates of lipid and protein molecules with a degree of structural organization. The term particle mass, rather than molecular mass, is used to describe their mass.

7.2.2 The apolipoproteins are the protein moieties that help to stabilize the lipid; they also provide specificity and direct the metabolism of the lipoproteins

At first a series of letters, A–E, was used to identify apolipoproteins as they were separated by electrophoresis (see Box 7.3), but it soon became apparent that most of these could be divided further into several individual proteins (Table 7.2). These are usually referred to in abbreviated form: apoAI, apoCIII, etc. The complete amino acid sequences of the ten major human apolipoproteins

are now known (AI, AII, AIV, AV, B (−48 and −100), CI, CII, CIII, D and E). The apolipoproteins other than apoB are sometimes referred to as soluble, or exchangeable, apolipoproteins. They may exist in lipid-free form in the plasma and they may exchange between lipoprotein particles. Apolipoproteins of the groups A, C and E have similar gene structures and some homologous stretches of sequence, and are believed to have evolved from a common ancestral gene, whereas the genes for apoB and apoD have distinct structures. The genes for apoAI, CIII, AIV and AV are arranged in one cluster spanning 60 kb on human chromosome 11, and the genes for apoCI, CII and E in a cluster on chromosome 19 (Table 7.2). ApoAV was only discovered in 2001 by bioinformatics analysis of what was, until then, known as the ApoAI/CIII/AIV gene cluster.

The apolipoproteins fulfil two main functions. First, they stabilize the lipid particles in the aqueous

Table 7.2 Characteristics of the human apolipoproteins.

Shorthand name	Gene name	Molecular mass (daltons)	Amino acid residues	Function	Major sites of synthesis	Chromosomal location
AI	APOA1	28,000	243	Cholesterol efflux from cells Activates LCAT	Liver Intestine	11q23-q24
AII	APOA2	17,000	154	May inhibit HL activity; Inhibits AI/LCAT	Liver Intestine	1q23.3
AIV	APOA4	44,500	376	Activates LCAT; Some evidence for AIV as a satiety factor	Intestine	11q23-qter
AV	APOA5	41,200 (nascent form)	366 (nascent form) 343 (mature form)	Gene variants are strongly associated with plasma TAG concentrations, but mechanism of action unclear: may be intracellular in liver (affecting TRL assembly) or effect on LPL-mediated TAG hydrolysis.	Liver	11q23-qter
B-48	APOB	241,000	2152	Structural component of chylomicrons	Intestine	2p23-p24
B-100	APOB	513,000	4536	Structural component of VLDL and LDL Binds to LDL-R	Liver	2p23-p24
CI	APOC1	6,600	57	Activates LCAT	Liver	19q13.2
CII	APOC2	8,800	79	Activates LPL (essential co-factor)	Liver	19q13.2
CIII	APOC3	8,800	79	Once thought to inhibit LPL; more likely involved in hepatic uptake of apoE-containing particles	Liver Intestine	11q23.3
D	APOD	22,000	169	Involved in cholesteryl ester transfer?	Brain, adipose tissue (role in brain may not be connected with lipid metabolism	3q26.2-qter
E	APOE	34,000	279	Ligand for LDL-R	Liver (60 – 80%) Other tissues including adipose tissue (remainder)	19q13.2

The functions of many of the apolipoproteins have not been fully clarified.

HL, hepatic lipase; LCAT, lecithin-cholesterol acyltransferase; LDL-R, LDL-receptor (B/E receptor); LPL, lipoprotein lipase; TAG, triacylglycerol; TRL, triacylglycerol-rich lipoprotein

environment of the blood, and maintain their structural integrity. Secondly, they are important in 'identifying' the lipoprotein and directing its metabolism in specific ways. The known metabolic functions of the apolipoproteins are shown in Table 7.2.

ApoB is common to the TAG-rich lipoprotein particles (chylomicrons, secreted from the small intestine, and VLDL secreted from the liver) and cholesterol-rich LDL. ApoB synthesis is intimately linked to the assembly of the TAG-rich lipoprotein particles: the particle forms around one molecule of the protein (Fig. 7.5 & Section 5.6.2), which never leaves the particle until the remnant particle is ultimately catabolized. There are two isoforms

of apoB in mammals encoded by the same gene. A full-length transcript produces a very large protein, of 4536 amino acids in humans, known as apoB100 (since it is 100% of the maximum length). A specific enzyme system present in some cells edits the mRNA coding for apoB to introduce a stop codon about halfway along. The editing enzyme is known as apolipoprotein B mRNA editing enzyme (gene APOBEC1) and is a cytosine deaminase, creating the codon UAA (termination) from CAA (glutamine). The edited transcript thus codes for a shorter protein, apoB48 (2152 amino acids in humans), so-called because it is 48% of the length of apoB100. In humans the intestine secretes apoB48 and the liver apoB100, and these proteins are therefore specific markers of chylomicrons and VLDL, respectively. In the rat, the liver secretes both apoB100- and apoB48-containing VLDL particles, and the intestine, as in humans, only apoB48.

7.2.3 The different classes of lipoprotein particles transport mainly triacylglycerols or cholesterol through the plasma

Chylomicrons are the largest and least dense of the lipoproteins and their function is to transport lipids of exogenous (or dietary) origin, as we saw in Section 7.1.3 above. Because the principal dietary fats are TAGs, this is true also of the chylomicrons. The TAG core is stabilized with a surface monolayer of phospholipids, unesterified cholesterol and proteins (Table 7.1). The core lipid also

contains some cholesteryl esters and minor fat-soluble substances absorbed with the dietary fats: fat-soluble vitamins, carotenoids and traces of environmental contaminants such as pesticides. It seems that the enterocyte continually secretes particles. Between meals these are small and poor in lipid, which is derived from biliary phospholipids or the lipids of cells shed from the gut mucosa, and from plasma NEFAs. When fat is being absorbed, the particles increase in size. Thus, increased lipid transport into the circulation after a fat-rich meal is achieved mainly by an increase in the size of the chylomicrons secreted, a mechanism that allows rapid changes in flux through this pathway. The large lipid-rich chylomicron particles scatter visible light, giving the blood plasma a 'milky' appearance after a fatty meal (Fig. 7.6).

VLDL particles, like chylomicrons, contain predominantly TAG (Table 7.1) and their function is to transport

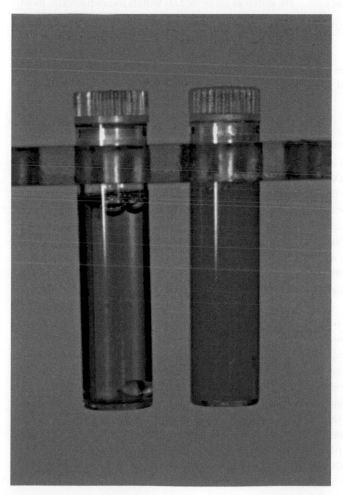

Fig. 7.6 Appearance of human blood plasma before (left) and after (right) a fatty meal. The cloudy (turbid) appearance after a fatty meal is due to the presence of TAG-rich chylomicrons, which are large enough to scatter visible light. (See the Colour Plates Section.)

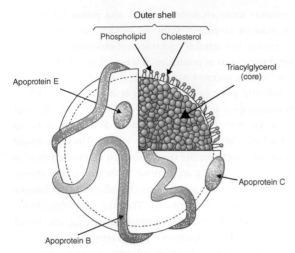

Fig. 7.7 Structure of a VLDL particle. (Other lipoprotein particles are similar in structure.) A monolayer of amphipathic molecules (phospholipids, unesterified cholesterol) stabilizes a droplet of hydrophobic lipids (triacylglycerols and cholesteryl esters) in the core. Apolipoprotein B wraps around the surface of the molecule, and other (soluble) apolipoproteins are present on the surface. Reproduced from GF Gibbons & D Wiggins (1995) Intracellular triacylglycerol lipase: its role in the assembly of hepatic very-low-density lipoprotein (VLDL). *Adv Enzyme Regul* **35**: 179–98, with permission of Elsevier.

TAGs of endogenous origin, synthesized in the liver. VLDL are spherical particles with a core consisting mainly of TAGs and cholesteryl esters, with unesterified cholesterol, phospholipids and protein mainly on the surface (Fig. 7.7). As with the chylomicrons, each VLDL particle is assembled around one molecule of apoB (in humans, apoB100), which remains with the particle throughout its lifetime. The particles also contain apoCI, CII, CIII and apoE in variable amounts, decreasing relative to apoB as particle density increases (i.e. as the particles become smaller).

As in chylomicron biosynthesis in the intestinal mucosa, nascent VLDL particles originate in the lumen of the ER, where they form around an emerging molecule of newly synthesized apoB100. They acquire phospholipids and further TAG, the latter by a process involving the MTP, known as 'second step' lipidation. A few individuals have been identified who have mutations rendering MTP inactive; such persons cannot secrete VLDL particles (see Section 10.5.2.1 for further discussion). The mature particles move to the Golgi apparatus where some of the apolipoproteins are

glycosylated. The Golgi vesicles then migrate to the cell surface where VLDL are exported by exocytosis into the subendothelial space, the space of Disse. Molecular details of these processes were described in more detail in Section 5.6.2.

The regulation of VLDL assembly and secretion is complex. A proportion of the apoB100 synthesized fails to associate with TAG and is degraded. Some nascent particles fail to acquire sufficient TAG to become 'competent' for further loading with TAG and are also degraded. Thus, increased availability of hepatic TAG may lead to a greater proportion of nascent VLDL particles reaching sufficient size for secretion.

The TAG-FAs secreted in VLDL originate from plasma NEFAs and TAG-FAs, taken up by the liver, and to a variable, but smaller, extent from *de novo* biosynthesis in the liver. However, before incorporation into VLDL they are stored as cytosolic LDs within the hepatocyte, mobilized (i.e. as FAs) from these stores, and then re-esterified to form the TAG that is incorporated into the VLDL particles. The purpose of this is presumably to allow a 'buffer' of hepatic TAG to cover fluctuating rates of VLDL secretion and also to confer additional points of control. The lipase responsible for mobilizaton of the cytosolic stores has not been identified for certain but an enzyme known simply as TAG hydrolase (gene name *CES1*, for carboxylesterase 1) may be responsible; another candidate is arylacetamide deacetylase (named for its ability to metabolize carcinogens and drugs; gene *AADAC*), which shares homology with hormone-sensitive lipase (HSL – see Section 4.2.2). Both TAG hydrolase and arylacetamide deacetylase are expressed predominantly in the liver. The G3P used in re-esterification may come either from glucose metabolism or from the phosphorylation of glycerol, taken up by the liver (released in turn mainly from adipose tissue), by the enzyme glycerol kinase.

LDL particles are the major carriers of plasma cholesterol in humans although this is not the case for all mammals (see Section 7.2.6). They are derived from VLDL in the plasma by a series of degradative steps that remove TAGs (described in more detail in Section 7.2.5), resulting in a series of particles that contain a progressively lower proportion of TAGs and are correspondingly richer in cholesterol, phospholipids and protein. The intermediate particles (IDL) are usually present at relatively low concentrations. During the transformations, the apoB component remains with the LDL particles and the apoC and E components are progressively

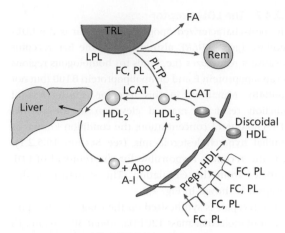

Fig. 7.8 The HDL pathway. Lipid-poor apoAI or pre-ß1 HDL acquires phospholipids (PL) and free cholesterol (FC) by interaction with cell membranes, forming discoidal HDL in the circulation. Through the action of lecithin-cholesterol acyltransferase (LCAT – see Fig. 7.14 below), and the acquisition of further lipids arising during the action of lipoprotein lipase (LPL) on triacylglycerol-rich particles (TRL) (phospholipid transfer protein, PLTP, is involved in this transfer), these particles swell and become spherical 'mature HDL'; smaller spherical particles are known as HDL₃, more lipid-rich particles as HDL₂. (Fatty acids, FA, are also produced by the action of LPL, and the smaller particles that remain are known as remnants (Rem).) These mature HDL particles can deliver their lipid content to the liver, possibly via the receptor SR-BI (see Section 5.2.4.4), to recycle as lipid-poor apoAI. An alternative route for delivery of lipid to the liver is through transfer of TRL remnants via the cholesteryl ester transfer protein (see Fig. 7.13). Based on PE Fielding & CJ Fielding (1996) Dynamics of lipoprotein transport in the circulatory system. In DE Vance & JE Vance, eds., *Biochemistry of Lipids, Lipoproteins and Membranes*, Elsevier, Amsterdam, pp. 495–516.

lost. Once only apoB remains, the particle is a mature LDL particle. ApoB has an important role in the recognition of LDL by cells since it must interact with specific cell-surface receptors before the LDL particle can be taken up and metabolized by the cell. Other receptors (for example on macrophages) recognize modified LDL (see Sections 7.2.4.4 & 10.5.1) and are responsible for the degradation of LDL particles that cannot be recognized by normal cell-surface LDL-receptors (see Section 7.2.4.2).

There is a variety of types of particle classified as HDL. They are very numerous in plasma: there are 10–20 times as many HDL particles in the circulation as all other lipoprotein particles combined. HDL precursor particles are not found within cells: HDL is formed in the

circulation. ApoAI has a very high affinity for phospholipid, a property that is essential to understanding HDL metabolism. Liver cells secrete apoAI, which becomes associated with some phospholipid molecules as it leaves the cell. The precursor particles in the circulation are described as pre-β HDL (from their migration pattern on electrophoresis). The smallest particles (pre-β1 HDL) are essentially apoAI with some phospholipid molecules. Further addition of phospholipid molecules leads to discoidal particles also called pre-β1 HDL. They acquire cholesterol from cells and from other lipoprotein particles (see Section 7.2.5) and mature into spherical, cholesterol-rich particles. These particles can give up their cholesterol (again by routes described later, see Section 7.2.5) and hence can recycle via lipid-poor forms (Fig. 7.8). Most HDL cholesterol is carried by the larger, spherical particles, which are divided by centrifugation into two subclasses, HDL₂ (larger particles) and HDL₃ (smaller).

A further class of lipoprotein particles is found to a variable extent, and is known as lipoprotein(a) (always read as 'lipoprotein-little a'). Lipoprotein(a), abbreviated Lp(a), is found at a somewhat higher density than LDL in the ultracentrifuge. Its structure is similar to that of an LDL particle, in which a large additional protein, apolipoprotein(a), has been covalently attached to the apoB100. Apolipoprotein(a) is related to the protein plasminogen, precursor of the enzyme plasmin in the fibrinolytic system. Its structure is characterized by multiple repeats of a 114 amino acid cysteine-rich domain known as a kringle (because its tertiary structure resembles the baked product of that name). The number of kringles varies from person to person and is genetically determined. Variations in the number of kringles give the Lp(a) particles a range of densities. The plasma concentration of Lp(a) particles varies considerably, but is of interest because of a positive relationship between Lp(a) concentration and risk of atherosclerosis.

7.2.4 Specific lipoprotein receptors mediate the cellular removal of lipoproteins and of lipids from the circulation

One function of the apolipoproteins has already been described as targeting of lipoproteins to specific destinations. This is achieved by the interaction of the apolipoproteins with specific cell-surface receptors. The TAG-rich lipoprotein particles, chylomicrons and VLDL, are largely confined to the vascular compartment by their size until they have undergone extensive lipolysis, allowing them to pass through the fenestrations in the endothelial lining of the capillaries. Some of the receptors involved are therefore expressed on endothelial cells, but most are expressed on parenchymal cells, and part of their selectivity may depend upon the exclusion, by virtue of size, of the larger particles. Before describing the major receptors involved in lipoprotein metabolism in further detail, it is useful to make some general points about membrane-bound receptors.

7.2.4.1 Membrane receptors

Membrane receptors are proteins that recognize specific molecules and bind them in preparation for the initiation of a biological process that takes place, initially, in that membrane. The process may be the transport of a molecule or particle (e.g. glucose, LDL) across the membrane or the triggering of a chemical message by a hormone or a growth factor. The recognition step demands that the receptor has specificity, like an enzyme for its substrate, known as a ligand. Binding is followed frequently, but not always, by internalization in which the membrane in the vicinity of the receptor forms a vesicle that encapsulates the receptor-ligand complex and enters the cell. Once inside, the receptor complex can be degraded by lysosomal enzymes and the receptor proteins may be recycled to be reinserted into the plasma membrane again. The number of membrane receptors is usually responsive to the availability of the ligand: as ligand availability increases, in general, so receptor expression decreases. This requires a mechanism for the control of the biosynthesis of the receptor. Receptors are usually glycoproteins, as described for the LDL-receptor below. The functional properties of receptors in membranes appear to be intimately related to the microenvironment provided by membrane lipids. Membrane proteins are on average associated with 30–40 molecules of phospholipid per molecule of peptide. These annular phospholipids are required for functional activity or for the stabilization of protein conformation. The structural and compositional changes of the lipid bilayer, which provides a fluid matrix for proteins, can induce alterations in functional properties of proteins in the intact membrane, for example by allowing changes in protein conformation or the diffusion or position of proteins in the membrane. This may be an important means by which diet – affecting the nature of the membrane phospholipid FAs – can affect hormone action, for instance.

7.2.4.2 The LDL-receptor

The best-characterized lipoprotein receptor is the LDL-receptor (gene *LDLR*), also known as the B/E receptor because it recognizes (i.e. binds to) homologous regions on apolipoprotein E and on apolipoprotein B100 (but not apoB48). Mutations in this receptor causing loss of function result in marked elevation in the plasma LDL-cholesterol concentration, the condition known as familial hypercholesterolaemia (see Section 10.5.2.1). It is the receptor responsible for the removal of LDL particles from the circulation. The structure of this receptor is shown in Fig. 7.9.

The receptor is synthesized on the rough ER as a precursor of molecular mass 120 kDa. About 30 min after its synthesis, the protein is modified to a mature receptor with an apparent mass of 160 kDa. The increase in molecular mass coincides with extensive modifications of the carbohydrate chains. The precursor molecule contains up to 18 N-acetylglucosamine molecules attached in O-linkage to serine and threonine residues (Fig. 7.9) as well as two high mannose-containing N-linked chains. The latter are modified extensively and the O-linked chains are elongated by addition of one galactose and two sialic acid residues to each N-acetylglucosamine during post-translational processing in the Golgi network.

The receptor is exported to specialized regions of the cell membrane known as coated pits, where the cytoplasmic leaflet is coated with the protein clathrin. After binding of an LDL particle through interaction of its apolipoprotein B100 molecule with the ligand-binding domain of the LDL-receptor, the receptor-LDL complex is internalized by endocytosis of the coated pit (Fig. 7.10). The coated vesicle so formed loses its clathrin coating, and its interior becomes acidified (an endosome). The LDL particle and receptor dissociate in the low pH environment of the endosome and the receptor is recycled back to the cell surface. The remainder of the particle is transferred to lysosomes, where the cholesteryl esters are hydrolysed by lysosomal acid hydrolases and the cholesterol adds to the cellular cholesterol pool. Cholesteryl esters may also be resynthesized by the action of acyl-CoA: cholesterol acyltransferase (ACAT, described earlier, see Section 4.9.7). The size of this cellular cholesterol pool in turn regulates expression of the LDL-receptor so that cellular cholesterol content is strongly regulated, as described in Section 7.3.1.

LDL-receptors are expressed by most cell types. There is some debate, however, about their biological function.

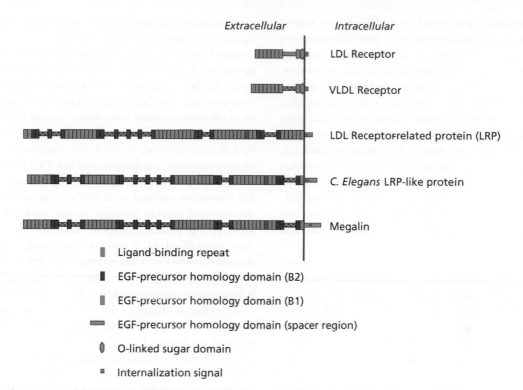

Fig. 7.9 The structure of the LDL receptor and related receptors. Each of the receptors in this family is built up from repeats of simpler units, some named according to their homology with the epidermal growth factor (EGF) receptor. The ligand binding domains are responsible for binding to the apolipoproteins of the target lipoprotein particles. The internalization signal domain is responsible for internalization of the receptor after the ligand is bound. (The apoE receptor 2 is not shown but is similar to LDL and VLDL receptors.) Based upon WJ Schneider (2008) Lipoprotein receptors. In DE Vance & J Vance, eds., *Biochemistry of Lipids, Lipoproteins and Membranes*, 5th edn, Elsevier, Amsterdam, pp. 555–78, Fig. 4.

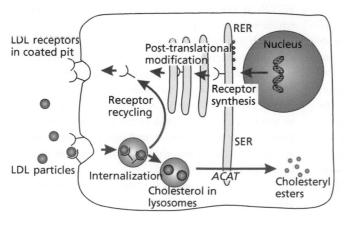

Fig. 7.10 Cell biology of the LDL receptor. ACAT is acyl-CoA cholesterol acyltransferase, which synthesizes cholesteryl esters that may be stored in the cell. The synthesis of the LDL receptor is closely regulated at the level of gene expression by the cellular cholesterol content (see Fig. 7.15). Redrawn from KN Frayn (2010) *Metabolic Regulation: a Human Perspective*, 3rd edn, Wiley-Blackwell, Oxford, and M Krieger (1999) Charting the fate of the 'good cholesterol': identification and characterization of the high-density lipoprotein receptor SR-BI, *Ann Rev Biochem* **68**: 523–58.

They represent a means by which cells can acquire cholesterol exported from the liver. Most cells expressing LDL-receptors, however, also have the enzymic capability of synthesizing cholesterol *de novo*, from acetyl-CoA. The alternative view is that LDL-receptors are there to remove cholesterol from the plasma. The fact that they are highly expressed in steroidogenic tissues such as the adrenal cortex argues for the former. However, in whole-body terms, the main site of LDL-receptor-mediated LDL removal is the liver. Cholesterol delivered to the liver can be removed from the body either by excretion directly in the bile, or after conversion to bile salts. Increasing LDL-receptor activity with the 'statin' drugs (see Section 7.3.1) can lead to marked reduction in plasma cholesterol concentration.

7.2.4.3 The LDL-receptor-related protein and other members of the LDL-receptor family

Because of its ability also to bind apoE, the LDL-receptor may play a role in removal of chylomicron remnants from the circulation. But it cannot be the major receptor involved in this process, because humans or animals lacking functional LDL-receptors (as in familial hypercholesterolaemia) do not show accumulation of chylomicron remnants. A related receptor, known as the LDL-receptor-related protein or LRP (gene *LRP1*), is the most likely candidate at present for the major chylomicron remnant receptor. The LDL-receptor and LRP, together with the other receptors (described below), form a family of receptors with structural homologies (Fig. 7.9). It should be noted that LRP undoubtedly has functions other than lipoprotein metabolism: it is also, for instance, a receptor for α2-macroglobulin.

LRP binds to apoE, in which chylomicron remnants are relatively enriched. It is expressed in most tissues, with the liver, brain and placenta as major sites of expression on a whole-body basis. Its expression is not regulated by cellular cholesterol content as is that of the LDL-receptor. However, in adipocytes, where it is also expressed, its concentration on the cell surface is rapidly increased by insulin. This increase in receptor appearance on the cell membrane is too rapid to be accounted for by an effect of insulin on protein synthesis, leading to the suggestion that the LRP, like the glucose transporter GLUT4, is present within the cells and translocates to the cell membrane when stimulated by insulin. This would represent a mechanism for directing some chylomicron remnants to adipose tissue in the period following a meal, although the major site of chylomicron remnant uptake is still considered to be the liver.

Both hepatocytes and adipocytes also produce apolipoprotein E, which may be expressed on the cell surface. It has been suggested that, in the liver, chylomicron remnants, once they become small enough via the activity of LPL in the capillary beds of other tissues, may enter the subendothelial space (the space of Disse). Here they bind to hepatic lipase, which is also expressed on the hepatocyte plasma membrane. Further TAG hydrolysis by hepatic lipase alters the particle composition so that cell membrane-attached apoE binds to the particle. As the remnant particle becomes further enriched with apoE by this means, it binds to adjacent LRP and is then internalized by a process very similar to that described for the LDL-receptor (Fig. 7.10). LPL also binds to the LRP. Since LPL is not synthesized in the adult liver, it is presumed that remnant particles carry this enzyme from peripheral tissues to the liver, where it assists in binding of the particles to LRP. Unlike the case for LDL-receptor, no human LRP deficiency state has yet been found. Attempts to create mice lacking the LRP have not been successful – the phenotype is lethal at an early embryonic stage. It may therefore be an essential protein for life.

A related receptor protein was discovered in 1992 by searching for genes with homology to the LDL-receptor gene. This receptor binds to, and internalizes, VLDL particles particularly avidly. It is expressed mainly in tissues outside the liver that might be expected to have a requirement for FAs: heart, skeletal muscle and adipose tissue. Within those tissues the so-called VLDL-receptor (gene *VLDLR*) is expressed mainly by the endothelial cells, which suits its supposed role as a receptor for large, TAG-rich particles that cannot cross the endothelial barrier. The precise physiological role of the VLDL-receptor is unknown, however. Its structure is shown in Fig. 7.9.

A further member of the family in mammals is known as megalin or LRP-2 (gene *LRP2*; Fig. 7.9). It is expressed in absorptive epithelial cells of the proximal tubules of the kidney. It has been suggested that its function is the re-uptake of fat-soluble vitamins that would otherwise be lost by urinary excretion.

This family of receptors is related also to the vitellogenin receptor responsible for delivery of lipid to the developing avian egg yolk, and there are closely related proteins in the fly *Drosophila melanogaster* and the nematode *Caenorhabditis elegans* (Fig. 7.9).

7.2.4.4 Scavenger receptors

Unrelated structurally to the LDL-receptor family are several receptors with the generic name of 'scavenger receptors' (SR). These were first identified as receptors in macrophages that mediate the uptake of LDL particles that have been chemically modified so that their affinity for the LDL-receptor is reduced. One such modification is oxidation, the significance of which will be discussed in Section 10.5.1. There are several families of SR, grouped according to structural features (eight families, A to H, are commonly recognized). They are characterized by a broad substrate specificity, and they may well have roles in macrophage function in host defence, removal of foreign substances, etc., beyond their role in lipid metabolism. Two closely related receptors in class A, the type I and type II SR (SR-AI, SR-AII) are expressed on the macrophage cell surface in clathrin-coated pits, similar to the LDL-receptor (Fig. 7.10). They bind and internalize modified LDL particles, which are then degraded by similar processes to those shown in Fig. 7.10. However, a crucial difference from the LDL-receptor system is that the cellular cholesterol content does not regulate expression of the SR. Thus, their activity can lead to unregulated accumulation of cholesterol in macrophages, an event that may initiate the series of changes that lead to atherosclerosis (see Section 10.5.1).

Class B SRs include the receptor known as SR Class B type 1 or SR-BI, and a structurally similar protein, with about 30% homology, that had been previously studied as a widely expressed cell-surface antigen given the name CD36. SR-BI was found to bind HDL particles with high affinity, and is now recognized as an HDL receptor. Its function in this role is covered in more detail in the following section. CD36 appears to have multiple roles. Amongst other physiological roles, it has been suggested as a long- and very long-chain FA transporter and also given the name 'fatty acid translocase' (FAT), discussed previously in Section 7.1.2 above.

7.2.5 The lipoprotein particles transport lipids between tissues but they interact and are extensively remodelled in the plasma compartment

It is convenient to describe lipoprotein metabolism as consisting of three pathways, but it will become clear that these are interrelated and interact with one another.

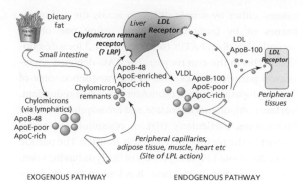

Fig. 7.11 Overview of the exogenous and endogenous pathways of lipoprotein metabolism. Lipolysis of particles by lipoprotein lipase (LPL) in capillaries of extrahepatic tissues is simplified: VLDL particles, in particular, may go through several cycles of lipolysis, and there is an intermediate class of particle known as Intermediate Density Lipoprotein (IDL) formed before final conversion to LDL particles. VLDL particles themselves may also be removed by VLDL receptors expressed in peripheral tissues. Interactions with HDL are shown in part in Fig. 7.13, but HDL is also an important donor of apolipoproteins (e.g. apoCII to nascent chylomicrons). Loosely based on J Herz (1998) Low-density lipoprotein receptor-related protein. In DJ Betteridge, DR Illingworth & I Shepherd eds., *Lipoproteins in Health and Disease*, Arnold, pp 333–59.

The pathway for transport of dietary fat is known as the *exogenous pathway* (Fig. 7.11). As the TAG-rich chylomicron particles, secreted from enterocytes, enter the plasma, they interact with other particles (presumably by simple physical contact) and acquire other exchangeable apolipoproteins, especially CII, CIII and apoE. ApoCII is essential for their further metabolism. The chylomicron particles come into contact with the enzyme LPL, which is expressed in a number of extra-hepatic tissues that can use FAs, including adipose tissue, skeletal and cardiac muscle and mammary gland. LPL was described in Section 4.2.2. It is anchored to the luminal aspect of the endothelial cells lining the capillaries in these tissues by binding to heparan sulphate proteoglycans. As chylomicrons pass through the capillaries, they bind to LPL and to the proteoglycan chains. LPL, with apoCII as an essential activator (cofactor), hydrolyses the core TAG in the particle. The rate of lipolysis is rapid, and it has been estimated that somewhere between 10 and 40 LPL molecules must act simultaneously on a chylomicron. The NEFAs that are generated can diffuse into the adjacent tissues, either by simple diffusion across the cell membranes or by facilitated diffusion, using one of the recently described FATPs (see Section 7.1.2 for a description of this in the enterocyte). The TAG-depleted chylomicron particle shrinks, and as a consequence some of the amphipathic surface monolayer becomes redundant. Surface components dissociate and are acquired by other lipoproteins, particularly HDL. A phospholipid-transfer protein in the plasma mediates this transfer. The particle dissociates from LPL (the process of lipolysis having taken perhaps a matter of minutes). It is known as a chylomicron remnant. It may undergo a further round of lipolysis in another tissue bed, but after loss of about 80% of its original content of TAG it will be removed rapidly from plasma. Shrinkage of the particle allows it to cross the endothelial barrier and interact with cell-surface receptors as described above. In addition, conformational changes to the apolipoproteins caused by the shrinkage of the particle expose different regions, which bind to specific cell-surface receptors, probably mainly LRP (see Section 7.2.4.3 above). The half-life of chylomicron-TAG in the circulation is about 5 min. The half-life of a chylomicron particle has been estimated at 13–14 min.

The exogenous pathway is regulated mainly by the rate of entry of dietary fat, but the tissue disposition of the dietary FAs is regulated by tissue-specific regulation of LPL activity. Adipose tissue LPL activity is up-regulated by insulin and therefore plays an important role in removal of dietary FAs in the period following a mixed meal (i.e. a meal that contains both fat and carbohydrate, since the latter will stimulate insulin secretion). The mechanism for activation of adipose tissue LPL by insulin is complex. Whilst insulin increases transcription of the LPL gene in adipose tissue, the major point of regulation in the short term (the hours following a meal) appears to be a diversion within the adipocyte of LPL between a degradative pathway, and secretion in an active, dimeric form for export to the endothelium. A key event may be dissociation of the active LPL dimer into inactive monomers, which are then degraded. Angiopoietin-like protein-4 (Angptl4, gene *ANGPTL4*) may carry out this role. Expression of the *ANGPTL4* gene in adipose tissue is highly regulated by nutritional state in the opposite way to LPL: it is increased in fasting. Since, like LPL, the enzymes of TAG biosynthesis in adipocytes are also activated in the fed state, the net effect is that dietary FAs tend to be readily deposited as adipocyte TAGs. In contrast, LPL in skeletal, and particularly cardiac, muscle is down-regulated by insulin, and increases in activity during fasting, and in skeletal muscle with exercise training. In these conditions, therefore, FAs tend to be directed to the tissues that need them for oxidation (Fig. 7.12).

The secretion of VLDL from the liver, and its further metabolism, is described as the *endogenous pathway* (Fig. 7.11). It is very similar initially to the exogenous pathway. VLDL particles also interact with LPL in capillaries and their core TAG is hydrolysed. The half-life of VLDL particles in the circulation is typically 12 h, but their fates are heterogeneous. As the particles lose their TAG through successive interactions with LPL, they also lose their surface apolipoproteins as described above, until they become LDL particles. However, along the way some of the particles are removed intact by endothelial VLDL-receptors. LDL particles have an even longer half-life in the circulation, typically 2–2.5 days, and they are removed by LDL-receptors on cell surfaces (see Section 7.2.4.3 & Fig. 7.10 above) in the subendothelial space. The major site for removal of LDL particles is normally the liver, although the LDL-receptor is a mechanism for delivering cholesterol to most cells. The expression of LDL-receptors is closely regulated to maintain cellular cholesterol homeostasis (see Section 7.3.1). Thus, the endogenous pathway is a means of delivering FAs to tissues via LPL and perhaps the VLDL-receptor, and cholesterol, via the LDL-receptor.

The third pathway is that of HDL metabolism (Fig. 7.13). As described above, HDL particles may not be secreted as such: rather, the basic components of apoAI and phosphoglycerides are secreted and the particles cycle through various stages in the plasma. If the LDL pathway of cholesterol delivery from liver to extrahepatic tissues is regarded as 'forward cholesterol transport', the HDL pathway represents 'reverse cholesterol transport' or the movement of cholesterol out of tissues and its transport to the liver and small intestine for ultimate excretion. These are the only mammalian organs that can excrete cholesterol from the body. The liver secretes cholesterol in the bile, both in unchanged form and after conversion to bile salts. Bile salt biosynthesis is highly regulated (see Section 7.3.3 below and Box 7.1). Excretion of cholesterol from the small intestinal mucosal cells was mentioned earlier (see Section 7.1.2). This is a newly recognized pathway known as trans-intestinal cholesterol efflux (TICE). Although its

(a)

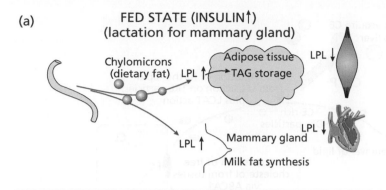

FED STATE (INSULIN↑)
(lactation for mammary gland)

Chylomicrons
(dietary fat)   LPL ↑

Adipose tissue
→TAG storage   LPL ↓

LPL ↑   Mammary gland
Milk fat synthesis

LPL ↓

(b)

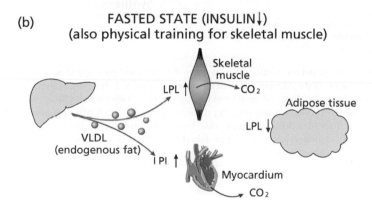

FASTED STATE (INSULIN↓)
(also physical training for skeletal muscle)

Skeletal
muscle
LPL ↑   →CO₂

Adipose tissue

LPL ↓

VLDL
(endogenous fat)

↓Pl ↑

Myocardium
CO₂

Fig. 7.12 Tissue-specific regulation of lipoprotein lipase (LPL) in relation to nutritional state. (a) In the fed state, insulin up-regulation of LPL activity directs the flow of dietary triacylglycerol-fatty acids to adipose tissue (and mammary gland during lactation). (b) In the fasting state, LPL in these tissues is down-regulated and endogenous (VLDL) triacylglycerol-fatty acids are directed more to heart and skeletal muscle. Physical activity up-regulates skeletal muscle LPL activity more than does fasting, however.

contribution to total cholesterol secretion is less than hepatic secretion (in mice TICE contributes around 30% of total cholesterol excretion), it is highly inducible, and is a promising target for drugs to lower plasma cholesterol concentrations.

As described earlier (see Section 7.2.3 and Fig. 7.8), immature lipid-poor HDL particles (pre-β1 HDL) interact with cell membranes to acquire unesterified cholesterol. This process is receptor-mediated. The receptor was discovered through identification of the gene defect in Tangier disease, a rare inherited disease in which cholesterol accumulates in tissues. The gene responsible for Tangier disease (see Section 10.5.2.1) encodes a protein that is a member of a family of cell-membrane-associated transporter proteins that have the property of binding ATP on their cytoplasmic side through an ATP-binding cassette (or protein 'motif'), ABC: this particular protein is known as ABCA1 (gene *ABCA1*). Hydrolysis of ATP can then power the transfer of substances across membranes. In the case of cholesterol efflux, ABCA1 seems to mediate

the transfer of cholesterol from cholesterol rafts in the cell membrane to the lipid-poor HDL particles (where it binds to apoAI). The unesterified cholesterol acquired by the particles becomes esterified with long- and very long-chain FAs through the action of the enzyme lecithin-cholesterol acyltransferase (LCAT), which is associated with HDL particles. This enzyme catalyses the transfer of a FA from PtdCho (originally called 'lecithin') to cholesterol to form a cholesteryl ester (Fig. 7.14). The phospholipid substrate for the reaction is present in the HDL particle. The hydrophobic cholesteryl ester moves to the core of the particle, which swells and becomes spherical as more cholesterol is acquired and esterified. The other product, lysoPtdCho, is transferred to plasma albumin from which it is rapidly removed from blood and reacylated. This is probably the origin of the bulk of HDL-cholesterol in plasma, although some is also acquired from the surface components of chylomicrons and VLDL particles when their TAG is hydrolysed by LPL, as described above.

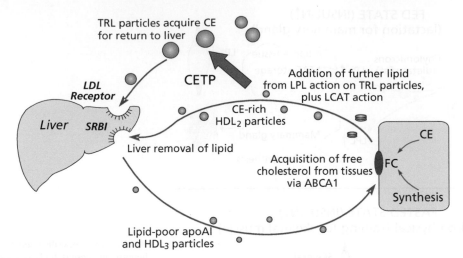

Fig. 7.13 The pathway of reverse cholesterol transport. For more details of HDL metabolism see Fig. 7.8. Lipid-poor apoAI acquires free cholesterol (FC) and phospholipid from peripheral tissues via the ATP-binding cassette protein-1 (ABCA1). The lipid content of the HDL particles is increased by transfer of additional cholesterol from the lipolysis of triacylglycerol-rich lipoprotein (TRL) particles. Esterification of this cholesterol by lecithin-cholesterol acyltransferase (LCAT) produces mature, lipid-rich HDL$_2$ particles. These can return their cholesterol to the liver in two ways. They may deliver it directly via a hepatic receptor (SR-BI) or they may exchange it for triacylglycerol from the TRL via the action of cholesteryl ester transfer protein (CETP). The TRL particles then carry this cholesterol to the liver for receptor-mediated uptake. CE, cholesteryl esters.

To complete the pathway of reverse cholesterol transport, HDL-cholesterol must reach the liver or small intestine. There are at least two routes by which this can occur. First, the SR-BI (see Section 7.2.4.4) is expressed in the liver and small intestine, and also in steroidogenic tissues (e.g. adrenal gland and ovary). 'Docking' of HDL particles with SR-BI is followed by off-loading their cholesteryl ester content. The cholesteryl esters enter the cellular pool and may be hydrolysed by lysosomal acid hydrolases as shown for LDL-receptor-mediated uptake (Fig. 7.10). This process is fundamentally different from the uptake of

LDL particles by the LDL-receptor, however, and has been called 'selective lipid uptake'. The difference is that the particle itself is not internalized and the cholesterol-depleted particle leaves the receptor to reenter the cycle of the HDL pathway.

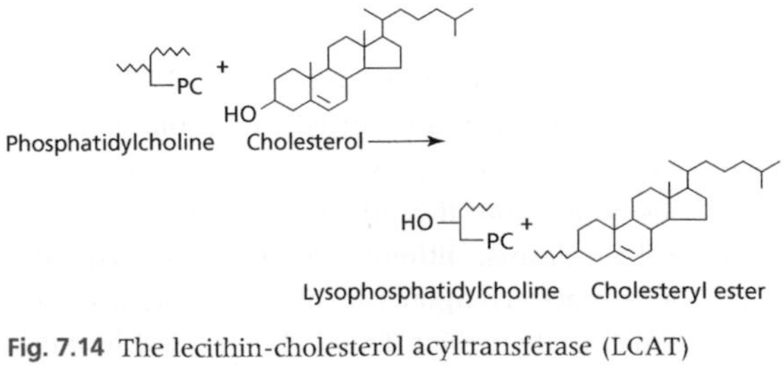

The second mechanism for delivery of HDL-cholesterol to the liver, and possibly intestine, brings us to an important way in which the lipoprotein pathways interact. Human plasma contains a protein known as cholesteryl ester transfer protein (CETP), which is secreted from the liver and from adipose tissue. CETP mediates the exchange of nonpolar lipids (TAGs and cholesteryl esters) between particles, according to concentration gradients. When HDL particles become enriched in cholesteryl esters, they may exchange these esters for TAGs carried in the TAG-rich lipoproteins, i.e. chylomicrons and VLDL. Thus, chylomicrons and VLDL acquire cholesterol that stays with the particle since it is not removed by LPL and it is eventually removed when the remnant particle (or LDL particle) is taken up by receptors, which, as we have seen, occurs mainly in the liver. Thus cellular cholesterol in extra-hepatic tissues, setting out for the liver in an HDL particle, may 'change vehicles' part way

Fig. 7.14 The lecithin-cholesterol acyltransferase (LCAT) reaction.

Phosphatidylcholine   Cholesterol →

Lysophosphatidylcholine   Cholesteryl ester

and end up being carried in a remnant particle. The HDL particles acquire TAGs in this process and these are hydrolysed by the enzyme hepatic lipase attached to the endothelial cells lining the hepatic sinusoids. What remains is a lipid-poor HDL particle, ready to recycle via acquisition of cellular cholesterol.

7.2.6 Lipid metabolism has many similar features across the animal kingdom, although there are some differences

Lipid and lipoprotein metabolism and their anatomical locations are different in groups of animals other than mammals. Insects have an organ known as the fat body, which combines the functions of the liver and adipose tissue. As well as detoxification and metabolic regulation, the fat body stores lipid as well as the carbohydrate glycogen, which is mobilized as trehalose (two glucose molecules linked by α 1–1 bonds). Fat is mobilized into the haemolymph in the form of DAGs rather than TAGs or NEFAs. DAGs are transported in lipoproteins, known as lipophorins. The major class of lipophorins has a density similar to mammalian HDL and is sometimes called HDLp. HDLp has two distinct nonexchangeable apolipoproteins, apolipophorin I and II, which are derived from cleavage of a precursor protein, apolipo-phorin II/I (apoLp-II/I). ApoLp-II/I is a member of the large family of lipid transfer proteins which includes mammalian apoB in addition to invertebrate and verte-brate vitellogenins (see below).

Egg yolk proteins fall into two major families, the vitellogenins (found in the gametes of some insects, many other invertebrates and oviparous vertebrates including most fish, amphibians, reptiles and birds) and the 'yolk proteins' found in many dipteran flies ('true flies'). The vitellogenins have homology to mam-malian apolipoprotein B (see above and Section 7.2.1) whilst the yolk proteins have homologies with the mam-malian lipase family that includes LPL.

Poikilothermic (cold-blooded) vertebrates (fish, amphibians and reptiles) generally store lipid in multiple tissues including liver and muscle as well as true adipose tissue. Adult amphibians and reptiles have fat bodies, usually paired masses of adipose tissue in the abdomen but in some reptiles adipose tissue is more widely distrib-uted including in the tail. Mechanisms of lipid deposition and mobilization, however, are basically similar to those in mammals, with transport of dietary fat in chylomicrons, and uptake mediated by LPL; mobilization of stored fat is in the form of NEFAs.

Mammalian species differ quantitatively in their pat-terns of lipoprotein metabolism. In some omnivores/herbivores, including man, guinea-pig and pig, lipopro-teins of the LDL type (in which apolipoprotein B pre-dominates) account for more than 50% of the total substances of density < 1.21 g/ml. These are the 'LDL mammals'. In the majority of mammals, however, HDLs are the predominant class and may account for up to 80% of plasma lipids. Herbivores, with the exception of guinea-pigs, camels and rhinos, and carnivores are 'HDL mammals'. It is worth noting that, although rats are often used for the study of lipid biochemistry, their lipoprotein pattern is of the HDL type and very different from that of man. Another distinct difference is that the rat liver secretes VLDL particles that contain either apoB48 or apoB100, whereas in humans the liver secretes only apoB100-containing particles. This almost certainly relates to the fact that rat plasma LDL-cholesterol con-centrations are relatively low, since apoB48-containing particles tend to be cleared from the circulation rapidly (as with chylomicrons in man). More generally, there are distinct differences between the lipid metabolism of ruminant and nonruminant mammals. In the former, most ingested carbohydrates are digested by microbes to short-chain FAs in the rumen, so glucose derived directly from the diet is limited; FAs and ketone bodies are more widely utilized as energy-producing fuels. Horses, rhinoceros and many rodents are not ruminants, but have abundant caecal and colonic microbes. Glucose is absorbed from the small intestine but short-chain FAs, absorbed from the colon, make a substantial contribution to fuel supply.

7.3 The coordination of lipid metabolism in the body

The flux of FAs through the pathways described in this chapter can vary enormously. The fat content of the diet varies in different parts of the world, between different people within one country, and even from day to day in one individual. The rate at which we transport FAs through the plasma varies considerably during a normal day. For NEFAs it is high at night during fasting, sup-pressed after meals, and stimulated to very high levels during aerobic exercise; for TAG-FAs it is roughly the converse. This requires coordination between metabolic pathways in different organs, in relation to both the influx of dietary lipids and the body's requirement for lipids. In Section 4.3 we saw how this coordination can be brought about by hormones, and in particular by insulin whose secretion from the pancreas is stimulated in the 'fed' state. Insulin acts both by affecting the activity of enzymes on a short-term basis (e.g. by reversible phosphorylation/dephosphorylation) and on a longer-term basis by regulation of gene expression. However, the main stimulus for insulin secretion is a rise in the plasma glucose concentration, not a change in lipid concentrations. It would seem sensible that the body should also have means for homeostasis of lipid metabolism that do not depend upon the simultaneous ingestion of carbohydrate in appropriate amounts.

Over the past few years there have been great advances in our understanding of how this is achieved. There are several interrelated systems that regulate gene expression to control the flux through various pathways of lipid metabolism, and whose regulators are themselves lipids or products derived from lipids.

7.3.1 The sterol regulatory element binding protein system controls pathways of cholesterol accumulation in cells and may also control fatty acid biosynthesis

The American biochemists and 1985 Nobel Prize winners, Mike Brown and Joseph Goldstein, showed that cellular cholesterol content is regulated by two parallel

mechanisms. When the content of unesterified cholesterol in cells increases, the expression of the LDL-receptor protein decreases. In addition, the key enzymes of cholesterol biosynthesis (hydroxymethylglutaryl [HMG]-CoA synthase, HMG-CoA reductase, squalene synthase, farnesyl diphosphate synthase; see Section 4.9.1) are repressed. Thus, any further increase in cellular cholesterol is minimized. Conversely, if the cellular unesterified cholesterol content falls, these pathways are activated (Fig. 7.15). The genes coding for all these proteins contain an upstream sequence known as the sterol regulatory element-1 (SRE-1). The protein that binds to SRE-1, and activates expression of the gene, is part of a larger protein called sterol regulatory element binding protein (SREBP). SREBP is normally localized in the ER. In the absence of cholesterol, a specific protease releases a peptide from SREBP, which migrates to the nucleus and binds to SRE-1, activating transcription. When cellular cholesterol concentrations are high, this protease is inhibited and gene expression is repressed. The natural regulator of this system appears not to be cholesterol itself but a hydroxylated derivative of cholesterol, an oxysterol. The working of the system is illustrated in Fig. 7.16.

There are three isoforms of SREBP: SREBP-1a, SREBP-1c and SREBP-2. They are members of a family of basic-helix-loop-helix-leucine zipper (bHLH-LZ) transcription factors. The first two arise from differential splicing of the transcript from one gene (*SREBF1*), whereas SREBP-2 is the product of a separate gene (*SREBF2*). SREBP-1a and 1c are concerned more with regulation of genes involved

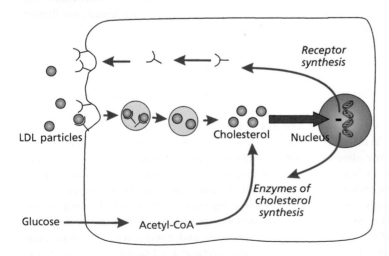

Fig. 7.15 Regulation of the cellular cholesterol content. When the content of unesterified cholesterol in cells increases, the expression of the LDL-receptor protein and of the enzymes of cholesterol biosynthesis is repressed.

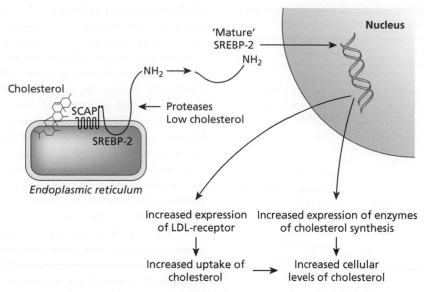

Fig. 7.16 The Sterol Regulatory Element Binding Protein (SREBP) system. The full-length SREBP protein is located in the endoplasmic reticulum (ER, a system of membranous cavities within the cytoplasm). It is associated with the SREBP cleavage-activating protein (SCAP), which 'senses' the level of cholesterol, or related sterols, within the membrane of the ER. When the cellular cholesterol content is low, the SCAP-SREBP complex migrates to the Golgi complex (not shown), where specific proteases cleave SREBP to release the N-terminal portion, 'mature' SREBP. Mature SREBP moves to the nucleus where it binds to sterol response elements in the promoter regions of many genes. SREBP-2 (as shown in the figure) mainly regulates genes concerned with cholesterol metabolism (LDL receptor, enzymes of cholesterol synthesis). SREBP-1 is regulated more by expression of the full-length protein (which is increased by insulin); its proteolytic cleavage is also stimulated by insulin independently of cholesterol. SREBP-1c increases the expression of genes concerned with fat storage (including acetyl-CoA carboxylase and fatty acid synthase). Reproduced from KN Frayn (2010) *Metabolic Regulation: A Human Perspective*, 3rd edn. Wiley-Blackwell, Oxford, with permission of John Wiley & Sons.

in FA metabolism including fatty acid synthase (FAS), acetyl-CoA carboxylase (ACC) and stearoyl-CoA (9-) desaturase (see Sections 3.1.3.1 & 3.1.8.1), whereas SREBP-2 regulates transcription of genes involved in cholesterol metabolism (see Section 4.9.1). SREBP-1c in adipose tissue is also known as adipocyte determination and differentiation factor-1, ADD-1.

Expression of SREBP-1c is itself under regulation by insulin. When insulin concentrations are high, more SREBP-1c is produced and FA biosynthesis is up-regulated. In physiological terms, it is probable that regulation is more relevant in times of fasting (low insulin), when SREBP-1c expression is severely reduced, and FA biosynthesis cannot be activated by the SREBP pathway. SREBP-1c acts coordinately with the carbohydrate response element binding protein, ChREBP, which responds to sugars.

Some of the genes whose expression is regulated by the SREBP system are shown in Table 7.3.

The SREBP system has been manipulated to alter cholesterol metabolism. Certain fungal metabolites, compactin from *Penicillium* spp. and mevinolin from *Aspergillus terreus*, were found in the 1970s to inhibit HMG-CoA reductase. Synthetic derivatives of these molecules are now widely used as drugs (the 'statins') to lower plasma cholesterol concentrations (see also Section 10.5.2.2). The main mode of action is that inhibition of hepatic cholesterol biosynthesis reduces hepatocyte unesterified cholesterol concentrations and this results in up-regulation of hepatic LDL-receptor expression via the action of SREBP-2. Hence more LDL particles are removed from the circulation and excess cholesterol may be excreted.

Table 7.3 Genes whose expression is increased by the sterol regulatory element binding protein (SREBP) system.

Gene	Function in cell
SREBP-1c	
Glucokinase	Glucose metabolism
Acetyl CoA carboxylase	*De novo* fatty acid biosynthesis
Fatty acid synthase	
ATP-citrate lyase	
Malic enzyme	
Stearoyl CoA desaturase	Synthesis of monounsaturated fatty acids
Glycerol phosphate acyl transferase	Triacylglycerol and phospholipid synthesis
Lipoprotein lipase	Import of lipoprotein-triacylglycerol fatty acids
PPAR-γ	Induction of adipocyte differentiation
SREBP-2	
LDL receptor	Import of LDL particles
HMG-CoA synthase	
HMG-CoA reductase	
Squalene synthase	*De novo* cholesterol biosynthesis
Farnesyl diphosphate synthase	
Lanosterol 14α demethylase	

Activation of the SREBP-1c system involves insulin and in the case of lipogenic enzymes, SREPB-1c acts in concert with the carbohydrate response element binding protein, ChREBP. Activation of SREBP-2 occurs in response to low cellular cholesterol content.

Based on H Shimano (2009) SREBPs: physiology and pathophysiology of the SREBP family. *FEBS J* **276**: 616–21 and H Shimano (2001) Sterol regulatory element-binding proteins (SREBPs): transcriptional regulators of lipid synthetic genes. *Prog Lipid Res* **40**: 439–52.

7.3.2 The peroxisome proliferator-activated receptor system regulates fatty acid metabolism in liver and adipose tissue

A number of apparently unrelated chemicals, such as pesticides, can induce the proliferation of the subcellular oxidative organelles called peroxisomes, especially in the livers of rodents. These agents were suggested to work through a common receptor called the peroxisome proliferator-activated receptor (PPAR). It is now realized that there is a family of PPARs and that a major role is to regulate FA metabolism. Like the SREBP system, the

PPARs are activated by lipids (in this case, FAs or their derivatives) and regulate the disposal of FA in different tissues. PPARs belong to the superfamily of nuclear receptors (see Section 6.2.3.1 & Box 6.1), and are considered members of group C of the subfamily 1 of the superfamily. In the case of PPARs, after binding their ligand, they must dimerize with another receptor, the retinoid X receptor (RXR), which must itself be activated by binding 9-*cis*-retinoic acid (see Section 6.2.3.1). The regulatory purpose of this dimerization is not known. The dimer, in complex with other regulatory proteins, then binds to peroxisome proliferator response elements (PPREs), in the promoter region of genes concerned with lipid metabolism (Table 7.4).

PPAR-α (gene *PPARA*) is expressed mainly in the liver. In rodents, its major function is to increase fat oxidation via proliferation of peroxisomes, and hence increased peroxisomal FA oxidation. In humans, this effect is minor or nonexistent. Instead, PPAR-α activation increases mitochondrial fatty acid oxidation and also up-regulates the secretion of the apolipoproteins forming HDL (apoAI, apoAII), thus raising HDL-cholesterol concentrations in plasma. Activation of PPAR-α by pharmacological agents (discussed below) reduces plasma TAG concentrations. In rodents this effect appears to involve up-regulation of hepatic expression of LPL, which is normally suppressed in the liver after birth. There is no evidence for this mechanism in humans, although the TAG-lowering effect is marked. Suppression of the expression of apoCIII may be involved, as apoCIII has been suggested to inhibit LPL and the hepatic uptake of TAG-rich remnants particles by hepatic receptors. Some genes, whose regulation is altered by PPAR-α, are listed in Table 7.4. In general, hepatic PPAR-α activation can be seen to increase FA oxidation and hence reduce excess lipid load in the body.

PPAR-δ or PPAR-β refer to the same protein, also known as NUC-1 and FAAR (fatty acid-activated receptor) (gene *PPARD*). PPAR-δ is ubiquitously expressed but a major site of expression, which distinguishes it from the other PPARs, is skeletal muscle (also small intestine and colon). PPAR-δ is more highly expressed in oxidative type I muscle fibres than in glycolytic type II muscle fibres. Activation of PPAR-δ has a number of effects that all tend to increase muscle fat oxidation. It increases the number of oxidative type I muscle fibres and activates the expression of most genes involved in FA oxidation, including those of mitochondrial biogenesis. Mice with muscle-specific overexpression of PPAR-δ have greatly

Table 7.4 Peroxisome proliferator-activated receptors (PPARs): tissue distribution and effects of activation.

Receptor	Other names	Main tissue distribution	Genes whose expression is increased by PPAR activation	Genes whose expression is suppressed by PPAR activation
PPAR-α		Liver (main site) Also kidney, heart, muscle, brown adipose tissue	Apolipoprotein AI Apolipoprotein AII Enzymes of peroxisomal fatty acid oxidation Liver FABP CPT-1 Enzymes of mitochondrial fatty acid oxidation	Apolipoprotein CIII
PPAR-δ	PPAR-β, NUC 1, FAAR (fatty-acid activated receptor)	Widespread, especially skeletal muscle, adipose tissue, liver, intestine	Enzymes of mitochondrial fatty acid oxidation. HDL concentrations increase with activation perhaps of ABCA1.	Not known
PPAR-γ1		Widespread at low levels		
PPAR-γ2		Adipose tissue	Factors involved in adipocyte differentiation Adipose tissue FABP (also known as aP2) Lipoprotein lipase Fatty acid transport protein Acyl CoA synthase GLUT4 Phosphoenolpyruvate carboxykinase Adiponectin (protein secreted by adipocytes)	Leptin

Collated from several sources. See **Further reading** for more information on PPARs.

increased exercise capacity and have been nicknamed 'marathon mice', because of increased capacity to oxidize FAs in skeletal muscle.

The third PPAR isoform, PPAR-γ (gene *PPARG*), is expressed mainly in adipose tissue and its major role in that tissue is the regulation of fat storage. Activation of PPAR-γ increases the expression of genes involved in adipocyte differentiation from precursor cells (pre-adipocytes) as well as increased FA deposition by up-regulation of the expression of genes regulating fat storage. These include components of the pathway of glucose uptake and biosynthesis, GLUT4 and phospho-enolpyruvate carboxykinase (an enzyme of gluco-neogenesis). Presumably their function is to increase availability of G3P for TAG biosynthesis. Therefore, in times of excess FA availability, this system ensures the coordinated storage of the excess FAs in adipose tissue, both by recruitment of new adipocytes and by up-regulation of fat storage pathways.

There is cross-talk between the SREBP and PPAR-γ pathways. One of the SREBP isoforms, SREBP-1c (ADD-1), is expressed in adipocytes. As described above,

SREBP-1c activation in adipose tissue increases expression of FAS and thus potentially the generation of ligands for PPAR γ (see below). There is also a more direct link in that SREBP-1 activation directly increases PPAR-γ expression.

The coordinated working of the PPAR-α, -δ and -γ systems is illustrated in Fig. 7.17. All can be seen to increase the disposal (either by oxidation or by storage) of excess FAs. This important action underlies the importance of the PPARs as pharmacological targets. Just as the SREBP system has been manipulated pharmacologically to alter cholesterol metabolism, so the PPAR system is the target for clinically useful drugs. A group of agonists of the PPAR-α receptor, known as fibric acid derivatives or fibrates, are useful drugs in the treatment of patients with elevated plasma TAG concentrations and low HDL-cholesterol concentrations. Their effects are apparently brought about by up-regulation of LPL and hepatic FA oxidation, thus allowing the disposal of excess FAs and also the secretion of apolipoproteins AI and AII, leading to increased HDL-cholesterol concentrations. No drugs are in clinical use to target PPAR-δ but such agents have

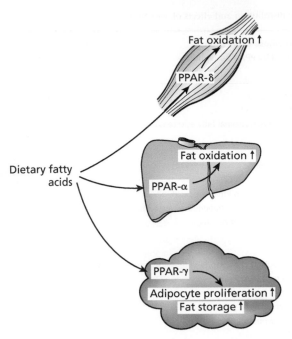

Fig. 7.17 The peroxisome proliferator-activated receptor (PPAR) system and regulation of fatty acid disposition. This is a simplification because expression of the PPAR isoforms is not so tissue-unique: e.g. PPAR-δ is also expressed, and regulates fatty acid oxidation, in liver.

been developed and tested in man. Generally they seem to improve the lipid profile and act against so-called 'metabolic syndrome' (see Section 10.5.2.4), with evidence that this reflects increased oxidation of FAs in muscle and perhaps liver. PPAR-γ activators have been in use for many years in the treatment of type 2 (maturity-onset) diabetes, which is usually associated with obesity. In this condition, the ability of tissues to respond to insulin is reduced ('insulin resistance'). PPAR-γ activators are known as thiazolidinediones or 'glitazones', and they improve the sensitivity of the body to insulin. A strong consensus view would be that this improvement comes about at least in part because of the induction of differentiation of new adipocytes, which can then sequester excess FAs, hence removing fat (TAG) from other tissues such as liver, skeletal muscle and pancreatic islets (see Section 10.4.1.3 for further discussion).

There has been much debate, however, about the endogenous ligands for the PPARs. They are activated by unsaturated long-chain FAs (less strongly by SFA) and by a number of other lipids. The actual ligand or ligands may be a metabolic derivative of a FA such as a member of the prostaglandin family. One candidate is 15-deoxy-12,14-prostaglandin J_2, which has been shown to be particularly active in cellular systems where PPARs are used as reporter genes. Another suggestion is that the endogenous ligand of PPARs is a molecular species of PtdCho, 1-palmitoyl-2-oleoyl-*sn*-glycerol-3-phosphocholine. It seems very likely that the natural ligand is generated intracellularly. Mice with liver-specific deletion of FAS suffer from metabolic abnormalities (especially impaired tolerance of fasting) similar to those seen in PPAR-α deficient mice and these changes are reversed when the mice are given a PPAR-α agonist. Similarly, mice with a heart-specific deletion of adipose triglyceride lipase (ATGL, named before TAG became the accepted term) suffer from impaired cardiac function, which is reversed when they are given a PPAR-α agonist. The suggestion in both cases is that the pathways concerned (*de novo* lipogenesis [DNL] and TAG hydrolysis, respectively) normally generate the ligands for activation of PPAR-α.

7.3.3 Other nuclear receptors that are activated by lipids regulate hepatic metabolism

Nuclear receptors were described briefly above and in previous sections (see Section 6.2.3.1 & Box 6.1). Two that are important in liver metabolism are known as FXR (the farnesoid-activated receptor) and LXR (liver X-receptor). Both heterodimerize with the retinoid X-receptor, RXR, as described for the PPARs (see Section 7.3.2).

FXR is misleadingly named as its main ligands are bile acids, such as chenodeoxycholic acid. FXR may be thought of as the endogenous bile acid sensor. Amongst pathways regulated by FXR is the pathway for bile acid biosynthesis from cholesterol (see Section 4.9.6). A key enzyme in this pathway is cholesterol 7-α hydroxylase, CYP7A1 (Box 7.1). CYP7A1 expression is suppressed by FXR activation, providing a means of regulating cholesterol and bile acid homeostasis. FXR also regulates many aspects of hepatic metabolism. FXR-deficient mice have elevated plasma lipid concentrations, which is somewhat counter-intuitive but probably reflects the fact that bile acids themselves play important roles in regulating lipid metabolism. FXR-deficient mice also tend to be insulin-resistant, leading to the idea that FXR could be a pharmacological target to improve lipid and glucose metabolism, an area that is being intensively pursued.

LXR was so named when it had been cloned but its ligand was not known. It is now known to bind oxysterols, oxidized metabolites of cholesterol, and is thus a cholesterol sensor. There are two isoforms, LXRα and LXRβ, produced from separate genes (*NR1H3*, for nuclear receptor subfamily 1, group H, member 3 and *NR1H2* respectively). LXRα is mainly expressed in the liver and to a smaller extent in small intestine and adipose tissue, whilst LXRβ is expressed ubiquitously. A major effect of hepatic LXRα activation in rodents is to up-regulate the CYP7A1 pathway of bile acid biosynthesis described in Box 7.1: thus, cholesterol is diverted into bile acid biosynthesis and may be excreted from the body. LXRα-deficient mice have, accordingly, high 'toxic' levels of cholesterol in the liver. However, this mechanism does not seem to operate in humans. LXR activation also increases expression of a number of other components of the 'reverse cholesterol transport' pathway (see Section 7.2.5), including the cholesterol transporter ABCA1. However, the therapeutic potential of LXR activation is limited by the fact that LXR also induces lipogenic pathways, in part directly and in part by increased expression of SREBP-1c (see Section 7.3.1). This can lead to excessive fat accumulation in the liver.

7.3.4 G protein-coupled receptors activated by lipids

Nuclear receptors, described above, reside within the cell: they may bind their ligands in the cytosol but their action is in the nucleus. An obvious feature of these receptors is that their ligands must be able to enter the cell, either by diffusion across the cell membrane, or, as in most cases, by carrier-mediated transport. In contrast, there is a large family of receptors that reside in the cell membrane and are placed to respond to extracellular signals. These have traditionally been thought of as receptors for hormones: for instance, the receptors for insulin, glucagon and adrenaline (US: epinephrine) all reside on the cell surface. But increasingly we are recognizing that many cell-surface receptors also respond to small molecules that we would think of more as 'metabolites' than hormones.

One family of such receptors is known as the G protein-coupled receptor (GPCR) family. GPCRs constitute the largest family of receptors, with around 800 GPCRs encoded in the human genome, of which the ligands for about 200 have been identified. GPCRs have a structure comprising an extracellular domain (the N-terminal domain), then seven trans-membrane domains,

giving this family of receptors the alternative name 7TMs (7 transmembrane), with a cytosolic C-terminal region. Ligands bind to the extracellular domain comprising the N-terminal domain and the extracellular loops connecting the membrane-spanning regions, although receptors for small molecules such as adrenaline have a short N-terminal domain, and the ligand-binding pocket is deep within the plane of the membrane.

GPCRs are so-named because their intracellular (C-terminal) domains interact with one of a family of proteins, the heterotrimeric G proteins that bind and hydrolyse guanosine trisphosphate (GTP). G-proteins themselves are formed of three subunits, one of which (the α subunit) may bind and hydrolyse GTP to GDP. When the ligand is bound, a conformational change in the GPCR leads the α subunit to exchange its GDP for GTP, which is then hydrolysed, returning the α subunit to an inactive state: but in this process, the G protein interacts with further signalling molecules. There are three major forms of G protein that interact with different signal chains. Gs *stimulates* adenylate cyclase, which produces cyclic AMP. Gi, in contrast, *inhibits* adenylate cyclase, whilst Gq activates phospholipase C (PLC, see Sections 4.6.4 & 8.1.3). In the mammalian photoreceptor there is a further form, 'transducin' (Gt) (see Section 6.2.3.1).

Several GPCRs are activated by FAs and other lipids. Some of these are described in Table 7.5. Several of them are involved in sensing fat, as FAs, in the gastrointestinal tract and enhancing the secretion of gut hormones such as glucagon-like peptide 1 (GLP-1), which in turn acts at the pancreatic β-cell to enhance insulin secretion. GPR40 is expressed in the β-cell and fulfils the same purpose. Several of them present attractive therapeutic targets and are being actively pursued in the pharmaceutical industry.

7.3.5 Adipose tissue secretes hormones and other factors that may themselves play a role in regulation of fat storage

The role of white adipose tissue as an organ of TAG storage will be described further in Section 9.1.1. In recent years it has become clear that adipocytes are also active in secretion of a number of compounds that may in turn help to regulate both lipid metabolism and other functions. Some of these are listed in Table 7.6. Most of these are proteins or peptides, and some clearly true hormones, but they will be described briefly here in

Table 7.5 G Protein Coupled Receptors (GPCRs) activated by lipids and related molecules.

GPCR number	Other names	Gene name	Ligand	Tissue expression (major tissues)	Physiological role and comments
GPR40	FFA1, FFAR1 (free fatty acid receptor 1)	FFAR1	FAs with chain-lengths in the range 12C-16C	Pancreatic β-cells	Potentiates glucose-stimulated insulin secretion
GPR41	FFA3, FFAR3	FFAR3	Short-chain FAs	Adipose tissue, GI tract (enteroendocrine cells)	Stimulation of leptin production; stimulation of gut hormone secretion
GPR43	FFA2, FFAR2	FFAR2	Short-chain FAs	Adipose tissue, GI tract (enteroendocrine cells)	Adipogenesis, reduction of lipolysis; stimulation of gut hormone secretion
GPR109A	HM74A, NIACR1	HCAR2 (hydroxycarboxylic acid receptor 2)	Probably 3-hydroxybutyrate (ketone body)	Adipocytes	Identified initially as the receptor for nicotinic acid (a component of the B-vitamin niacin), used in large doses to treat high triacylglycerol concentrations. Only known metabolic function is to suppress adipocyte lipolysis. Since 3-hydroxybutyrate is a product of hepatic fatty acid oxidation, this could provide a feedback loop.
GPR119	Oleoylethanolamide receptor	GPR119	Oleoylethanolamide and other lipids containing oleic acid, e.g. 2-oleoyl glycerol	Pancreatic β-cells, GI tract	Oleoylethanolamide has appetite-suppressing activity (although not entirely via GPR119). It is related to the endogenous cannabinoids (Section 4.6.6). 2-monoacylglycerol stimulation in the GI tract may enhance GLP-1 secretion (together with GPR40).
GPR120		FFAR4	n-3 PUFAs	Macrophages, GI tract, adipose tissue, brain (hypothalamus)	Has been suggested to modulate anti-inflammatory effects of n-3 PUFAs. Human genetic variation associated with obesity and insulin resistance.
GPR131	GPBAR1 (G protein-coupled bile acid receptor 1), TGR5	GPBAR1	bile acids	Liver, adipose tissue, intestine, gall bladder	Regulates gall bladder filling with bile, gut motility, and secretion of GI tract hormones.
	CB_1, CB_2	CNR1, CNR2	Endogenous cannabinoids	CB_1: brain, neurons. CB_2, immune cells. Also, both: adipose tissue and muscle.	Respond to endogenous cannabinoids (Section 4.6.6) released from nearby cells (acting in a paracrine or autocrine fashion). Activation of CB_1 was shown to reduce obesity but had unwanted side-effects.
GPR26 (LPA_1), GPR23 (LPA_4), GPR92 (LPA_5)	$LPAR_{1-6}$	LPAR1 – LPAR6	LPA	Widespread, including blood and immune cells, cells of blood vessel walls, fibroblasts	There may be nine LPARs in total. They respond to LPA produced locally from lysophosphatidylcholine. Biological effects include immune activation and many others. (See Section 8.3).
	$S1P_{1-5}$	S1PR1 – S1PR5	Sphingosine 1-phosphate	Widespread, including vascular and immune cells.	S1PRs respond to sphingosine 1-phosphate produced locally (intracellularly) by the phosphorylation of sphingosine, derived from the deacylation of ceramide. Widespread biological effects (see Section 8.3).

GI tract, gastrointestinal tract; LPA, Lysophosphatidic acid.
When each GPCR was first discovered, it was given a sequential number. However, as their ligands have been identified, GPCRs have mostly been given other names relating to their function. Some have never been given numbers.
Note also that most receptors for eicosanoids are GPCRs (not listed here).

Table 7.6 Proteins and other factors secreted by adipocytes

Secreted product	Potential role
Proteins	
Lipoprotein lipase	Transport to endothelium; uptake of circulating triacylglycerol fatty acids
Complement components D (adipsin), B, C3	Formation of C3a-desarg or acylation stimulating protein: stimulation of fatty acid storage
Cholesteryl ester transfer protein (CETP)	Systemic participation in reverse cholesterol transport; adverse role in generation of atherogenic lipoprotein phenotype in obesity/insulin resistance. See Section 10.5.2.4.
Apolipoprotein-E	Lipoprotein metabolism
Angiotensinogen	Regulation of blood pressure (after conversion to angiotensin-II)
Adiponectin	Protein secreted that appears to act as a hormone, increasing sensitivity to insulin and improving cardiovascular risk profile. Secretion decreases as adipose stores expand.
Leptin	Hormone signalling size of fat stores, playing widespread role in metabolic regulation. See Section 10.4.1.1.
Retinol binding protein 4 (RBP4)	Binds vitamin A (see Section 6.2.3.1). Appears to act on liver to produce adverse metabolic profile.
Plasminogen activator inhibitor-1 (PAI-1)	Increases coagulability of blood
Tissue factor	Coagulation pathway
Tumour necrosis factor-α	Cytokine, perhaps reducing fat storage; systemically, may reduce sensitivity to insulin (although there is some doubt about adipose tissue's contribution to systemic concentration)
Interleukin-6	Cytokine with role in inflammatory processes
Other factors	
Prostaglandins PGE$_2$, PGI$_2$ (prostacyclin)	Probably local role in regulation of blood flow and lipolysis
Non-esterified fatty acids, glycerol	Products of lipolysis; distribution to other tissues

The list is not exhaustive and is expanding all the time.

so far as they relate to lipid metabolism. Because of their similarity in function to cytokines, hormones secreted from cells of the immune system, they are collectively called adipokines.

The hormone leptin was discovered in 1994 in the laboratory of Jeffrey Friedman at Rockefeller University in New York, by positional cloning of the ob gene responsible for gross, spontaneous obesity in a mutant strain of laboratory mice. Mice homozygous for the mutation (ob/ob mice) show the obese phenotype. Leptin is expressed almost exclusively in white adipose tissue. It carries a signal from adipose tissue to the brain that the fat stores are enlarged and that it is appropriate to reduce food intake and increase energy expenditure. It also signals to other organs and tissues, and widespread roles of leptin in regulation of energy metabolism, as well as reproductive and immune status, have been shown. Leptin signals to the reproductive system that the fat stores are sufficient to support pregnancy and birth. This probably explains the observation that extreme thinness (arising, for instance, either from eating disorders or from intense physical training) is associated with amenorrhoea (absence of menstruation). Although the main factor regulating leptin secretion from the adipocyte is

cell size (i.e. TAG content), there is additional modulation by the short-term feeding state. This appears to be mediated largely by insulin, which acts through ADD-1 (or SREBP-1c, see Section 7.3.1), to increase leptin expression. Thus, leptin secretion increases on feeding (acting to reduce further food intake) and is suppressed during fasting (so acting to increase food intake). The leptin system also operates in humans, although it became clear soon after the discovery of leptin that the vast majority of obese humans have high, not low plasma leptin concentrations in their circulation, as appropriate for their enlarged fat stores. Thus, the majority of human obesity does not result from a failure of adipose tissue to secrete leptin, but from a failure of the brain to respond to it. Clinical trials of recombinant leptin as a treatment for human obesity have been disappointing, attributed generally to a phenomenon of 'leptin resistance'. Low leptin concentrations are probably a more important signal than high levels. A small number of individuals has been identified, who carry a mutation rendering leptin ineffective and show remarkable obesity from a very early age. Some of these people have been treated with recombinant human leptin, which has normalized their eating behaviour and led for the first time to weight loss. See Section 10.4.1.1 for further discussion of this condition.

Other proteins secreted from adipocytes that have a role in lipid metabolism include LPL (secreted for transport to the capillary endothelium), apolipoprotein E and CETP. Adipocytes also secrete a number of peptide components of the complement system, which is important in host defence against infection. These include the factors known as D, B and C3, all components of the alternative (as distinct from the classic) complement pathway. Factors D (also known as adipsin, to reflect its origin), B and C3 interact to produce a peptide of 76 amino acids that is known to immunologists as C3a-desarg (formed by removal of a C-terminal arginine from the fragment C3a). C3a-desarg has been identified as a potent stimulator of FA esterification in adipocytes and termed acylation stimulating protein (ASP). Production of ASP by adipocytes is stimulated by the presence of chylomicrons. This suggests a coordinated system of regulation, whereby the arrival in adipose tissue capillaries of dietary fat in the form of chylomicron-TAG triggers ASP secretion, which in turn stimulates the storage of dietary FAs as adipocyte-TAG after their release from chylomicrons by LPL

KEY POINTS

- Dietary fat consists mainly of TAGs with some cholesterol and phospholipids (Chapter 6). In order for these compounds to be used by the body, they must be digested within the intestinal tract, absorbed into the epithelial cells (enterocytes) lining the intestine, secreted into the bloodstream, and distributed appropriately to tissues.

Digestion and absorption of dietary lipid species

- Digestion involves partial hydrolysis of TAGs, mainly by pancreatic lipase, producing 2-MAGs and NEFA. Cholesteryl esters are hydrolysed by carboxyl ester lipase to produce free cholesterol and NEFA and phosphoglycerides by a specific PLA_2 which produces lysophospholipids and, again, NEFA.

- These products are 'solubilized' or emulsified with emulsifiers that are partly the products of hydrolysis (e.g. MAGs, lysophospholipids, long- and very long-chain FAs), but also the bile salts produced from cholesterol in the liver.

- Absorption of the components of the emulsion droplets, or mixed micelles, is mainly by specific transporter mechanisms. Specific FA transporter proteins are involved. Cholesterol absorption is partial (around 50%) and complex, involving transport into the enterocyte followed by later efflux of some cholesterol and also other plant-derived phytosterols; each step involves specific transporter proteins.

- Within the enterocytes the products of digestion are recombined into multimolecular complexes, with specific proteins (apolipoproteins), known as lipoprotein particles. In the enterocytes these are known as chylomicrons and are the largest and most lipid-rich (mainly TAGs) of the various classes of lipoprotein particles in the circulation.

- Chylomicrons enter the circulation via the lymphatic drainage of the small intestine (thus bypassing the liver). Some other products of digestion, especially the short- and medium-chain FAs, enter the hepatic portal blood directly and reach the liver. Chylomicrons also carry fat-soluble vitamins and other lipid-soluble components of the diet (e.g. xenobiotics).

Pathways of lipid transport around the body

- FAs are delivered to tissues from chylomicron-TAGs by the action of LPL in the capillaries of the tissues. LPL is regulated in a tissue-specific manner: it is activated in adipose tissue after a meal, so delivering dietary FAs for storage. In skeletal muscle, it is activated after endurance exercise. This pathway for delivery of dietary FAs to tissues is known as the exogenous pathway of lipoprotein metabolism.

- The liver also secretes lipoprotein particles, VLDL, that, like chylomicrons, carry mainly TAGs. These also deliver FAs to tissues via LPL. This pathway is known as the endogenous pathway of lipoprotein metabolism.

- Cholesterol (mainly as long-chain cholesteryl esters) is another component of VLDL particles. As they lose their TAGs through the action of LPL, VLDL become relatively richer in cholesterol and eventually become smaller particles which are carriers mainly of cholesterol. These are the LDL that can deliver cholesterol to tissues via a specific receptor, the LDL receptor.

- There is also a pathway for removal of excess cholesterol from tissues, for excretion by the liver (in the bile) or through the small intestinal enterocytes. This involves the lipoprotein particles, the HDL, which mainly carry cholesterol and phospholipids. The pathway is known as reverse cholesterol transport.

- The pathways described above operate in mammals. In other animals the pathways may be related but often differ in specifics. For instance, in insects lipid transport between cells is in the form of DAGs rather than TAGs.

Coordination of lipid metabolism in the body

- Pathways of lipid metabolism are highly regulated. The cholesterol content of cells is tightly controlled through the operation of the SREBP system, which regulates mRNA expression of enzymes of cholesterol synthesis, and of the LDL receptor.

- FA metabolism is regulated on a short-term (hour to hour) basis largely by insulin, but over longer periods by a system of nuclear receptors/transcription factors known as PPARs.

- Other nuclear receptors (especially LXR and FXR) regulate bile salt synthesis from cholesterol and other aspects of lipid metabolism mainly in the liver. There are also many GPCRs that respond to FAs and other lipid species and regulate specific aspects of lipid metabolism.

- Adipose tissue itself secretes hormones and other peptides regulating, or involved in, lipid metabolism. A key example is the hormone leptin, whose secretion from adipocytes is increased as they increase in size and number, and signals to the brain to reduce appetite, hence creating a feedback loop to regulate body fat stores.

Further reading

Digestion and absorption

Abumrad NA & Davidson NO (2012) Role of the gut in lipid homeostasis, *Physiol Rev* **92**:1061–85.

Armand M (2007) Lipases and lipolysis in the human digestive tract: where do we stand?, *Curr Opin Clin Nutr Metab Care* **10**:156–64.

Azer SA (2004) Do recommended textbooks contain adequate information about bile salt transporters for medical students?, *Adv Physiol Educ* **28**:36–43.

Davis HR Jr & Altmann SW (2009) Niemann-Pick C1 Like 1 (NPC1L1) an intestinal sterol transporter, *Biochim Biophys Acta* **1791**:679–83.

Soupene E & Kuypers FA (2008) Mammalian long-chain acyl-CoA synthetases, *Exp Biol Med (Maywood)* **233**:507–21.

Lipid transport

Coleman RA & Mashek DG (2011) Mammalian triacylglycerol metabolism: synthesis, lipolysis, and signaling, *Chem Rev* **111**:6359–86.

Kzhyshkowska J, Neyen C & Gordon S (2012) Role of macrophage scavenger receptors in atherosclerosis, *Immunobiology* **217**:492–502.

Rosenson RS, Brewer HB Jr, Davidson WS *et al.* (2012) Cholesterol efflux and atheroprotection: advancing the concept of reverse cholesterol transport, *Circulation* **125**:1905–19.

Sundaram, M & Yao Z (2012) Intrahepatic role of exchangeable apolipoproteins in lipoprotein assembly and secretion, *Arterioscler Thromb Vasc Biol* **32**:1073–8.

Xiao C, Hsieh J, Adeli K & Lewis GF (2011) Gut-liver interaction in triglyceride-rich lipoprotein metabolism, *Am J Physiol Endocrinol Metab* **301**:E429–E446.

Coordination of lipid metabolism

Blad CC, Tang C & Offermanns S (2012) G protein-coupled receptors for energy metabolites as new therapeutic targets, *Nat Rev Drug Discov* **11**:603–19.

Ehrenborg E & Krook A (2009) Regulation of skeletal muscle physiology and metabolism by peroxisome proliferator-activated receptor delta, *Pharmacol Rev* **61**:373–93.

Feige JN, Gelman L, Michalik L, Desvergne B & Wahli W (2006) From molecular action to physiological outputs: peroxisome proliferator-activated receptors are nuclear receptors at the crossroads of key cellular functions, *Progr Lipid Res* **45**:120–59.

Ferré P & Foufelle F (2007) SREBP-1c transcription factor and lipid homeostasis: clinical perspective, *Horm Res* **68**:72–82.

Jeon T-I & Osborne TF (2012) SREBPs: metabolic integrators in physiology and metabolism, *Trends Endocrinol Metab* **23**:65–72.

Kalaany NY & Mangelsdorf DJ (2006) LXRS and FXR: the yin and yang of cholesterol and fat metabolism, *Annu Rev Physiol* **68**:159–91.

Miyauchi S, Hirasawa A, Ichimura A, Hara T & Tsujimoto G (2010) New frontiers in gut nutrient sensor research: free fatty acid sensing in the gastrointestinal tract, *J Pharmacol Sci* **112**:19–24.

Mutoh T, Rivera R & Chun J (2012) Insights into the pharmacological relevance of lysophospholipid receptors, *Br J Pharmacol* **165**:829–44.

Oosterveer MH, Grefhorst A, Groen AK & Kuipers F (2010) The liver X receptor: control of cellular lipid homeostasis and beyond: Implications for drug design, *Prog Lipid Res* **49**:343–52.

Porez G, Prawitt J, Gross B & Staels B (2012) Bile acid receptors as targets for the treatment of dyslipidemia and cardiovascular disease, *J Lipid Res* **53**:1723–37.

Shimano H (2009) SREBPs: physiology and pathophysiology of the SREBP family, *FEBS J* **276**:616–21.

Wahli W & Michalik L (2012) PPARs at the crossroads of lipid signaling and inflammation, *Trends Endocrinol Metab* **23**:351–63.

CHAPTER 8
Lipids in transmembrane signalling and cell regulation

Until recently, lipids were seen as serving three main biological functions. They are the core structural components of cellular membranes, they are versatile energy stores and they provide subcutaneous insulation in warm-blooded animals. It is clear from the early chapters of this book that cells make many different types of membrane phospholipids. But why do they do this, since most of the membrane proteins that have been critically examined can function efficiently in lipid bilayer membranes that contain only a few types of amphiphilic lipids?

The last few decades have seen a radical change in this view of membrane phospholipids as being fairly inert. First it became clear that rapid metabolic turnover of membrane lipids is a normal part of the homeostatic maintenance of membranes (see Sections 4.5–4.9 & 5.4). And then it was gradually realized that some membrane lipids – and their metabolites – are amongst the most important and dynamic regulators of cell behaviour and of membrane trafficking processes within cells. Some lipids are extracellular signals that control cell behaviour, some participate in the cellular processes by which cells respond to extracellular signals, and some contribute to the identities of intracellular compartments and to membrane traffic between these compartments.

First it was discovered that prostaglandins and other metabolites of arachidonic acid (20:4, *n*-6) and of other polyunsaturated fatty acids (PUFAs) are crucial regulators in the reproductive, inflammatory and circulatory systems (see Section 3.5). This led to a Nobel Prize for Bengt Samuelsson, Sune Bergström and John Vane in 1982, and whole 'new' families of biologically active fatty acid (FA) metabolites, such as resolvins and protectins, are still being discovered (see Section 3.5.13). The plot thickened when it was realized that immune stimulation of blood leucocytes provokes them to release a very potent 'platelet

activating factor' (PAF) with the solubility properties of a lipid. PAF turned out to be a structurally unusual phospholipid that acts as an extremely potent extracellular signal (see Sections 4.5.11 & 10.3.7.2).

These discoveries established the principle that very low concentrations of specialist lipids in the extracellular environment can dramatically regulate the behaviour of cells in many ways, and that these responses depend on highly selective interactions between the extracellular lipids and cell surface receptors that have evolved to recognize them. There have since been other discoveries of this type, for example with the identification and characterization of leukotrienes (see Sections 3.5.9–11). More recently, it was realized that blood plasma contains minuscule amounts of sphingosine 1-phosphate (S1P) and lysophosphatidic acid (LysoPtdOH) and that these regulate the function of the endothelial barrier between tissues and the blood, the formation and function of platelets and the activation status of the immune system (see Sections 4.8.7, 8.3 & 10.3). Membrane phospholipids also turned out to be precursors of the endogenous ligands of the receptors that respond to the psychoactive constituents of cannabis (see Section 8.2).

Membrane lipid molecules in the target cells also play central roles in signal transduction, the suite of transmembrane processes by which information from extracellular signals is transmitted to the machinery of the cell interior. In the 1980s it became clear that receptor-controlled hydrolysis of phosphatidylinositol 4,5-bisphosphate (PtdIns(4,5)P_2) liberates two second messengers: *myo*inositol 1,4,5-trisphosphate (Ins(1,4,5)P_3) regulates cytosolic Ca^{2+} ion concentration and *sn*-1,2-diacyglycerol (DAG) activates protein kinase C (PKC). And then there was a second major signalling system that is involved in insulin action, control of cell survival and

---

Lipids: Biochemistry, Biotechnology and Health, Sixth Edition. Michael I. Gurr, John L. Harwood, Keith N. Frayn, Denis J. Murphy and Robert H. Michell.
© 2016 John Wiley & Sons, Ltd. Published 2016 by John Wiley & Sons, Ltd.

other processes (see Section 8.1.4.). These receptors control a phosphoinositide 3-kinase (which is generally termed PI 3-kinase, abbreviated to PI3K) that makes the second messenger PtdIns$(3,4,5)P_3$. It has since been realized that cells make several more inositol-based glycerophospholipids, generically termed polyphosphoinositides (PPIn), and that seven interrelated PPIn are essential players in a remarkably diverse set of cell functions. These include roles for PPIn as molecular identifiers of intracellular compartments and as regulators of vesicular traffic amongst these compartments (see Sections 8.1.5–8.1.8). As a result of these developments, intervention in these and other lipid-based elements of cell regulation has become a focus in the development of drugs for a variety of conditions.

8.1 Phosphoinositides have diverse roles in cell signalling and cell compartmentation

By the 1950s it was known that the polyol *myo*inositol (Ins) is a constituent of some of the phospholipids of animal cell membranes, that the brain contains a lipid with a headgroup that has at least two esterified phosphate groups per Ins, and that yeast and mammals that are rigorously deprived of Ins do not thrive – they sometimes do not survive. It was clear that Ins-containing phospholipids do something special in eukaryotic cells.

Now, more than half a century later, we know that the various PPIn have a striking variety of specific functions in all eukaryotic cells. Several key concepts underpin our knowledge of their cellular roles:

• Phosphatidylinositol (PtdIns), a 'standard' phospholipid with an *sn*-1,2-diacylglycerol 3-phosphate backbone, makes up a few per cent of the phospholipids of all eukaryotic cells. Archaea contain phospholipids that have the same Ins1*P* headgroup as PtdIns, but these have a mirror image *sn*-2,3-diradylglycerol 1-phosphate backbone, in which most of the side-chains are polyisoprenoid ethers (Fig. 8.1; see also Section 2.3.5). Ancient archaea were probably the first organisms to make Ins and Ins phospholipids (e.g. archaetidylinositol, ArcIns), and one of these organisms probably passed these capabilities to its eukaryotic descendants. Relatively few bacteria make Ins and Ins-based phospholipids, and the ancestors of those

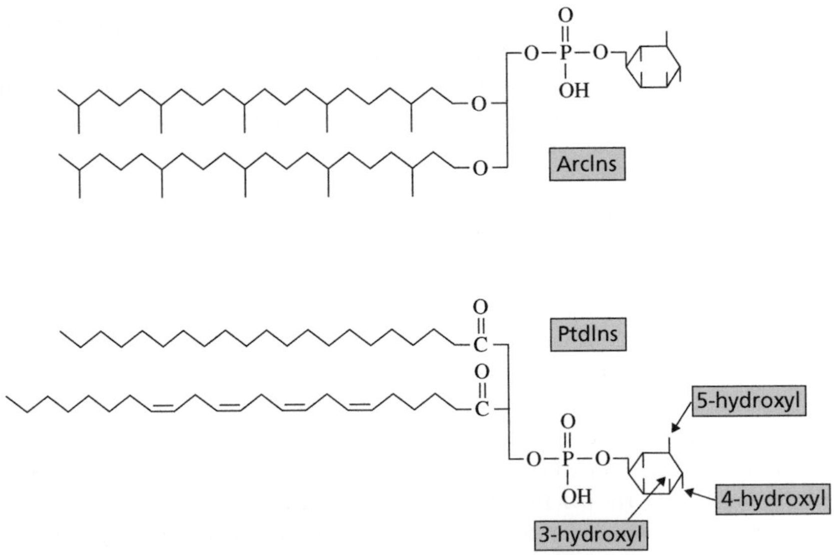

Fig 8.1 Structures of PtdIns and of archaetidylinositol (ArcIns). Note that the stereochemical configurations of the diradylglycerol backbones of the two lipids are opposite: PtdIns is *sn*-1,2-diacylglycerol-3-phosphoryl-D-1-myoinositol, and ArcIns is *sn*-2,3-diphytanylglycerol-1-phosphoryl-D-1-myoinositol (see section 5.2.2 for a discussion of the structural differences between the phosphoglycerides of eukaryotes and bacteria and of archaea). The 3, 4 and 5-hydroxyl groups of PtdIns are labelled, to indicate the positions at which additional phosphate groups are added to make the various PPIn.

that do probably gained these abilities as a result of ancient lateral gene transfers from archaea.

- Every eukaryotic cell converts small proportions of its PtdIns into at least four – and sometimes as many as seven – PPIn. These have one, two or three monoester phosphate groups attached to the 3 and/or 4 and/or 5-position(s) of the inositol ring of PtdIns (Fig. 8.1). The PPIn are the key to the functional versatility of the phosphoinositides. PtdIns is synthesized on the endoplasmic reticulum (ER) and is the ultimate substrate for the biosynthesis of all PPIn. How newly synthesized PtdIns moves from the ER to the other cell membranes in which it is converted into the various PPIn is still not well understood.

- The majority – sometimes 80% or more – of the PtdIns and PPIn molecules in mammalian tissues have a single FA combination: they are sn-1-stearoyl(18:0)-

2-arachidonyl(20:4) molecular species. This homogeneity is in marked contrast to the heterogeneous mixtures of acyl chain pairings in other membrane phospholipids. The phosphoinositides of nonmammalian organisms, such as yeast and *Dictyostelium* have the same pairing of acyl and/or alkyl substituents on the 1- and 2-positions of their glycerol backbone.even when they do not replicate this 'mammalian' acyl composition.

- PtdIns, the parent lipid of all PPIn, is usually metabolically renewed with a half-life of a few hours, significantly faster than most other membrane phospholipids. However, the monoester phosphate groups of the PPIn, which come directly from the γ–phosphate of ATP, turn over extremely quickly (the relevant pathways are outlined in Fig. 8.2). For example, when cells are allowed to incorporate radioactive inorganic phosphate, there is no substantial delay between the

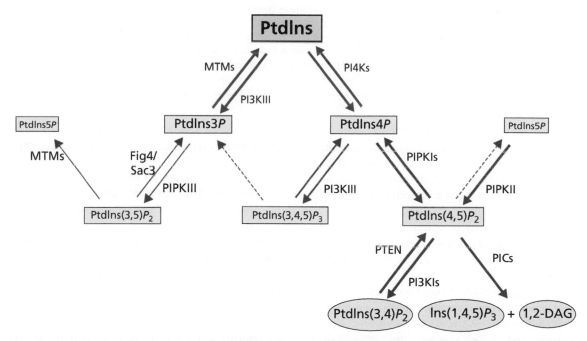

Fig. 8.2 The pathways by which PtdIns and the various PPIn are interconverted in metazoan cells. The pathways shown are those used by mammalian and other metazoan cells. Most other eukaryotes do not make PtdIns(3,4,5)P_2 or PtdIns(3,4)P_2, at least by the PI3K routes. The relative sizes and densities of the lettering and of the arrows give some indication of the relative concentrations of the various PPIn and the relative rates of the reactions that interconvert them The abbreviations used represent the most common names used for the enzymes that catalyse the indicated reaction: **Fig4/Sac3**, PtdIns(3,5)P_2 5-phosphatase; **MTMs**, PPIn 3-phosphatases of the myotubularin family; **PIC**, phosphoinositidase C (or phospholipase C); **PI4K**, PtdIns 4-kinase; **PI3K1**, Type IA or IB PI3K; **PI3KII**, Type II PI3K; **PI3KIII**, Type III PI3K; **PIPkI**, PtdIns4P 5-kinase; **PIPkII**, PtdIns5P 4-kinase; **PIPkIII**, PtdIns3P 5-kinase (Fab1 or PIKfyve); **PTEN**, PtdIns(3,4,5)P_3 3-phosphatase. The three compounds in shaded ovals are those that are now known to have widespread biological roles as second messengers.

Box 8.1 The classification and naming of functional domains in proteins.

As the sequencing of proteins and determination of their folded structures became more commonplace, it was realized that: (a) proteins with related functions often have at least partially similar amino acid sequences; (b) the most conserved parts of their sequences usually occupy the same spaces in their similar 3D structures; and (c) each of these conserved organizational motifs has its own specific function(s). Conserved parts of proteins have functions of many types – they may be elements of an enzyme active centre, surface patches that interact with partner proteins, clefts into which activating ligands bind... and so on. Sometimes a single polypeptide chain sequentially encodes several functional modules. For example, mammalian fatty acid synthase is a homodimer of 270 kDa polypeptide chains, each subunit of which includes a thioester 'acyl carrier' site for the growing acyl chain and five catalytic enzyme modules (see Chapter 3). Remember, however, that the native protein is folded into a compact tertiary structure.

During the last couple of decades, the sequencing of genes has become routine, as have the cloning and expression of the proteins they encode and targetted mutagenesis of these expressed proteins. As a result, researchers have access to the native proteins, to individual domains therefrom and to variant forms of these with changed amino acid sequences. This bonanza of protein-coding sequences and expressed proteins has led to ever-greater appreciation of the multidomain nature of many proteins. For example, the PICs and PI3Ks shown in Figs. 8.3 and 8.5 make up families of multidomain enzymes that include catalytic domains and also domains that mediate interactions with phosphoinositides, with anionic membrane surfaces and with regulatory partner proteins.

Nowadays, the hypothetical protein that a newly sequenced gene predicts is often first described in terms of the recognizable protein domains that it includes – and what their presence suggests the protein might do. Three large online databases – SMART, Pfam and the NCBI Conserved Domain Database – summarize information on the currently known families of proteins and protein domains.

The names of protein domains, including those that bind to PPIn, are commonly acronyms that are historic in origin. Their derivations are frequently opaque and have little meaning for most readers (see the next paragraph). These acronyms now simply serve as unambiguous names, but some scientific journals continue to require every paper to spell out these historical derivations at first use.

Let us illustrate with two examples. PH (**P**leckstrin **H**omology) domains constitute a large domain family that has become particularly well known because interactions between these PH domains and one or other of the PPIn are essential to the functions of their host proteins. However, a PH domain is a particular type of folded protein module, and it is often forgotten that many PH domains – including the original domain in the platelet protein pleckstrin on which their name is based – do not bind selectively to PPIn. And the name given to **FYVE** domains, most or all of which bind PtdIns3*P*, was simply constructed from the initials of **F**ab1, **Y**OTB, **V**ac1 and **E**EA1, the four proteins whose sequence similarity first allowed the domain to be recognized.

incorporation of radioactivity into the γ-phosphate of ATP and the 4- and 5-phosphate groups of PtdIns(4,5)P_2. Complex regulation of the activities of the kinases and phosphatases that attach and remove the phosphate groups of PPIn allows rapid and precise control of PPIn concentrations in cells.

- Each of the PPIn is unevenly distributed between the various organelles of eukaryotic cells. It is made, exerts its functions and is destroyed at one or more characteristic sites in a cell.
- Each PPIn regulates certain cell function(s) as a result of its interactions with specific PPIn-binding domains in target proteins (see Box 8.1 for a comment on such domains and their naming). The PPIn-binding domains are commonly members of a few structural families. Amongst these are some (though not necessarily all) domains in the FYVE, PH, PX, Tubby, and ENTH families. The PPIn-binding domains are structurally very diverse, but in each type of domain the PPIn binding site includes a domain-specific conserved motif of basic amino acid residues through which the

protein interacts with the negatively charged phosphate arrays on the headgroups of their partner PPIn.

8.1.1 The 'PI response': from stimulated phosphatidylinositol turnover to inositol (1,4,5)P₃-activated Ca²⁺ mobilization

In the 1950s, Mabel and Lowell Hokin, working mainly in Madison, Wisconsin, discovered that when acetycholine stimulates the exocrine pancreas to secrete digestive enzymes it also provokes rapid metabolic turnover of the tissue's phospholipids. This accelerated turnover is largely restricted to PtdIns and its precursor phosphatidate (PtdOH) and was christened 'the PI response'. (PI was an earlier abbreviation for phosphatidylinositol.) A similar 'PI response' occurs in many stimulated cells, from the salt glands of seabirds to human blood platelets, when they respond to external stimuli that interact with cell surface receptors (for some examples, see Table 8.1). The effective stimuli, and the receptors through which they act, are multifarious – they include some hormones

Table 8.1 Examples of receptors that signal through PIC activation.

Stimulus	Receptor type	Tissue responses include:
Acetylcholine	Muscarinic (M_1, M_3)	Smooth muscle contraction, salivary secretion
Noradrenaline	α_1-adrenergic	Smooth muscle contraction
Histamine	H_1	Allergic reactions, including bronchoconstriction
Serotonin	$5\text{-}HT_2$	Blowfly salivary gland secretion
Glutamate	$MGluR_1$	Modulation of neurotransmission
Vasopressin	V_1	Hepatic glycogenolysis, vasoconstriction
Oxytocin	OT-R	Uterine smooth muscle contraction
Angiotensin	AT_1	Vascular smooth muscle contraction, aldosterone secretion
Thyroliberin	TRHR	TSH secretion from anterior pituitary thyrotroph cells
Thrombin	PAR1*	Platelet activation
Platelet-derived growth factor	PDGFR1, PDGFR2	Proliferation
Sperm		Mammalian egg fertilization

*PAR: protease-activated receptor

and neurotransmitters (activating processes such as exocytotic secretion and smooth muscle contraction), antigens that activate immune cells and activators of platelet aggregation and secretion.

The effective receptors are highly specific: when one extracellular stimulus acts through two or more types of receptor, only some types of receptor evoke the 'PI response'. For example, the hormone vasopressin (also known as antidiuretic hormone) has two types of receptor (V_1 and V_2). V_1 receptor stimulation causes constriction of arteries, so increasing blood pressure, and stimulates glycogen breakdown in the liver – and it always evokes a robust 'PI response', whatever is the physiological response being displayed by the tissue being studied. By contrast, V_2 receptor stimulation activates adenylate cyclase and regulates fluid output through the kidney, but has no effect on PtdIns metabolism.

Observations such as these established that the stimulation only of some types of cell surface receptors evokes a 'PI response' and suggested that stimulation of these receptors might always provoke this response. It became apparent that such stimulation activates a phospholipase C (PLC), which removes the headgroup of a phosphoinositide, and that the increased phosphatidate (PtdOH) and PtdIns biosynthesis reincorporates at least some of the liberated 1,2-diacylglycerol (DAG) into phosphoinositides. Thus, the classic 'PI response' reflects the metabolic replenishment of the phosphoinositides that have been depleted by the action of a phosphoinositide-

specific PLC. The name phosphoinositidase C (PIC) is preferable for this limited family of eukaryotic enzymes (see Sections 8.1.1–8.1.3). It clearly distinguishes them from the many other PLCs from diverse sources that have been studied: a few of these are specific for PtdIns, but many can hydrolyse a variety of phospholipids (see Section 4.6.4).

So why do some stimuli stimulate phosphoinositide hydrolysis in this way? In the 1970s Bob Michell, in Birmingham, UK, recognized that stimulation of a PIC by cell surface receptors is always associated with an increase in the cytosolic free Ca^{2+} ($[Ca^{2+}_i]$) concentration in the stimulated cells, and he suggested that it might cause the increase in $[Ca^{2+}_i]$ that triggers physiological responses such as exocytotic secretion and smooth muscle contraction. Michael Berridge and John Fain, working in Cambridge, UK, confirmed that an Ins-requiring process is indeed involved when 5-hydroxytryptamine elevates $[Ca^{2+}_i]$ in the salivary glands of blowflies. Michell's group then discovered that intense vasopressin stimulation of hepatocytes provokes very rapid $PtdIns(4,5)P_2$ hydrolysis – at an initial rate of about one per cent of the total per second – and suggested that this $PtdIns(4,5)P_2$ hydrolysis is the reaction that initiates Ca^{2+} mobilization in cells. This set the scene for Berridge and his collaborators to discover that receptor-activated PICs liberate $Ins(1,4,5)P_3$, which diffuses to the ER and triggers rapid Ca^{2+} release from this ATP-loaded membrane compartment (see Section 8.1.3).

8.1.2 After the 1980s: yet more polyphosphoinositides, with multifarious functions in signalling and membrane trafficking

For a long time it was assumed that eukaryotic cells made only three phosphoinositides – quite a lot of PtdIns and small amounts of PtdIns4P and PtdIns(4,5)P_2 – but then the situation changed dramatically. First, Lew Cantley's Boston group found a PI3K activity that was associated with activated growth factor receptors and some oncogenes – and they realized that PtdIns3P is a normal constituent of unstimulated animal cells. They then discovered that their PI3K could phosphorylate all of the then-known phosphoinositides – making PtdIns3P, PtdIns(3,4)P_2 and PtdIns(3,4,5)P_3.

During the 1990s the combined work of several groups established that the PtdIns(3,4,5)P_3 which receptor-controlled PI3Ks make very rapidly in stimulated cells acts as a membrane-localized second messenger that controls multiple downstream signalling pathways (see Section 8.1.4). Next it emerged that PtdIns3P and PtdIns(3,4)P_2 have their own distinctive functions, notably in regulating membrane trafficking processes (see Sections 5.4.2–5.4.3, 8.1.7 & 8.1.8). And, finally, discovery of PtdIns5P and of PtdIns(3,5)P_2 completed the currently known set of PPIn. PtdIns(3,5)P_2 has recently been garnering numerous functions (see Section 8.1.9), and roles for PtdIns5P are starting to emerge (see Section 8.1.10).

8.1.3 Signalling through receptor activation of phosphoinositidase C-catalysed phosphatidylinositol 4,5-*bis*phosphate hydrolysis

Receptor regulation of cAMP formation by adenylate cyclase was defined in the 1970s as the prototype of cell signalling processes in which an *extracellular* 'first messenger', such as a hormone, stimulates a transmembrane signal-transducing apparatus in the plasma membrane to control an enzyme that makes an *intracellular* 'second messenger'. This second messenger then regulates intracellular processes. The same principles are in play during signalling through the PIC and PI3K pathways: plasma membrane receptors responding to extracellular stimuli activate signal transduction mechanisms and these activate enzymic reactions – PtdIns(4,5)P_2 hydrolysis by PIC and phosphorylation by PI3K – at the inner face of the plasma membrane.

Early work, initially on mammalian erythrocytes – in which the plasma membrane is the only membrane – had established that PtdIns, PtdIns4P and PtdIns(4,5)P_2 are rapidly interconverted at the inner face of the plasma membrane by a subset of the pathways shown in Fig. 8.2.

Eukaryotic cells contain several types of PIC, all of which share key structural features (Fig. 8.3). However, only some of these have been proved to be involved in receptor-activated signalling. PICs typically have a PPIn-binding PH domain near their N-terminus, a nearby Ca^{2+}-binding EF hand domain, and a C2 domain towards the C-terminus that is responsible for relatively non-specific binding to the phospholipid surface of membranes. In between, there are two half-domains, designated X and Y, that fold together in the protein to form the catalytic domain that cleaves PtdIns(4,5)P_2 into Ins(1,4,5)P_3 and 1,2-DAG.

The stimuli that activate PtdIns(4,5)P_2 hydrolysis operate mainly through two functional classes of cell surface receptors (see Fig. 8.4 & Table 8.1), each of which stimulates the activity of particular types of PIC. Many of these receptors, such as V$_1$-vasopressin receptors or M1-muscarinic acetylcholine receptors, are members of

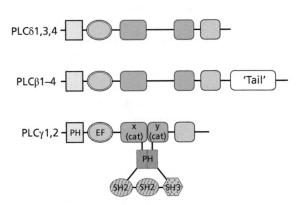

Fig. 8.3 A simple domain organization depiction of three types of PIC. PICδs are present in almost all eukaryotes, and PICβs and PICγs are mainly confined to metazoans, in which they are regulated by activation of cell surface receptors (see the text). Conserved domains are as identified in the PLCγ illustration. Domains are denoted as: PH, pleckstrin homology domain (PLCγs include a PH domain that is formed from two noncontiguous sequences that fold together); EF (Ca^{2+}-binding EF hand); SH2 (*src* homology Type II domain, interacts with phosphotyrosine motifs on activated RTKs); SH3, *src* homology Type III domain; C2, membrane-binding C2 domain.

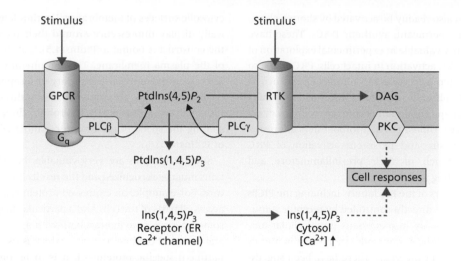

Fig. 8.4 A summary depiction of the PIC signalling pathway. Many cell stimuli activate one or more of the many PICs in metazoan cells, leading to simultaneous liberation of Ins(1,4,5)P_3 and DAG. The resulting increases in cytosolic [Ca^{2+}] and activation of multiple types of PKC are responsible, either directly or through intermediate target proteins, for regulating most of the cell responses to the relevant pharmacological subsets of receptors for these stimuli (example of such receptors include α_1-adrenergic, V_1-vasopressin, H_1-histamine). Such responses include exocytotic secretion by a variety of cells, contraction of a variety of small muscles, and pro-thrombotic activation of platelets.

the very large family of mammalian G-protein-coupled receptors (GPCRs – see Section 7.3.4). These activate guanine nucleotide exchange on coupled heterotrimeric G-proteins (G_Q and/or G_{11}) at the cytoplasmic surface of the plasma membrane, and the GTP-ligated and activated $G\alpha$ is liberated and activates PICs of the β-subtype. Sometimes the released $G\beta\gamma$ dimer provides additional, synergistic, activation of PICβ. Other effective receptors, such as that for platelet-derived growth factor (PDGF), a polypeptide growth factor, are members of the Receptor Tyrosine Kinase (RTK) family, and activate PICs of the γ-subtype. The aminoacid sequences of PICγs include paired SH2 (*src* homology 2) domains which bind to PIC-activating tyrosine-phosphorylated motifs in the activated RTKs. Activation of RTKs, and sometimes of GPCRs, also brings about downstream activation of the small GTPase Ras. PICγs (and some other PICs) include Ras response domains, and these are additionally activated by GTP-ligated – and thus active – Ras.

Once Ins(1,4,5)P_3-triggered release of Ca^{2+} from an ATP-loaded store in the ER had been observed in many cells in response to diverse stimuli, it was recognized that PIC-generated Ins(1,4,5)P_3 has a widespread role as a Ca^{2+}-mobilizing intracellular second messenger, at least in animal cells. Katsuhiko Mikoshiba in Osaka then

identified a very large tetrameric Ca^{2+} channel localized in the ER as the Ins(1,4,5)P_3 receptor.

Perhaps more surprising was the discovery that *sn*-1,2-DAG, the other product of PtdIns(4,5)P_2 hydrolysis, is also a second messenger: the PIC signalling system was the first that was known to generate two second messengers from the same molecule. The released DAG activates PKC, discovered in brain by Yasutomi Nishizuka's group in Kobe, Japan. DAG was the first second messenger that was known to be retained in the membrane in which it is formed rather than being a freely diffusible cytosolic molecule. Nishizuka's initial PKC was the first of a substantial family of related protein kinases, most of which are activated by DAGs, that are now collectively termed the PKCs. The so-called 'conventional' DAG-activated PKCs, notably PKCβ, require a trace of Ca^{2+}, have a native affinity for the anionic lipid surface at the cytosolic surface of the plasma membrane and possess a DAG-binding C1 domain. PKCs are mainly cytosolic in unstimulated cells and the DAG that is released in the plasma membrane by PIC activation both 'attracts' and activates PKCs.

The PKC-activating DAG that is liberated from PtdIns(4,5)P_2 in mammalian cells is largely the 1-stearoyl,2-arachidonyl molecular species. PKCs in intact cells can also readily be activated by shorter-chain, and membrane-permeant, synthetic DAGs. These have been particularly valuable in experimental exploration of the effects of PKC activation in intact cells. PKCs are also potently activated by a variety of phorbol esters such as tetradecanoyl phorbol acetate (TPA). These plant-derived tumour-promoting molecules are complex diterpene derivatives. Unlike DAG, phorbol esters are resistant to metabolism, and persistent activation of PKC accounts for much of their pro-inflammatory and tumour-promoting activity.

Several members of the PIC family, including the PICβs and PICγs that are directly controlled by receptor stimulation, are found only in metazoans (multicellular animals), but PICs of the δ-subfamily are present in almost all eukaryotes. Over the years, there have been suggestions of PIC-initiated Ca^{2+} mobilization in many other eukaryotes, such as green plants, despite the fact that genomic studies have failed to find $Ins(1,4,5)P_3$-receptor Ca^{2+} channels in these organisms. As a result, it is still not clear what role(s) these ubiquitous PICδs play in non-metazoan eukaryotes.

8.1.4 Polyphosphoinositide-binding domains as sensors of polyphosphoinositide distribution in living cells

As mentioned earlier, the various PPIn exert their biological actions through their interactions with a multitude of proteins that include domains that interact with varying degrees of specificity with one or more of the PPIn (see Box 8.1). This property has also made the more PPIn-specific of these domains into valuable tools with which cell biologists can visualize the distributions of individual PPIn within living, fixed or sectioned cells. Such techniques allow cell biologists to map the distributions and movements of the various PPIn within living or fixed cells by expressing a variety of tagged domains that selectively bind to one or other of the PPIn.

This principle can be illustrated by reference to the use of the PH domain of PICδ as a tool for examining the intracellular distribution of $PtdIns(4,5)P_2$ by fluorescence microscopy. Cells are transfected with a genetic construct that makes them express a hybrid protein consisting of the PH domain of PICδ coupled to a fluorescent protein such as the green fluorescent protein (GFP) from jellyfish. The PH domain of the expressed GFP-PICδPH construct then binds selectively to $PtdIns(4,5)P_2$ at the cytosolic surfaces of membranes. Unstimulated cells typically display fluorescence around their margin, where the construct is bound to $PtdIns(4,5)P_2$ at the inner face of the plasma membrane. This peripheral GFP-PICδPH fluorescence decreases in intensity or disappears within seconds when cells are exposed to a stimulus that triggers PIC-catalysed $PtdIns(4,5)P_2$ hydrolysis, directly demonstrating the speed of the receptor-stimulated hydrolysis of $PtdIns(4,5)P_2$.

Such techniques are very valuable, but their limitations must be recognized and the results interpreted with care. For example, an expressed protein may not freely access all parts of the cell. And a particular tagged protein sometimes needs to interact both with a particular PPIn and a second binding partner – which is usually a compartment-specific protein – if it is to become securely associated with its 'home' intracellular structure. As a result, binding of such sensors at particular sites usually indicates that the target PPIn is present there, but failure to see a fluorescence signal at other sites does not always mean that the PPIn is absent.

A complementary approach has been to raise antibodies to an individual PPIn and use these to map its intracellular distributions in fixed and sectioned cells – the PPIn-bound antibodies are detected by secondary antibodies that may be fluorescently labelled (for light microscopy) or tagged with electron-dense labels such as gold particles (for electron microscopy).

8.1.5 Signalling through phosphoinositide 3-kinase-catalysed phosphatidylinositol 3,4,5-*tris*phosphate formation

The three different types of PI3K (see Fig. 8.5) make the various 3-phosphorylated PPIn. Class I PI3Ks (PI3K-Is) make $PtdIns(3,4,5)P_3$ in response to receptor activation; Class II PI3Ks (PI3K-IIs) make at least some of the cellular $PtdIns(3,4)P_2$; and Class III PI3K (PI3K-III) makes $PtdIns3P$.

$PtdIns(3,4,5)P_3$ is made from plasma membrane $PtdIns(4,5)P_2$ only by receptor-activated PI3K-Is, of which there are four in mammals (Fig. 8.5). The PI3K-IAs are heterodimeric complexes of a catalytic subunit (designated p110α, β, or δ) and a p85 or p55 regulatory subunit. They are primarily activated as a result of interactions between pairs of SH2 domains in the p85 or p55 regulatory subunits and PI3K-specific phosphotyrosine-containing motifs that are formed either: (a) in the

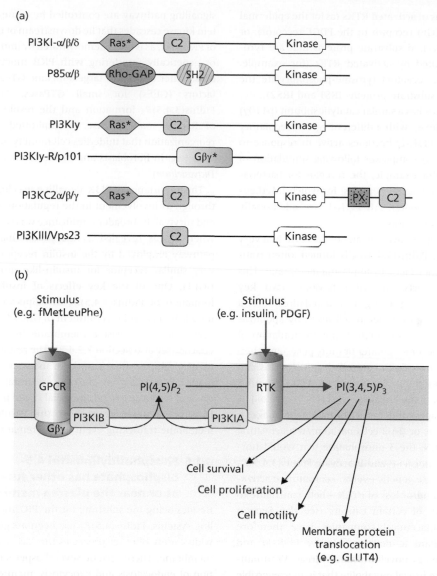

Fig. 8.5 Signalling through receptor activation of PI3K. (a) The top two pairs of schema depict, respectively, the catalytic (above) and regulatory (below) subunits of Type IA and Type IB PI3Ks. The bottom two depict a Type II PI3K (above) and Type III PI3K (bottom). Domains are denoted thus: Ras∗, domains that interact with activated RAS proteins, contributing to activation of the PI3K; C2, membrane interaction domain; Kinase, catalytic domain; SH2 (*src* homology Type II domain, interacts with phosphotyrosine motifs on activated RTKs); Rho-GAP, domain homologous to those in other proteins that activate the Rho GTPase; Gβγ∗, domain that mediates activation by the Gβγ dimers from activated heterotrimeric G proteins; PX, phagocytic oxidase homology (phox) domain. (b) A schematic depiction of the membrane events that initiate signalling through Type IB (left) and Type IA (right) PI3Ks. *f*MetLeuPhe is a peptide ligand that activates receptors on phagocytic leucocytes that initiate chemotaxis towards bacteria that release *f*Met-initiated signal peptides as a result of the synthesis of secreted proteins. PDGF is platelet-derived growth factor. Signalling through Type I PI3Ks stimulates many cell-specific responses, some of which are indicated. Most are initiated by interactions between PtdIns$(3,4,5)P_2$ and PtdIns$(3,4,5)P_2$-specific PH domains in target protein, notably the protein kinases PDK1 and Akt.

cytosolic domains of activated RTKs (as for the epidermal growth factor (EGF) receptor or the PDGF receptor); or (b) in RTK-associated substrate proteins that are tyrosine-phosphorylated by activated RTKs (for example, activated insulin receptors tyrosine-phosphorylate the insulin receptor substrate proteins IRS1 and IRS2).

The one PI3K-IB has a similar catalytic subunit (p110γ) but this is complexed with a different type of regulatory subunit (p105). PI3K-Iγ becomes active in response to release of free G$\beta\gamma$ complexes following stimulation of certain GPCRs (for example, the receptor for bacteria-derived formylated signal peptides, for which fMetLeu-Phe serves as a model). All PI3K-Is are additionally activated by activated Ras.

Whenever receptors activate any of the PI3K-Is, a very small amount of PtdIns$(3,4,5)P_3$ is formed, often transiently, at the inner face of the plasma membrane. This has profound effects on cell behaviour. Two key responses in many growing cell populations are a decreased tendency to undergo cell death by apoptosis and acceleration of cell growth. If the accumulation of PtdIns$(3,4,5)P_3$ becomes greater or more persistent, then cells may tend towards uncontrolled growth. This can happen, for example, if: (a) one of the PI3K-Is (either p110α, in cells of many lineages, or p110δ, in myeloid or lymphocyte lineages) undergoes a mutation that makes it persistently active; or (b) if cells harbour an inactivating mutation in PTEN, the 3-phosphatase that would normally 3-dephosphorylate and inactivate PtdIns$(3,4,5)P_3$ (see Fig. 8.2). These genetic events – constitutive activation of a PI3K-I and/or loss of PTEN – both contribute to the development of certain cancers (see also Section 10.2.2.3). Pharmaceutical companies have therefore been working hard to develop selective PI3K-Iα and PI3K-Iδ inhibitors as novel anticancer drugs. Wortmannin, a steroid-like fungal metabolite that is an irreversible inhibitor of PI3Ks, was very valuable in the initial identification of biological processes that are regulated through PI3Ks. However, wortmannin also 'unspecifically' inhibits a number of protein kinases, and it is increasingly being superseded by these more selective inhibitors.

The PtdIns$(3,4,5)P_3$ that is made in stimulated cells has many direct molecular targets, most of which associate with it through PtdIns$(3,4,5)P_3$-selective PH domains. Notable amongst these are two cooperatively acting protein kinases, phosphoinositide-dependent protein kinase-1 (PDK1) and protein kinase B (PKB, also known as Akt). Many of the events regulated through the PI3K signalling pathway are controlled by tissue-specific protein kinase cascades that lie downstream of the activation of PKB/Akt, as is often demonstrated by immunologically or genetically interfering with PKB function in intact cells. Other important targets include GTPase-activating factors (GEFs) for small GTPases. For example, PtdIns$(3,4,5)P_3$ formation and the resulting activation of Rac, a small GTPase, is implicated in the actin reorganization that underlies cell motility during chemotaxis (e.g. in fibroblasts and in the cellular slime mould *Dictyostelium*).

The importance of PI3K signalling was first recognized through its involvement in the regulation of cell growth and survival. Its broader significance was recognized only when it was revealed as the predominant signalling pathway employed by the insulin receptor and by the very similar receptor for insulin-like growth factor-1 (IGF1). One of the key effects of insulin-stimulated formation of PtdIns$(3,4,5)P_3$ in adipocytes or skeletal muscle is very rapid recruitment of GLUT4 glucose transporters into the plasma membrane from intracellular vesicles (see also Section 7.2.4.3). The resulting enhancement of glucose flux into these tissues causes the dramatic fall in circulating glucose concentration that is the classical response to a postprandial rise in plasma insulin. However, knowledge of how this insulin-stimulated membrane trafficking event occurs remains incomplete.

8.1.6 Phosphatidylinositol 4,5-*bis*phosphate has other functions at or near the plasma membrane

Besides being the substrate for the PIC and PI3K signalling systems, PtdIns$(4,5)P_2$ has been assigned roles in a wide variety of other processes that occur at the plasma membrane. These include several aspects of the regulation of endocytosis and exocytosis, maintenance of the activity of diverse ion channels for K$^+$, Ca^{2+} and other ions, and control of various aspects of cytoskeletal function.

8.1.7 Phosphatidylinositol 4-phosphate in anterograde traffic through the Golgi complex

PtdIns4P and PtdIns$(4,5)P_2$ were first discovered in brain myelin and in erythrocyte membranes, and PtdIns4P was initially regarded just as an intermediate metabolite *en route* to PtdIns$(4,5)P_2$ at the plasma membrane. However, eukaryotic cells always express two or three separate PtdIns 4-kinases, at least one of which is localized to the *trans*-Golgi. This localization requires the GTPase ARF1, and maintenance of this pool of Golgi PtdIns4*P* is essential to ensure the integrity and function of the Golgi complex.

Once PtdIns4*P* has been made at the *trans*-Golgi, it is recognized by several lipid transfer proteins – including CERT, OSBP and FAPP2 – each of which includes both a PtdIns4*P*-binding PH domain through which it recognizes the *trans*-Golgi, and an ER recognition domain. The ceramide-binding protein CERT transfers newly synthesized ceramide (N-acyl-sphingosine) from the ER to the PtdIns4*P*-rich *trans*-Golgi, where the ceramide is used for sphingomyelin biosynthesis (see Sections 2.3.3 & 4.8.1). OSBP and FAPP2 play similar roles in trafficking other newly synthesized membrane lipids – cholesterol and glucosylceramide, respectively – from the ER to the *trans*-Golgi, from which these lipids (or their modified products) can move on to post-Golgi membrane compartments such as endosomes, lysosomes and the plasma membrane. OSBP, as an example, has a remarkable ability to bind and transfer, in a mutually exclusive manner, *either* cholesterol or PtdIns4*P*. It is hypothesized that: (a) OSBP binds simultaneously to closely apposed elements of the ER and *trans*-Golgi; (b) it then reciprocally transfers cholesterol from the ER to the *trans*-Golgi and PtdIns4*P* in the opposite direction; and, finally, (c) ER PtdIns4*P* is dephosphorylated to PtdIns by the ER PPIn 4-phosphatase Sac1, so rendering these lipid transfers irreversible.

8.1.8 Phosphatidylinositol 3-phosphate in regulation of membrane trafficking

PtdIns3*P* is a normal constituent of eukaryotic cells of all types, and is made by ubiquitous PI3K-IIIs. The prototypic PI3K-III is encoded by the *vps34* gene of the yeast *Saccharomyces cerevisiae* – *vps* identifies the product of this gene as implicated in 'vacuolar protein sorting'. PI3K-III functions in several large hetero-oligomeric complexes that always include both PI3K-III and a core regulatory subunit (encoded in yeast by *vps15*). Genetic studies showed that PtdIns3*P* formation is necessary for the trafficking of proteins within the cell, and it has become clear that PtdIns3*P* executes many key functions in endosomal compartments. In particular, many newly synthesized proteins traffic through the Golgi to endosomes *en route* to lysosomes (or in yeast to the vacuole); and endocytosed proteins such as growth factor receptors, low density lipoprotein receptors or transferrin (with their associated ligands or cargoes) traffic to endosomes where they are sorted. Some are returned to the cell surface, whilst others are sent to lysosomes for destruction.

Cells express many proteins that selectively bind both to PtdIns3*P*, often through FYVE or PX domains, and to a second, and compartment-defining, molecule – this process is known as 'coincidence detection'. For example, the early endosomal protein EEA1 has a FYVE domain that binds to PtdIns3*P* and a second domain that simultaneously associates with the endosomal GTPase Rab5, and the combined affinities for these two molecules anchors EEA1 to early endosomes. (The affinity of a single FYVE domain for PtdIns3*P* is not usually sufficient to achieve stable binding of a FYVE domain-containing protein to its 'home' organelle, so the mapping of PtdIns3*P* distribution in cells is usually done with a fluorescent protein construct that expresses two or more copies of a FYVE domain in tandem.)

PtdIns3*P* has been implicated in a variety of processes by which proteins passing through endosomal compartments are sorted and routed to other parts of the cell. These include: the return to the Golgi complex of receptor proteins (such as the mannose 6-phosphate receptor) that are needed for continued flux of newly assembled lysosomal proteins out of the Golgi; several steps in the trafficking of proteins to the lysosomal interior; and the return of some endocytosed receptors to the plasma membrane.

PtdIns3*P* is also an essential player in the initiation of macroautophagy. During this process, it is made by a different PI3K-III-containing complex that is regulated by the protein Beclin-1 (Atg6 in yeast). Macroautophagy is a process in which cells enclose substantial volumes of their cytoplasm, often including organelles such as mitochondria, into an autophagic vacuole. The contents of the autophagic vacuole are then trafficked to lysosomes where they are dismantled to provide supplies for ongoing cell metabolism. One important function of autophagy is to allow cells to weather periods of nutritional deprivation. For example, newborn mice undergo a brief period of starvation during the transition from placental support *in utero* to their first milk feed and they are dependent on autophagy to survive this stress. Autophagy is also employed during the structural and functional remodelling that occurs during insect metamorphosis and cell differentiation.

8.1.9 Type II phosphatidylinositol 3-kinases, phosphatidylinositol 3,4,-*bis*phosphate and endocytosis

The three PI3K-IIs in mammals have been enigmatic – they are widely expressed but they participate neither in the PI3K-I signalling pathway nor in the PtdIns3P-dependent regulation of endocytic processes. Combinations of cell biological and genetic investigations have implicated the PI3K-IIs in a variety of cell functions, including insulin-stimulated glucose uptake, exocytotic secretion by neurosecretory cells and resistance to cell death by apoptosis, though in still mysterious ways. For example, genetic knockout of one of these, PI3K-C2α, yields underweight mice that have a reduced lifespan.

A reasonably well characterized action is the role of PI3K-C2α in the early stages of clathrin-mediated endocytosis of macromolecules such as the activated EGF receptor. In this process, proteins that are to be internalized become clustered in the plasma membrane and a 'cage' of clathrin molecules then assembles on the cytoplasmic surface of the membrane in a PtdIns(4,5)P_2-dependent manner. A clathrin-coated pit is thus formed in the plasma membrane. The next step of the endocytic process is invagination and pinching off of endocytic vesicles that will travel to intracellular endosomes. For this to happen, PtdIns(3,4)P_2 must be made by PI3K-C2α in the membrane of the coated pit. Recruitment to the membrane of SNX9, a PtdIns(3,4)P_2 effector protein that induces membrane curvature, seems to be a key step in this invagination process.

8.1.10 Phosphatidylinositol 3,5-*bis*phosphate, a regulator of late endosomal and lysosomal processes

By the mid-1990s it was assumed that the PPIn had all been identified – and then PtdIns(3,5)P_2 and PtdIns5P were discovered as normal cell constituents of strikingly low abundance. PtdIns(3,5)P_2 is made by 5-phosphorylation of PtdIns3P (Fig. 8.2). This reaction is catalysed by a large and complex PtdIns3P 5-kinase that is known as FAB1 in yeast and PIKfyve in animals, and this functions in cells in an even larger complex organized around an essential 'scaffold protein' (VAC14 in yeast). Cells in which FAB1 or VAC14 is inactivated (either mutationally or pharmacologically) cannot make PtdIns(3,5)P_2 and they develop large intracellular vacuoles. In mice, this leads to a characteristic form of neurodegeneration in which neurones are vacuolated. These late endosomal and/or lysosomal vacuolar structures normally receive constant inputs of membrane by vesicular trafficking, both by delivery of newly synthesized organelle components via the Golgi complex and as endocytosed material that is destined for lysosomal destruction. When these lysosomal vacuoles cannot recycle this excess membrane back to the Golgi complex in a normal manner, they become engorged and their functions are compromised. The yeast β-propeller protein Atg18, the first PtdIns(3,5)P_2 effector to be identified, is necessary for this retrograde trafficking of membrane from the endo-lysosomal compartments to the *trans*-Golgi.

Mutational and pharmacological studies have implicated PtdIns(3,5)P_2 in a remarkably disparate spectrum of important cell processes, some of which are summarized in Fig. 8.6. A single theme has emerged from all of these studies: PtdIns(3,5)P_2 is made near the functional interface between late endosomal and lysosomal compartments, and the various cellular functions that it initiates are all launched from there.

8.1.11 Phosphatidylinositol 5-phosphate functions are starting to emerge

The fact that PtdIns5P is present in animal cells at extremely low concentrations makes it especially difficult to study, but recent evidence has implicated it in normal cell regulation and in the responses of cells to some pathogens. PtdIns5P can be produced in cells by at least two pathways, shown in Fig. 8.2. One is through the action of PPIn 3-phosphatases of the myotubularin family on PtdIns(3,5)P_2. Much of the work on PtdIns5P has focussed on its possible functions in the nucleus, where it binds to the PHD domains of ING2, a protein that senses chromatin modifications and regulates cell growth.

The other pathway to PtdIns5P is employed during infection with the diarrhoea-producing pathogen *Shigella flexneri*. This organism introduces the virulence factor IpgD – a PtdIns(4,5)P_2 4-phosphatase – into infected cells. The result is that PtdIns5P is made near the plasma membrane, where it interacts with a PtdIns5P-selective PH domain in Tiam1, a guanine nucleotide exchange factor for the cytoskeletal regulator protein Rac1 The activated Rac1, in turn, stimulates invasive cell motility of the type employed by metastasizing cells.

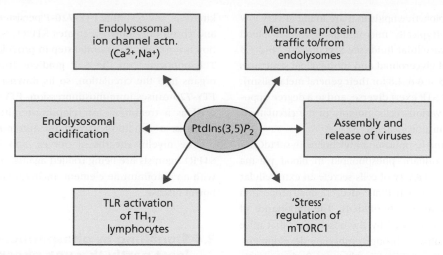

Fig. 8.6 A summary of the functions of PtdIns(3,5)P$_2$. PtdIns(3,5)P_2 is made in the dynamic endosomal/lysosomal regions of eukaryotic cells by the Fab1/PIKfyve complex (see the text), and it is likely that all of its diverse functions are initiated by PtdIns(3,5)P_2 that is formed and functions in this region of cells. The PtdIns(3,5)P_2-binding effector proteins that initiate many of these cell functions are yet to be identified. One subset of functions (top left of figure) relies on regulation of the activities of ion channels and pumps in the membranes of endolysosomes. Another subset (top right of figure) are consequences of regulated vesicular membrane traffic through endosomes: these include retrograde protein traffic from endolysosomes, controlled delivery of specific proteins to particular regions of the plasma membrane, and virus assembly and release. Yet others (bottom of figure) include metabolic switching in response to changes in nutritional status and the development, in response to activation of Toll-like receptors (TLRs), of the TH$_{17}$ lymphocytes that are involved in some aspects of innate immunity and autoimmunity.

8.2 Endocannabinoid signalling

The psychoactive properties of *Cannabis sativa* have long been known, and it became clear during the 1990s that mammals and other vertebrates express at least two types of closely related G-protein-coupled cannabinoid receptors (CB1 and CB2 receptors), especially in the nervous system (CB1) and immune system (CB2) (see Section 7.3.4 for a brief discussion of GPCRs). Both receptors are potently activated by tetrahydrocannabinol, the main psychoactive constituent of cannabis.

These discoveries raised an intriguing question: what are the endogenous ligands that activate these receptors *in vivo*? Two well characterized 'endocannabinoids' emerged from this search: 2-arachidonyl-glycerol and a series of N-acyl derivatives of ethanolamine, particularly N-arachidonyl-ethanolamine (which is also known as anandamide). Their synthesis and degradation are described in Section 4.6.6.

There is growing evidence that these molecules have important regulatory roles, particularly at synapses in the nervous and immune systems. The picture is clearest at glutamatergic synapses in the striatum. Here excitatory transmission stimulates postsynaptic PtdIns(4,5)P_2 hydrolysis by PIC (see Section 8.1.3) and a DAG lipase (DAGLα) removes the 1-acyl group from the released DAG. Being derived from a phosphoinositide, the liberated monoacylglycerol is mainly 2-arachidonyl-glycerol (see Section 8.1). This 2-arachidonyl-glycerol is released from the postsynaptic neurone, crosses the synaptic cleft and activates CB1 receptors on the presynaptic terminal and/or axon, exerting feedback inhibition on subsequent neurotransmitter release. This brings about 'postsynaptic depression', a sustained suppression of excitatory signalling through the synapse. 2-arachidonyl-glycerol is degraded by a MAG lipase that is mainly presynaptic.

8.3 Lysophosphatidate and sphingosine 1-phosphate in the circulation regulate cell motility and proliferation

LysoPtdOH (the now accepted abbreviation for lysophosphatidate which is often abbreviated to LPA) and sphingosine 1-phosphate (S1P) are two physically quite similar and important bioactive lipids that are found at very low concentrations (typically micromolar or less) in blood plasma and extracellular fluid (see Sections 2.3.2 & 2.3.3 for discussion of glycerolipid and sphingolipid structures and Sections 4.5, 4.6 & 4.8 for their general metabolism). LysoPtdOH and S1P exert diverse, and to a degree opposite, effects on various cells, notably in the circulatory, nervous and immune systems.

Albumin-bound lysophosphatidylcholine (lysoPtdCho) is the most abundant phospholipid in blood plasma ($\leq$200 μM), and a variety of cells secrete an extracellular PLD, known as autotaxin, that converts a small proportion of this lysoPtdCho into lysoPtdOH. The importance of autotaxin is demonstrated by the observation that mice that lack it die halfway through embryonic development. S1P is made by two sphingosine kinases. Both are intracellular, and embryonic lethality is again the result when both are deleted. S1P destined for plasma is made inside cells, particularly endothelial cells, from which at least some of it exits on a transporter named Spns2.

LysoPtdOH and S1P have half-lives in the circulation of only a few minutes, and they exert their effects on target cells by activating specific cell surface GPCRs. There are several lysoPtdOH receptors (at least 6) and 5 S1P receptors (S1PR1-5), with a small degree of cross-sensitivity at some of these. LysoPtdOH and S1P are inactivated, at least in part, by 2 lipid phosphate phosphatases (LPP1 and LPP3) that have their active sites exposed on the external surface of cells.

In the circulatory system, S1P protects the integrity of the permeability barrier presented by the endothelium, counteracting increased vascular permeability that is provoked, for example, by histamine or PAF. In contrast, lysoPtdOH provokes disruption of the tight junctions between endothelial cells and increases vascular permeability. Local ceramide production also tends to prejudice endothelial barrier function, at least in part by promoting apoptosis of endothelial cells (see Section 8.5).

Some cancers have properties that chronically expose them to elevated concentrations of lysoPtdOH: they secrete more autotaxin and/or express less LPP1 and/or LPP3 than normal cells. The lysoPtdOH receptor-driven consequences include enhanced growth, survival and motility – and greater resistance to chemotherapy.

S1P also has an important role in immune regulation. This was emphasized when it was discovered that the immunosuppressant FTY-720 (fingolimod) is a sphingosine mimic that is phosphorylated by sphingosine kinase-2. The resulting FTY-720-P persistently activates and chronically down-regulates S1PR1. S1PR1 activation is normally an essential step in provoking egress of T lymphocytes up the S1P gradient from lymphoid organs into the circulation, so its down-regulation by FTY-720 causes immunosuppression. FTY-720 is being used as a treatment for relapsing-remitting multiple sclerosis, a condition caused by autoimmune attack on the myelin sheaths of nerves, and it and other S1PR1 agonists are being trialled against other diseases with an autoimmune element, including inflammatory bowel disease.

8.4 Signalling by phospholipase D, at least partly through phosphatidate

Historically, PLDs have long been known as potent activities from plants that are useful in the laboratory for making stereochemically correct phosphatidate (PtdOH) from biological substrates such as egg-yolk PtdCho (see Section 4.6). However, PLDs were not generally regarded as constituents of animal tissues until the 1980s. A PLD activity against PtdCho, which was stimulated by PKC-activating phorbol esters, was then detected in mammalian cells, and yeast and mammalian PLDs were soon cloned by Frohman's group at Stony Brook, USA.

Activation of PLD in cells has two immediate results: an increase in the concentration of PtdOH, a lipid that is already present in cells as a central biosynthetic intermediate; and a very slight decrease in the concentration of abundant PtdCho. These changes are very difficult to measure unambiguously, which makes experimental analysis of PLD function technically difficult. However, PLDs have a catalytic property that is biologically irrelevant but technically convenient – they catalyse transphosphatidylation, the transfer of a phosphatidyl group from PtdCho onto a small primary alcohol. For example, in the presence of a small concentration of n-butanol PLDs make the 'novel' phosphoglyeride phosphatidylbutanol, and PLD activity can therefore be monitored by incorporation of radioactively labelled n-butanol into lipids.

Mammals have two major PLDs – PLD1 and PLD2 – with similar overall structures. The activity of both requires interactions with PtdIns(4,5)P_2 through a PH domain and a second site. PLD1 and PLD2 also include a PPIn-binding PX domain of less well defined selectivity and function. PLD1 is generally associated with intracellular membranes along the Golgi/secretory vesicle/endosome/lysosome continuum, and PLD2 with the plasma membrane. However, their intracellular localizations often change during changes in cell state (e.g. during adherence to surfaces) or following various types of cell stimulation. The difficulties mentioned above have meant that attempts to understand the regulatory roles of PLD1 and PLD2 have had to rely on combinations of two complementary experimental approaches: (a) gene transfection techniques in cells and gene ablation studies in intact animals; and (b) studies of the effects of added PtdOH on cell functions, some of which have aimed to determine whether added PtdOH can substitute for PLD activity when PLD expression or activity in cells has been suppressed.

Initial observations suggested that PLD1$^{-/-}$ and PLD2$^{-/-}$ mice were fairly normal, which seemed to question whether these enzymes were of any importance in vivo. However, closer inspection revealed subtle but important changes in the biology of these mutant animals. For example, starvation normally stimulates macroautophagy in the liver (see Section 8.1.7), and this did not happen in PLD1$^{-/-}$ mice; and β-amyloid-derived peptides interfere with synaptic function in a mouse model of Alzheimer's disease, and this effect was ablated in PLD2$^{-/-}$ animals.

The most compelling evidence for functions of PLDs and the PtdOH they produce relates to a complex hierarchy of effects on cell shape and motility. Activity of PLD2 appears to be especially important when fairly rounded cells in suspension start to adhere to a substratum and to take on a more spread and potentially motile configuration. During this process, 'inside-out' signalling by PLD2-generated PtdOH in the cells in suspension seems to expedite the transition of cell-surface adhesion receptors known as 'integrins' to a more adherent state and to activate a PtdIns4P 5-kinase, so increasing the local PtdIn(4,5)P_2 concentration at those parts of the plasma membrane that are in contract with the substratum. Once the cells have become adherent, PLD2 activity decreases.

Many other cellular effects, particularly enhancement of various aspects of cell motility, chemotaxis and survival, are regulated in some way by changes in the function of PLDs. These often involve direct effects of the PLDs themselves and/or of the PtdOH they produce on the activities of small GTPases such as Rac1 that are direct regulators of cytoskeletal function. PtdOH is likely to be involved as a signalling molecule in many of these processes. However, the complexity of these systems has been emphasized by the discovery that at least part of the effect of PLD2 on cytoskeletal control by Rac2a is a

consequence of a direct regulatory protein–protein interaction between PLD2 and Rac2.

8.5 Ceramide regulates apoptosis and other cell responses

Some simple sphingolipids other than S1P – including sphingosine, its N-acylated derivative ceramide and ceramide 1-phosphate – exert complex regulatory effects on cell function. The best characterized of these actions is the induction of apoptosis, a widespread mechanism of 'programmed cell death', by ceramide that is liberated in cells that are exposed to stresses such as ionizing radiation or treatment with cytotoxic anticancer drugs (e.g. taxol, doxorubicin).

Cells contain a complex mixture of ceramide molecules with a variety of N-acyl chain lengths – from 14C to 26C – and functions. These serve as biosynthetic precursors of sphingomyelin and diverse glycosphingolipids (Figs. 4.26 & 4.28 & Section 4.8) and as important cell regulators. Some ceramides are synthesized from sphingosine and acyl CoAs by a family of ceramide synthases

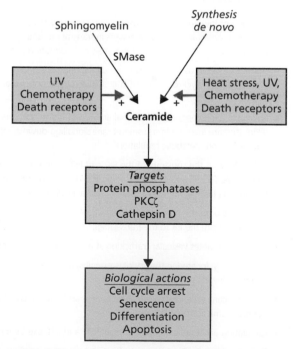

Fig. 8.7 Multiple roles of ceramides in cell regulation. As indicated in the text, 'signalling' ceramides may be formed by multiple routes, at multiple sites in the cell, and have multiple target proteins.

located at the ER, whilst others are formed by the degradation of more complex sphingolipids – for example, by lysosomal sphingomyelin-specific PLCs known as sphingomyelinases. Many investigations of ceramide actions on cells have employed ceramides with short FA chains that readily enter cells from the culture medium. This is a convenient experimental approach, but it is difficult to relate its results to the events within cells that are directly regulated by endogenously formed ceramides – which are diverse populations of molecules, made by several pathways, with multiple molecular targets. Some of the events in signalling by ceramides are summarized schematically in Fig. 8.7.

The sequence of cellular events that leads to apoptosis can be initiated in a variety of ways. Ligation of 'death receptors', such as that for tumour necrosis factor-α (TNF-α), initiate the so-called 'extrinsic' pathway and cell insults, such as radiation injury or DNA damage, trigger the 'intrinsic' pathway. Whatever the provocation, these pro-apoptotic provocations switch on a cascade of intracellular proteases known as caspases, with caspase-9 in a central role.

The permeabilization of mitochondria, causing release of cytochrome c and consequent caspase-9 activation, is a major mode of initiation of apoptosis following the intracellular formation of ceramides. This happens by at least two mechanisms. In one, ceramides somehow have a direct pore-forming effect on the mitochondria. In the second, they serve as activators of protein phosphatases, notably protein phosphatase 2A (PP2A), so activating the dephosphorylation of certain apoptotic regulatory proteins – for example, the activation of pro-apoptotic Bax and inactivation of anti-apoptotic Bcl-2.

Another important action of ceramides seems to be activation of PKCζ, one of the so-called 'atypical' members of the PKC family (see Section 8.1.3). This appears to play a role, for example, in inhibiting the activity of PKB/Akt downstream of insulin (see Section 8.1.5) and provoking insulin resistance.

KEY POINTS

- A variety of low abundance phospholipids serve key regulatory functions in eukaryotic cells: as extracellular or intracellular signalling molecules; as central players in transmembrane signalling processes; as identifiers of intracellular compartments; and as regulators of the exchange of materials or information between cell compartments.

- Lipids that serve as extracellular stimuli include prostanoids, PAF, leukotrienes, lysoPtdOH, S1P, and the endocannabinoid 2-arachidonoylglycerol. Most of the cellular actions of these diverse compounds are initiated by their binding to cognate GPCRs.

- Eukaryotic cells contain a substantial amount of PtdIns and small quantities of up to seven phosphorylated derivatives of PtdIns (PPIn) that are involved in transmembrane signalling downstream of cell surface receptors and in many aspects of membrane trafficking and metabolic regulation.

- PtdIns(4,5)P_2 is the substrate of the PIC and PI3K signalling systems. The various PIC and PI3K isoenzymes are variously activated by multiple GPCRs and by receptor tyrosine kinases and receptor-associated tyrosine kinases. Activated PICs liberate the second messengers Ins(1,4,5)P_3 and 1,2-DAG, and activated Type I PI3Ks synthesize the messenger PtdIns(3,4,5)P_3.

- PtdIns4P is both the precursor of PtdIns(4,5)P_2, notably at the plasma membrane, and a regulator of protein and sphingolipid trafficking from the ER to the trans-Golgi.

- PtdIns3P regulates vesicular trafficking at endosomal compartments and during autophagy.

- PtdIns(3,5)P_2, located in the endolysosomal system, exerts diverse regulatory effects on membrane trafficking and permeability and on cell-wide metabolic switching.

- The endocannabinoids N-acyl-ethanolamine and 2-arachidonoylglycerol are modulators of immune cell trafficking and neurotransmission.

- Circulating and locally generated lysoPtdOH and S1P exert generally opposite effects on the permeability of the vasculature.

- PLD-generated PtdOH has major effects on cell adhesion, shape and motility.

- Intracellular formation of ceramide, particularly in stressed cells, tends to be pro-apoptotic.

Further reading

Brindley DN & Pilquil C (2009) Lipid phosphate phosphatases and signaling, *J Lipid Res* **50**:S225–S230.

Brindley DN, Lin FT & Tigyi GJ (2013) Role of the autotaxin-lysophosphatidate axis in cancer resistance to chemotherapy and radiotherapy, *Biochim Biophys Acta* **1831**:74–85.

Bunney TD & Katan M (2011) PLC regulation: emerging pictures for molecular mechanisms, *Trends Biochem Sci* **36**:88–96.

Chavez JA & Summers SA (2012) A ceramide-centric view of insulin resistance, *Cell Metab* **15**:585–94.

Di Paolo G & De Camilli P (2006) Phosphoinositides in cell regulation and membrane dynamics, *Nature* **443**:651–7.

Dove SK, Dong K, Kobayashi T, Williams FK & Michell RH (2009) Phosphatidylinositol 3,5-bisphosphate and Fab1p/PIKfyve underPPIn endo-lysosome function, *Biochem J* **419**:1–13.

Elphick MR (2012) The evolution and comparative neurobiology of endocannabinoid signalling, *Phil Trans Roy Soc B* **367**:3201–15.

Falasca M & Maffucci T (2012) Regulation and cellular functions of class II phosphoinositide 3-kinases, *Biochem J* **443**:587–601.

Gomez-Cabronero J (2011) The exquisite regulation of PLD2 by a wealth of interacting proteins: S6K, Grb2, Sos, WASp and Rac2 (and a surprise discovery: PLD2 is a GEF), *Cell Signalling* **23**:1885–95.

Hannun YA & Obeid LM (2008) Principles of bioactive lipid signalling: lessons from sphingolipids, *Nat Rev Mol Cell Biol* **9**:139–50.

Ho CY, Alghamdi TA & Botelho RJ (2012) Phosphatidylinositol-3,5-bisphosphate: no longer the poor PIP2, *Traffic* **13**:1–8.

Holthuis JCM & Menon AK (2014) Lipid landscapes and pipelines in membrane homeostasis, *Nature* **510**:48–57.

Kok K, Geering B & Vanhaesebroeck B (2009) Regulation of phosphoinositide 3-kinase expression in health and disease, *Trends Biochem Sci* **34**:115–27.

Maceyka M & Speigel S (2014) Sphingolipid metabolites in inflammatory disease, *Nature* **510**:58–67.

Mayinger P (2012) Phosphoinositides and vesicular membrane traffic, *Biochim Biophys Acta* **1821**:1104–13.

McCartney AJ, Zhang Y & Weisman LS (2014) Phosphatidylinositol 3,5-bisphosphate: low abundance, high significance, *Bioessays* **36**:52–64.

McCrea HJ & De Camilli P (2009) Mutations in phosphoinositide metabolizing enzymes and human disease, *Physiology (Bethesda)* **24**:8–16.

Michell RH (2008) Inositol derivatives: evolution and functions, *Nat Rev Mol Cell Biol* **9**:151–61.

Marat AL & Haucke V (2016) Phosphatidylinositol 3-phosphates – at the interface between cell signalling and membrane traffic. *EMBO J* **35**:561–579.

Rudge SA & Wakelam MJO (2009) Inter-regulatory dynamics of phospholipase D and the actin cytoskeleton, *Biochim Biophys Acta* **1791**:856–61.

Schink KO, Raiborg C & Stenmark H (2013) Phosphatidylinositol 3-phosphate, a lipid that regulates membrane dynamics, protein sorting and cell signalling, *Bioessays* **35**:900–12.

Viaud J, Lagarrigue F1, Ramel D, *et al.* (2014) Phosphatidylinositol 5-phosphate regulates invasion though binding and activation of Tiam1, *Nature Commun* doi 10.1038/ncomms5080.

Young MM, Kester M & Wang H (2013) Sphingolipids: regulators of crosstalk between apoptosis and autophagy, *J Lipid Res* **54**:5–19.

CHAPTER 9

The storage of triacylglycerols in animals and plants

Both plants and animals have evolved to store energy in the form of triacylglycerols (TAGs). TAGs provide a concentrated form of metabolic energy, with a metabolizable energy value of 37 kJ/g. The stored TAGs are almost anhydrous. The fat content of human adipose tissue, for example, is typically 80–90% by weight. This contrasts with energy storage as carbohydrate (glycogen or starch), for which the intrinsic metabolizable energy value is lower (17 kJ/g), and which is stored in hydrated form (glycogen is stored with three times its own weight of water).

Many cell types contain some TAGs in the form of lipid droplets (LD, see Section 5.5), but these are not stored as an energy supply for the whole organism. TAGs that are primarily an energy store for the whole organism are normally stored in specialized cells in animals and plants. There are highly regulated pathways of storage and mobilization that control the amount stored and deliver fat as needed for energy requirements.

9.1 White adipose tissue depots and triacylglycerol storage in animals

Typical energy stores in an adult human of normal weight are around 400–500 g of glycogen (7–8 MJ of energy supply, sufficient for about 24 hours), and around 15 kg of TAG (550 MJ of energy, sufficient for nearly 2 months). Furthermore, the glycogen stores, which are present mostly in skeletal muscle and in liver, are limited (see Section 4.3.1). In contrast, the store of TAG can apparently expand almost without limit.

TAGs are stored in specialized cells known as adipocytes, which make up most of the volume of adipose tissue. There are two types of adipose tissue in mammals,

distinguished histologically, known as 'white' and 'brown' (Fig. 9.1). White adipose tissue is the more abundant and is the main tissue involved in the storage of body fat. Brown adipose tissue has a more specialized function in energy metabolism (see Section 9.2).

White adipose tissue is widely distributed throughout the body. In humans, a large proportion is located just beneath the skin (subcutaneous adipose tissue) and is the tissue that influences the contours of the body. It also provides an insulating and protective layer. Fat contributes a larger proportion of the body weight in women than in men and their subcutaneous adipose tissue is correspondingly more abundant. The tissue is also located internally, surrounding some organs (e.g. perirenal adipose tissue around the kidneys, pericardial adipose tissue around the heart) and along the intestinal tract (mesenteric adipose tissue), and in the omentum. (The mesentery and omentum are folds in the lining of the abdominal cavity, the peritoneum, which seem to hang down and suspend the intestinal tract.) These internal depots are sometimes called 'visceral fat', although a stricter definition of visceral fat is usually taken to be those adipose depots whose blood drains into the hepatic portal vein (mesenteric and omental adipose tissue).

Although adipose tissue contains many types of cells, the ones responsible for fat storage are the adipocytes, often simply called 'fat cells', which are bound together with connective tissue and supplied by an extensive network of blood vessels (Fig. 9.1). The tissue also contains smooth muscle and endothelial cells (in blood vessels), cells of the immune system, and diverse other cells, all of which are collectively known as the stromavascular fraction (SVF). In the laboratory, the SVF is separated from adipocytes after digestion of the

Lipids: Biochemistry, Biotechnology and Health, Sixth Edition. Michael I. Gurr, John L. Harwood, Keith N. Frayn, Denis J. Murphy and Robert H. Michell.
© 2016 John Wiley & Sons, Ltd. Published 2016 by John Wiley & Sons, Ltd.

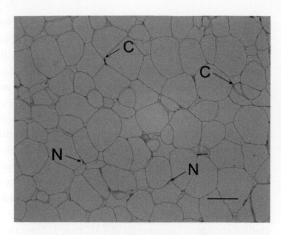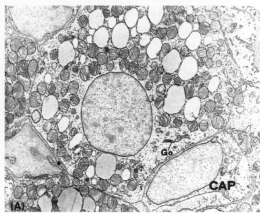

Fig. 9.1 Histology of white and brown adipose tissues. Left, white adipose tissue under the light microscope. Each cell consists of a large lipid droplet (white) surrounded by a narrow layer of cytoplasm. The nucleus (N) can be seen in some cells. There are capillaries (C) at the intersections of the cells: some are marked. The scale bar represents 100 μm (0.1 mm). Picture courtesy of Rachel Roberts. Right, an electron micrograph of brown adipose tissue. In this high-powered view, one adipocyte nearly fills the picture. Unlike the white adipocytes shown above, it has multiple lipid droplets (white areas) and many mitochondria (white adipocytes also have mitochondria, but not so densely packed). CAP is a capillary adjacent to the cell, Go the Golgi apparatus. The picture represents a width of about 14 μm (i.e. it is about 14 times more enlarged than the left hand picture). From S Cinti (2001) The adipose organ: morphological perspectives of adipose tissues. *Proc Nutr Soc* **60**: 319–28, reproduced with permission from Cambridge University Press. The combined picture is reproduced from KN Frayn (2010) *Metabolic Regulation: A Human Perspective*, 3rd edn. Wiley-Blackwell, Oxford. (See the Colour Plates section.)

connective tissue with collagenase (adipocytes float in the centrifuge, whereas other cells sink). The SVF contains adipocyte precursor cells, known as preadipocytes, which can be stimulated to differentiate into new adipocytes. It is also recognized that mature adipocytes may die, so that there is a continuous turnover of adipocytes; in adult humans, about 10% of fat cells are renewed annually (the average age of a fat cell is about 10 years).

In a mature adipocyte, most of the cell volume is occupied by the large LDs consisting of TAGs with some cholesterol and a phospholipid outer monolayer (see Section 5.5.4). White adipocytes are unusual in being able to expand to many times their original size by increasing their content of stored TAGs. They are in general the largest cells in the human body other than ova, commonly reaching 0.1 mm in diameter.

9.1.1 Adipocyte triacylglycerol is regulated in accordance with energy balance

The TAG content of adipocytes is strictly regulated and reflects whole-body energy balance. The energetics of the body are governed by the first law of thermodynamics, which says in effect that if there is any difference

between energy intake and energy expenditure, that difference is reflected in a change in the body's energy stores. In the very short term (minutes), that might mean a change in body temperature, but in the rather longer term (hours) it will reflect changes in glycogen stores, and over longer periods still – say of the order of a day or more – it will be reflected in changes in the amount of TAGs stored in adipocytes. This is summarized in Fig. 9.2.

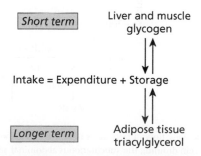

Fig. 9.2 Energy balance and the body's TAG store. The first law of thermodynamics dictates that any difference between energy intake and energy expenditure is reflected in a change in the body's energy stores. In the short term, this means changes in glycogen stores, but over longer periods of time it is translated exactly into changes in the body's store of TAG.

Note that if energy intake exceeds energy expenditure, then the 'storage' term is positive: TAG stores increase. But if the opposite is true ('negative energy balance', as during fasting or dieting or prolonged periods of exercise) then TAG stores will decrease.

Because of the role of adipocytes in long-term energy storage, they are sometimes described as 'inert'. This is far from the truth. The TAG content of adipocytes changes rapidly during each typical day, albeit by only a tiny percentage of the total. During the overnight fast, TAG stores are mobilized as described below (see Section 9.1.2) by hydrolysis to liberate nonesterified fatty acids (NEFAs). NEFAs are transported in the plasma (bound to albumin), to other tissues (e.g. liver, skeletal muscle), where they constitute an important metabolic fuel when dietary glucose is not available. After each meal, the action of lipoprotein lipase (LPL) in the capillaries of adipose tissue, whose activity is up-regulated by insulin (see Section 7.2.5 & Fig. 7.12) liberates fatty acids (FAs) from circulating TAGs (mainly in the chylomicron fraction,

carrying dietary fat: see Section 7.2.5), which flow into the adipocytes and are re-esterified to rebuild TAG stores. This flow of FAs in and out of adipocytes (across the capillary wall into the plasma) during the normal daily cycle has been termed 'transcapillary fatty acid flux' and has been measured in healthy human volunteers (Fig. 9.3). During a typical day, the total area under the curve above the baseline (representing fat storage) must almost exactly match the total area under the curve below baseline (fat mobilization), and so the TAG stores remain constant from day to day. However, a slight imbalance of energy intake and expenditure will shift these relative proportions above and below the baseline, resulting in a gradual change in the amount of TAG stored. Also potentially superimposed on these changes resulting from FA uptake and release is the generation of new FAs by the process of *de novo* lipogenesis (DNL) from glucose (mainly) and amino acids (for pathway see Section 3.1.3).

This dynamic view of adipocyte function is reinforced by measurements of the turnover of the total body TAG

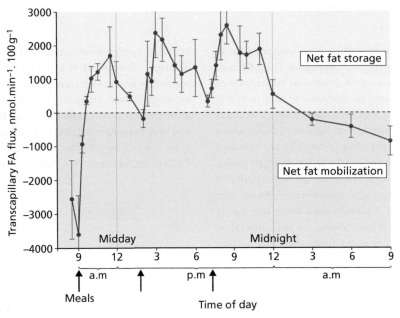

Fig. 9.3 The daily flow of fatty acids in and out of adipose tissue. Measurements were made in healthy volunteers by catheterization of a small vein draining the subcutaneous abdominal adipose tissue (and comparison with the arterial inflow). The volunteers fasted overnight before the experiment, then were fed three meals (arrows) of equal nutrient content, to match their energy requirements. Values < 0 represent a net loss of fatty acids from adipocytes (fat mobilization); values > 0 represent FAs flowing into the adipocytes (net fat storage). Notice a state of marked fat mobilization in the initial fasting state, switching to a state of fat storage after meals. Note also that the state of fat storage is maintained throughout most of 24 h. Based on data in T Ruge, L Hodson, J Cheesema *et al.* (2009) Fasted to fed trafficking of FAs in human adipose tissue reveals a novel regulatory step for enhanced fat storage. *J Clin Endocrinol Metab* **94**: 1781–8.

pool using radioactive tracer measurements. These indicate that the body's TAG pool turns over (i.e. FAs are lost to oxidation and replaced by new fat storage) about once every year or two. The TAG store of each of the adipocytes (which, as noted above, are continuously being replaced) is itself replaced about six times during the life of the cell.

These individual pathways of fat storage and mobilization are described in more detail in the following sections and in Chapter 4.

9.1.2 Pathways for fat storage and mobilization in white adipose tissue and their regulation

9.1.2.1 Uptake of dietary fatty acids by the lipoprotein lipase pathway

As was described in Section 6.1.2.2, the FA composition of an animal's adipose tissue TAGs reflects that of its diet. This has been studied extensively in humans, mainly in the hope that measurement of adipose tissue TAG composition (which is easily done with a tiny needle-biopsy through the skin) will be a biomarker for dietary FA composition. When dietary records of FA intake are compared with adipose tissue composition, there are strong correlations for the essential fatty acids (EFA; Section 6.2.2) (e.g. correlation coefficients of ~ 0.5–0.7 for linoleic acid, $18{:}2n$-6), whereas correlations for FAs that may be synthesized *de novo* in humans are weaker (typically 0.1–0.6 for palmitic acid ($16{:}0$) for example – see Hodson *et al.* (2008) in **Further reading** for more information). The implication is that dietary fat may be stored directly in adipose tissue, but adipose tissue TAG composition also reflects biosynthesis *de novo* within the body.

For people eating typical Western diets (which tend to be rich in fat and easily digestible carbohydrates), the major route for deposition of adipose tissue TAGs is that mediated by LPL in the capillaries (see Section s 9.1.1 & 7.2.5). Adipose tissue LPL is specifically up-regulated by insulin (at various levels from transcriptional to activation or degradation), whereas insulin tends to suppress LPL activity in muscle (heart and skeletal muscle) (see Fig. 7.12). The mechanisms were described in Section 7.2.5. This directs dietary FAs, carried in the chylomicrons (see Section 7.2.5), to adipose tissue for storage. Furthermore, LPL preferentially acts upon chylomicron-TAGs compared with VLDL-TAGs. The reason for this is not entirely clear although it may relate to the greater

size of chylomicrons: it has been estimated that 30–40 LPL molecules act simultaneously on a chylomicron during TAG hydrolysis. LPL hydrolyses the TAGs in the chylomicrons, releasing FAs that are taken up into adipocytes. As discussed for intestinal absorption of FAs (see 7.1.2), FA transport across the cell membrane may involve passive diffusion by 'flip-flop' across the adipocyte membrane but is now thought to occur mainly by means of transporter proteins. In the adipocyte these would be CD36 (fatty acid translocase, FAT) and the fatty acid transport protein (FATP) family members FATP1 and FATP4 (see Section 3.1.1 & 7.1.2).

FAs inside the cell are rapidly esterified or 'activated' to form acyl-CoA (see Section 3.1.1), and this is the starting point for their esterification to glycerol 3-phosphate (G3P) in the pathway of TAG biosynthesis (see Section 4.1.1). This pathway of esterification is itself up-regulated after meals. Insulin stimulates FA esterification in adipocytes. The locus of action is not absolutely clear and it may act at several steps. In addition, insulin may increase the provision of G3P by stimulating glucose uptake and glycolysis. There may be additional stimulation of this pathway through the action of Acylation Stimulating Protein (ASP) as described in Section 7.3.5.

9.1.2.2 *De novo* lipogenesis and adipose tissue triacylglycerols

There has been much controversy over the importance of DNL in the storage of TAGs in adipose tissue. At a whole-body level, DNL does not seem to contribute to net fat storage under normal conditions in humans eating typical 'Western' diets that contain plenty of fat. The evidence for that is largely from measurements of the whole-body respiratory exchange ratio (RER; often called respiratory quotient, RQ). RER is the ratio of the rate of expiration of CO_2 to the rate of consumption of O_2. According to simple stoichiometric equations, the ratio is 1.00 for the oxidation of sugars, but 0.70 for the oxidation of fats. However, DNL results in a ratio > 1.00 (one reason is that 3-carbon units, e.g. pyruvate, are converted into 2-carbon units, acetyl-CoA, for biosynthesis of FAs, with loss of 1 CO_2 per 2 carbons added to the chain). At a whole-body level, an RER > 1.0 implies a net synthesis of fat: i.e. fat synthesis at a greater rate than fat oxidation. Lower values of RER do not mean that no DNL occurs, just that fat oxidation occurs at a greater rate than DNL. RER values consistently > 1.0 are not normally seen in people on typical 'Western' diets.

A series of studies by Acheson and colleagues at the Institute of Physiology in Lausanne, Switzerland, in the 1980s set DNL in perspective. They fed healthy volunteers a single very large carbohydrate load (480 g) in an attempt to show conversion into fat, but RER values were only ever transiently > 1.0, and over the next 10 h there was, on average, a net oxidation of 17 g fat, together with a net gain of 346 g glycogen. In later experiments, the same investigators fed volunteers for one week diets that provided more energy than their energy requirements (i.e. overfeeding), with the excess energy mainly as carbohydrate. Under those conditions, RERs soon rose to values consistently > 1.0 and remained so until the diet was stopped. This shows the physiological role of the pathway of DNL. When there is an excess of carbohydrate above energy requirements, the excess is channelled at first into glycogen: but, as noted earlier, glycogen stores are limited to a few hundred grams and then the only option for the body to dispose of excess carbohydrate is to convert it into fat. The same principle is used in the production of *foie gras*. Ducks or geese are overfed starch, in the form of corn, which leads to massive fat deposition, mainly in the liver.

It should be noted that similar studies have not been conducted in people who habitually eat diets with a very low fat content: it may be that DNL plays a bigger role than in people on a 'Western' diet. A similar consideration may explain why DNL is more readily demonstrated in rodents. Laboratory rats and mice are usually maintained on diets with a very low fat content, so DNL is a necessary pathway.

However, although the pathway of DNL may contribute to adipose tissue TAG stores under some conditions, it does not necessarily occur in the adipose tissue. In fact the liver is likely to be the main site of DNL (although this is difficult to confirm in humans), and the FAs so produced may then be exported as very low density lipoprotein (VLDL) TAGs (see Section 7.2.3) and taken up by adipose tissue by the LPL pathway.

Rat adipocytes incubated in vitro with radiolabelled glucose and insulin will readily convert the glucose into lipid. Much of the radiolabel in lipid will be in the glycerol moiety of TAGs (often misleadingly called DNL in the literature) but some will be in the fatty acyl moieties, representing true DNL. However, this is much more difficult to demonstrate in mature human adipocytes. Furthermore, mature human adipocytes have low levels of expression of some of the enzymes of DNL, for instance ATP: citrate lyase (see Section 3.1.3). In contrast, as human adipocytes differentiate from preadipocytes, they readily make TAGs from glucose or amino acids. This might be a key to the role of adipocyte DNL in humans: it could be a pathway that initiates the storage of TAGs, before the tissue architecture (blood vessels etc.) necessary for the LPL pathway is operative. However, that is still speculative.

9.1.2.3 Fat mobilization from adipose tissue

TAGs are stored to be used as an energy source when required. In normal daily life, that means during the overnight fast (see Fig. 9.3) or during exercise. In other circumstances it might mean during prolonged energy deficit, as during fasting for whatever reason or during illness, when intake is reduced (and energy expenditure may be high). TAGs stored in adipose tissue are delivered to other tissues by hydrolysis (lipolysis) to release NEFAs, which are released into the plasma bound to albumin.

The pathway for adipose tissue fat mobilization, or lipolysis, was described in outline in Section 4.2.2. More details are given in Box 9.1. This pathway is stimulated by hormonal and nervous stimuli, but it is also markedly suppressed by insulin. The effect of insulin is the predominant one during normal periods of feeding and fasting overnight (see Fig. 9.3), suppressing fat mobilization after meals when fat storage is the predominant pathway. Fat mobilization is stimulated by catecholamines: either circulating adrenaline (US: epinephrine), or noradrenaline (US: norepinephrine) released from sympathetic nerve endings in close proximity to the adipocytes. Catecholamines act via β-adrenergic receptors of the G protein-coupled receptor family (GPCRs – see Section 7.3.4) to increase cyclic AMP concentrations. These activate protein kinase A and activate the lipolysis cascade as described in Box 9.1. Insulin, in contrast, acts through the insulin receptor to activate a particular phosphodiesterase (PDE3B), which reduces cyclic AMP concentrations. Constitutively active phosphatases presumably then dephosphorylate the key regulatory proteins, leading to a suppression of lipolysis: i.e. lipolysis is a pathway that normally runs at a low rate until stimulated.

Catecholamines are released during stress of various sorts. There have been several reports of very high plasma NEFA concentrations in 'stress' situations, for instance in racing drivers before a race. It had long been assumed that catecholamines are also the main

Box 9.1 The pathway of triacylglycerol (TAG) hydrolysis in white adipocytes

In Chapter 4, one intracellular lipase, known as hormone-sensitive lipase (HSL, see Section 4.2.2), was mentioned. HSL was for many years thought to be the principal TAG lipase of adipocytes. However, in about the year 2000, mice were bred that were deficient in HSL. Their phenotype was relatively normal and isolated adipocytes showed some residual activity of lipolysis, with an accumulation of diacylglycerols (DAGs), implying that HSL acts as a DAG lipase. These observations led to identification of an additional lipase, now called adipose triglyceride [triacylglycerol] lipase (ATGL). Current understanding is that these lipases act sequentially, with a monoacylglycerol (MAG) lipase (MGL) that was characterized in the 1970s. Each releases one FA, and MGL also releases free glycerol. Free glycerol is exported from the adipocyte and will eventually be taken up by the liver (the activity of glycerol kinase is normally very low in white adipocytes, so little of the glycerol is converted into G3P and used in re-esterification of FAs).

The pathway of adipocyte lipolysis is highly regulated. HSL is activated on a very short-term basis by reversible phosphorylation by protein kinases A and G (see text). This is brought about by changes in the intracellular concentrations of cyclic AMP and cyclic GMP. The regulation of ATGL is less direct. ATGL has an essential coactivator protein called CGI-58 (comparative gene identification member 58). There is additional regulation at the surface of the LDs where these lipases must act. As described in Section 5.5.4, LDs are coated with specific proteins, a major one of which in the adipocyte is perilipin A (or perilipin 1, human gene *PLIN1*). Perilipin A is itself subject to phosphorylation, under the same conditions as HSL, and this appears to cause a conformational change that allows lipases to access the lipid droplet. CGI-58 is bound to perilipin in the unstimulated state, but released, and hence is free to activate ATGL, upon perilipin phosphorylation.

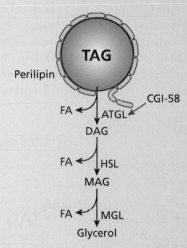

The figure shows a lipid droplet surrounded by perilipin. Upon phosphorylation of perilipin, and also of HSL, a conformational change in perilipin (i) allows the lipases access to the LD and (ii) allows CGI-58 to dissociate and thus to activate ATGL. NB: The lipases act much closer to the surface of the LD than is shown here. Fatty acids (FA on diagram) may be released from the cell for transport to other tissues, although a proportion is always reactivated (thio-esterified to CoA) and re-esterified to form TAGs. Glycerol is exported from the cell via an aquaglyceroporin channel.

stimulators for lipolysis during exercise. This is certainly so in rodents. In humans, exercise-induced lipolysis is blocked to a considerable extent by administration of a β-adrenergic antagonist ('β-blocker'), but not entirely so. However, in the year 2000 Max Lafontan and colleagues in Toulouse, France, discovered a new pathway for activation of human lipolysis during exercise. It is mediated by the natriuretic peptides, atrial and brain natriuretic peptide (ANP, BNP). These hormones normally regulate renal sodium excretion. However, they are released from the heart (both ANP and BNP, despite the latter's name) during the stress of exercise, and act also on specific receptors on adipocytes. These receptors are coupled to guanylate cyclase (alternative name guanylyl cyclase), which produces cyclic GMP, which in turn activates protein kinase (PK) G, and brings about similar changes to those induced by PKA. This pathway is specific to humans so far as is known at present.

There are also catecholamine-induced inhibitory pathways. α-Adrenergic receptors are present on adipocytes and are coupled through Gi proteins to inhibit adenylate cyclase, so leading to an inhibition of lipolysis. This effect is predominant when catecholamine concentrations are very high: it has been demonstrated in dogs and there is some evidence for its operation in humans with extremely high catecholamine concentrations after physical injury, when circulating NEFA concentrations may be relatively low. It also becomes predominant as fat cells increase in size (the expression of α-adrenergic receptors relative to β-adrenergic receptors increases as fat cells expand) and this may be one reason why it is difficult to mobilize fat once it has been stored in excess. The relative expression of α-adrenergic and β-adrenergic receptors also differs between different adipose depots. In humans, there is a clear predominance of α-adrenergic effects in the fat of the legs and buttocks (lower-body, or gluteofemoral fat: see Section 10.4.1.2 for more information), whereas in abdominal fat β-adrenergic effects predominate.

The regulation of fat deposition and mobilization in white adipose tissue is summarized in Fig. 9.4.

9.2 Brown adipose tissue and its role in thermogenesis

9.2.1 Brown adipose tissue as a mammalian organ of thermogenesis

Brown adipose tissue is so-named because, unlike white adipose tissue, it is rich in mitochondria and iron-containing pigments (hence its brown colour) involved in O_2 transport (Fig. 9.1). It is clearly a tissue that is specialized in oxidation (in fact, of FAs) but it is also a site of TAG storage, and hence is classed as an adipose tissue. Its role is to oxidize FAs in order to produce heat, which can then be distributed to other organs via the circulation.

Brown adipose tissue is found only in mammals and it has been argued that it gave mammals an evolutionary advantage in being able to survive in cold conditions. It is present in smaller animals and neonates of larger animals that have a large body surface area (through which heat is lost) in relation to their body mass (in which heat is generated). It plays an important role in hibernating animals, generating heat to allow them to warm up from their hibernating state.

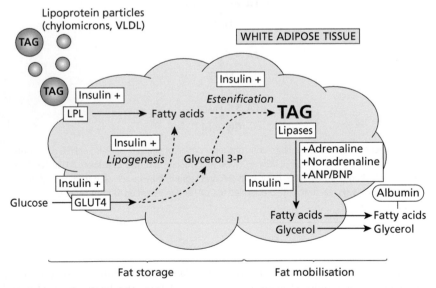

Fig. 9.4 Regulation of fat storage and mobilization in white adipose tissue. Fat storage is the process of biosynthesis of TAGs; fat mobilization (or lipolysis) is the process of hydrolysis of the stored TAGs to release NEFAs into the plasma (bound to the carrier protein albumin), so that they can be taken up by other tissues. ANP/BNP, atrial and brain natriuretic peptides; LPL, lipoprotein lipase; G3P, glycerol 3-phosphate; VLDL, very-low-density lipoprotein. GLUT4 is the insulin-regulated glucose transporter expressed in white adipocytes. The major pathways and main sites of hormonal regulation are shown: a plus sign indicates stimulation, a minus sign inhibition. Dashed lines show multiple enzymic steps. NB: Adrenaline/noradrenaline (US: epinephrine/norepinephrine), ANP/BNP and insulin regulate the lipolysis pathway by modulation of cyclic AMP and cyclic GMP concentrations (not shown here). More details of the lipases involved in fat mobilization are given in Box 9.1. Reproduced from KN Frayn (2010) *Metabolic Regulation: A Human Perspective*, 3rd edn. Wiley-Blackwell, Oxford.

The heat is generated by oxidation of FAs, which are mainly obtained from the tissue's own stores of TAGs (see Section 9.2.2). The TAGs are stored in multiple, small LDs, 1–3 µm in diameter (Fig. 9.1). The cells contain a large number of specialized mitochondria, adapted for oxidizing the FAs from the LDs that they surround (e.g. they are densely packed with cristae, increasing their oxidative capacity).

The tissue is activated by release of noradrenaline from sympathetic nerves, with which it is richly supplied. This leads to activation of β-adrenergic receptors, which are coupled to both TAG hydrolysis (as in white adipose tissue; see Section 9.1.2.3) and regulation of blood flow: both lipolysis and blood flow increase on sympathetic activation. The increase in blood flow is important as the heat generated from FA oxidation needs to be transported out of the brown adipose tissue depot to the rest of the body.

It was believed for many years that brown adipose tissue is absent from adult humans. That view changed in the early 2000s with increased use of the technique of positron-emission tomography (PET) for scanning patients to look for tumours. This use of PET depends on the tumour, as a site of high metabolic activity, taking up a radioactive tracer that is an analogue of glucose ([18F]fluorodeoxyglucose). Sites of tracer uptake can then be imaged externally. Radiologists using this technique began to notice that many patients apparently displayed 'tumours' in the region of the neck and chest that had the unexpected property of being arranged symmetrically around the body's midline (Fig. 9.5). Eventually it was recognized that these are not tumours, but depots of brown adipose tissue, also taking up glucose

because they are so metabolically active. The identification of brown adipose tissue has been confirmed by showing that these depots are activated in the cold, and also by direct biopsy.

This finding is of great interest in discussions about how to manage, or treat, obesity (see Section 10.4.1). Brown adipose tissue is the one tissue in which excess calories can be 'burned off', since (as will be described below) FA oxidation is not coupled to ATP biosynthesis. In the 1980s and early 1990s there were many attempts by the pharmaceutical industry to activate brown adipose tissue to oxidize excess FAs, mostly with novel β-adrenergic agonists that would, it was hoped, specifically target brown adipocytes. This field of research came to a halt for two reasons: all the agents tested clinically had adverse side effects, coupled with very low efficacy; and, as noted above, it was gradually accepted at this time that brown adipose tissue was not present in significant quantities in adult humans. However, the research may now be resurrected with a greatly improved understanding of the physiology and cell biology of brown adipose tissue.

Brown adipocytes, like white adipocytes, differentiate from precursor cells in the tissue. However, they have a different cellular origin from white adipocytes. They come from a precursor more closely related to muscle cell precursors; in the laboratory, it is possible to isolate such cells and differentiate them into either brown adipocytes or skeletal muscle cells depending upon culture conditions. They are present in specific anatomical depots. In rodents the most prominent depots are between the shoulder blades (interscapular depots) but in hibernating mammals there are extensive depots wrapped around the

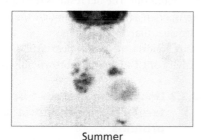
Summer

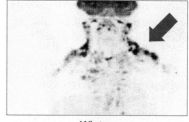
Winter

Fig. 9.5 Brown adipose tissue (BAT) depots around the neck region displayed by Positron Emission Tomography (PET). The figure shows two scans on one individual, in the summer (left) and in the winter (right), when the depots are more prominent. The BAT depots show up as dark areas of high metabolic activity (arrow shows one such area). In this individual they are particularly present in the supraclavicular/cervical areas. Pictures courtesy of Professor ME Symonds, University of Nottingham, UK and representative of patients studied in the paper: ITH Au-Yong, N Thorn, R Ganatra, AC Perkins & ME Symonds (2009) Brown adipose tissue and seasonal variation in humans. *Diabetes* **58**: 2583–7.

abdominal aorta, whose function is to warm the blood reaching all other organs. In humans, the most prominent depots are in the upper chest and neck region (Fig. 9.5), and specifically these surround arteries: the neck arteries supplying the brain, the vertebral arteries, and there are smaller depots around the renal arteries.

It is now understood that there is a further type of adipocyte that is of the same cellular origin as white adipocytes, but which may develop characteristics of brown adipocytes: increased mitochondrial density and expression of uncoupling protein-1 (UCP1: see Section 9.2.2). In fact these adipocytes may 'transdifferentiate' between brown and white phenotypes. They are known as 'brown in white' or 'brite' adipocytes, or as 'beige' adipocytes. (Another terminology is that these are 'recruit-able' brown adipocytes, whereas those of the interscapular depot in rodents are known as 'constitutive' brown adi-pocytes.) These 'brite' adipocytes exist within normal white adipose tissue depots. In rodents their formation may be induced by treatment with a peroxisomal prolif-erator activated receptor (PPAR)-γ agonist. Their existence in humans is still under investigation.

Up to half the metabolic rate of small mammals at normal ambient temperatures may be accounted for by brown adipose tissue, and considerably more at low temperatures. It is the only tissue involved in the phe-nomenon of nonshivering thermogenesis: the ability of an animal to increase heat production in the cold without the muscular contractions associated with shivering. There is, of course, tremendous interest in knowing whether, and to what extent, brown adipose tissue might contribute to metabolic rate in adult humans. One esti-mate, based on imaging the tissue and making various assumptions about its metabolism, was that it contrib-uted about 5% to metabolic rate on average. That would mean a greater contribution in some individuals, but more precise information is lacking at present.

9.2.2 Uncoupling proteins dissociate fatty acid oxidation from ATP generation

The unique feature of brown adipose tissue is the presence, in its mitochondrial membrane, of the 'uncoupling pro-tein-1' (UCP-1; gene name *UCP1*) or, in older literature, thermogenin. This is the molecular key to thermogenesis. Normally the operation of the electron transport chain (carrying reducing equivalents from the oxidation of FAs or other substrates) leads to the pumping of protons from the mitochondrial matrix across the inner mitochondrial membrane. This builds up a 'proton gradient' that then provides the energy needed for ATP synthesis, i.e. phos-phorylation of ADP by ATP synthase. UCP1 functions as a proton transporter and is situated in the inner mitochon-drial membrane. Therefore, in the mitochondria of brown adipose tissue, UCP 'short-circuits' the mechanism for ATP synthesis by allowing the protons to leak back into the mitochondrial matrix. Thus, the energy liberated by oxi-dation of substrates is not 'captured' in ATP for chemical purposes and is released instead as heat.

It should be noted that ATP hydrolysis in metabolic processes ultimately leads to heat generation in all tis-sues, unless the energy associated with ATP is captured in a biosynthetic reaction. What is different in brown adi-pose tissue is that oxidative pathways are not restricted by accumulation of ATP; we should envisage uncoupling by UCP1 as unblocking oxidative pathways so that they can run in an unregulated manner.

Noradrenaline does not activate UCP1 directly, although it does activate pathways for UCP1 gene expression. Rather, UCP1 activity is controlled in the short term by long-chain FAs. Therefore, as TAG hydrol-ysis is increased under noradrenaline stimulation, FAs will be released that will both activate UCP1 and act as substrates for oxidation.

The involvement of FAs in UCP1 action may be more intimate even than that. One hypothesis for UCP1 action is that it transports not protons, but FA anions, across the mitochondrial membrane. This has been observed exper-imentally. In this theory, protonated (uncharged) FAs diffuse into the mitochondrial matrix by the flip-flop mechanism (see Sections 3.1.1 & 7.1.2); UCP1 transports FA anions out. The net result is the equivalent of inward transport of a proton. Although there is experimental evidence that this mechanism can operate in isolated mitochondria, it is not clear whether it explains the activity of UCP1 in intact brown adipose tissue.

9.2.3 Uncoupling protein-1 belongs to a family of mitochondrial transporter proteins

The UCP1 of brown adipose tissue is a member of the large family of mitochondrial anion transporter proteins. Within this family, several closely related proteins have been found. The mammalian proteins UCP2 (expressed in many tissues) and UCP3 (expressed mainly in skeletal muscle) have close sequence homology with UCP1.

More distantly related are UCP4 and UCP5 (the latter also known as brain mitochondrial carrier protein-1, BMCP1), which are predominantly expressed in brain and nervous tissues. These other proteins are not thought to play roles in whole-body thermogenesis and their physiological function may be more to do with FA transport. For instance, UCP3 expression increases in conditions of increased FA availability, such as starvation. In starvation, mechanisms that lead to energy dissipation are unlikely to become prominent and hence it is generally considered that this is not the function of UCP3. However, the exact physiological roles of the other UCPs are not clear.

Members of this family are also found in plants and fungi, indicating an early origin in the eukaryote evolutionary tree. Their functions are not known, although it has been suggested that one closely related protein found in some plants, including potatoes (in which it is known as *Solanum tuberosum* uncoupling protein, StUCP), may play a role in warming up tubers for sprouting.

9.3 Lipid storage in plants

Many plant species accumulate storage lipids in their principal reproductive propagules, namely their fruits, seeds and pollen grains. In contrast, plants do not normally accumulate long term storage lipids in their vegetative tissues, such as leaves, stems and roots. Although the majority of vegetative plant cells contain small numbers of cytosolic and plastidial nonpolar LDs, recent evidence suggests that these droplets have a variety of nonstorage functions that include signalling, stress responses and involvement in overall lipid and membrane trafficking. Efforts are now underway to increase the TAG content of some vegetative tissues, especially leaves, so that they might be used as more efficient sources of renewable biomass-based fuels. Each of the major sites of lipid storage is discussed in more detail below.

9.3.1 Major sites of lipid storage
9.3.1.1 Fruits
A fruit is a maternally derived structure that normally encloses the seeds of a plant. In some plants, such as the coleoptiles of cereal species, the fruit is reduced to a single cell layer on the outside of the seed or grain. In other cases the fruit is a more substantial structure containing a skin or pericarp that encloses a relatively large fleshy mesocarp and an inner endocarp within which the seeds are located. When it is fully ripe, the fleshy mesocarp acts as a nutritious bait that attracts animals which consume

the fruit. The seeds generally pass through the gut of the animal and are thereby dispersed well away from the mother plant. While most fleshy fruits tend to accumulate starches and/or simple sugars as attractants, in some plants the mesocarp becomes highly enriched in storage lipids. Examples include commercially important edible fruits such as olives, oil palm, and avocado. The mesocarp oil of the oil palm plant, *Elaeis guineensis*, is the most important vegetable oil in terms of its global production and is widely used either as an edible oil or for a variety of industrial uses (see Chapter 11).

Olive oil is principally derived from the oil-rich fruit mesocarp of the olive tree, *Olea europea*. Olive oil is especially prized for the unique organoleptic qualities that give rise to the distinctive bouquet of flavours and aromas that are characteristic of cold-pressed or virgin oil. Many of these flavour and volatile aroma compounds are derived from partial oxidative breakdown of the oil as the olive fruits ripen. This process is mediated by lipoxygenases (see Sections 3.2.5 & 3.2.6) that are produced in the final stages of ripening and the volatile lipid derivatives act as a signal to animals indicating that the fruits are ready for consumption.

Although oleogenic fruits can accumulate large amounts of TAGs, these have not evolved for long term storage for the use of the plant, and therefore their mechanism of accumulation differs somewhat from true lipid stores in seeds and pollen grains. Hence the TAGs in fruit tissues are produced in the form of LDs, as described in Chapter 5, Section 5.5, but they do not contain a proteinaceous boundary layer. This means that the droplets tend to fuse with one another to produce large irregular structures in the region of 5–10 μm diameter that eventually take up much of the cytosolic volume. As the fruit ripens, mesocarp cell walls break down and the TAGs are released with some of them becoming oxidized as discussed above. The result is that, in oily fruits such as olive or avocado, the ripe mesocarp turns into a relatively soft oily mush that is highly attractive to most animals including humans.

9.3.1.2 Seeds
In contrast to fruits, seeds accumulate storage compounds for the benefit of the germinating embryo as it seeks to establish itself as a new plant. Most seeds accumulate a mixture of lipids, carbohydrates and proteins, although the proportions of these storage compounds vary considerably in different species. In the major commercial oilseeds, such as rapeseed (called 'canola' in Canada) and sunflower, storage lipids typically make up almost half of total seed weight. In plants

with large oil-rich seeds, such as nuts, storage lipids can account for as much as 75% of total seed weight. In contrast, nonoleogenic seeds such as peas and beans are highly enriched in carbohydrates and/or proteins and can contain as little as 2–5% w/w lipid.

Within a given seed, storage lipids can accumulate within the developing embryo itself and/or in the surrounding maternally–derived endosperm layer. In rapeseed and sunflower seeds the endosperm is consumed during seed development and therefore all of the storage lipids are accumulated in the cotyledons of the developing embryo that eventually constitutes over 90% of the volume of a mature seed. In mature castor beans the embryo remains as a tiny structure and most of the storage lipids accumulate in the large endosperm that makes up most of the seed. In some oil-rich cereals such as maize, the relatively large endosperm mostly accumulates carbohydrates, while most of the storage lipids accumulate in the much smaller embryo.

As described in Chapter 5, Section 5.5, storage lipids in most seeds are accumulated as small droplets of about 1 μm diameter that are enclosed in a protein-rich coat that mostly consists of oleosins and caleosins. These proteins stabilize the small LDs and enable them to the withstand the rigours of seed desiccation, a prolonged dormancy period that can last for many years, and the subsequent sudden shock of germination, when the LDs are rapidly broken down to provide energy for the newly emerging seedling. The mobilization of storage LDs is a highly efficient process that begins with the proteolytic digestion of oleosins. The removal of the oleosin coat enables lipases to gain access to the TAG core of the droplets. TAG lipolysis results in the release of large quantities of NEFAs that are channelled for β-oxidation to specialized organelles, called glyoxysomes, that line up immediately adjacent to the disintegrating LDs.

9.3.1.3   Pollen grains

The pollen grains of higher plants transport the male gametes and, like seeds, require dispersal as propagules in order to reach a receptive female flower. Many of the more advanced plant species produce sticky pollen grains that adhere to insect vectors such as bees and moths, thus ensuring their efficient propagation. The sticky covering on the external wall of such pollen grains is a lipid-rich mixture produced by the nutritive maternal cells of the tapetum during pollen development. The tapetal cells initially accumulate specialized cytoplasmic lipid storage structures termed tapetosomes and these are released from the lysing cells following apoptosis of the tapetum. The released lipids form a sticky mixture that coats the pollen grains shortly before they finish their development. Failure to produce the lipidic pollen coat results in male sterility for such plants.

In addition to their external lipid coat, entomophilous or insect-propagated pollen grains often accumulate large numbers of small oleosin-enclosed storage LDs in their cytosol in a manner very similar to that in seeds. Pollen grains have a much shorter lifespan than seeds, typically measured in hours rather than months or years, but they are also propagules that must be dispersed into the environment and subsequently germinate as part of the fertilization process. Once a pollen grain lands on a receptive female flower, it produces a long tube that grows into the female stigma tissue until it reaches an ovary where the male nuclei are delivered for fertilization. The lipid stores in pollen grains provide much of the energy required for germination and the subsequent establishment of the pollen tube. Many of the enzymes involved in storage lipid biosynthesis and mobilization in pollen grains are encoded by genes that are closely related to those involved in seed lipid metabolism, the only significant difference being in their tissue-specific expression.

KEY POINTS

Carbohydrates and lipids as energy stores

- Both plants and animals store energy for use when energy supply from the environment is low.

- Although both plants and animals produce polysaccharides as an energy store (starch and glycogen respectively), the major long-term store of energy is as TAGs. These can be stored in almost anhydrous form and represent a very efficient way to store excess energy.

Adipose tissue as the main energy storage organ in animals

- Most vertebrates store TAGs as an energy reserve in specialized cells, adipocytes. Adipocytes are grouped in the tissue called adipose tissue, along with other cell types. There are two types of adipose tissue, distinguished histologically but also in terms of cellular origin and function: white and brown.

- White adipose tissue contains the largest energy store. TAGs are synthesized in adipocytes either from preformed FAs, that are taken up into the cell following the action of LPL on lipoprotein TAGs in the capillaries (major route), or from FAs synthesized *de novo* from glucose or amino acids (minor route in humans under most conditions).

- White adipose tissue TAG stores are mobilized (hydrolysed to release FAs and glycerol) when energy is required elsewhere in the body. This occurs between meals, after an overnight fast, to a greater extent during prolonged fasting, and during periods of increased energy expenditure (e.g. physical activity). Both fat deposition and fat mobilization are highly regulated, insulin being a key regulator. They are regulated in such a way that the body's store of TAGs in white adipocytes quantitatively reflects the difference between energy intake and energy expenditure over a long period.

- Brown adipose tissue is a specialized tissue for generating heat. It is prominent in small mammals or newborn human infants, less so in adult humans: but there is intriguing evidence that it is still present in many adult humans. Heat is released from oxidation of FAs that is not coupled to ATP production. The molecular process involves a protein, UCP1, that short-circuits the normal build-up of a proton gradient across the inner mitochondrial membrane that drives ATP biosynthesis.

Plant lipid storage

- Many plants accumulate storage lipids in their principal reproductive propagules, namely fruits, seeds and pollen grains. In contrast, plants do not normally accumulate long-term storage lipids in their vegetative tissues, such as leaves, stems and roots.

- In some plants the mesocarp (fleshy part of the fruit) becomes highly enriched in storage lipids. Examples include olives, oil palm and avocado. The fleshy mesocarp acts as a nutritious bait that attracts animals which consume the fruit, leading to dispersal of the seeds which generally pass through the gut of the animal.

- In contrast, seeds accumulate storage compounds for the benefit of the germinating embryo as it seeks to establish itself as a new plant. In the major commercial oilseeds, such as rapeseed and sunflower, storage lipids typically make up almost half of total seed weight. In plants with large oil-rich seeds, such as nuts, storage lipids can account for as much as 75% of total seed weight.

- Many advanced plant species produce sticky pollen grains that adhere to insect vectors. The sticky covering is a lipid-rich mixture produced by the nutritive maternal cells of the tapetum during pollen development. In addition to their external lipid coat, entomophilous or insect-propagated pollen grains often accumulate large numbers of small oleosin-enclosed storage LDs in their cytosol in a similar manner to seeds. The lipid stores in pollen grains provide much of the energy required for germination and the subsequent establishment of the pollen tube.

- Many of the enzymes involved in storage lipid biosynthesis and mobilization in pollen grains are encoded by genes that are closely related to those involved in seed lipid metabolism, the only significant difference being in their tissue-specific expression.

Further reading

Pathways of fat storage and mobilization

Arner P, Bernard S, Salehpour M, *et al.* (2011) Dynamics of human adipose lipid turnover in health and metabolic disease, *Nature* **478**: 110–13.

Hodson L, Skeaff CM & Fielding BA (2008) Fatty acid composition of adipose tissue and blood in humans and its use as a biomarker of dietary intake, *Prog Lipid Res* **47**: 348–80.

Lafontan M & Langin D (2009) Lipolysis and lipid mobilization in human adipose tissue, *Prog Lipid Res* **48**: 275–97.

Spalding KL, Arner E, Westermark PO, *et al.* (2008) Dynamics of fat cell turnover in humans, *Nature* **453**: 783–7.

Brown adipose tissue

Cannon B & Nedergaard J (2004) Brown adipose tissue: function and physiological significance, *Physiol Rev* **84**: 277–359.

Peirce V, Carobbio S & Vidal-Puig A (2014) The different shades of fat, *Nature* **510**: 76–83.

Townsend KL & Tseng YH (2014) Brown fat fuel utilization and thermogenesis, *Trends Endocr Metab* **25**: 168–77.

Virtanen KA & Nuutila P (2011) Brown adipose tissue in humans, *Curr Opin Lipidol* **22**: 49–54.

Wu J, Cohen P & Spiegelman BM (2013) Adaptive thermogenesis in adipocytes: is beige the new brown?, *Genes Dev* **27**: 234–50.

Plant lipid storage

Gunstone FD, Harwood JL, Dijkstra AJ, eds., (2007) *The Lipid Handbook*, 3rd edn. Taylor & Francis, UK, pp. 703–82.

Leprince O, van Aelst AC, Pritchard HW & Murphy DJ (1998) Oleosins prevent oil-body coalescence during seed imbibition as suggested by a low-temperature scanning electron microscope study of desiccation-tolerant and -sensitive oilseeds, *Planta* **204**: 109–19.

Lersten NR, Czlapinski AR, Curtis JD, Freckmann R & Horner HT (2006) Oil bodies in leaf mesophyll cells of angiosperms: overview and a selected survey, *Am J Bot* **93**: 1731–9.

Murphy DJ, ed., (2005) *Plant Lipids: Biology, Utilisation and Manipulation*. Blackwell, Oxford, UK.

Winichayakul S, Scott RW, Roldan M, *et al.* (2013) In vivo packaging of triacylglycerols enhances arabidopsis leaf biomass and energy density, *Plant Physiol* **162**: 626–39.

CHAPTER 10

Lipids in health and disease

Many diseases are associated with disorders of lipid metabolism. The choice of amounts and types of lipids in the diet can play an important role in maintaining good health. In some cases treatment with certain lipids, usually by addition to the diet, can be useful in reducing risk of disease, or in managing or even treating diseases. This chapter will be concerned mainly with animals, in fact almost exclusively humans, but there are also aspects of plant health that will be touched upon.

There are many disorders of human metabolism that result from alterations in a single gene, usually grouped as 'inborn errors of metabolism'. These are inherited in a Mendelian fashion. Very often they are evident early in life, for instance because they result in failure to thrive in infancy. Others may not be manifest until later in life. One example is that of familial hypercholesterolaemia (defective low density lipoprotein (LDL)-receptor: see Section 10.5.2.1). It used to be the case that this was recognized when an adult had a heart attack (myocardial infarction) at a young age. Nowadays, there is widespread screening of families of carriers, so it is often picked up and treated before symptoms develop.

There are also many conditions that result from the action of multiple genes (polygenic conditions). There is a wide range of conditions that result in altered (usually elevated) concentrations of lipoproteins in the circulation. Most of these have a genetic component but cannot be ascribed to the effect of a single defective gene. A common end-result of altered plasma lipoprotein concentrations is cardiovascular disease, meaning myocardial infarction (heart attack), stroke or peripheral vascular disease, with the underlying pathology of the blood vessel wall described as atherosclerosis.

In this chapter we will look at some of these conditions, beginning with the Mendelian disorders of lipid metabolism, and the chapter will finish with a description of lipoprotein disorders and the process of atherosclerosis. Lipids are also important in plant health, not just as energy stores (see Section 9.3) but also as hormones, for instance the jasmonates or 'jasmonins' (see Section 3.4.1), derived mainly from α-linolenic acid ($18:3n$-3).

10.1 Inborn errors of lipid metabolism

'Inborn error of metabolism' is a term used to describe the condition that arises from an inherited mutation in a gene that affects the activity of the protein that it encodes (usually rendering it inactive). The protein concerned may be, for instance, an enzyme, a carrier protein or a transporter, a receptor or a transcription factor. Because the term is used to refer to single-gene mutations ('monogenic' conditions), these conditions are inherited in a Mendelian fashion. They are catalogued in the freely available resource: Online Mendelian Inheritance in Man (OMIM) (© Johns Hopkins University) and always referred to by their OMIM catalogue number. There is also a commercial catalogue, Scriver's *The Online Metabolic & Molecular Bases of Inherited Disease* (see **Further reading**).

Naturally, many of these conditions affect lipid metabolism. Some are covered later in this chapter (e.g. disorders of lipoprotein metabolism, including Type 1 hyperlipoproteinaemia and familial hypercholesterolaemia, see Section 10.5.2.1). Here we will discuss the large group of disorders that are related to impaired breakdown (hydrolysis or oxidation) of various lipid classes. These conditions may lead to the accumulation of specific lipids, and are then generally called 'lipidoses'. These usually arise because the affected enzyme is needed for the breakdown of a specific lipid molecule. Since the biosynthesis of these lipids is not impaired, the result of

Lipids: Biochemistry, Biotechnology and Health, Sixth Edition. Michael I. Gurr, John L. Harwood, Keith N. Frayn, Denis J. Murphy and Robert H. Michell.

the enzyme deficiency is the gradual accumulation of lipids in the tissues. Other defects, for instance most of those in fatty acid (FA) oxidation, do not result in lipid accumulation but affect other pathways (often the generation of metabolic energy). These conditions are rare but usually extremely severe, and often fatal at an early age. Some options for treatment are available but gene therapy, which might be considered the ultimate cure, is still largely a research area.

10.1.1 Disorders of sphingolipid metabolism

There is a range of inherited disorders of sphingolipid catabolism (see also Section 4.8.7), in which various sphingolipids or glycosphingolipids accumulate. These conditions are sometimes called the sphingolipidoses. Because the hydrolysis of sphingolipids occurs in lysosomes, they are also classified as a subgroup of the lysosomal storage diseases. Lipid accumulation is primarily in the lysosome, but it is unlikely that this is the primary cause of the disease; rather, lipids 'spill over' into other cellular compartments, where they affect cellular function. Many of these conditions involve lipid accumulation in cells of the central nervous system, and consequent impairment of neural function, or indeed neurodegeneration.

These conditions arise from mutations in genes encoding enzymes of sphingolipid breakdown or in genes for activator proteins for these enzymes. More than 40 such diseases are known. Their combined incidence is about 1 in 20,000 births, so they are rare conditions. The most common is Gaucher disease with an incidence of 1 in 59,000. Some of the most important are summarized in Table 10.1. These conditions are frequently fatal at an early age, which indicates how important it is that the amounts and types of lipids in membranes are strictly controlled to preserve biological function.

It should be noted from Table 10.1 that mutations in one gene may give rise to various manifestations of disease (e.g. age of onset and severity), largely dependent upon which part of the enzyme that it encodes is affected. In general, the greater is the residual activity, the less the severity of the condition. These conditions are all recessive: that is, one functional allele is sufficient to prevent occurrence of the disease. An alternative description is that they are autosomal recessive conditions, with the exception of (X-linked)-adrenoleukodystrophy, the gene for which (*ABCD1*) is on the X-chromosome.

Most of the lipids involved in these disorders are readily synthesized in the body (see Sections 4.8.1–4.8.6) so that treatment by elimination of particular lipids from the diet is ineffective. As noted earlier, gene therapy is in its infancy and not yet in clinical use in any of these conditions. Another treatment under investigation in animal models, but not yet in humans, is cell-based therapy, in which stem cells are manipulated to overexpress the necessary enzymes and then implanted. Direct replacement of the affected enzyme seems the most logical approach, but its applicability is limited because enzymes given intravenously (they would be digested if given orally) cannot access the central nervous system. Enzyme replacement therapy using recombinant enzymes (produced in bacteria) is, however, very successful in the nonneuropathic form of Gaucher disease (Type I: see Table 10.1) as the accumulation of glucosylceramide is mainly in cells of the reticuloendothelial system, which are readily accessible from the bloodstream. Enzyme replacement therapy has been used in other sphingolipidoses including Fabry disease, for which two recombinant enzyme products are available. However, its effectiveness is limited by access of the enzyme to the site where it is needed, and of course biological rejection of a foreign protein.

As with all genetic diseases, families with the defect may be offered genetic counselling. At one time, early diagnosis relied on measurements of enzyme activity, for instance in white blood cells, but now detection usually depends on screening for mutations in the gene concerned.

10.1.2 Disorders of fatty acid oxidation

FA oxidation occurs in mitochondria and peroxisomes (also called microbodies; see Section 3.2.1.1). Inherited defects in almost any one of the many proteins involved in the pathway have been observed and constitute an important group of inherited metabolic disorders.

Like the sphingolipidoses, these are rare conditions usually with a recessive inheritance, but they are more common than disorders of sphingolipid metabolism. The overall prevalence is population-dependent but is around 1 in 5,000 to 1 in 10,000 births in several populations, although much less in Chinese people. The most common disorders are medium-chain acyl-CoA dehydrogenase deficiency, with a prevalence of around 1 in 10,000 to 1 in 15,000 in European/American populations, and very long-chain acyl-CoA dehydrogenase

Table 10.1 Enzyme deficiencies and accumulating lipids in the main sphingolipidoses.

Disease	OMIM number*	Clinical features	Major lipid accumulation	Enzyme defect	Gene affected
Farber lipogranulomatosis (Farber disease)	228000	Very early hoarseness, dermatitis, skeletal deformation, mental retardation, death usually before 2 years.	Ceramide	Acid ceramidase	*ASAH1*
Generalized gangliosidosis (GM$_1$ gangliosidosis)	230500 (Type I) 230600 (Type II) 230650 (Type III)	Type I is infantile, severe with mental retardation, liver enlargement, skeletal deformities, red spot in retina; Type II is late-infantile/juvenile, Type III adult/chronic.	Ganglioside GM$_1$ (monosialotetrahexosylganglioside)	β-Galactosidase-1	*GLB1*
Gaucher diease	230800 (Type I) 230900 (Type II) 231000 (Type III).	Spleen and liver enlargement, erosion of long bones and pelvis; Types II and III involve nerve damage also; usually diagnosed in childhood but may be later.	Glucosylceramide (glucocerebroside)	Acid beta-glucosidase (glucocerebrosidase) (cleaves the beta-glucosidic linkage of glycosylceramide)	*GBA*
Krabbe disease (globoid cell leukodystrophy)	245200	Mental retardation, almost total absence of myelin, globoid bodies in white matter of brain; usual presentation by 6 months, death by 2 years.	Galactocerebroside	Galactosylceramidase (Galactocerebrosidase)	*GALC*
Metachromatic leukodystrophy (two forms)	250100	Motor symptoms, rigidity, mental deterioration, and sometimes convulsions, onset in second year of life and death usually before 5 years; there are also later-onset forms including adult with slower progression.	Galactosphingosulphatides	Aryl sulphatase A (hydrolyzes cerebroside sulphate)	*ARSA*
Niemann-Pick disease Types A and B	257200 (Type A) 607616 (Type B)	As Type C (below) but ranging from a severe infantile form with neurologic degeneration and death by 3 years (type A) to a later-onset non-neurologic form (type B) compatible with survival into adulthood.	Sphingomyelin	Sphingomyelin phosphodiesterase-1 (acid sphingomyelinase).	*SMPD1*
Niemann-Pick disease Type C**	257220	Usual onset 2 - 4 years of age. Neurologic abnormalities (ataxia, seizures, loss of speech), spasticity, dementia, psychiatric manifestations.	Cholesterol (various cellular locations) and glycosphingolipids (in lysosomes)	Niemann-Pick disease, type C1, a sterol transporter (related to NPC1L1; see Box 7.2)	*NPC1* (95% of cases) *NPC2* (5% of cases)
Tay-Sachs disease	272800	Developmental retardation, paralysis, dementia and blindness; death in the second or third year. There are subtypes with more or less rapid onset.	Ganglioside GM$_2$ (a structural variant of GM$_1$) (both are monosialotetrahexosylgangliosides)	α-subunit of Hexosaminidase A	*HEXA*

(Continued)

Table 10.1 *(Continued)*

Disease	OMIM number[*]	Clinical features	Major lipid accumulation	Enzyme defect	Gene affected
Other GM2-gangliosidoses	268800	Similar phenotype to Tay-Sachs disease		Sandhoff disease: β-subunit of Hexosaminidase A	*HEXB*
	272750			AB variant: GM2A activator protein	*GM2A*
Fabry disease (also called Anderson-Fabry disease)	301500	Reddish purple skin rash, kidney failure, pain in lower extremities, one variant has cardiac disease, often seen in adulthood and may be slowly-progressing.	Gal-Gal-Glu-ceramide (globotriaosylceramide)	α-Galactosidase A	*GLA*

[*] OMIM: Online Mendelian Inheritance in Man (http://www.omim.org/ and see text).

[**] Note that Niemann-Pick disease Type C is not strictly a disorder of sphingolipid catabolism, but that there is a secondary accumulation of sphingolipids.

Table based on, and updated from EF Neufeld (1991) Lysosomal storage diseases. *Annu Rev Biochem* **60**:257–80 and M Eckhardt (2010) Pathology and current treatment of neurodegenerative sphingolipidoses. *Neuromol Med* **12**:362–82.

See Chapter 2 for more on structures of lipids involved and Chapter 4 for more on their metabolism.

---

deficiency at around 1 in 30,000 births. As with all autosomal recessive conditions, the disease risk is much augmented in populations in which consanguineous marriage (e.g. between first cousins) is a custom.

The clinical manifestation of a defect in an enzyme of FA oxidation is not necessarily lipid accumulation. FA oxidation and the energy it provides underlie many processes. Like the sphingolipidoses, defects in FA oxidation may become apparent at different ages according to the severity of the defect. Low blood glucose concentration with low ketone body concentrations (hypoketotic hypoglycaemia) is a common presentation in the neonatal period, particularly in response to metabolic stress (e.g. fasting, even overnight, or illness). This reflects the fact that hepatic gluconeogenesis, needed to provide plasma glucose at a time when the diet is mostly fat-based (i.e. milk), requires energy derived from FA oxidation. The low glucose and ketone body concentrations deprive the brain of an energy source, and patients may die rapidly. Onset in later life is usually associated with muscle problems – pain, weakness, and muscle breakdown with excretion of breakdown products in the urine. The heart (myocardium) is also affected (cardiomyopathy), reflecting the high metabolic demands of the myocardium, a substantial proportion of which is normally met by FA oxidation. Another consequence, seen especially in long-chain acyl-CoA dehydrogenase deficiency, is that a woman who is

heterozygous for a defect in FA oxidation, but carrying a homozygous foetus, may suffer a build-up of lipids and metabolites in her circulation, resulting in fat accumulation in the liver ('acute fatty liver of pregnancy') and major systemic metabolic complications (see Section 10.4.1.3). Disorders of the mitochondrial trifunctional protein complex (responsible for β-oxidation of FAs) are also associated with progressive nerve and retinal damage and are resistant to treatment. Interestingly, deficiency of the short-chain 3-hydroxyacyl-CoA dehydrogenase leads to oversecretion of insulin (and again hypoglycaemia), implying a role for FA oxidation in regulation of insulin secretion.

Some of the major FA oxidation deficiencies are listed on Table 10.2. It will be seen that many steps in the pathway of FA oxidation may be affected. Not shown are some rare cases where the defect appears to be in cell membrane transport of FAs. CD36 (also known as Fatty Acid Translocase (FAT), see Sections 3.1.1 & 7.1.2) deficiency is also known, and indeed relatively common in Japanese people and some African and African American populations. It does not have a clear phenotype, probably reflecting the fact that there are multiple systems for transport of FAs across the cell membrane. There is some evidence that people with CD36 deficiency are 'insulin resistant' and a suggestion that this genetic alteration might increase susceptibility to development

Table 10.2 Inborn errors of fatty acid oxidation.

System affected	Disease and enzyme deficiency	OMIM number*	Gene affected
Mitochondrial fatty acid transport			
	Systemic carnitine deficiency (resulting from a defect in carnitine transport into muscle cells through mutation in the transporter).	212140	*SLC22A5*
	CPT-1A (liver-specific) deficiency; deficiencies in CPT-1B (muscle type) and CPT-1C (brain type) have not been reported.	255120	*CPT1A*
	CPT-2 deficiency	600649 (infantile), 255110 (adult form)	*CPT2*
Mitochondrial fatty acid oxidation			
	VLCAD deficiency	201475	*ACADVL*
	MCAD deficiency (may be homozygous)	201450	*ACADM*
	SCAD deficiency	201470	*ACADS*
	LCHAD deficiency	609016	*HADHA*
	Mitochondrial trifunctional protein deficiency – may be caused by mutations in the α-subunit (includes LCHAD activity) or the β-subunit (LCKAT activity)	609015	*HADHA* (α-subunit) *HADHB* (β-subunit)
Peroxisomal fatty acid oxidation			
	Adrenoleukodystrophy (also known as X-linked adrenoleukodystrophy); deficiency of ABCD1	300100	*ABCD1*
	Acyl-CoA oxidase deficiency	264470	*ACOX1*
	Phytanoyl-CoA hydroxylase deficiency, Refsum disease	266500	*PHYH*

There are more than 30 defects identified in the pathway of fatty acid oxidation, and this table is representative of the steps affected rather than exhaustive. See **Further reading** for more information and Section 3.2.1 for details of FA oxidation.

Based partly on P Rinaldo, D Matern & MJ Bennett (2002) Fatty acid oxidation disorders, *Annu Rev Physiol* **64**:477–502; RJA Wanders, J Komen & S Kemp (2011) Fatty acid omega-oxidation as a rescue pathway for fatty acid oxidation disorders in humans. *FEBS J* **278**:182–94.

ABCD1, an ATP-binding cassette family transporter: see Section 7.2.5 for further information on this family; CPT, carnitine palmitoyltransferase; LCHAD, long-chain 3-hydroxyacylacyl-CoA dehydrogenase; LCKAT, long-chain 3-ketothiolase; MCAD, medium-chain acyl-CoA dehydrogenase; SCAD, short-chain acyl-CoA dehydrogenase; VLCAD, very long-chain acyl-CoA dehydrogenase.

of metabolic syndrome (see Section 10.5.2.4 for further explanation).

Another very rare condition that could be called a FA oxidation defect is malonyl-CoA decarboxylase deficiency. Malonyl-CoA decarboxylase breaks down malonyl-CoA, an intermediate in the pathway of lipogenesis that is a very potent inhibitor of carnitine palmitoyltransferase-1 (CPT-1) and hence of FA oxidation (see Section 3.2.1.6). Thus, without removal of malonyl-CoA, FA oxidation is inhibited and the signs are similar to those of the classic FA oxidation defects.

Unlike the sphingolipidoses, in many cases dietary treatments can be effective in patients with FA oxidation defects. Affected infants can be rescued by frequent feeding to avoid fasting, provision of energy in the form of glucose and, in the case of disorders of long- and very long-chain FA oxidation, administration of fat as medium-chain triacylglycerols (TAGs). Medium-chain TAGs consist of FAs of chain length 8–12 carbons (esterified as usual to glycerol). These FAs are readily absorbed directly into the hepatic portal system (see Section 7.1.3) and rapidly oxidized in the liver.

There are some special examples of dietary treatment in cases of peroxisomal FA oxidation defects. Adrenoleukodystrophy arises from lack of the carrier ABCD1 responsible for entry of very long-chain FAs into peroxisomes for oxidation. Saturated very long-chain FAs such as 26:0 and 24:0 accumulate, and cause damage, in plasma and tissues. Patients have been treated with restriction of very long-chain FA intake, and administration of monounsaturated fatty acids (MUFA) including oleic ($18:1n$-9) and erucic ($22:1n$-9) with beneficial results (lowering of plasma very long-chain FA concentrations, and improvement in symptoms). This was illustrated in the film *Lorenzo's Oil*. Lorenzo's Oil, patented by Lorenzo Odone's father, Augusto Odone, is a 4:1 mixture of trioleoylglycerol and trierucoylglycerol, which is combined with moderate reduction of fat in the diet.

Refsum disease is a deficiency of phytanoyl-CoA hydroxylase, leading to failure of α-oxidation of the branched-chain FA, phytanic acid (see Section 3.2.2). In patients with the disease, there is a characteristic build-up of phytanic acid in the blood where it may represent 30% of the total fatty acids. Patients are usually identified in childhood/adolescence with nerve and retinal damage. Phytanic acid is formed in the gut of ruminant animals from phytol, a constituent of chlorophyll and hence a universal constituent of green plants. Patients must therefore follow a diet low in phytanic acid, meaning in practice low in fats from ruminant animals and some fish.

Because disorders of FA oxidation may be 'silent' until a metabolic stress precipitates acute illness and even death, there are now widespread programmes for screening newborns for these defects. These mostly rely on tandem mass-spectrometric methods for measurement of acyl-carnitines in dried blood spots. The pattern of acyl-carnitines is characteristically altered in the various conditions: for instance, blood octanoylcarnitine concentrations are elevated in medium-chain acyl-CoA dehydrogenase deficiency, whereas free carnitine concentrations are elevated in CPT-1A deficiency. However, as with most screening programmes, there is debate about the effectiveness, and it is only in medium-chain acyl-CoA dehydrogenase deficiency that benefits are clear.

10.1.3 Disorders of triacylglycerol storage

A disorder in which TAGs accumulate in tissues was recognized in the 1960s, but its molecular basis was not understood until recently. A diagnostic characteristic is the accumulation of TAGs in vacuoles in leukocytes (which can be observed in a blood sample), known as

Jordan's anomaly after the person who first described it. The condition is called neutral (nonpolar) lipid storage disease (NLSD). There are two forms of NLSD. One involves muscle damage (myopathy) but the skin is normal (neutral lipid storage disease with myopathy, NLSDM, OMIM number 610717). In the other, the skin is also affected by a scaly condition known as ichthyosis (neutral lipid storage disease with ichthyosis, NLSDI or Chanarin-Dorfman syndrome after the people who characterized it in the 1970s; OMIM number 275630). Both these conditions are now known to result from defects in hydrolysis of TAGs. NLSDM represents lack of the enzyme adipose tissue triglyceride lipase (ATGL, human gene *PNPLA2*: see Box 9.1). Chanarin-Dorfman syndrome or NLSDI results from lack of the protein CGI-58 (the name comes from *comparative gene identification-58*; gene *ABHD5*), which, as described in Box 9.1, is an essential coactivator protein for ATGL. These are nonlysosomal, nonperoxisomal and nonmitochondrial diseases since hydrolysis of TAGs occurs in the cytosolic compartment of cells.

One theory for the adverse effects of these conditions is that ATGL-mediated hydrolysis of TAGs is necessary to generate ligands for the peroxisome proliferator activated receptor (PPAR)-α (see Section 7.3.2). One function of activation of PPAR-α is to induce the enzymes of fat oxidation. The metabolic state of mice lacking ATGL is rescued by administration of bezafibrate, an agonist of PPAR-α (restoring heart function and preventing early death). Two patients with NLSDM have been treated with bezafibrate for 6 months. There was a reduction in the amount of TAG seen in muscle biopsies, but no obvious clinical benefit, although the study was small and a longer period of treatment may be needed.

Human mutations rendering the other enzymes of TAG hydrolysis (hormone-sensitive lipase and monoacylglycerol [MAG] lipase) inactive have not been described. Deficiency of perilipin-1 (gene *PLIN1*), a protein that coats lipid droplets (LDs – see Section 5.5) and acts in concert with hormone-sensitive lipase (HSL), leads to a form of lipodystrophy and is discussed later (see Section 10.4.2).

10.1.4 Disorders of lipid biosynthesis

Defects in lipid *catabolism* have been recognized for decades. More recently there has been an appreciation that a large group of conditions may result from defects in lipid *biosynthesis*. Inherited defects have been identified in many steps of the synthetic pathways of glycerophospholipids, sphingolipids and FAs (elongation in particular), linked mainly to neurological and muscular disorders. This is an emerging field and the reader is referred to the review by Lamari *et al.* (2013) in **Further reading** for more information.

10.2 Lipids and cancer

Cancer results from the uncontrolled proliferation of cells. It arises from mutations in genes that normally control cell division or promote programmed cell death (apoptosis). The development of cancer involves a number of stages. Initiation is the initial process of DNA mutation; promotion is a stage at which a potentially cancerous cell, through accumulation of further mutations and also epigenetic changes, becomes a fully cancerous cell. It is now recognized that many mutations have generally accumulated before a cell becomes cancerous, and sometimes there is an underlying genetic predisposition. Then there is a stage of progression, or development by cell division, following which the cancerous cells may spread and invade other tissues (metastasis).

Lipids are involved in cancer in many ways. Dietary lipids, including micronutrients such as fat-soluble antioxidants, may play a role in predisposing to, or protecting from, cancer. The development of cancer involves cellular changes that include alterations to lipid components of the cell (and this may be relevant to the design of new treatments). The growth of a tumour requires new blood vessels to be formed (angiogenesis), and this and the proliferation of cells both require a supply of FAs for synthesis of membrane lipids. Finally, there may be a role for certain dietary lipids in the control or treatment of certain cancers or aspects of cancer.

There are a few proven pharmaceutical means to reduce incidence of certain cancers, each of which may mediate their effects through changes in lipid metabolism. Daily low-dose (typically 75 mg) aspirin has been prescribed in many clinical studies as a prophylactic treatment against cardiovascular disease (especially in secondary prevention). An analysis of these studies showed an unexpected, but highly significant, reduction (of around 40%) in deaths from colorectal cancer amongst those taking aspirin. Further analysis of many studies has confirmed effects of aspirin against both incidence of, and deaths from, a range of cancers including oesophageal, gastric, biliary and breast cancer. Various mechanisms have been proposed. These low doses of aspirin are considered to act mainly via inhibition of cyclo-oxygenase (COX)-1 (see Section 3.5.2) and inhibition of platelet activation. Platelet activation may be involved in induction of paracrine signalling between cells in developing tumours. However, another suggestion is that aspirin has an additional mechanism of action, activation of AMP-activated protein kinase (AMPK). Interestingly, another known pharmaceutical prophylactic against cancer is the drug metformin, a member of the biguanide family of drugs used in treatment of Type 2 diabetes. Metformin is also an activator of AMPK. AMPK activation signals to change cellular metabolism in a 'catabolic' direction, with inhibition of lipogenesis and activation of fat mobilization and oxidation. Given the role of increased *de novo* lipogenesis (DNL) in cancer discussed below (see Section 10.2.2.5), it is possible that this is one mechanism by which these agents exert their anticancer effects. Other agents with a proven role in cancer prevention are the 'selective oestrogen (US: estrogen) receptor modulators' including tamoxifen and raloxifene. These drugs inhibit oestrogen signalling through the oestrogen receptor, and are protective against (and useful in the treatment of) breast cancers that are oestrogen-responsive. Since oestrogens are derivatives of cholesterol, this has a basis in lipid biochemistry if not exactly metabolism. There is interest in the use of another class of compounds in this context, the aromatase inhibitors, which inhibit the conversion of androgens into oestrogens. In postmenopausal women, aromatase action is the main source of circulating oestrogens (see Section 10.4.1 for more on this pathway, in connection with obesity). Finally it should also be pointed out that weight loss, achieved through abdominal surgery ('bariatric surgery' to reduce the size of the stomach or reduce absorption of nutrients), also reduces cancer incidence. This is perhaps not surprising in view of the strong link between obesity and cancer risk (see Section 10.4.1).

10.2.1 Dietary lipids and cancer

There is a large body of evidence on the role of dietary lipids in cancer. It is based on epidemiological studies of various sorts in humans, and on feeding studies in animals. It should be stressed that the evidence is often conflicting. In 1997–98, two major reports on diet and cancer were published in which all existing evidence was reviewed. One of these was the UK's Committee on Medical Aspects of Food and Nutrition report *Nutritional Aspects of the Development of Cancer* (HMSO 1998). The other was the World Cancer Research Fund/American Institute for Cancer Research (WCRF/AICR) report, which was updated in 2007 and their website presents up to date information (see **Further reading**).

The WCRF/AICR expert group assessed the literature and graded the evidence for associations between dietary components and specific cancers (in each case either protective or risk-inducing) on a scale from 'convincing', through 'probable', 'limited – suggestive' to 'substantial effect on risk unlikely'. For total dietary fat, no convincing or probable evidence of a link with any cancer was found, although there was 'limited – suggestive' evidence of increased risk for lung and breast cancer. It should be noted that evidence is adjusted for total energy intake: body fatness has 'convincing' associations with risk of several cancers including oesophagus, pancreas, colorectal, breast [postmenopausal], endometrial and kidney.

For 'foods containing beta-carotene' (a vitamin A precursor: see Section 6.2.3.1) there was evidence for 'probable' protection against oesophageal cancer; for 'foods containing vitamin D', there was 'limited-suggestive' evidence for protection against colorectal cancer; and for 'foods containing vitamin E', limited-suggestive evidence for protection against oesophageal and prostate cancer. For one specific dietary antioxidant, lycopene (the red carotenoid pigment of tomatoes), there was 'probable' protection from prostate cancer.

However, it should be emphasized that cancer prevention may not be as simple as supplementation with the antioxidant vitamins at high doses. Several supplementation studies have been carried out, none so far with positive results, and in one study of β-carotene supplementation, there was an excess of mortality from lung cancer (in those already at high risk).

The WCRF/AICR report indicates 'limited-suggestive' evidence for fish consumption as protective against colorectal cancer. In a separate meta-analysis of data from 21 prospective studies of breast cancer, involving a total of 688,000 participants, dietary intake of marine very long-chain *n*-3 polyunsaturated fatty acids (PUFAs) was found to protect against breast cancer, with a 5% reduction in risk for every 0.1 g/day increment in consumption. The relationship was not found for fish consumption itself.

Cancer of the colon is common, but its prevalence varies widely between countries. In comparisons between countries, colon cancer prevalence is correlated with meat and fat consumption. However, the interpretation of such studies needs caution as many other factors may be different between countries. The evidence from other types of study, such as case-control studies (where each 'case', or person suffering from the disease, is matched with a nonsufferer with otherwise similar characteristics), for the link between meat consumption and colon cancer is very conflicting, with some studies showing clear positive links and others the opposite. There is reasonably consistent evidence (meaning that it has been found in most, but not all studies), however, for a protective effect of dietary fibre intake against colorectal cancer (the WCRF/AICR report rates it as 'probable' evidence). Whilst dietary fibre (or nonstarch polysaccharide) is not a dietary lipid, the association is relevant to this book. The mechanism responsible is believed to be that fermentation of nonstarch polysaccharides by bacteria in the colon produces the short-chain FAs, acetate, propionate and butyrate. Butyrate is an important and preferred fuel for the colonic mucosa and there is also strong evidence that it exerts antiproliferative effects on the mucosal cells, which could protect from cancer development.

10.2.2 Cellular lipid changes in cancer and their use in treatment

There are many changes in lipid metabolism in cancer cells, and opportunities to target lipid-related processes in treatment, as summarized in Fig. 10.1.

10.2.2.1 Cell surface glycosphingolipids

Cells express a number of carbohydrate and lipid derivatives on their surfaces. Amongst these are glycosphingolipids (GSLs), which are involved in cellular recognition, cell adhesion and regulation of cell growth: the carbohydrate chains of GSLs are exposed to the external environment and are part of the 'glycome' (a far more complex set of molecules than even the genome because of the many permutations of carbohydrate chains possible). The cell glycome, including its GSL complement, is an important aspect of its antigenicity: i.e. specific antibodies may be directed against the carbohydrate moities. The ABO blood group antigens, which cause an immune reaction if blood of one group is infused into someone of another blood group, are of

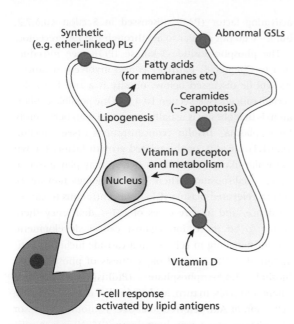

Fig. 10.1 Lipid-related therapeutic targets in cancer. Abbreviations: GSLs, glycosphingolipids; PLs, phospholipids. For further information see the text.

this type, although most of the sugar molecules in that case are attached to proteins with a minority attached to GSLs.

Cancer cells express variant forms of GSLs, often in larger amounts than normal. Typically there is a return of expression of GSLs normally expressed early in development ('developmental antigens'), and not in differentiated tissues. These GSLs may play a role in the uncontrolled cell growth of cancer. There are characteristic changes in GSLs in different cancers.

GSLs perform many functions on the surface of cells. They form clusters known as 'glycosynapses' together with signal transducing proteins and other membrane proteins. Glycosynapses mediate cell adhesion and signal transduction. The microenvironment formed by glycosynapses of interacting cells is critical to cell adhesion, as occurs during development and in cancer progression.

These cell-surface antigens are released to some extent into the circulation and may therefore provide a means of diagnosis of particular cancers. Equally exciting is the idea that specific antibodies against these antigens might be used to target drugs to the cancer cells: either drugs could be coupled to the antibodies, or the presence of the

antigens might be used as the basis for development of a specific vaccine. These ideas have been promoted consistently by the Japanese biochemist, Senitiroh Hakomori, now working in the US, who has been active in this field for more than 50 years.

α-Galactosylceramide (also known as KRN7000) is a GSL originally derived from the marine sponge *Agelas mauritanius*, and is a ligand for the cell-surface receptor CD1d expressed by antigen-presenting cells of the immune system (see Section 10.3.1 below). α-Galactosylceramide, when presented attached to CD1d, is recognized by the T-cell receptor on a particular set of T-lymphocytes, invariant natural killer T-cells. When activated, these invariant natural killer T-cells rapidly produce a panel of cytokines, including interferon-gamma (IF-γ) and interleukin (IL)-4, and bring about antitumour, antiviral and anti-inflammatory responses. KRN7000 was discovered because of its antitumour activity in mice. KRN7000 has since been tested in clinical trials in advanced cancer and in hepatitis B and C. It has to be delivered intravenously. Results so far have not been clear, but further work optimizing methods of delivery is under way.

10.2.2.2 Ceramide metabolism

Ceramides are sphingolipids (see Sections 2.3.3 & 4.8.1) that are often associated with cell death through the apoptosis pathway. This could be seen as beneficial in cancer. However, in cancer cells there is an up-regulation of pathways of ceramide removal, including glucosylceramide synthase, a glucosyltransferase that produces glucosylceramide. Furthermore, different ceramide species – differing in the acyl chain attached to sphingosine – may have different effects. For instance, 16C-ceramide, which is generated by the enzyme ceramide synthase 6, has been implicated in cancer cell proliferation, whereas 18C-ceramide mediates cell death. Thus, there has been considerable interest in the enzymes involved in ceramide synthesis and removal as therapeutic targets in cancer treatment. Exogenous short-chain ceramides, such as 6C-ceramide, which is permeable to cells, are also under investigation. Both targeting of pathways and administration of exogenous ceramides have been shown to be effective in some animal cancer models. Human trials of the drug fenretinide are under way in many types of cancer. Fenretinide is a retinoid (related to vitamin A: see Section 6.2.3.1) which causes ceramide to accumulate in tumour cells and is associated with generation of reactive oxygen species (ROS), resulting in cell death through apoptosis and/or necrosis. Because of its fat-solubility, fenretinide accumulates preferentially in fatty tissue such as the breast, potentially contributing to its effectiveness against breast cancer. Tamoxifen, which is an antagonist of the oestrogen receptor in breast tissue with established anticancer properties (see introduction to Section 10.2), is also a potent inhibitor of ceramide glycosylation, so leading to ceramide accumulation and hence death of tumour cells.

It is probable that multiple lines of therapy will need to be combined, including targeting of ceramide pathways in combination with more conventional chemotherapy aimed at cell division.

10.2.2.3 Phospholipid-related pathways

New potential cancer treatments based on phospholipid analogues have been developed and in some cases tested in clinical trials. These compounds are intended to become incorporated into target membranes, either the cell membrane or intracellular membranes, such as the endoplasmic reticulum (ER). Thus, their mode of action differs from conventional anticancer drugs that target DNA replication. Several types of compounds have been tested, although the most promising are alkyl analogues of phosphatidylcholine (PtdCho) or of lysophosphatidylcholine (lyso-PtdCho). These molecules accumulate in the cell membrane, since they are resistant to normal pathways of degradation. Here they disrupt signalling processes, they may inhibit phosphocholine cytidylyltransferase leading to inhibition of PtdCho biosynthesis, and they disturb lipid microdomains. Ultimately they might lead to cell death through the apoptosis pathway.

The first such compound synthesized, with ether-linked substitutions at glycerol positions sn-1 and sn-2, was 1-O-octadecyl-2-O-methyl-glycero-3-phosphocholine, known as edelfosine. This has potent antitumour effects in cellular models and some animal models but has found limited applicability in human disease because of adverse effects such as pulmonary oedema and impairment of hepatic function. Perifosine is a more recent derivative in which the choline headgroup is substituted by a heterocyclic methylated piperidyl residue. It is considered to inhibit Akt (also known as protein kinase B [PKB]) activity in the cell membrane. It has been tested in various human cancers, but whilst its toxicity is low, its therapeutic effect is also not sufficient for it to be used as a single treatment. These compounds are related structurally to platelet

activating factor (PAF), discussed in Section 10.3.7.2, but they do not appear to act through the PAF receptor.

The phosphoinositide-3-kinase (PI3K; see introduction to Chapter 8) pathway is also involved in cancer. As will be discussed below, obesity is a risk factor for certain cancers (see Section 10.4.1). One possible explanation is that obesity is usually associated with persistently high plasma insulin concentrations (see Section 10.4.1.3). Insulin, or the related growth factors known as insulin-like growth factors (IGFs), may play a role in cancer predisposition. There is epidemiological evidence relating elevated plasma IGF-1 concentrations to cancer incidence, and in rare cases of IGF-1 deficiency there seems to be protection against cancer development. Insulin signalling in cells is via a cascade including activation of PI3K-catalysed biosynthesis of phosphatidylinositol 3,4,5-triphosphate (PtdIns(3,4,5)P_3) (see Chapter 8). This, in turn, leads to further events including activation of Akt, referred to above. Tumour-suppressor phosphatase and tensin homologue (PTEN) is a specific PtdIns(3,4,5)P_3 3-phosphatase and mutations of the *PTEN* gene, which increase mitogenic signalling through the PI3K-Akt signalling pathway, commonly occur during cancer development. Rare cases of inherited mutations of *PTEN* cause a syndrome known as Cowden syndrome (OMIM number 158350) which includes cancer predisposition.

10.2.2.4 Vitamin D and cancer

As noted in Section 6.2.3.2, low vitamin D status has been linked with increased risk of some cancers. 1,25 (OH)$_2$D$_3$ (calcitriol, the active form of vitamin D) and its analogues inhibit proliferation, angiogenesis, migration/invasion and induce differentiation (reversion to normal cell type) and apoptosis in cancer cell lines. As tumour cells dedifferentiate, the expression of the vitamin D receptor and of the enzyme CYP27B1 (which makes calcitriol) decreases, while the expression of CYP24A1 (which makes the inactive form, 24,25(OH)$_2$D$_3$) strongly increases, so depriving tumour cells of active vitamin D. Therefore there has been interest in the use of vitamin D, or analogues thereof, in cancer treatment.

Observations on randomized trials where vitamin D has been given, such as the Women's Health Initiative, show equivocal results (in this case complicated by women in the placebo group taking supplements), and other studies of vitamin D administration in various cancers have not shown antitumour activity (see

Leyssens *et al.* (2013) in **Further reading** for more information).

The epidemiological evidence is strong for a link between low vitamin D status and prostate cancer. Administration of high doses of $1,25(OH)_2D_3$ is limited because of problems with hypercalcaemia. 1α-Hydroxyvitamin D_2 is a vitamin D analogue with less calcaemic activity, which has shown promise in cellular and rodent studies. However, it has not proved to be as effective as hoped against human prostate cancer.

10.2.2.5 *De novo* lipogenesis in tumour cells

The German biochemist Otto Warburg demonstrated in 1929 that cancer cells have high rates of glucose uptake, but with glycolysis, rather than glucose oxidation, the predominant pathway of glucose utilization (the Warburg effect). Only more recently was another typical characteristic discovered: that cancer cells have high rates of FA biosynthesis by the DNL pathway of *de novo* lipogenesis (DNL – see Section 3.1.3). Just as the description of the Warburg effect led to many attempts to block glycolysis as a treatment for cancer, much attention is now focussed on the possibility of blocking FA biosynthesis.

The need for FA biosynthesis in cancer is obvious. Cell proliferation and tumour growth, with associated angiogenesis (new blood vessel growth), requires membrane lipids. There is a specific up-regulation of the activity of fatty acid synthase (FAS, human gene *FASN*) in many types of cancer, which correlates with poor prognosis, leading to the description of *FASN* as a 'a metabolic oncogene' (i.e. gene causing cancer). The increase in FAS activity is a result of increased mRNA and protein expression, driven by a variety of growth-inducers, as well as, in some cancers (e.g. of the prostate), gene duplication leading to 'copy number gain'. Perhaps surprisingly, FAS can be secreted from cells and detected in the bloodstream: the concentration of FAS in serum is strongly associated with tumour stage and survival of patients with a variety of cancers, suggesting a use in diagnosis and prognosis.

Other enzymes of lipogenesis are also up-regulated, including ATP:citrate lyase and acetyl-CoA carboxylase 1. Not surprisingly, all have been seen as possible drug targets. A variety of inhibitors of these enzymes had been tested over the years as weight-gain inhibitors, and some of the same drugs have been tested in preclinical models for anticancer activity. However, many have undesirable side-effects (such as weight loss, in the context of cancer).

This is a rapid area of research: see **Further reading** for more information.

Cancer cells may also obtain FAs from their environment. In particular, close relationships have been described between cancer cells and adipocytes, and it has been suggested that cancer cells may stimulate FA release from neighbouring adipocytes. Increased cancer cell growth and metastasis (spread) have been linked with cancer cell proximity to adipocytes.

10.2.3 Dietary lipids and the treatment of cancer

Cancer causes death through a number of mechanisms including direct invasion of critical organs. However, a common and life-threatening feature of many cancers is a marked loss of body mass, both fat and muscle, known as cachexia. Fat loss predicts survival in patients with advanced cancer.

There have been many investigations of the cause of cancer cachexia. One theory is that a circulating factor, known as cachectin, leads to inhibition of adipose tissue lipoprotein lipase (LPL) and consequent failure of fat storage (so that there is continuous net fat loss). Cachectin is now known to be identical to tumour necrosis factor-α (TNF-α), a cytokine produced by many cell types including macrophages and adipocytes. Another theory is that zinc-α_2-glycoprotein, a protein released by adipocytes, causes cachexia by stimulating adipocyte fat mobilization. The close relationship between cancer cells and adipocytes (see Section 10.2.2.5) could also imply that cancer cells 'suck' FAs out of adipocytes.

The evidence for such mechanisms has largely been based on animal studies. In humans, measurements of energy intake have shown that the loss of body fat is mainly a problem of energy balance. Patients lose appetite and, in many cases involving cancers of the gastrointestinal tract, have difficulty eating.

There have been some interesting developments in the treatment of cancer cachexia by the oral administration of relatively large amounts of very long-chain *n*-3 PUFAs (1–2 g/day of eicosapentaenoic acid, 20:5*n*-3), together with a conventional nutritional supplement. Patients who were previously losing weight have shown an increase in body weight during this treatment. It is likely that the mechanism relates to a general anti-inflammatory effect of large doses of the very long-chain *n*-3 PUFAs, mediated in part by their role as precursors of eicosanoids of the 3-series rather than the 2-series

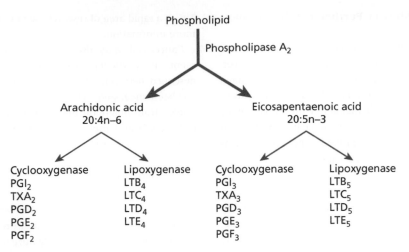

Fig. 10.2 Synthesis of eicosanoids from PUFAs of the *n*-6 and *n*-3 families. Abbreviations: PG, prostaglandin; TX, thromboxane; LT, leukotriene. Reproduced, with permission of John Wiley & Sons, from P Yaqoob & PC Calder (2011) The immune and inflammatory systems. In SA Lanhan-New, IA Macdonald & HM Roche, eds., *Nutrition and Metabolism,* 2nd edn. Wiley-Blackwell, Oxford, pp. 312–38.

(Fig. 10.2, Section 3.5.13, Table 3.11 & Fig. 3.54). The very long-chain *n*-3 PUFAs also tend to reduce the production of pro-inflammatory cytokines including TNF-α and IL-6.

10.3 Lipids and immune function

10.3.1 Involvement of lipids in the immune system

The immune system protects the body from infectious agents and other noxious insults. Life would hardly be possible without this well-integrated system but failure to regulate these processes appropriately can lead to damage to the body's own tissues, as seen in severe inflammation and in autoimmune diseases. There has long been an understanding that diet has an important influence on immunity. Energy malnutrition and deficiencies in specific nutrients result in increased susceptibility to infection and increased mortality, especially in children. There is also much evidence, albeit not always consistent, that the activity of the immune system may be modified (either amplified or suppressed) by specific types of dietary FAs.

The immune system has two general components (Fig. 10.3). The *innate* immune system consists of physical barriers and phagocytic cells (granulocytes, monocytes and macrophages, and natural killer cells). It is not very specific in its action and it has no 'memory'. The *acquired* immune system is more complex. It involves the specific recognition of molecules foreign to the host. Lymphocytes are key players in the acquired immune system. B-lymphocytes produce antibodies against specific antigens located on the surface of invading organisms (humoral immunity). T-lymphocytes do not produce antibodies. They recognize peptide antigens attached to major histocompatibility complex proteins on the surfaces of so-called antigen-presenting cells (cell-mediated immunity). In response to antigenic stimulation, they secrete cytokines (e.g. ILs, interferon (IF), transforming growth factor, etc.), whose function is to promote the proliferation, differentiation and activation of T-lymphocytes and other cells of the immune system. These include B-lymphocytes, monocytes, neutrophils, natural killer cells and macrophages, thus linking activation of the acquired immune system to activation of innate immunity. Once the system has been activated, there is a proliferation of the lymphocytes capable of recognizing, and responding to, the specific antigen involved. This expanded specific lymphocyte population may persist for a long time, giving rise to 'memory' of previous infections; B lymphocytes also contribute to immunologic memory. Natural killer cells comprise a third type of lymphocyte. They do not express surface markers identifying them as either B- or T-lymphocytes; they have a role in lysing infected cells and in graft rejection.

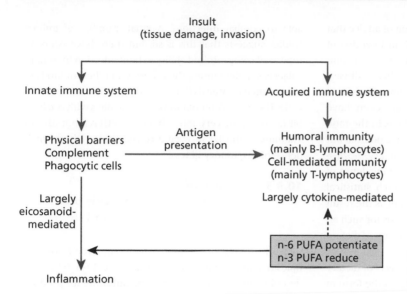

Fig. 10.3 Schematic view of the immune system and the effects of polyunsaturated fatty acids (PUFAs).

Inflammation is a manifestation of activation of the innate immune system. Classically, the signs of inflammation were described as pain, heat, redness, swelling and loss of function. Inflammation may be acute, as in response to tissue damage, or chronic. Inflammation may be seen as a useful part of the response to an invading organism or other noxious stimulus, but it can also become damaging when it appears excessive, or excessively prolonged. It is now recognized that many diseases have an underlying inflammatory component, sometimes called subclinical (or low-grade) inflammation, detectable by measurement of inflammatory markers in the blood (for instance the C-reactive protein, produced by the liver in response to inflammation). Therefore there is much interest in possible nutritional means of modulating inflammation (and other aspects of the immune response), sometimes called 'immunonutrition'.

Lipids are involved in immunity in several ways. Because of their importance as a source of energy, dietary lipids may help to improve the depressed immune function seen in malnourished individuals. Individual lipids, including some fat-soluble vitamins and different types of FAs, play more specific roles through their contribution to membrane structure and integrity (see Section 5.3), their metabolism to eicosanoids and other lipid mediators (see Section 3.5), their participation in processes of lipid peroxidation (see Section 3.3), or their influence on gene expression (see Section 7.3). In addition, some glycosphingolipids serve as cell-surface antigens as was discussed in relation to lipids and cancer (see Section 10.2.2.1). Lipid mediator production is clearly a central feature of inflammatory processes and the use of COX inhibitors, the nonsteroidal anti-inflammatory drugs (NSAIDS) that specifically target the prostanoid family of eicosanoids, is a cornerstone of the treatment of inflammatory disorders (see Section 3.5.3).

10.3.2 Dietary lipids and immunity

In the 1970s it was discovered that diets rich in PUFAs (mainly of the *n*-6 family) could prolong the survival in mice of skin allografts (grafts from genetically nonidentical mice). Such diets were subsequently employed as adjuncts to conventional immunosuppressive therapy to reduce rejection of transplanted human kidney. However, later studies revealed that the effects were not long-lasting. Such PUFA-rich diets also appeared beneficial in treating patients with multiple sclerosis, an autoimmune disease in which an inappropriate immune response to one of the body's own proteins damages the myelin sheaths of nerve cells. Again, however, these apparent benefits have not been substantiated in later studies.

More recently, interest has centred on diets rich in PUFAs of the *n*-3 series, as found in fish oils, especially eicosapentaenoic acid (EPA, 20:5*n*-3) and docosahexaenoic acid (DHA, 22:6*n*-3). In many industrialized countries, increases in intakes of linoleic acid

(18:2*n*-6) have occurred mainly as a result of advice that such dietary changes are likely to result in lowering of plasma LDL-cholesterol concentration (see Section 10.5.4.2) and a reduction in coronary heart disease incidence and mortality. At the same time, intakes of *n*-3 PUFAs have not increased, and may even have decreased, resulting in significant reductions in the ratio of dietary *n*-3/*n*-6 PUFAs. Many people have pointed out the correlation between the fall in the dietary *n*-3/*n*-6 PUFA ratio and the concomitant rise in the incidence of inflammatory diseases such as asthma. Such statistical correlations do not, of course, show that one is the result of the other, but there is some biological basis for such an association. See Box 10.1 for further discussion of dietary *n*-3/*n*-6 PUFA ratios.

There have been many studies of the effects of dietary supplementation with *n*-3 PUFAs, usually in the form of capsules of fish oil or of isolated very long-chain *n*-3 PUFAs. Positive benefits of these FAs have been shown in rheumatoid arthritis, an auto-immune disease with a strong inflammatory component, with reductions in joint swelling and pain, easing of morning stiffness and a lessening of disease activity.

In many animal models of inflammation, dietary very long-chain *n*-3 PUFAs have been shown to reduce inflammation. Very long-chain *n*-3 PUFAs are now commonly added to cat and dog foods because of research showing beneficial effects on osteoarthritis in dogs and, although less well investigated, cognitive development in puppies. However, investigations in humans of very long-chain *n*-3 PUFA supplementation in inflammatory conditions other than rheumatoid arthritis have not been so positive. These include psoriasis, inflammatory bowel disease (ulcerative colitis and Crohn's disease) and asthma. Although some human studies may show positive results, overall no clear benefits have been demonstrated when all available data are analysed together (meta-analysis). However, the authors of most of these meta-analyses comment that the studies reviewed are often small in scale or of unsound design, so there is keen interest in conducting further studies in these conditions. (Effects of these FAs on cardiovascular disease risk factors will be considered later: see Sections 10.5.4.2 & 10.5.4.3).

As noted above, inflammation is not necessarily always adverse: it is an important part of the normal immune process. Given that some PUFAs may suppress immune responsiveness, an important question arises. Are animals that habitually consume diets rich in unsaturated lipids less able to combat infection? A small number of animal studies suggests that this is so, but there have been no reports of compromised ability to fight infection in human subjects supplementing their diets with different kinds of unsaturated fatty acids (UFAs). Indeed, in contrast, a study in the US of supplementation of infants' diets with cod liver oil (a source of very long-chain *n*-3 PUFAs, but also of vitamins A and D) showed a considerably reduced incidence of respiratory illness.

10.3.3 Influence of dietary polyunsaturated fatty acids on target cell composition and function

Cells of the immune system have membrane phospholipids that are rich in arachidonic acid, 20:4*n*-6, which, as described in Section 3.5, is a major substrate for production of eicosanoids. Typically, human immune cells isolated from the bloodstream contain around 20% arachidonic acid and about 1% EPA and 2.5% DHA. The composition can be changed by dietary manipulation. In particular, it has been shown consistently in cellular studies as well as in intact animals and humans that diets (or culture media) supplemented with very long-chain PUFAs of the *n*-3 series (as found in fish oils) lead to enrichment of immune cell phospholipids with the respective FAs, and a consequent decrease in arachidonic acid content.

One mechanism for the change in FA composition may relate to the intimate links between immune cells and adipose tissue. Lymph nodes, which are components of the immune system packed with macrophages and lymphocytes, are generally surrounded by adipose tissue. There is strong evidence that specific FAs, such as very long-chain *n*-3 PUFAs, may be transferred between nearby adipocytes and immune cells, probably stimulated by passage of cytokines in the other direction. Dendritic cells are antigen-presenting cells that migrate through tissues including adipose tissue. After initial recognition and uptake of an antigen (activation), they migrate to the lymph nodes where they interact with T- and B-lymphocytes to initiate their roles in the immune response. During their migration through adipose tissue, they have been shown to acquire specific FAs, which become incorporated into their membrane phospholipids. It has been suggested that the close interaction between adipocytes and immune cells allows the latter to acquire specific FAs needed to fight against invasion,

Box 10.1 Dietary polyunsaturated fatty acids and the n-3:n-6 balance

Dietary PUFAs are essential to health: indeed, specific FAs of the n-3 and n-6 families (α-linolenic acid, 18:3n-3 and linoleic acid, 18:2n-6) constitute the essential dietary FAs for mammals (Section 6.2.2).

When PUFAs replace SFAs in the diet, some beneficial changes are observed:
- lowering of circulating lipid (cholesterol and/or TAG) concentrations, lowering risk of cardiovascular disease;
- potentially improvement in sensitivity to insulin, lowering diabetes risk.

But how much of each of these families of PUFA should we consume? That is rather controversial.

n-6 PUFA, when replacing SFA or carbohydrates:
- potentially increase inflammatory responses;
- potentially improve sensitivity to insulin (thus diabetes risk)

n-3 PUFA, when replacing SFA or carbohydrates:
- reduce platelet aggregation and prolong bleeding time (in large amounts);
- stabilize the heart rhythm (shown in experimental animals, not clearly in humans);
- potentially reduce ('dampen down') inflammatory responses

(in each case, effects are listed in approximate order of the strength of the evidence in humans). (Note that effects on plasma lipid concentrations are covered elsewhere: see Box 10.4.)

In addition, very long-chain n-3 PUFAs (or fish intake) have been shown to improve symptoms in rheumatoid and osteoarthritis (Section 10.3.2), to reduce risk of breast cancer (Section 10.2.1), and to benefit some aspects of cardiovascular disease risk (Section 10.5.4.3). In animal models, very long-chain n-3 PUFAs have several additional benefits including protection against onset of Alzheimer's disease. The evidence for the latter in man is mixed, although a recent meta-analysis suggested that treatment periods have been too short to show benefit.

Analysis of the diets of our palaeolithic hunter-gatherer ancestors by archaeologists suggests that they ate a dietary n-3:n-6 ratio of ~1 : 1. The n-3 PUFA came mainly from plant sources (hence much of it was α-linolenic acid), although in some geographical locations sea-food (fish and shellfish) could have contributed quite large amounts of EPA, 20:5n-3, and DHA, 22:6n-3. There is an argument that this is the period when our genes evolved and therefore we would be healthy if we returned to this type of diet.

Current diets in industrialized countries have a n-3:n-6 ratio more like 1 : 5 or 1 : 10 (the National Diet and Nutrition Survey in the UK found a ratio of 1 : 5.4 in adult men and women in 2000/2001; in the US, Kris-Etherton et al. (2000) reckoned the dietary n-3 : n-6 ratio to have changed from around 1 : 8 in the 1930s to 1 : 10 in the 1990s). Should we therefore try to change this ratio in favour of n-3 PUFA?

One argument in favour of changing the ratio in favour of n-3 PUFA is that there is competition between n-3 and n-6 PUFA for desaturation via the 6-desaturase and elongation (see Section 3.5.13, Fig. 3.54 and Section 6.2.2.2). Therefore, more n-6 PUFA in the diet will decrease the extent of conversion of α-linolenic acid to EPA and DHA. This can be shown in cellular systems, but is not clearly borne out in human feeding trials.

Most nutritionists agree that it would be beneficial to increase intake of oily fish, and therefore EPA and DHA intake. This would, of course, change the n-3 : n-6 ratio in favour of n-3 PUFAs.

But actually a number of expert reviews now consider that the n-3 : n-6 PUFA ratio is not, in itself, a valuable concept. A major reason is that this ratio assumes that all n-6 PUFAs are identical in their effects, and likewise for n-3 PUFA including α-linolenic acid, whereas this is manifestly not the case.

Information in this box is based on:

Kris-Etherton PM, Taylor DS, Yu-Poth S et al. (2000) Polyunsaturated fatty acids in the food chain in the United States. *Am J Clin Nutr* **71**(Suppl):179S–188S.

Food and Agriculture Organization (FAO) of the United Nations (2010) *Fats and Fatty Acids in Human Nutrition*. Report of an Expert Consultation. FAO Food and Nutrition Paper 91, FAO, Rome (Final report).

Goyens PLL, Spilker ME, Zock PI, Katan MB & Mensink RP (2006) Conversion of α-linolenic acid in humans is influenced by the absolute amounts of α-linolenic acid and linoleic acid in the diet and not by their ratio. *Am J Clin Nutr* **84**:44–53.

Further reading

Cunnane SC (2003). Problems with essential fatty acids: time for a new paradigm? *Prog Lipid Res* **42**:544–68.

Stanley JC, Elsom RL, Calder PC et al (2007). UK Food Standards Agency Workshop Report: the effects of the dietary n-6:n-3 fatty acid ratio on cardiovascular health. *Br J Nutr* **98**:1305–10.

Griffin BA (2008) How relevant is the ratio of dietary n-6 to n-3 polyunsaturated fatty acids to cardiovascular disease risk? Evidence from the OPTILIP study. *Curr Opin Lipidol* **19**:57–62.

Kuipers RS, Luxwolda MF, Janneke Dijck-Brouwer DA et al (2010) Estimated macronutrient and fatty acid intakes from an East African Paleolithic diet. *Br J Nutr* **104**:1666–87.

without a need for dietary change. (See Pond (2005) in **Further reading** for more information.)

A change in membrane phospholipid FA composition may affect membrane physical properties (see Section 5.3.1). Changes in membrane fluidity may modify activities of membrane-bound enzymes (see Section 5.3.2 for discussion of membrane proteins). For example, G-protein activity changes can result in changes in adenylate cyclase, phospholipase A_2 (PLA$_2$) and phospholipase C (PLC) activities. Alterations in membrane phospholipids can influence the production of lipid mediators such as diacylglycerols (DAGs; see Section 8.1.3).

More importantly, however, the pattern of production of eicosanoids and other related lipid mediators will be modified by changes in the nature of COX and lipoxygenase (LOX) substrates (e.g. arachidonic acid replaced by EPA). In general, eicosanoids formed from arachidonic acid (n-6) are regarded as pro-inflammatory, whereas those formed from very long-chain PUFAs of the n-3 family reduce inflammation (Table 3.11; Fig. 10.2). In addition, n-3 PUFAs give rise to mediators involved in resolving inflammation and protecting against damage, the so-called resolvins and protectins (see Section 3.5.13). For instance, a short period of dietary fish oil supplementation in healthy volunteers (containing 3–4 g EPA/day for 4 weeks) led to an increase in neutrophil EPA content, a concomitant reduction in neutrophil arachidonic acid content, and suppression of neutrophil leukotriene B$_4$ (LTB$_4$, see Sections 3.5.9 & 3.5.10) biosynthesis. (Neutrophils are part of the innate immune system, and are the most abundant type of white blood cells in human blood and so are easily isolated.) A suppression of neutrophil chemotactic responsiveness to LTB$_4$ (i.e. the ability of neutrophils to sense and move towards a source of LTB$_4$) was also observed.

10.3.4 Influence on other aspects of immune function

Proliferation of lymphocytes is an essential component of the acquired immune response. Animal studies indicate that, in general, lymphocytes from animals given high-fat diets proliferate less vigorously when stimulated than those given low-fat diets. Among high-fat diets the order of potency is: fats rich in saturated fatty acids (SFAs) < oils rich in n-6 PUFAs < olive oil < linseed oil < fish oil. Cytotoxic T-lymphocytes and natural killer cell activities also tend to be suppressed by high-fat diets in a manner similar to the lymphocyte proliferation. Some, but by no means all, human studies using cells removed from volunteers replicate these results.

Several studies in animals have described lessening of various aspects of the immune/inflammatory response by n-3 PUFAs (either fed in the diet or added to cell culture). For instance, inflammatory signs including macrophage infiltration in the adipose tissue of obese mice are lessened on a fish oil-rich diet; antibody production in response to various stimuli is reduced by very long-chain n-3 PUFAs in stimulated rat spleen lymphocytes and in mice challenged with the antigen ovalbumin. Curiously, there are several reports of enhanced immunity in chickens given supplemental very long-chain n-3 PUFAs. Human studies have generally not found any such effects, and, as noted earlier, there is no evidence that dietary UFAs of any type compromise the ability to fight infection in humans.

10.3.5 Availability of vitamin E

As discussed in Section 3.3, PUFAs are highly susceptible to peroxidation unless adequate vitamin E is available as an antioxidant. When diets that contain high levels of n-3 PUFAs from fish oils are given, some diminution in the concentration of vitamin E in plasma and some tissues may be observed. This is because such diets contain relatively low levels of vitamin E. Concomitantly, some increase in the concentration of the products of lipid peroxidation may be measured in the plasma and tissues. Such changes do not occur when the sources of n-3 PUFAs are plant oils, because they are richer in vitamin E than fish oils. There is limited evidence that a diet supplemented with vitamin E reduces the suppression of antigen-stimulated lymphocyte proliferation by n-3 PUFAs. Functions such as neutrophil phagocytosis are suppressed by vitamin E deficiency and enhanced by vitamin E supplementation. However, vitamin E supplementation decreases bactericidal activity, probably because such activity is promoted by ROS, which are suppressed by vitamin E. There is evidence for a reduction in respiratory infections in elderly people given vitamin E supplementation, although no convincing evidence in younger people who are adequately nourished.

10.3.6 Lipids and gene expression

In addition to effects mediated via incorporation into membrane phospholipids, PUFAs of the n-3 family may modify gene expression. These n-3 PUFAs are ligands for the PPARs (see Sections 3.1.1 & 7.3.2), and PPARγ activation inhibits the expression of inflammatory cytokines, and directs the differentiation of immune cells towards anti-inflammatory phenotypes. Another transcription factor central to immune responses is nuclear factor-κB (NF-κB). The inhibitor of NF-κB (I-κB) and NF-κB form a heterotrimer in the cytosol (two molecules of NF-κB with one of I-κB). Phosphorylation of I-κB allows NF-κB dimers to dissociate from I-κB, translocate to the nucleus, and increase the transcription of a variety of 'pro-inflammatory' genes, including those for COX-2, TNF and IL-6. The translocation of NF-κB is regulated both positively and negatively by various PUFAs. Arachidonic acid stimulates NF-κB translocation and thus stimulates transcription of 'pro-inflammatory' genes and the PUFAs EPA and DHA have the opposite effect. The mechanisms by which EPA and DHA inhibit NF-κB translocation are not understood but may involve the kinases implicated in I-κB phosphorylation.

FAs may also act via the family of toll-like receptors (TLRs), either directly or by altering the lipid composition of the membranes in which they are located. TLRs are 'pattern recognition receptors' expressed on cells of the innate immune system that recognize conserved molecular structures from a wide variety of bacteria and viruses. Following recognition of ligands, TLRs recruit signalling adapters to initiate a pro-inflammatory signalling cascade, leading to the production of cytokines such as TNF and IL-1β. There is some evidence that the TLR system is activated by FA, especially SFAs, and inhibited by very long-chain n-3 PUFAs.

FAs also signal via several G protein-coupled receptors (GPCRs): see Sections 7.3.4 and Table 7.5 for more information.

10.3.7 Other lipids with relevance to the immune system

10.3.7.1 Lipopolysaccharide (endotoxin) in the cell envelope of Gram-negative bacteria is responsible for toxic effects in the mammalian host

Infection with Gram-negative bacteria such as *Escherichia coli* can have devastating effects on the host, including the phenomenon of toxic or septic shock. The component of the bacteria responsible for these effects was termed 'endotoxin' at the end of the 19th century. In the 1950s, the chemical components of endotoxin were discovered to contain a lipid moiety known as Lipid A. We now recognize that endotoxin is a lipopolysaccharide component of the outer cell membrane of Gram-negative

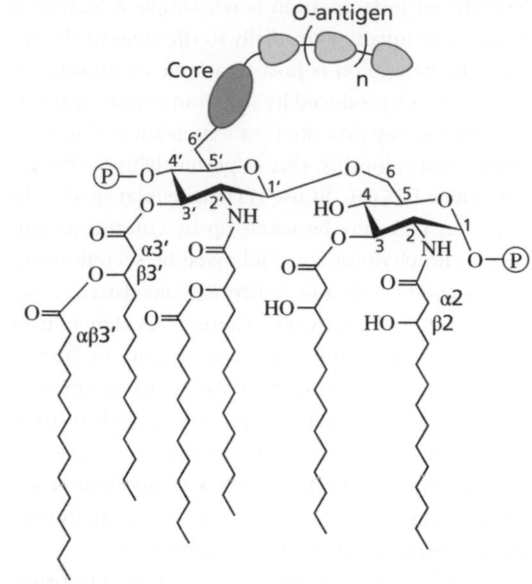

Fig. 10.4 Structure of bacterial lipopolysaccharide (also called endotoxin). The O-antigen region may contain many repeating monosaccharides. The Core region usually contains 10–15 monosaccharides. Lipid A is represented by the acyl chains at the bottom of the figure, which are derivatives of myristic acid (14:0), 3-hydroxymyristate and 3-myristoxymyristate. According to bacterial species and conditions, lauroyl (12:0) and other saturated fatty acids, and palmitoleoyl (16:1) acyl chains may also be present. The acyl chains are attached to glucosamine residues (shown), and these are in turn attached to 3-deoxy-D-manno-octulosonic acid residues (shown here simply as part of the Core). Reproduced, with permission of Elsevier, from Xwang & PJ Quinn (2010) Lipopolysaccharide: Biosynthetic pathway and structure modification, *Prog Lipid Res* **49**:97–107.

bacteria, and it is often called simply lipopolysaccharide or LPS.

Lipopolysaccharide is a complex polymer comprising three parts (Fig. 10.4). It was described in more detail in Section 5.6.1.

The core of Lipid A is a heavily substituted disaccharide of glucosamines (Fig. 10.4). The amino groups are substituted with 3-hydroxymyristate while the hydroxyl groups contain saturated (12C–16C) acids and 3-myristoxymyristate. Palmitoleoyl (16:1n-7) acyl chains may also be present depending on bacterial species and conditions.

Lipid A is the component of LPS mainly responsible for its toxicity, and it interacts with TLRs (see Section 10.3.6), especially TLR 4, which is expressed by many cells of the innate immune system. (See Peri *et al.* (2010) in **Further reading** for more information on this pathway.)

LPS-induced inflammation is not simply a feature of sepsis (bacteria invading normally sterile areas the body). The mammalian colon is host to a large community of bacteria, and LPS produced by the Gram-negative bacteria can find its way into the host's circulation. There are two mechanisms for this. One is 'permeability' of the gut wall, in which LPS can diffuse via paracellular spaces; the other is that LPS may be taken up by enterocytes and packaged into chylomicrons, followed by secretion into the lymphatic vessels and hence the bloodstream (see Section 7.1.2 for pathway). There is evidence from rodent studies, confirmed to some extent in human volunteers, that high-fat diets increase entry of intestinal LPS into the circulation by either one or both of these routes. The inflammation of adipose tissue that is seen in obesity (see Section 10.4.1.3) has a component representing LPS-stimulation of pro-inflammatory pathways, believed to originate from intestinal bacteria.

Two closely-related proteins are involved in transport of LPS in the circulation: LPS-binding protein (LBP) and bactericidal/permeability-increasing protein (BPI). LBP is secreted from hepatocytes, and transports LPS to the cell-surface receptor CD14, expressed on monocytes and myeloid cells (granulocytes), part of the process of delivery of LPS to the TLR 4. Thus, LBP could be seen to have a pro-inflammatory role, activating host defences to infection. BPI, on the other hand, which is produced by neutrophils, has bactericidal properties, and binds LPS avidly, so preventing its transfer via LBP to CD14. LPS in the circulation is also bound by lipoproteins (LBP may facilitate the transfer of intestinally absorbed LPS to lipoproteins) by incorporation of Lipid A into the surface phospholipid monolayer. This seems effectively to neutralize LPS: indeed, mice given Gram-negative infections have been saved by infusion of a lipid emulsion. One marked response to sepsis is a rise in concentrations of TAG-rich lipoproteins (this is brought about by cytokines, including TNF-α), and it has been suggested that this is a protective response and that TAG-rich lipoproteins might be considered part of the innate immune system.

10.3.7.2 Platelet activating factor: a biologically active phosphoglyceride

Platelet activating factor (PAF – introduced in Section 4.5.11) is a phosphoglyceride that acts as a mediator of many biological processes. PAF is found in many living organisms (including plants), but in mammals it has roles that include effects on immune function and in inflammation; hence its inclusion in this section. PAF is a 1-O-alkyl-2-acetyl-sn-glycero-3-phosphocholine (Fig. 10.5) and is produced by many cells involved in host defence. It acts like a hormone, signalling to other cells. Its concentration is regulated both by its cellular production rate and by its degradation.

There are two routes for the biosynthesis of PAF: the *remodelling* pathway (Fig. 10.5) and the *de novo* pathway (see also Section 4.5.11). Remodelling is the pathway used in acute reactions and is up-regulated in response to a variety of stimuli associated with inflammatory stress. It involves removal of a 2-arachidonoyl group from 1-O-alkyl-2-arachidonoyl-sn-glycero-3-phosphocholine to form 1-O-alkyl-sn-glycero-3-phosphocholine (lyso-PAF), which is then acetylated by a lyso-PAF:acetyl-CoA acetyltransferase using acetyl-CoA as substrate. Biosynthesis *de novo* is not so acutely regulated and is responsible for constitutive production of PAF. It begins with 1-alkyl-glycero-3-phosphate and involves acetylation of the 2-carbon followed by the action of a cholinephosphotransferase.

PAF degradation involves removal of the 2-acetyl group, to form lyso-PAF, by a Ca^{2+}-independent phospholipase A_2, known as PAF acetylhydrolase (PAF-AH; Fig. 10.5). The term PAF-AH actually refers to a small family of intracellular and secreted enzymes. The two intracellular PAF-AH isoforms and one secreted (plasma) isoform exhibit different biochemical characteristics. A loss-of-function mutation in the plasma PAF-AH has been found to cause various inflammatory diseases. The mutation is common in Japanese people (4% homozygous, 27% heterozygous) and increases the risk of inflammatory conditions, including asthma, atherosclerosis (see Section 10.5.1) and some forms of kidney disease. The secreted form is now more commonly known as lipoprotein-associated PLA$_2$ (Lp-PLA$_2$). (The human gene is *PLA2G7* as it belongs to group VII of the phospholipase A_2 family – see Section 4.6.2.) Lp-PLA$_2$ is produced mainly by macrophages. It is found in human plasma, associated with low-density and high-density lipoproteins (see Section 7.2.3). The activity of Lp-PLA$_2$ in plasma correlates with risk of cardiovascular disease (see Section 10.5.1) and inhibitors of this enzyme are in clinical trials with the aim of reducing the incidence of cardiovascular disease. This seems counterintuitive, given that it inactivates PAF, a mediator with strong pro-inflammatory actions, and the outcome of these trials is awaited with interest.

PAF has several unusual features: (1) it was the first well-documented example of a biologically active

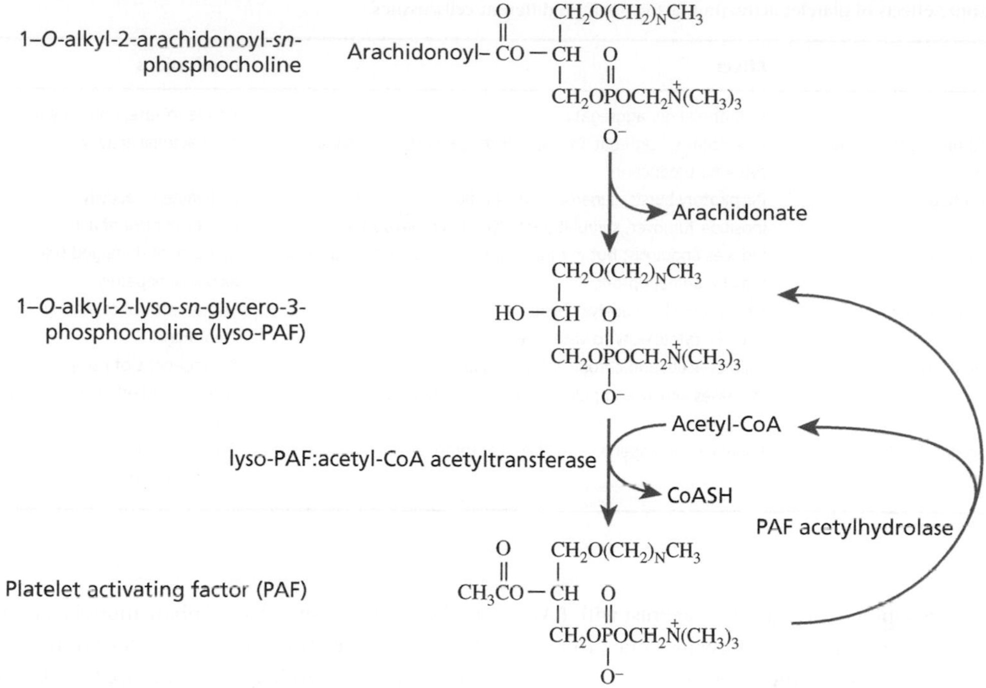

1–*O*-alkyl-2-arachidonoyl-*sn*-phosphocholine

$$\text{Arachidonoyl}-\overset{\overset{\text{O}}{\|}}{\text{CO}}-\overset{\overset{\text{CH}_2\text{O(CH}_2)_N\text{CH}_3}{|}}{\underset{|}{\text{CH}}}\quad\overset{\text{O}}{\underset{|}{\text{CH}_2\text{OPOCH}_2\overset{+}{\text{N}}(\text{CH}_3)_3}}$$

Arachidonate

1–*O*-alkyl-2-lyso-*sn*-glycero-3-phosphocholine (lyso-PAF)

lyso-PAF:acetyl-CoA acetyltransferase

Acetyl-CoA

CoASH

PAF acetylhydrolase

Platelet activating factor (PAF)

Fig. 10.5 Platelet activating factor and its biosynthesis and degradation. The pathway of synthesis *de novo* is not shown.

phosphoglyceride; (2) it has an ether link at position 1 (fatty acyl derivatives have 1/300th the activity); and (3) an acetyl moiety is present at the 2-position (butyryl substitution decreases activity some 1000-fold). PAF is a very potent molecule having biological activity at 5×10^{-11} M concentrations. Some of its actions are listed in Table 10.3, where it will be seen that this molecule has wide-ranging effects in many tissues.

PAF acts through a specific GPCR (human gene *PTAFR*) that is expressed in platelets and a variety of other cells, including those of the immune system (Table 10.3). Binding of PAF to its receptor initiates complex intracellular signalling events. These include: stimulation of the phosphoinositidase C signalling pathway (see Section 8.1.3), so increasing intracellular Ca^{2+} mobilization and increasing protein phosphorylation via PKC and other protein tyrosine kinases; activation of the mitogen-activated protein (MAP) kinase pathway, which in turn can lead to activation of the pro-inflammatory NF-κB system (see Section 10.3.6) and to activation of immediate-early genes such as *c-fos* and *c-jun* (which are involved in initiating the transition of cells from quiescence to proliferation); and by increasing arachidonate turnover and the formation of arachidonate metabolites (see Section 3.5). MAP kinase signalling stimulates cell proliferation in many healthy tissues and in tumours. PAF can affect inflammatory responses by altering expression of cytokines, such as TNF-α.

As well as effects on the immune cells (Table 10.3), PAF has potent angiogenic effects: it stimulates new blood vessel growth (angiogenesis), which is important for tumour growth (see Section 10.2.2.5). The PAF/PAF receptor system has therefore been seen as a potential therapeutic target in cancer and also in inflammatory conditions including sepsis, asthma and rhinitis (allergic inflammation of the nasal passages). Some chemically synthesized derivatives of PAF (e.g. edelfosine, with a methoxy group at position 2) show highly selective antitumour properties, apparently by inhibiting the cytidylyltransferase of the CDP-choline pathway (see Section 4.5.5) and limiting membrane formation. They do not appear to act via the PAF receptor. (The possible antitumour activity of these compounds was discussed previously in Section 10.2.2.3.) Trials of a PAF-receptor antagonist in Gram-negative sepsis have shown little or no benefit and trials in other conditions are on-going.

Table 10.3 Some effects of platelet activating factor (PAF) in different cells/tissues.

Cell/tissue	Effect	Implications
Platelet	Degranulation, aggregation	Amine release, coronary thrombosis
Neutrophil and other cells of the immune system	Chemotaxis, aggregation, superoxide generation; enhances cytokine production	Antibacterial activity
Alveolar macrophage	Respiratory bursts, superoxide generation	Antibacterial activity
Liver	Inositide turnover, stimulationof glycogen breakdown	Overall control of activity
Hepatocytes, neurons	Induces apoptosis; but in other cell types, e.g. epidermal cells, may be anti-apoptotic	Removal of damaged cells, e.g. in alcoholic hepatitis.
Exocrine secretory glands	Similar effects to acetylcholine	Overall control of activity
Leukaemic cells	Specific cytotoxicity towards the cells	Treatment?
Vascular permeability	Mimics acute and chronic inflammation	Pathogenesis of psoriasis?
Lungs	Increases airway and pulmonary oedema; decreases compliance	Asthma and other respiratory conditions
Tumours, Endothelial cells	Enhanced angiogenesis and vascular remodelling	Could contribute towards tumour growth and spread

Rupatidine, a histamine H_I-receptor antagonist with PAF receptor antagonist properties, is showing some promise in inflammatory diseases of the airways.

10.3.7.3 Pulmonary surfactant

Every time we breathe out, our lungs are prevented from collapsing by a unique lipoprotein mixture termed the 'pulmonary surfactant'. Pulmonary surfactant is adsorbed at the alveolar air/liquid interface where it lowers the surface tension and, therefore, reduces contractile forces at the surface and the work needed to expand the lungs. As the alveolar surface area decreases during expiration, the surface film is compressed and undergoes a phase transition from a fluid to a solid gel state, preventing collapse of the deflated lungs. A dramatic example of its importance is provided by a disease known as respiratory distress syndrome (RDS) of the newborn. In this condition, premature infants who do not have sufficient (endogenous) pulmonary surfactant are unable to expand their lungs properly and suffer from various complications.

Pulmonary surfactant, which can be isolated by carefully washing out lungs repeatedly with isotonic saline, has a unique composition, summarized in Table 10.4. In contrast to serum lipoproteins, surfactant comprises mainly phospholipids. PtdCho accounts for approximately 80 % of total and in humans about 70% of this is a single molecular species – dipalmitoylPtdCho. Thus, the lung surfactant lipids have a composition particularly suited

to the formation of stable gel phase monolayers at the air–liquid interface. The unsaturated PtdCho molecular species are believed to help in the intracellular trafficking in the type II epithelial cells of the lung where surfactant is synthesized, assembled into organelles know as lamellar bodies, and secreted into the alveolar subphase by exocytosis. The unsaturated components also contribute to the rapid spreading of surfactant in the alveoli at body temperatures. Phosphatidylglycerol (PtdGro), which is unusual in mammalian tissues, is also present and may assist in the dissolution of lamellar bodies to form the surface film.

Table 10.4 Composition of alveolar surfactant.

	(% w/w)		
	Human	**Rabbit**	**Rat**
Total protein	15	8	10
Total lipid	85	91	89
	(% w/w of total lipid)		
Phosphatidylcholine	74	83	73
(dipalmitoylphosphatidylcholine)	(68)*	(63)*	(82)*
Phosphatidylglycerol	6	3	5
Sphingomyelin	4	2	2
Other phospholipids	7	3	9
Cholesterol (free and esterified)	4	3	3
Other lipids	5	6	8

* Percentage of phosphatidylcholine.

Four specific proteins are present in human surfactant (surfactant protein-A (SP-A), SP-B, SP-C and SP-D). SP- B and SP-C assist in the spreading of surfactant over the air–liquid interface in the lung. SP-A and SP-D contribute to the host innate immune defence system and aid in the phagocytosis of microorganisms including bacteria, viruses and moulds (US: molds) and in other immune functions. SP-B and SP-C are small and remarkably hydrophobic. In fact, they copurify with lipid in normal extraction systems!

It is currently thought that surfactant forms a film through structures containing SP-B or SP-C. Furthermore, as the alveolar surface film is compressed during expiration, the unsaturated phospholipids are selectively squeezed out of the surface monolayer via the adsorption structures into bilayer stacks, leaving a monolayer highly enriched in saturated lipids. The resulting disaturated phospholipid-enriched monolayer can withstand sufficient surface pressure to reduce surface tension to near zero, thus stabilizing the alveoli at end-expiration. During inspiration, the unsaturated lipids flow back onto the surface through the adsorption structures thus eliminating the need to apply fresh surfactant with each breath.

Detailed studies of lipid metabolism in type II cells have been made by the use of normal cells (isolated from whole lungs by collagenase treatment to separate them carefully from the more abundant type I cells and the underlying tissue). Attention has been focused on the formation of dipalmitoylPtdCho. The PtdCho in surfactant is made by the CDP-base pathway (see Section 4.5.5). However, the operation of this route results in a number of PtdCho molecular species, containing UFAs at the *sn*-2-position as in other tissues such as liver. Additional dipalmitoyl species are provided by the remodelling of PtdCho by removing the UFAs at the *sn*-2-position with PLA_2 and replacing them by acylation with palmitoyl-CoA i.e. Lands type remodelling (see Section 4.6.2 & Fig. 4.23). Careful experiments by the Dutch workers Batenburg and van Golde have shown that about half of the dipalmitoyl PtdCho in rat surfactant is produced *de novo* and half through deacylation/reacylation.

The type II cell is important not only for surfactant biosynthesis but also seems to be involved in its recycling. In much the same way as it would be wasteful for an animal continually to produce bile salts without any reabsorption, so pulmonary surfactant is constantly taken up by alveolar cells. Little is known of this process (and still less of its control) but it appears that all major lipid components are recycled. The careful balance of secretion and re-uptake of surfactant ensures that optimal amounts are always available to function in the alveolar spaces.

There would be little point to a young foetus in producing surfactant until near the time for birth and active breathing, so foetuses start to make large amounts of surfactant only towards the end of pregnancy – after about 32 weeks in humans. After this time the typical surfactant components can be found, and tested for, in the amniotic fluid. Various tests have been used of which the PtdCho/sphingomyelin ratio is the most common. A ratio of at least 2 is taken to indicate significant surfactant production. Nevertheless this test is of little use for most infants likely to suffer neonatal RDS (below) because they simply haven't produced enough surfactant. It may be useful, however, for diabetic mothers whose babies used to have a significant death-rate. The problem can be much alleviated by careful nursing and treatment.

Neonatal RDS is a developmental disorder which is caused by immaturity of the baby's lungs. The lack of adequate amounts of surfactant causes morphological alterations to the hyaline membranes and lung collapse (atelactasis) as well as physiological changes (decreased lung compliance, hypoxaemia) and used to be a major cause of death in premature infants. At one time about 25% of such babies died and others were left with handicaps. There have been three approaches to the treatment of the disease. The first has been to employ ventilators which maintain a positive air pressure in the lungs and thus counteract alveolar collapse. The second has been to hasten the development of the foetal lungs by the use of hormones, namely corticosteroids. Such treatments obviously require some prewarning of a possible Caesarean delivery or premature natural birth and are not helpful in all cases due to pregnancy complications. The third treatment is to instil surfactant mixtures into the baby's lungs in order to reproduce the effects of natural surfactant until such time as the baby can synthesize its own. Various preparations are in current use and they usually are made from extracts of animal lungs which contain both phospholipids and the surfactant proteins, SP-B and SP-C. Use of these clinical surfactants has been very successful and has succeeded in limiting the effects of RDS significantly and also in allowing younger and younger babies to survive. The use of such clinical surfactants is now routine in the UK for babies of less than 32 weeks' gestation.

Although respiratory distress of the newborn is a very tragic and dramatic disease, that of adults known as acute lung injury or acute respiratory distress is far more prevalent and has a worse prognosis. At least 60,000 people are affected each year in the USA alone, with a mortality of 40–50%. These conditions arise from inflammation due to a number of causes but not by lung immaturity. Hormone therapy is therefore of limited use and surfactant therapy has not been effective in decreasing mortality.

Several other diseases or conditions are known to affect surfactant metabolism and, hence, lung function. For example, alveolar proteinosis is a disease arising from a number of causes, including industrial work-related diseases such as silicosis, and can lead to a massive accumulation of surfactant (up to 40-times normal amounts) to the extent that it impairs gas exchange and breathing. The opposite occurs in poisoning with the herbicide, paraquat. The type II cells no longer produce surfactant and the victim makes painful efforts to breathe but may die within a few days.

10.4 Effects of too much or too little adipose tissue: obesity and lipodystrophies

10.4.1 Obesity and its health consequences

As discussed in the previous chapter, Section 9.1.1, the store of TAGs in adipocytes expands when energy intake exceeds energy expenditure. If this continues for a prolonged period, the body's fat stores become excessive, with adverse consequences for health. This is the condition known as obesity.

Obesity is defined by the World Health Organization (WHO) as 'abnormal or excessive fat accumulation that presents a risk to health'. It is classified by the WHO in terms of the body mass index (BMI), which is calculated as body mass (kg) divided by square of the height (m). Overweight is defined as a BMI of 25–30 kg/m², and obesity as a BMI > 30 kg/m². Some authorities define morbid obesity (with serious health consequences) as a BMI > 40 kg/m². BMI is not a perfect measure of 'adiposity' (TAG stores) because in principle a person with big musculature could have a high BMI. However, amongst the general population there is a strong relationship between body mass and adiposity (i.e. fat content) (Fig. 10.6).

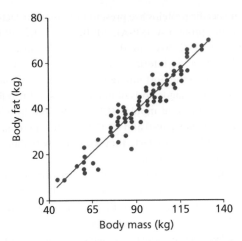

Fig. 10.6 Relationship between body fat content and body mass. The data are from measurements made in 104 healthy women. They show a strong relationship between body fat content (essentially TAG in adipocytes) and body mass (correlation coefficient 0.96). Redrawn from JD Webster, R Hesp & JS Garrow (1984) The composition of excess weight in obese women estimated by body density, total body water and total body potassium. *Human Nutr: Clin Nutr* **38C**:299–306.

In principle, the adipocyte TAG stores may increase in two ways. One is that individual adipocytes may increase their fat store, and swell (hypertrophy). The other is that new adipocytes may be stimulated to differentiate from precursor cells (preadipocytes) and begin to store TAGs (hyperplasia). In practice hypertrophy is almost universal in obesity, but to differing degrees; and the same is true for hyperplasia. Estimates of the total number of adipocytes in the body, made by taking a biopsy, measuring average cell size, and relating that to total fat stores, show that obese people generally have more adipocytes than lean people (for adults, typical figures in lean people are 4×10^{10} cells; in obese, $6 - 8 \times 10^{10}$ cells: see Spalding *et al.* (2008) in **Further reading**). Nevertheless, the degrees of hypertrophy and hyperplasia differ from one obese person to another, and may relate to health consequences (see below). It has been argued that some people become obese because they have more fat cells that tend to fill up, but this argument has to be carefully made to avoid contravening the first law of thermodynamics. An alternative explanation is that obese people have generated more fat cells to store the excess. (Both may be true to some extent.)

The process of differentiating new fat cells from preadipocytes depends on activation of a cascade of transcription factors, key amongst which is the nuclear receptor PPAR-γ. Ligands for PPAR γ (which were discussed in Section 7.3.2) may be generated by the processes involved in fat cell expansion, so that there is a natural tendency for increased fat storage to generate new adipocytes. Synthetic ligands for PPAR-γ have been developed by the pharmaceutical industry and are in use as treatments for type 2 diabetes mellitus (see Section 10.4.1.3). Preadipocytes can be stimulated to differentiate into new adipocytes in cell culture by adding a cocktail including a PPAR-γ ligand, thyroid hormone and insulin. It is not necessary to add FAs: provided that there is a substrate for DNL, developing adipocytes (even human) will produce a range of FA by a combination of DNL, chain elongation and desaturation.

Obesity carries serious health consequences. The Prospective Studies Collaboration collated data on 900,000 people from 57 different prospective studies of mortality, and estimated that all-cause mortality increased approximately 30% for each 5 kg/m^2 increase in BMI: so the risk of dying at any given age is more than double in a person of BMI 40 kg/m^2 compared with a person of 25 kg/m^2. There was increased mortality from many causes: vascular disease (mainly coronary heart disease and stroke), diabetes, kidney and liver disease, cancer and respiratory disease. They found, in common with many other studies, that the lowest mortality rates were seen in people with BMI around 25 kg/m^2. A lower BMI was also associated with increased mortality. There are many difficulties with confounding variables in that analysis: thin people may smoke more, or people may be thin because they have an underlying disease, although the authors corrected for these variables as well as they could. There is some debate about the health consequences of mild overweight (BMI 25–30 kg/m^3). In the Prospective Studies Collaboration data, overweight people displayed increased mortality but in large studies collated at the Centers for Disease Control and Prevention in the US, overweight was associated with lower mortality than normal weight (defined as BMI 18.5–25 kg/m^2).

Obesity is not solely a bad thing: it has an evolutionary basis. Some animal species become naturally fat in a seasonal manner in preparation for long-distance flights, for childbearing or for hibernation. There is no doubt that periods of food shortage have shaped human evolution, and those who can lay down fat stores in times of plenty will have a survival advantage when food is scarce. There are reports in the literature of obese people who have been completely starved (other than water and vitamin and mineral supplements) for periods of up to 17–18 weeks in order to lose weight. In contrast, people of initially normal body weight cannot survive more than 6–8 weeks without food. As noted earlier (see Section 7.3.5), adipose tissue secretes hormones and one of its important functions is to convert androgens (male sex hormones), produced in the adrenal glands, into oestrogens (US: estrogens – female sex hormones). Most oestrogens are secreted directly from the ovaries in premenopausal women but this ceases at menopause. After this, synthesis from adrenal androgens, catalysed by aromatase in adipose tissue, is the only source of oestrogens. Oestrogens are important for several functions including maintenance of bone mass. Therefore, thin postmenopausal women are at increased risk of osteoporosis, whereas obese women are protected.

10.4.1.1 Causes of obesity

A detailed discussion of the causes of obesity is beyond the scope of this book (see **Further reading** for more information). However, a few points are relevant so far as they pertain to lipid metabolism.

In one sense, the causes of obesity (and how to reverse it) are well known and incontrovertible. Accumulation of energy stores in the form of TAGs in adipocytes reflects a period when energy intake has exceeded energy expenditure. The situation will be reversed during a period when energy intake is less than energy expenditure. But that statement raises more questions than it answers. To put the real question in its simplest form: does someone become obese because of (1) a greater energy intake than someone who remains of normal weight, or (2) a lesser energy expenditure?

There are some medical conditions in which the second of these is the answer. The most clear-cut is hypothyroidism (low concentrations of thyroid hormones, which 'set' metabolic rate throughout the body). In fact, the diagnosis of thyroid disorders used to be based on measurement of 'basal metabolic rate'.

But the first of these is the answer for most obese people: their energy intake is higher than 'normal'. This has been demonstrated clearly in a large number of studies that measured either basal metabolic rate or daily energy expenditure (sometimes over several days). Energy expenditure in obese people was almost always greater than that of lean people. This finding must imply that obesity is maintained by increased energy intake; it does not necessarily tell us how obesity began to develop, but as John Garrow, a British nutritionist who spent many years studying and helping obese people, remarked in 1985 after measuring energy expenditure in 100 women: 'It is difficult to believe that fat people with a high metabolic rate relative to fat-free mass once had a low metabolic rate.'

Obesity is a highly inherited condition and 70–80% of the variation in BMI is attributable to underlying genetic variation, a figure very similar to that for height. Recent years have seen an enormous number of studies of the genes involved. These help us to understand the pathways involved.

There are a few well-defined single-gene mutations that cause obesity. Most prominent is homozygous mutation in the *LEP* gene encoding the peptide hormone leptin (OMIM 614962), which, as described in Section 7.3.5, regulates energy intake in relation to body fat stores. Leptin deficiency causes massive obesity, with highly abnormal eating behaviour (enormous drive to eat), from soon after birth. Some of the small number of people described with this condition have been treated with human leptin produced by recombinant DNA technology, with remarkable restoration of normal eating behaviour and many aspects of health. Even rarer, but causing a similar phenotype, is a mutation in the *LEPR* gene encoding the leptin receptor, and this condition (OMIM 614963) does not, of course, respond to leptin treatment. More common, but with

a less pronounced phenotype, is mutation of the receptor for melanocortin, melanocortin receptor 4 (gene *MC4R*). This is part of the system in the hypothalamus that regulates appetite. It has been estimated that around 1 in 200 obese people has a mutation in this receptor (the condition is listed generally in OMIM under Obesity, 601665). Different mutations cause various degrees of malfunction, and the activity of the receptor in test systems correlates with the energy intake of the person concerned, measured with a test meal, a quite remarkable link between biological activity and a behavioural trait (see O'Rahilly & Farooqi (2008) in **Further reading**). Several other genes responsible for familial obesity have been identified. All are involved with regulation of energy intake (appetite) rather than of energy expenditure or metabolic rate. These genetic studies bear out the conclusion from physiological measurements. Large genome-wide association studies have identified many genetic loci associated with the much more common *polygenic* pattern of inheritance of obesity. These include one or two genetic variations that might be involved with regulation of peripheral metabolism: for instance, a relatively common polymorphism in the PPAR-γ receptor that controls adipocyte differentiation (see Section 7.3.2). As with *monogenic* inheritance, many of the loci that have been identified, so far as their function is known, relate more to energy intake than to expenditure.

Study of the causes of obesity has led to development of drug treatments. These are summarized in Box 10.2.

Box 10.2 Lipid-related drug targets in obesity

There has been much interest in a pharmaceutical treatment for obesity. In the 1930s, 2,4-dinitrophenol, a mitochondrial uncoupler (so mimicking the action of the uncoupling protein of brown adipose tissue: see Section 9.2.2), was used to increase metabolic rate, but proved fatal in many cases. According to the UK's National Health Service website (http://www.nhs.uk/Conditions/Obesity/Pages/Treatment.aspx), more than 120 different drug treatments for obesity have been tried over the years.

There have been a number of attempts to 'activate' brown adipose tissue to 'burn off' excess calories. These have not as yet led to useful drug treatments (discussed also in Section 9.2.1).

At present only two drugs are in clinical use. Phentermine is a drug of the phenethylamine class, similar to amphetamine. Like most of the drugs that have been used in obesity, it targets hypothalamic pathways regulating appetite, in recognition of the role that increased energy intake plays in obesity.

The other drug in use at present targets a different pathway: fatty acid absorption from the small intestine. A bacterial metabolite known as tetrahydrolipstatin (orlistat is the generic drug name) is a potent irreversible inhibitor of pancreatic lipase in the small intestine, and so reduces the absorption of dietary fat. It is taken as a pill. At its most effective, it reduces fat absorption by about 30%. Patients taking this medication have to keep to a relatively low-fat diet as an unpleasant side-effect is 'leak' of unabsorbed fatty material. Orlistat has been shown, in a randomized clinical trial, to reduce the incidence of type 2 diabetes in obese people.

10.4.1.2 The health risks of excess adiposity depend upon where the excess is stored

It is a long-standing observation that the health risks of obesity vary according to the distribution of fat over the body. The observation of different patterns of fat distribution is an old one. A 25,000-year-old carved figurine, the Venus von Willendorf, depicts a female, perhaps thought to be especially fertile, with large fat deposits around her hips and breasts; Shakespeare in *Comedy of Errors* has Dromio referring to Nell, who clearly carried her fat around her midriff, with the words: 'No longer from head to foot than from hip to hip: she is spherical, like a globe.'

This topic of fat distribution was first investigated systematically by the French physician Jean Vague in Marseilles in the 1940s. In a classic paper published in 1947 Vague drew attention to two broad patterns of fat distribution, one typically female with fat mainly on the lower body (often thought of as pear-shaped) and one typically male with abdominal and upper-body fat accumulation (often envisaged as apple-shaped; Fig. 10.7). He pointed out that the male pattern can also occur in women, and vice versa, and that irrespective of sex, the upper-body (apple) type of fat distribution was associated

with a much higher incidence of several diseases including coronary heart disease and type 2 diabetes.

Larger studies have since confirmed powerful adverse effects on health of fat accumulation in the upper body (generally referred to as abdominal obesity), but many studies have shown that fat accumulation in the lower part of the body (gluteal and femoral regions, sometimes called gluteofemoral fat, loosely called hip fat) is not only less adverse than abdominal obesity, but is actually protective against a range of health problems. The INTERHEART study of 27,000 people from 52 countries compared the characteristics of people who had suffered a myocardial infarction (MI, or heart attack: see Section 10.5.1) with those of healthy people. There was a strong positive correlation between waist circumference, as a proxy measure of abdominal fat deposition, and risk of MI. In contrast, the relationship between hip circumference and risk of MI was strongly negative, implying a protective effect of hip fat. The same has been found for incidence of type 2 diabetes and various other adverse effects that are associated with abdominal fat deposition (see **Further reading**).

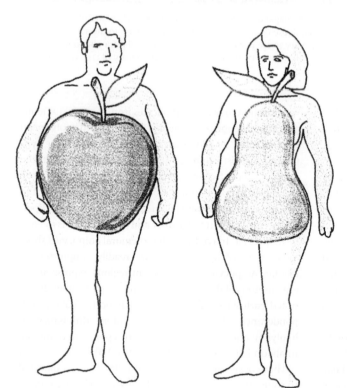

Fig. 10.7 Characterization of the two extremes of body fat distribution as apple- and pear-shaped. Reproduced from KN Frayn (1997) Obesity: a weighty problem, *Biological Sciences Review* **10**:17–20, with permission.

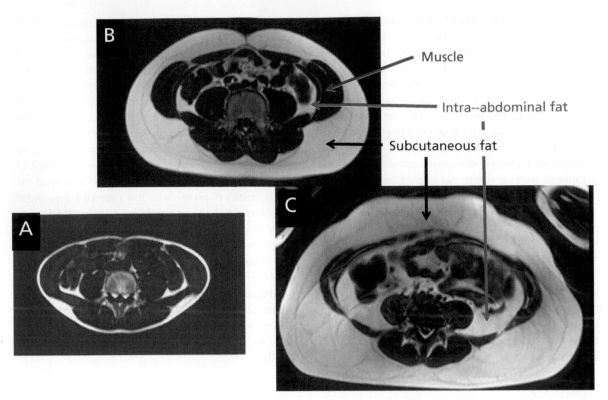

Fig. 10.8 Subcutaneous and intra-abdominal adipose tissue imaged by magnetic resonance imaging (MRI). These are MRI scans of the abdomen in people with different amounts and distributions of adipose tissue, which shows white. All are to the same scale. (a) lean person; (b) obese person with mainly subcutaneous fat; (c) very obese person with accumulation of intra-abdominal ('visceral') fat (this person also has liver fat accumulation). Pictures courtesy of Dr Rajarshi Banerjee, University of Oxford Centre for Clinical Magnetic Resonance Research.

A further development has been the understanding that abdominal fat accumulation involves both subcutaneous fat and intra-abdominal fat, often called in this context visceral fat (see Section 9.1.1 for a stricter definition of visceral fat). This has been made possible because of imaging techniques such as computed tomography (CT, an X-ray technique) and magnetic resonance imaging (MRI), which enable the visceral fat to be visualized and quantified (Fig. 10.8). Many studies suggest that the amount of visceral fat is more strongly associated with adverse health effects than is subcutaneous abdominal fat. However, subcutaneous abdominal fat is certainly not 'innocent' and in any case this distinction is difficult to make, because amounts of visceral and subcutaneous abdominal fat are highly correlated (correlation coefficients in many studies are around 0.7 or greater).

Studies of adipocytes isolated from different body regions have shed some light on these different relationships to health. Adipocytes from the subcutaneous abdominal depot can be incubated in vitro and lipolysis can be stimulated by adding a catecholamine (often the synthetic β-agonist isoprenaline – US: isoproterenol). In the presence of the β-agonist, lipolysis can then be suppressed by increasing concentrations of insulin. When visceral and subcutaneous abdominal adipocytes are compared, lipolysis occurs faster in the former and more insulin is needed to suppress it. In contrast, gluteofemoral adipocytes show lower rates of lipolysis that are readily suppressed by insulin. Adipocytes from different regions express different amounts of the relevant receptors and enzymes. It was noted earlier that gluteofemoral adipocytes have a preponderance of α-adrenergic receptors (inhibitory to lipolysis) and that β-adrenergic receptors predominate in abdominal depots (see Section 9.1.2.3).

These observations made in vitro have been supported to some extent by studies made in vivo (which are more difficult to perform). They have led to a hypothesis that gluteofemoral fat is a long-term storage depot, a view reinforced by studies of weight loss, which show that the gluteofemoral fat is lost more slowly than other depots (summed up in the commonly used phrase 'a moment on the lips: a lifetime on the hips'). In contrast, the lipid stores in abdominal fat (and visceral fat especially) turn over more quickly, with greater release of nonesterified fatty acids (NEFAs) into the circulation, and as discussed below (see Section 10.4.1.3), this might have adverse metabolic effects. It was noted earlier (see Section 9.1.1) that blood from the true visceral fat depots (omental and mesenteric) drains directly into the hepatic portal vein, and so might have particular effects on liver metabolism. A complementary, but different, view is that the differing relationships to health reflect the release of adipose tissue-derived hormones or 'adipokines' (see Section 7.3.5), some of which may have adverse or beneficial effects on metabolism of other organs. The question of how adipose tissue fat deposition relates to two particular aspects of health is discussed further in the following sections.

10.4.1.3 Obesity and the risk of developing type 2 diabetes

Diabetes mellitus is a condition in which blood glucose concentrations are abnormally raised, leading to 'overspill' of glucose into the urine (mellitus refers to sweet urine). There are two major types of diabetes mellitus, type 1 and type 2. Type 1 (in older literature, insulin-dependent or juvenile-onset diabetes) arises because of an autoimmune process that destroys the insulin-secreting β-cells of the pancreas. It almost always develops in childhood and requires insulin for treatment. Its aetiology (US: etiology) is therefore not directly concerned with lipid metabolism. People with type 1 diabetes are often thin and their insulin resistance (described below) is less marked than in type 2.

Type 2 diabetes usually comes on later in life (usually after 40). Older literature called it maturity-onset or noninsulin-dependent diabetes mellitus (because insulin is not usually necessary for treatment). It is strongly associated with obesity. In a large study of men aged 50–60, the risk of developing type 2 diabetes was increased 5–15-fold in overweight men (BMI 25–30 kg/m^2) and 40–50-fold increased in men with BMI > 33 kg/m^2, compared with lean men.

Two metabolic changes underlie the development of type 2 diabetes: insulin resistance and impairment of insulin secretion. Insulin resistance is a condition in which greater concentrations of insulin than normal are required for normal regulation of metabolic processes. (It can also be defined as a condition in which normal concentrations of insulin produce less-than-normal metabolic changes.) Insulin resistance is strongly associated with abdominal obesity (Fig. 10.9). Insulin-resistant people may be able to maintain normal blood glucose concentrations, because the pancreatic β-cells can secrete sufficient additional insulin, as shown in Fig. 10.9. However, this does not mean that all is normal metabolically. Insulin resistance is strongly associated with raised plasma TAG concentrations and other adverse health effects such as raised blood pressure; mechanisms will be considered below.

In general, obesity and insulin resistance develop in parallel, requiring the pancreatic β-cells to secrete more and more insulin. At some point, the pancreatic β-cells will be unable to do so. Then insulin concentrations will be insufficient to maintain normal glucose concentrations, and the pathway to type 2 diabetes is established. Thus, both insulin resistance and impaired insulin secretion are required for the development of type 2 diabetes. As will be described below, these conditions may have common origins in lipid metabolism. Earlier writings on type 2 diabetes tended to emphasize the insulin resistance. However, since the year 2000, a series of increasingly powerful genome-wide association studies has seen the identification of many genetic variants that contribute to increased risk of type 2 diabetes. (In its most common form, it has a strong inheritance, but is polygenic in origin.) Most of the contributing genes relate to the pathway of insulin secretion from the pancreas, and considerably fewer to pathways of insulin resistance.

A common feature of insulin resistance is fat deposition in several organs. This is prominent in skeletal muscle, the TAG content of which is closely related to insulin resistance. (There is a paradox in that highly trained athletes also have relatively high muscle TAG content, but there may be a qualitative difference in the location or nature of the LDs: see **Further reading** and Section 5.5 for description of LDs.) When fat accumulates in the liver without excessive alcohol intake, it is known as nonalcoholic fatty liver disease (NAFLD). NAFLD may progress through stages of increasing severity eventually to liver cirrhosis, a very serious condition. Liver fat accumulation (as in muscle) is strongly associated with insulin resistance and an adverse pattern of metabolism.

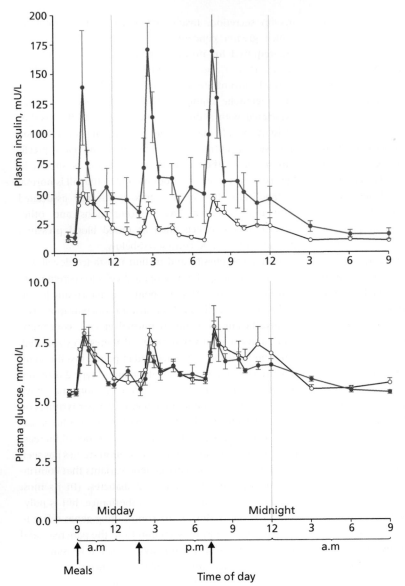

Fig. 10.9 Insulin resistance in abdominally-obese men. The experiments were the same as described in Fig. 9.3. Here data are shown for abdominally obese men (blue solid circles) and lean men of similar age (black open circles). Upper panel, plasma concentrations (in arterial blood) of insulin; lower panel, plasma glucose concentrations, in response to three meals during a 24 h period. Note that the abdominally obese men have identical plasma glucose concentrations to the lean men, but markedly increased insulin concentrations, especially in response to each meal ('insulin resistance'). Based on data in SE McQuaid, L Hodson, MJ Neville *et al.* (2011) Down-regulation of adipose tissue fatty acid trafficking in obesity: a driver for ectopic fat deposition? *Diabetes* **60**:47–55.

For example, one striking feature is that liver fat correlates closely with the secretion of large TAG-rich very low density lipoprotein (VLDL) particles (thus raising plasma TAG concentrations). Fat also accumulates in the pancreas. The pancreas mainly consists of exocrine tissue, producing and secreting digestive juices, and this can become infiltrated with adipocytes in obesity. But the islets of Langerhans, which contain the insulin-secreting β-cells, can also accumulate fat and the β-cells themselves may develop LDs and impaired insulin secretion. This storage of TAGs in tissues other than adipose tissue is known as 'ectopic fat deposition'. There is debate about whether the accumulation of TAGs in muscle and liver is a cause or a consequence of insulin resistance but there is a possibility that ectopic fat deposition underlies the development of both insulin resistance and impaired insulin secretion in type 2 diabetes.

Why does this ectopic fat accumulation accompany obesity? A prevalent idea is that elevated plasma concentrations of NEFAs may underlie the insulin resistance of obesity, perhaps via ectopic fat deposition. Adipose tissue releases NEFAs and it seems logical to suppose that, as adipose tissue mass increases, so will concentrations of NEFAs. High concentrations of NEFAs will produce insulin resistance, in the sense that major glucose-consuming organs such as skeletal muscle and liver will tend to oxidize FAs rather than glucose. This is part of normal metabolic fuel partitioning. But now it is becoming clear that severe insulin resistance can occur without elevated plasma NEFA concentrations. In fact, the high insulin concentrations associated with insulin resistance (as in Fig. 10.9) may well restrain lipolysis and so produce relatively normal circulating NEFA concentrations.

Current views of the relationship between obesity and insulin resistance build upon a more dynamic picture of lipid metabolism. As described earlier (see Section 9.1.2 & Fig. 9.3), adipose tissue takes up incoming dietary fats for storage in the period after meals, and releases them later in a controlled way when they are required by other tissues for energy generation. This is analogous to the role of the liver in glucose metabolism. When dietary glucose is available, the liver switches off glucose release, takes up glucose, and stores it as glycogen. During fasting or exercise, glycogen is hydrolysed and glucose is released for other tissues. Adipose tissue does the same with dietary fat, and thus protects other tissues (e.g. liver, skeletal muscle, pancreas) from what has been called 'excessive flux of FAs' (but note that FAs here may refer to TAG-FAs, e.g. in chylomicrons or chylomicron remnants, as well as NEFAs). This function of adipose tissue in the daily flow of FAs in and around the body has been described as 'buffering' (see **Further reading**).

When fat cells become enlarged, as in obesity, their metabolic functioning is impaired. They become more resistant to taking up FAs, and their capacity to release FAs upon catecholamine stimulation is reduced. This reflects down-regulation of the expression of the key lipases, ATGL and HSL (Box 9.1), as well as an increased ratio of α- to β-adrenergic receptors, so that the inhibitory effect of catecholamines begins to predominate (see Section 9.1.2.3). The normal 'buffering' is impaired and dietary fat tends to stay in the circulation for longer, potentially to be taken up by other tissues and to result in ectopic fat deposition. A simple way of looking at this is that the adipocytes 'fill up' to their limit and excess fat has to go somewhere else. This is one possible view of the changes that link obesity to insulin resistance and to impairment of insulin secretion.

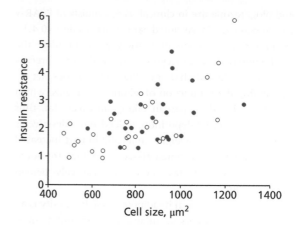

Fig. 10.10 Larger adipocytes are associated with insulin resistance. Data are from 59 healthy individuals. A fasting blood sample was taken to estimate insulin resistance from fasting glucose and insulin concentrations (using the so-called HOMA method), and a biopsy of subcutaneous abdominal adipose tissue was taken. The biopsy was processed for histology, and the average cell size in μm^2 was measured. Men are shown as solid points, women as open points. The correlation coefficient between insulin resistance and cell size is 0.38. Based on data in R Roberts, L Hodson, AL Dennis *et al.* (2009) Markers of de novo lipogenesis in adipose tissue: associations with small adipocytes and insulin sensitivity in humans. *Diabetologia* **52**:882–90.

Evidence for this view comes from the fact that people with a predominance of large adipocytes ('hypertrophic obesity') tend to have more severe adverse metabolic effects (e.g. insulin resistance) than people with larger numbers of smaller adipocytes ('hyperplasic obesity' – Fig. 10.10). The idea is that some people can more readily develop new adipocytes when needed to store more fat. TAGs stored in adipocytes are, in themselves, not harmful to other tissues: if the body has excess fat to store, this is the safest place for it to be. But other people cannot: their adipocytes expand, their capacity to absorb more fat is reached and insulin resistance and other adverse changes follow. This has also been called the 'adipose tissue expandability hypothesis' (see **Further reading**). Further evidence for this view of adipose tissue function and dysfunction comes from studies of the few people who lack adipose tissue – the condition of lipodystrophy (see Section 10.4.2).

One class of drugs used in the treatment of type 2 diabetes may work directly upon this mechanism. These are the thiazolidinediones or 'glitazones' (rosiglitazone and pioglitazone are in clinical use), agonists of PPAR-γ (see Section 7.3.2). As noted earlier (see Section 10.4.1), an important function of PPAR-γ is to stimulate the production of new adipocytes from preadipocyte precursors. It is believed that the thiazolidinediones act by stimulating the production of additional fat cells, which can then play their part in taking up excess fat, thus reducing ectopic fat deposition. Indeed, one side-effect is an increase in body weight, reflecting in part increased adipose stores but, as noted above, TAGs that are stored in adipocytes, and stay there, are relatively benign metabolically.

Another change seen in adipose tissue in obesity is an expression of markers of inflammation, including infiltration with macrophages. One possible explanation is that very large adipocytes may 'burst' and then macrophages are attracted and surround, and eventually engulf, the fragments: there are descriptions of macrophages surrounding damaged adipocytes, seen histologically as 'crown-like structures'. There is an idea that inflammatory changes lead to secretion of pro-inflammatory cytokines, such as TNF-α (see Section 10.2.3), that may lead to insulin resistance in other tissues. This mechanism seems to operate in rodents but evidence is not so clear in human obesity.

10.4.1.4 Ectopic fat deposition and insulin resistance: cause or effect?

It was noted above that there is some debate about whether ectopic fat deposition reflects cause or effect in the development of insulin resistance. This will be discussed here in terms of skeletal muscle, although similar arguments and observations apply to the liver.

Insulin resistant conditions (especially obesity) tend to be associated with impaired mitochondrial function. Therefore the capacity to oxidize, and hence dispose of, FAs in muscle may be reduced, and TAGs may accumulate. That would represent an effect, rather than a cause, of insulin resistance. Against that are many observations. Skeletal muscle TAG content can be increased over a period of a few hours by artificial means, namely intravenous infusion of a lipid emulsion to supply fat at a high rate: insulin resistance, measured in the whole body, develops in parallel with the development of muscle TAG stores. There is also strong cellular evidence for mechanisms that may link TAG accumulation to insulin resistance. However, it is generally agreed that TAGs themselves are not the 'culprits'. As was

argued above for TAGs stored in adipocytes, fat in LDs may be relatively inert, but something related to the lipid stores may cause problems. In the case of ectopic fat deposition, various lipid-related species have been suggested as the direct link to insulin resistance, including acyl-CoA species and intracellular NEFA. But the most consistent evidence currently implicates ceramides (see Section 4.8.1) as the active species. Ceramide accumulation is believed to activate signalling pathways involving certain isoforms of PKC, which may phosphorylate components of the insulin signalling chain in a way that adversely affects function.

10.4.1.5 Obesity and the risk of cardiovascular disease

Mortality from cardiovascular disease (mainly coronary heart disease and ischaemic stroke – see Section 10.5.1 for more details) is considerably increased in obese compared to lean individuals. Almost all this increase can be accounted for by alterations in plasma lipid and lipoprotein concentrations. Statistically, it is sometimes claimed that obesity per se is not a significant risk factor for cardiovascular disease, but this merely implies that it has no residual effect beyond what would be expected from the alterations in lipid concentrations.

The typical alterations in lipid concentrations in obesity include elevated plasma TAG concentrations and decreased concentrations of the 'protective' form of cholesterol-rich lipoprotein, HDL (sometimes called 'good cholesterol'). There is also a characteristic change in the nature of LDL particles. They increase in number (each includes one apoB molecule, so plasma apoB concentrations also increase – see Section 7.2.2) but decrease in size (the so-called 'small dense LDL'). These changes in plasma lipids are known to carry increased risk of cardiovascular disease and are described as the 'atherogenic lipoprotein phenotype' (see Section 10.5.2.4 for more details). Because concentrations of some lipoprotein species (particularly HDL) may be decreased, this is often described as a dyslipidaemia rather than a hyperlipidaemia.

These changes may all result primarily from an increased secretion of TAGs by the liver, in TAG-rich VLDL particles. As noted earlier, increased hepatic secretion of TAGs appears to be driven by increased liver fat stores: it can be seen as a move by the liver to reduce its fat stores. The origin of increased liver fat stores in obesity was described earlier (see Section 10.4.1.3).

10.4.2 Lipodystrophies

In Sections 10.4.1.2–10.4.1.5 above, we discussed the adverse effects on health when excess TAGs are stored in adipose tissue – the condition of obesity. But for some people, the problem is just the opposite: not having enough adipose tissue. The causes and consequences of this condition, known as lipodystrophy, help to illuminate the normal role of adipose tissue.

There are various conditions that result in a lack of adipose tissue. Some are genetic and may result in total (familial total (or generalized) lipodystrophy) or partial (familial partial lipodystrophy) lack of adipose tissue. Others are acquired. The most common of these is a form of lipodystrophy associated with human immunodeficiency virus (HIV) infection, usually known as HIV-associated lipodystrophy. Although HIV-associated lipodystrophy was observed before the widespread treatment of HIV infection became common, it is considerably more pronounced during treatment with highly active antiretroviral therapy, especially protease inhibitors.

Of the familial types of lipodystrophy, about half have a known genetic cause. Many of the genes identified encode proteins that are involved in adipocyte differentiation, TAG synthesis and LD formation. Some are listed in Table 10.5.

The mechanism of HIV-associated lipodystrophy is unclear. Various suggestions have been made. The HIV-protease (targeted by protease inhibitors) is homologous to cellular retinoic acid binding protein (CRABP) (see Section 6.2.3.1). CRABP is involved in the activation of retinoid X receptors (RXR), which form dimers with PPARs and so might affect adipocyte differentiation (see Sections 7.3.2 & 10.4.1), but other mechanisms have also been proposed.

There are obvious cosmetic consequences of lipodystrophy, arising from lack of subcutaneous fat. In many forms of partial lipodystrophy, the loss of fat is particularly pronounced on the extremities; hence the legs, for instance, may look 'muscled' and the veins stand out (the condition of phlebactasia, an identifying feature) (Fig. 10.11). The consequent 'masculine' appearance is one reason that the condition is more frequently diagnosed in women: it may not be so apparent in men. The fat depots in the cheeks are also affected in some conditions, giving the face a lean appearance, although they are spared in other types.

Whatever the origin of the lipodystrophy, the metabolic consequences are very consistent. Patients with lipodystrophy usually have high concentrations of lipids (especially TAGs) in the blood, are insulin-resistant (see Section 10.4.1.3), and many develop type 2 diabetes. (Congenital generalized lipodystrophy was at one time called 'lipoatrophic diabetes'.) This appears paradoxical: these are exactly the metabolic features associated with an excess of adipose tissue TAGs, i.e. obesity. But this paradox sheds light on the normal role of adipose tissue. Whenever we eat a meal that contains fat, TAGs enter the bloodstream and need to be removed. In the presence of well-functioning adipocytes, much of the dietary fat will be taken up and stored in adipose tissue (see Section 9.1.2 & Fig. 9.3). It was argued in Section 10.4.1.3 that adipose tissue in obesity is effectively over-full and its action in removal of dietary fat is impaired: hence fat accumulates in other tissues (ectopic fat deposition), where it is associated with insulin resistance and other adverse changes. Exactly the same would be expected in lipodystrophy: when there is too little adipose tissue for normal removal of dietary fat, it will go elsewhere with similar consequences. This was nicely illustrated in an experiment by David Savage and colleagues in Cambridge, UK, in which patients with lipodystrophy, and healthy volunteers, were given a large fat load (30% energy above requirements provided as fat during 40 h). The healthy volunteers stored most of the fat (presumably in adipose tissue) and did not change their energy expenditure. In contrast, the lipodystrophic patients increased their energy expenditure and fat oxidation, but in some of the patients there were also very large increases in plasma TAG concentration.

The adverse metabolic profile in subjects with lipodystrophy translates into a high cardiovascular disease risk, of a similar magnitude to subjects with familial hypercholesterolaemia (see Section 10.5.2.1). There is more than 100-fold higher risk of a myocardial infarction or coronary by-pass grafting in women with familial partial lipodystrophy than in healthy controls.

One other important feature of lack of adipose tissue is that the patients have low plasma concentrations of hormones secreted from that tissue, notably leptin (see Section 7.3.5). Some patients have been treated with recombinant leptin (as for patients with extreme obesity due to leptin deficiency) and this can be very effective in normalizing the metabolic profile.

Table 10.5 Some genetic causes of lipodystrophy.

Condition	OMIM* number	Characteristics	Defect	Gene affected
Congenital generalized lipodystrophy, or Berardinelli-Seip congenital lipodystrophy (various subtypes as below)		Near absence of adipose tissue from birth or early infancy and severe insulin resistance. Altered glucose tolerance or (usually) diabetes, and hypertriacylglycerolaemia.		
Type 1	608594		1-acylglycerol-3-phosphate O-acyltransferase-2, involved in triacylglycerol synthesis (Section 4.1.1)	AGPAT2
Type 2	269700	The most severe form: patients are born without any adipose tissue. (In the other subtypes, at least 'mechanical' adipose depots, in joints, are preserved.)	Seipin, a protein named after the condition, involved in lipid droplet formation in adipocytes.	BSCL2
Type 3	612526		Caveolin 1, a scaffolding protein involved in membrane function (Section 5.3.3).	CAV1
Type 4	613327	Includes muscular dystrophy	Polymerase I and transcript release factor (hence name of gene) but this protein, also called cavin, also plays a critical role in the formation of caveolae and the stabilization of caveolins.	PTRF
Familial partial lipodystrophy (FPLD) (various types as below)		Gradual loss of peripheral adipose tissue especially on the limbs and buttocks with preservation of subcutaneous fat on the trunk; often starts to become apparent around puberty.		
FPLD Type 2, Dunnigan type lipodystrophy	151660	The commonest form. Mutations elsewhere in the same gene cause muscular dystrophy. Heterozygous condition.	Lamin A/C is a member of the family of lamin proteins involved in nuclear stability.	LMNA
FPLD Type 3	604367	The next most common form. Severe insulin resistance. Heterozygous condition.	PPAR -γ (Section 7.3.2) is a key regulator of adipocyte differentiation.	PPARG
FPLD Type 4	613877	Rare cause of FPLD. Heterozygous condition.	Perilipin 1, a lipid droplet-associated protein involved in adipocyte lipolysis (Box 9.1).	PLIN1
FPLD Type 5	615238	One patient described. Adipocytes in affected patient have multilocular rather than unilocular lipid droplets.	Cell-death-inducing DNA fragmentation factor a-like effector c (CIDEC), a lipid droplet protein.	CIDEC
AKT2 mutations	Not listed	Very rare cause of severe insulin resistance, diabetes and lipodystrophy. Heterozygous condition.	Akt1 and Akt2 are protein serine/threonine kinases, also known as protein kinase B, involved in insulin signalling.	AKT2
CAV1 mutations	Not listed	'Atypical partial lipodystrophy' with hypertriacylglycerolaemia; only described in small number of patients. Fat loss from upper body only. Heterozygous condition.	Caveolin 1, mutations of which may also cause generalized lipodystrophy (see above; see also Section 5.3.3 for description of caveolins).	CAV1

*OMIM: Online Mendelian Inheritance in Man (http://www.omim.org/ and see Section 10.1).
PPAR-γ, peroxisome proliferator-activated receptor-γ.

Note: FPLD Type 1, or Kobberling-type lipodystrophy, does not have a defined genetic cause.
The classification of FPLD sub-types here is based on OMIM. Garg (2011) in **Further reading** numbers them differently.
Conditions noted as 'heterozygous' are inherited in an autosomal dominant manner; others are autosomal recessive (i.e. manifest in the homozygous form).

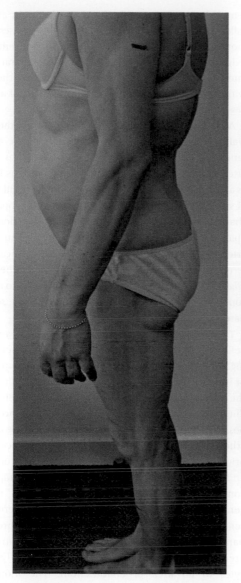
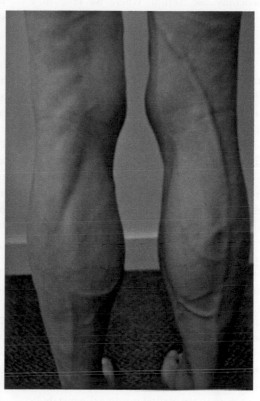

Fig. 10.11 Familial partial lipodystrophy. Left, typical appearance of a patient with familial partial lipodystrophy. Loss of subcutaneous adipose tissue is particularly apparent on the arms and legs, so that the muscles show through clearly. Right, phlebectasia – pronounced veins – seen on the legs. Photographs courtesy of Dr Sara Suliman, with consent of the patient. (See the Colour Plates section.)

10.5 Disorders of lipoprotein metabolism

Defects in a variety of lipid metabolic pathways can give rise to inappropriately high (or sometimes low) concentrations of lipoproteins in the blood. Cardiovascular disease is often a result of altered lipoprotein concentrations. Altered lipoprotein concentrations are commonly observed in a range of chronic conditions including obesity and diabetes.

A quick note about measurement of lipids is necessary here. Concentrations of cholesterol and other lipids are usually measured in serum, the liquid portion of clotted blood. Serum is essentially the same as blood plasma but without the protein fibrin. There is little difference between serum and plasma concentrations of lipids, and often the type of sample is not specified. Here we will generally use 'serum' where we refer to a measured concentration and 'plasma' where we are discussing more generally the transport of lipids in the circulation.

10.5.1 Atherosclerosis and cardiovascular disease

Cardiovascular disease is a broad term for a number of conditions involving pathological changes in blood vessels, including those of the heart and brain. The WHO considers that cardiovascular disease is the largest single cause of death worldwide, causing 30% of all deaths. The term cardiovascular disease includes coronary heart disease (CHD, also called ischaemic heart disease) and cerebrovascular disease (affecting blood flow to a region of the brain). Peripheral vascular disease, in which blood flow to the extremities (often the legs) is impaired, is another component of cardiovascular disease and a cause of considerable morbidity (ill-health but not necessarily death). The common, chronic pathology affecting blood vessels that underlies these conditions is atherosclerosis. In CHD, the coronary arteries that supply oxygen and nutrients to the heart muscle (myocardium) become narrowed by deposits in the arterial wall to such a degree as to prevent the coronary circulation meeting the metabolic demands of the heart. This may result in chest pain, especially on exertion, because the myocardium is deficient in oxygen, and lactic acid accumulates locally. This is the condition known as angina pectoris (or commonly just as angina). Local and systemic factors may also increase the likelihood of platelet aggregation and blood clot (thrombus) formation at the site of narrowing. This can result in complete blockage of an artery: a heart attack (or MI). The immediate effect is failure to supply a region of the myocardium with oxygen, which may lead to electrical disturbances and in severe cases cessation of beating (cardiac arrest), which has a high mortality. There is damage to, and 'scarring' of, the myocardium in patients who survive.

If the affected vessels are those supplying the brain, then the result of blockage is ischaemic stroke. (Ischaemic strokes account for around 80% of all strokes; the remainder result from bleeding into the brain and are called haemorrhagic strokes.) Stroke also arises commonly when a detached portion of clot (an embolus) lodges in a narrowed artery supplying the brain. The immediate effect is failure to supply a region of the brain with oxygen and nutrients. If the blockage is of a major vessel, this may be fatal.

Atherosclerosis is an irregular thickening of the inner wall of the artery that reduces the size of the arterial lumen (Fig. 10.12), particularly near junctions in the arterial tree. The thickening is caused by lesions known as atherosclerotic plaques. These consist of a proliferation of smooth muscle cells, connective tissue, mucopolysaccharides, fat-filled foam cells (in which the predominant lipid is cholesteryl ester) and deposits of calcium. The artery wall is locally thickened and loses elasticity. The term atherosclerosis was coined in 1904, from the Greek *athere*, meaning porridge or gruel and referring to the soft consistency of the core of the plaque, and *sclerosis*, hardening.

The first stage in atherosclerosis is the fatty streak, a yellowish, minimally raised spot (the spots later merging into streaks) in the arterial wall. Forty-five per cent of infants coming to post mortem examination during the first year of life have fatty streaks in their aorta. These therefore develop early in life (and are quite possibly reversible) and the development of the mature atherosclerotic plaque must be a long-term process. The origin of the fatty streak has mostly been studied in animal models of hyperlipidaemia (elevated lipoprotein concentrations). These are usually rabbits, with a genetic predisposition to high serum cholesterol concentrations, given a high-fat diet containing predominantly SFAs and cholesterol. The fatty streak is preceded by the adhesion of monocytes and T-lymphocytes to an area of endothelium. The monocytes migrate into the subendothelial space where they differentiate into macrophages, which then begin to engulf large amounts of lipid via the scavenger receptor pathway (see Section 7.2.4.4). Most of this lipid comes from LDL particles that have become chemically modified whilst trapped in the tissue (the 'response to retention' hypothesis). The modification probably arises through lipid peroxidation, which leads in turn to peroxidation of the apoB100 (see Section 3.3 for a description of peroxidation, which can also affect proteins). This reduces the affinity of the LDL particles for the LDL-receptor, and instead makes them ligands for

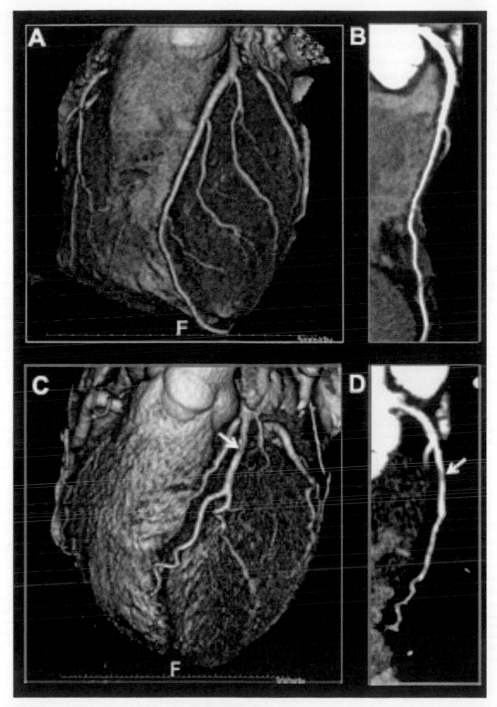

Fig. 10.12 Atherosclerosis demonstrated using the technique of computed tomography coronary angiography. This technique allows construction of a 3-dimensional image of the heart and the coronary tree (a, c) and a planar image (b, d). (a), (b) show a healthy heart; (c), (d) show a diseased (atherosclerotic) left anterior descending (LAD) artery. The white arrow denotes an atheromatous plaque resulting in a significant LAD stenosis (narrowing). Note also the diffuse vessel disease in the distal LAD (D) (mottling along the vessel). Picture courtesy of Drs Charalambos Antoniades and Alexios Antonopoulos. (See the Colour Plates section.)

the family of scavenger receptors. As we saw in Section 7.2.4.4, lipid uptake by the scavenger receptor pathway is not subject to feedback regulation by cellular cholesterol content; therefore, the macrophages may engulf large amounts of lipid, giving them a foamy appearance under the microscope. These 'foam cells' are characteristic of the atherosclerotic plaque.

Further development of the atherosclerotic plaque involves proliferation of smooth muscle cells of the arterial wall and the elaboration of a connective tissue matrix, forming a fibromuscular cap over the lesion. Within the lesion, macrophages may die and release their contents, forming a semiliquid pool of extracellular lipid. The lid of the lesion may remain firm, with the lesion protruding into the arterial lumen, obstructing flow, but not causing acute damage. However, in a process that may involve inflammatory cells, some plaque caps become unstable and are damaged. The contents of the plaque are then exposed, resulting in the normal response to damage to a vessel wall – thrombus formation.

The processes involved in the development of CHD and myocardial infarction, and the possible involvement of lipids, are illustrated in Fig. 10.13.

10.5.2 Hyperlipoproteinaemias (elevated circulating lipoprotein concentrations) are often associated with increased incidence of cardiovascular disease

Diseases that involve elevation of circulating lipid concentrations (hyperlipoproteinaemias or hyperlipidaemias) are often associated with increased incidence of atherosclerosis and myocardial infarction. They are often subdivided into primary and secondary hyperlipoproteinaemias. Primary hyperlipoproteinaemias are considered to arise directly from a genetic cause, while secondary hyperlipoproteinaemias arise from some other environmental or medical cause, such as inappropriate diet, diabetes or obesity. If the diabetes or obesity is controlled or cured, or the diet improved, the hyperlipoproteinaemia will disappear or lessen. However, the simplicity of this classification is now under question as we increasingly recognize that most diseases reflect interactions between genome and environment. For instance, even in familial hypercholesterolaemia, caused by a defect in the LDL-receptor (see Section 10.5.2.1), the degree of elevation of serum cholesterol concentration is

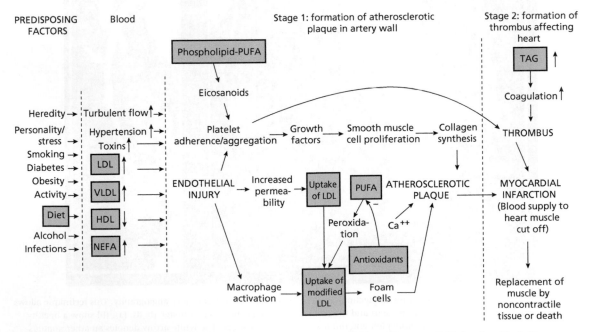

Fig. 10.13 Possible processes involved in the development of ischaemic heart disease and myocardial infarction. Shaded boxes indicate points at which lipids may be involved.

Table 10.6 The classification of hyperlipidaemias according to phenotype.

Type	Serum cholesterol	Serum triacylglycerol	Particles accumulating	Usual underlying defect	Treatment
I	+	+++	Chylomicrons	Lipoprotein lipase deficiency; apolipoprotein CII deficiency	Low-fat diet; medium-chain triacylglycerols to replace long-chain fatty acids
IIa	++	N	LDL	LDL receptor defect or LDL overproduction	Reduce dietary saturated fat and cholesterol; resins; 'statin' drugs (HMGCoA reductase inhibitors)
IIb	++	++	VLDL, LDL	VLDL or LDL overproduction or impaired clearance	Reduce dietary saturated fat and cholesterol; reduce weight; fibrate drugs (PPAR-α activators); statins
III	+	++	Chylomicron- and VLDL-remnants	Impaired remnant removal; may be due to particular isoform of apolipoprotein E, or apo-E deficiency	Reduce dietary saturated fat and cholesterol; fibrates; statins
IV	N or +	++	VLDL	VLDL overproduction or clearance defect	Reduce weight if obese, reduce alcohol intake; n-3 PUFA supplementation; fibrates
V	+	+++	Chylomicrons, VLDL and remnants	Lipoprotein lipase defect (not complete absence) or apolipoprotein CII deficiency	

This is known as the Fredrickson classification. N, normal; +, mildly raised; ++, moderately raised; +++, severely raised. Resins are nonabsorbed compounds that bind bile salts and cholesterol in the gastrointestinal tract so that they are excreted (see Section 7.1.2).

dependent upon other lifestyle and medical factors such as diet and thyroid hormone status.

A different classification is based upon the observed lipid phenotype. The classification normally used is that proposed in 1967 by the American clinician and biochemist, Donald Fredrickson (Table 10.6). Fredrickson believed that this was a classification of familial (i.e. primary) hyperlipidaemias, but it is now recognized to include many conditions that we would regard as secondary.

Secondary hyperlipoproteinaemias may be associated with many conditions. Among the most common are diabetes and obesity (see Section 10.5.2.4), thyroid disease (low thyroid hormone concentrations lead to hypercholesterolaemia and hypertriacylglycerolaemia), overconsumption of alcohol (hypertriacylglycerolaemia) and liver and kidney diseases. (Please note that in clinical parlance, 'triacylglycerolaemia' is usually called 'triglyceridaemia'.) Some drug treatments for other medical conditions may also lead to hyperlipoproteinaemias. A prominent example that is causing concern at present is a marked hypertriacylglycerolaemia that is commonly observed in patients with human immunodeficiency virus (HIV) infection treated with viral protease inhibitors, that is associated with lipodystrophy (see Section 10.4.2).

10.5.2.1 Single gene mutations affecting lipoprotein metabolism

The background to most disorders of lipoprotein metabolism is polygenic, combined with environmental factors, and the phenotypes are correspondingly variable. However, there are a few well-described single-gene mutations that lead to 'classic' primary hyperlipoproteinaemias. These are inborn errors of metabolism in the same way as those described earlier (see Section 10.1). In several cases, identification of the gene responsible has led to increased understanding of the pathway. Some will be described here as they illustrate features of the lipoprotein transport system. Further details are in Table 10.7. Other than familial hypercholesterolaemia, these are all autosomal recessive conditions: that is, they are manifest only in the homozygous state.

The most common example of a single-gene defect raising the concentration of a lipoprotein is mutation in the *LDLR* gene that encodes the LDL-receptor (see Section 7.2.4.2), rendering it completely or partially inactive and unable to bind to LDL. Therefore LDL particles, carrying mainly cholesterol, cannot be removed from the plasma by the usually major route, and the plasma LDL-cholesterol concentration increases. Since LDL carries most of the cholesterol in plasma, the total

Table 10.7 Inborn errors of lipoprotein metabolism.

Disease and enzyme deficiency	Characteristics	OMIM number[*]	Gene affected
Conditions raising lipoprotein concentrations			
Familial hypercholesterolaemia (LDL-receptor deficiency)	Heterozygous form leads to isolated raised serum cholesterol concentration (Fredrickson Type IIa) and, if not recognized and treated, often myocardial infarction in early middle-age. It is very readily treated with statin drugs (inhibitors of HMG-CoA reductase: Section 7.3.1), and adequately-treated patients have normal life expectancy.	143890	*LDLR*
Familial combined hyperlipidaemia	Elevation of both cholesterol and TAG concentrations in serum in a variable fashion.	144250	Heterogeneous: see text
Lipoprotein lipase deficiency or familial chylomicronaemia syndrome (Fredrickson Type I hyperlipoproteinaemia)	Accumulation of chylomicrons in the plasma gives creamy appearance (reflecting the normal role of LPL in hydrolysing or 'clearing' chylomicron-triacylglycerols: Section 7.2.5). In the heterozygous state there is sufficient LPL activity for almost normal triacylglycerol clearance but carriers may display features of familial combined hyperlipidaemia.	238600	*LPL*
Sitosterolaemia	Clinical manifestations similar to familial hypercholesterolaemia but with normal or moderately elevated serum cholesterol concentrations. Plant sterol concentrations in blood are markedly elevated.	210250	*ABCG5* or *ABCG8*
Conditions resulting in lower lipoprotein concentrations			
Chylomicron retention disease or Anderson (or Anderson's) disease	Severe fat malabsorption and failure to thrive in infancy. Deficiency of fat-soluble vitamins, low serum cholesterol concentrations, and a selective absence of chylomicrons from blood. Accumulation of chylomicron-like particles in membrane-bound compartments of enterocytes, which contain large cytosolic lipid droplets.	246700	*SAR1B.*
Abetalipoproteinaemia	Defect in microsomal triacylglycerol transfer protein which plays a critical role in VLDL assembly (Sections 5.6.2, 7.1.3 and Fig. 7.5). There may be signs of fat malabsorption, and neurological and visual impairment. A characteristic finding is abnormally shaped red blood cells (often star-shaped) reflecting abnormal membrane phospholipid composition, secondary to the changes in lipoprotein concentration and composition.	200100	*MTTP*
Tangier disease	Very low circulating high-density lipoprotein (HDL)-cholesterol concentrations. Clinically, a peculiar and distinguishing feature is large orange-yellow tonsils. In the absence of normal ABCA1-mediated cholesterol efflux, cholesterol-laden macrophages accumulate in lymphoid tissues such as the tonsils, and since their lipids are derived ultimately from the diet, they include fat-soluble pigments such as β-carotene.	205400	*ABCA1*
Lecithin-cholesterol acyltransferase deficiency (one variant called fish-eye disease)	Clinical manifestation differs according to the exact nature of the mutation. Fish-eye disease is so-called because of corneal opacities. Despite low HDL-cholesterol concentrations (see Section 10.5.2.3 below), patients do not seem to have increased risk of CHD.	245900 and 136120	*LCAT*

[*] OMIM: Online Mendelian Inheritance in Man (http://www.omim.org/ and see Section 10.1).

cholesterol concentration is also elevated. This is the condition of familial hypercholesterolaemia, which occurs in about 1 in 500 people worldwide. The homozygous form is much rarer and more difficult to treat (since statin treatment relies upon there being some LDL-receptor function to up-regulate: see Section 7.3.1). A given phenotype may result from different causes, and familial hypercholesterolaemia may also (much more rarely) result from mutations in the *APOB* gene, encoding apoB100, that affect the ability of apoB100 to bind to the LDL-receptor. *APOB* mutation is technically known as familial defective apolipoprotein B100 or FDB (OMIM 144010), but in most clinical situations would not be distinguished from familial hypercholesterolaemia. Another mutation that can give rise to the same phenotype is in the proprotein convertase subtilisin/kexin type

9 (PCSK9, gene *PCSK9*). PCSK9 is involved in intracellular degradation of the LDL receptor and, in this case, it is *gain of function* mutations that raise serum LDL-cholesterol concentrations. Loss of function mutations in *PCSK9* are also known and result in very low serum LDL-cholesterol concentrations and much reduced risk of CHD. (See also Box 10.3 for discussion of drug targets arising from these observations.)

Some patients with early CHD display elevated serum concentrations of both TAGs and cholesterol. The relative concentrations can vary widely between sequential observations in the same patient. The serum concentration of apoB100 is elevated, indicating a large number of LDL particles, with similarities to the atherogenic lipoprotein phenotype (see Section 10.5.2.4). This pattern is known as familial combined hyperlipidaemia (FCHL).

Box 10.3 Lipid-related drug targets for hyperlipidaemias

Many drug treatments are available for different types of hyperlipidaemias. The statin drugs (inhibitors of HMG-CoA reductase; Fig. 4.38 & Section 7.3.1) are effective at lowering elevated serum cholesterol concentrations in many patients and confer benefits in reduced risk of cardiovascular disease (Section 10.5.2.2). They also somewhat reduce elevated serum TAG concentrations, probably because lower availability of hepatocyte cholesterol reduces VLDL-TAGs secretion. Other steps in the process of cellular uptake of cholesterol from the circulation are being investigated as drug targets. One is the protein proprotein convertase subtilisin/kexin type 9 (PCSK9, gene *PCSK9*). As mentioned in Section 10.5.2.1, PCSK9 is involved in intracellular degradation of the LDL receptor. Inhibitors of PCSK9 have the potential to increase removal of cholesterol from the circulation and hence reduce circulating concentrations, and are in clinical trials.

Elevated serum TAG concentrations are targeted more effectively by the fibrate drugs, also called fibric acid derivatives. These act as activators of PPAR-α (Sections 7.3.2 & 10.5.2.3). Probably through their action in regulating apolipoprotein expression, they also raise serum HDL-cholesterol concentrations.

Niacin, also called nicotinic acid or vitamin B3, also lowers serum TAG concentrations and raises HDL-cholesterol. For these purposes it is given in large doses (typically 1–2 g/day; compare with its recommended intake as a vitamin for prevention of pellagra, which is < 20 mg/day). Niacin is a ligand for the G protein-coupled receptor GPR109A (Table 7.5), whose endogenous ligand is thought to be 3-hydroxybutyrate, a ketone body: i.e. a product of hepatic fatty acid oxidation. The only known metabolic action of niacin is to suppress adipocyte lipolysis, thus potentially reducing supply of nonesterified fatty acids to the liver and consequently hepatic TAG secretion. However, it is thought to have additional effects leading to substantial elevation of HDL-cholesterol concentration, reduction in Lp(a) concentration (see Section 7.2.3), and anti-inflammatory effects.

Agents that inhibit the action of CETP have marked effects in raising serum HDL-cholesterol concentrations but have, as yet, not proved useful in reducing CHD risk. Their use is described more in the main text.

Before the introduction of the statins, many patients with elevated cholesterol concentrations were managed with so-called resins, substances that make a gel with water and bind cholesterol in the intestinal tract. This targets the entero-hepatic circulation of cholesterol (Section 7.1.2). Nowadays, patients who do not benefit sufficiently from statin treatment often receive the drug ezetimibe. Ezetimibe is an inhibitor of the Niemann-Pick C1-like protein-1 involved in cholesterol absorption in the small intestine (Section 7.1.2 & Box 7.2). Note that dietary cholesterol itself is not a major regulator of plasma cholesterol concentrations (Section 10.5.4 & Box 10.4): but because these treatments interrupt the entero-hepatic circulation of cholesterol they may have greater effects than reduction in dietary cholesterol.

Certain plant sterols and stanols (phytosterols) can also interfere with the absorption of cholesterol of both dietary and biliary origin, and may have a useful cholesterol-lowering effect. They are marketed as components of spreads. It has long been believed that these phytosterols compete with cholesterol for incorporation into mixed micelles (Section 7.1.1), and tend to exclude cholesterol, hence reducing its absorption. However, more recent evidence suggests that they may increase the rate of cholesterol efflux from enterocytes back into the intestinal lumen.

There have been studies aimed at identifying a single gene mutation in FCHL but there is no consensus: candidates include *USF1* (encoding upstream transcription factor 1), a transcription factor controlling expression of several genes involved in lipid and glucose homeostasis, and LPL (gene *LPL*). It is generally agreed that FCHL is a genetically heterogeneous condition.

Type III hyperlipidaemia (see Table 10.6 for definition) is not shown in Table 10.7, because it is not inherited in Mendelian fashion. It arises from a defect in removal of remnant lipoprotein particles. A common underlying genetic basis is homozygosity for a particular form of apolipoprotein E, known as apoE$_2$. However, it appears that other genetic or acquired factors (e.g. obesity) must be present for Type III hyperlipidaemia to become manifest.

Complete lack of activity of LPL, due to mutation of the *LPL* gene, leads to familial chylomicronaemia syndrome or Fredrickson Type I hyperlipoproteinaemia. An identical phenotype is caused by a different mutation, in this case the *APOC2* gene encoding apoCII, and abolishing its ability to activate LPL (OMIM 207750). For clinical purposes it may not be important to distinguish these causes, although genetic analysis is increasingly being applied to these conditions. People with this condition must live on a very low-fat diet. Almost identical lipoprotein patterns are also seen in patients carrying mutations in the *GPIHBP1* or *APOA5* genes, which affect the activity of LPL. *GPIHBP1* encodes the glycosylphosphatidylinositol-anchored high density lipoprotein-binding protein 1, involved in export of LPL to its endothelial site of action (see Sections 4.2.2 & 7.2.5); *APOA5* encodes apolipoprotein AV (see Section 7.2.2 & Table 7.2).

Sitosterolaemia is a rare condition in which serum concentrations of plant-derived, and also shellfish-derived, sterols similar to cholesterol are elevated. Patients present with skin lesions similar to those seen in familial hypercholesterolaemia or with early signs of CHD. Serum cholesterol concentrations are not always particularly elevated. It arises from mutations in the genes for the transporters ABCG5 or ABCG8 involved in exporting these sterols from the enterocyte (Box 7.3).

There are also monogenic conditions that result in lowered lipoprotein concentrations. Chylomicron retention disease (Anderson disease) is caused by a mutation in the gene *SAR1B*, which encodes a GTPase involved in protein transport from the ER to the Golgi apparatus. Inactivation results in an inability to secrete chylomicrons from enterocytes.

Abetalipoproteinaemia is a complete absence of apoB100 from plasma: i.e. there are no VLDL or LDL particles. ('Beta' refers to the original electrophoretic separation of lipoproteins, wherein 'alpha' lipoproteins were predominantly those we now call HDL, 'beta' lipoproteins were those we now recognize as LDL. VLDLs were called prebetalipoproteins.) This condition arises from mutations in the *MTTP* gene encoding microsomal triacylglycerol transfer protein (MTP). This protein transfers TAGs to the maturing apoB-containing lipoprotein particle during the intracellular assembly process (see Section 7.2.3). Failure to assemble the lipoprotein particle leads to intracellular degradation and absence of cellular lipoprotein secretion. There are also conditions in which levels of apoB100 are low but not absent (hypobetalipoproteinaemia). These may represent the heterozygous form of abetalipoproteinaemia, or they may be separate conditions, associated with a mutation causing a truncation of apoB100. Defects in the *PCSK9* gene which result in very low serum LDL-cholesterol concentrations were mentioned above.

Tangier disease is a rare condition first described by Donald Fredrickson in people living on Tangier Island in Chesapeake Bay, USA. The gene for Tangier disease was cloned in 1999 and shown to be *ABCA1*, encoding the ABC-cassette transporter A1, which, as described in Section 7.2.5, facilitates the efflux of free cholesterol from cells to HDL in the circulation. Indeed, it was the mapping of this gene that led to understanding of the function of ABCA1. Another inherited defect leading to low HDL concentrations is deficiency of lecithin-cholesterol acyltransferase (LCAT, see Section 7.2.5), caused by mutations in the *LCAT* gene.

10.5.2.2 Low density lipoprotein cholesterol and risk of cardiovascular disease

Several specific aspects of lipid metabolism relate to the risk of atherosclerosis and CHD. The most clearly associated is an elevated LDL-cholesterol concentration. Over many years, epidemiological observations of the differences in CHD rates between different countries have shown a strong relationship with average (across the population) serum LDL-cholesterol concentrations. The same is also true when differences between members of the same population are studied. The Prospective Studies Collaboration (investigators from large-scale studies pooling their data) in 2007 analysed data from prospective, observational studies that included 12 million person-years of observation, and 55,000 deaths from cardiovascular disease. They concluded that for each 1 mmol/l difference in total serum cholesterol (which is mainly LDL-cholesterol), mortality from CHD changed by a factor 2 (i.e. 1 mmol/l greater serum cholesterol doubles mortality), in people aged 40–49. Effects were slightly smaller, but still large, in older age groups up to 89-year-olds. In addition, people with a defect in LDL removal, usually through a mutation in the LDL-receptor (familial hypercholesterolaemia; see Sections 7.2.4.2 & 10.5.2.1), have a very high risk of developing CHD at an early age if they are not treated appropriately.

The 'lipid hypothesis of CHD' – that elevated serum LDL-cholesterol concentrations play a direct role in CHD – was first proposed in its modern form in the 1950s, but even in the 1970s was the subject of considerable controversy. Now it is generally accepted. A major reason for this acceptance has been that several large prospective clinical trials have shown that lowering LDL-cholesterol concentrations with the 'statin' drugs (that inhibit HMG-CoA reductase – see Section 7.3.1) leads to marked reductions in mortality from CHD, in proportion to the reduction in LDL-cholesterol achieved. This applies in both primary (i.e. treating those who do not yet have CHD) and secondary (treating patients with established CHD) settings. A large meta-analysis of cholesterol-lowering trials, involving > 200,000 patients, concluded that for every 1 mmol/l decrease in serum cholesterol concentration, there was a 30% decrease in 'CHD events' and a 25% decrease in CHD-related mortality (see Gould *et al.* (2007) in **Further reading**). Box 10.3 gives more information on drug targets in lipoprotein disorders.

10.5.2.3 Low high-density lipoprotein cholesterol concentrations and risk of cardiovascular disease

Despite the clear role of LDL-cholesterol in CHD, many patients suffering an early myocardial infarction have a relatively normal blood concentration of LDL-cholesterol (Fig. 10.14). Increasingly, other abnormalities of lipid metabolism are being seen as associated with development of atherosclerosis.

The strongest relationship between CHD risk and non-LDL lipids is with the serum HDL-cholesterol concentration. In epidemiological studies, serum HDL-cholesterol concentration is strongly negatively related to CHD risk: i.e. an elevated HDL-cholesterol concentration appears protective. This is understandable if we consider serum HDL-cholesterol concentrations to reflect activity of the reverse cholesterol transport pathway, removing excess cholesterol from peripheral tissues, such as macrophages, to the liver for disposal (see Section 7.2.5). However, a causal relationship between serum HDL-cholesterol concentrations and CHD risk has recently been questioned. One method to investigate this relationship is to observe the effects of genetic variants that result in increases or decreases in serum HDL-cholesterol concentrations, a technique that is referred to as Mendelian randomization. Such studies have generally failed to find corresponding differences in CHD risk. It is possible that the strong negative associations between serum HDL-cholesterol concentrations and CHD arise indirectly from the negative association between HDL cholesterol and TAG concentrations (see Section 10.5.2.4 for explanation of this relationship). In contrast, the relationship of elevated serum TAG concentrations to CHD has been tested using

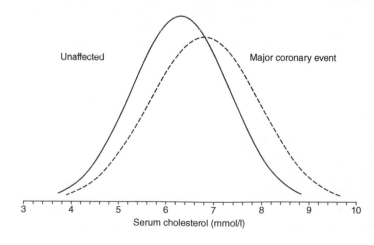

Fig. 10.14 Distribution of serum cholesterol concentrations in 438 men who had a major coronary event and 7252 unaffected men. Note that there is considerable overlap between the distributions. An elevated serum cholesterol concentration is certainly a major risk factor for development of coronary heart disease, but nevertheless many people experiencing the disease have relatively normal cholesterol concentrations. Data from the British Regional Heart Study: redrawn by DJ Wald (1992) in M Marmot & P Elliott, eds., *Coronary Heart Disease Epidemiology,* Oxford University Press, Oxford. Reproduced with permission.

Mendelian randomization and appears to be causal. For more on Mendelian randomization to investigate the role of serum lipids in CHD, see Holmes *et al.* (2014) in **Further reading**.

Epidemiological studies have to some extent been reinforced by clinical trials. Fibrate drugs lower serum TAG and moderately raise HDL-cholesterol concentrations (see Section 7.3.2 & Box 10.3). In clinical trials, they have provided mixed results in terms of reduction in incidence of CHD (measured by 'events' such as myocardial infarction), but there is a clear positive effect when these drugs are used in patients with raised serum TAG concentrations.

Another treatment with similar effects on serum lipids is niacin (Box 10.3). In some clinical trials, niacin has produced significant benefits in terms of measures of CHD (e.g. measures of plaque size by imaging) and in a few, reductions in mortality. However, the largest clinical trial of niacin in 26,000 patients (HPS2-THRIVE) reported in 2013 that no benefits had been observed.

The field of HDL-raising therapy has been further complicated by the results of trials of cholesteryl ester transfer protein (CETP)-inhibitors (see Section 7.2.5 for description of CETP). CETP mediates the transfer of cholesteryl esters from HDL particles into other lipoproteins. Inhibitors of this process have been developed, and in clinical trials they markedly increase serum HDL-cholesterol concentrations. However, the first large clinical trial of such an agent, called torcetrapib, showed a surprising increase in CHD events, despite a dramatic increase in HDL-cholesterol levels. Many investigators believed that this was due to so-called 'off-target effects' (i.e. unwanted side-effects not related to CETP inhibition), and further CETP inhibitors have followed into clinical trials. However, none of the trials has yet reported beneficial effects, and one, on the agent dalcetrapib, was terminated early, probably because interim analysis indicated no benefit. Results from a very large trial using anacetrapib (REVEAL) are currently awaited. Anacetrapib is a particularly potent CETP inhibitor showing a doubling of serum HDL-cholesterol concentrations but also a lowering of LDL-cholesterol.

The difficulty of showing clinical benefit, especially reduced mortality, with drugs elevating plasma HDL-cholesterol concentrations should make us think again about the role of HDL. As described in Section 7.2.5, HDL particles transfer cholesterol from peripheral tissues, including macrophages, to the liver and intestine for excretion. Possibly, in a population study, high serum concentrations of HDL-cholesterol indicate that the pathway is functioning well, and CHD risk will be low. But one component of the pathway, as described in Section 7.2.5, is that CETP transfers the cholesteryl esters from HDL into TAG-rich particles that can then be taken up by the liver and the cholesterol excreted. Blocking this process would raise HDL-cholesterol concentrations but not increase cholesterol excretion. See Box 10.3 for more on drug targets in lipoprotein disorders.

10.5.2.4 Atherogenic lipoprotein phenotype

Beyond the positive relationship of CHD with elevated LDL-cholesterol concentration and its negative relationship with HDL-cholesterol, there are other more subtle effects. Increasingly there is recognition of a phenotype that predisposes to CHD, known by a variety of names, including the atherogenic lipoprotein phenotype, the metabolic syndrome, syndrome X, or hyperapobetalipoproteinaemia (a high concentration of apoB in the circulation). The characteristics usually associated with the atherogenic lipoprotein phenotype are listed in Table 10.8. A brief description of its origin in terms of lipid metabolism may be given as follows. It may stem from overproduction of VLDL-TAGs from the liver. That, in turn, is closely related to the presence of excess fat in the liver (a component of ectopic fat deposition, discussed in Section 10.4.1.4). An increased circulating

Table 10.8 Features of the atherogenic lipoprotein phenotype.

Feature	Comment
Hypertriacylglycerolaemia (often called hypertriglyceridaemia)	May by relatively mild, but usually is more marked after meals
Low serum HDL-cholesterol concentration	Reflects impairment of triacylglycerol-rich lipoprotein metabolism
Small, dense LDL particles	Total LDL-cholesterol concentration may be normal, hence number of particles must be increased
Hyperapobetalipoproteinaemia (high concentration of apoB-100 in serum)	Reflects increased number of LDL particles, as above
Insulin resistance	These features are often associated with impaired action of insulin on metabolic processes

concentration of TAGs in VLDL particles will compete for clearance by LPL when chylomicrons become enriched with TAGs in the postprandial period. This will lead to a prolonged residence of partially lipolysed chylomicron and VLDL remnants in the circulation, something that has often been noted to be associated with increased CHD risk. If the circulating pool of TAG-rich lipoproteins is increased, there will be increased opportunity for CETP to exchange TAG from VLDL and chylomicron remnants for cholesteryl esters from HDL and LDL particles. VLDL and chylomicron remnant particles therefore become enriched in cholesteryl esters. Some people believe that these become atherogenic particles – that is, they may themselves penetrate the arterial wall and become engulfed by macrophages. (There is some evidence for the presence of apoB48 in arterial lesions, although this is difficult to demonstrate as apoB48 is enormously outnumbered in the circulation by apoB100.) But, in addition, LDL and HDL particles thereby lose cholesteryl esters. The TAGs that they have gained in exchange can be hydrolysed by hepatic lipase, which acts preferentially on these smaller particles. Therefore the particles have lost both TAGs and cholesterol. The result is a low HDL-cholesterol concentration (associated with increased risk of CHD as discussed above) and a predominance of small, relatively lipid-poor LDL particles.

These so-called small, dense LDL particles have attracted much attention, and may well be the atherogenic particles of the atherogenic lipoprotein phenotype. There is always a population of LDL particles of a variety of sizes in the circulation, but some people tend to have a predominance of large, buoyant cholesterol-rich LDL particles (known as pattern A) and others a predominance of smaller, denser relatively lipid-poor LDL particles (pattern B). There is some evidence for a genetic component to this LDL phenotype, but it is clearly modifiable by diet and by other influences on the lipoprotein phenotype. Pattern B is associated with much higher risk of CHD than is pattern A, even though the total LDL-cholesterol concentration may be the same. The increased number of LDL particles that must be present in pattern B (to give the same total LDL-cholesterol concentration) accounts for the term hyperapobetalipoproteinaemia, which covers the same, or a closely related, phenotype. It is postulated that smaller LDL particles may more readily cross the endothelial wall and hence become substrates for macrophage uptake. In addition, there is experimental evidence that small,

dense LDL particles are more susceptible to peroxidation than are larger, more buoyant particles, and hence more likely to be substrates for macrophage uptake.

10.5.2.5 Nonesterified fatty acids and the heart

Myocardial infarction may lead to death. If it does, this is not usually because of complete obstruction of blood flow to the myocardium, but because the electrical rhythm of the heart becomes grossly disturbed, leading to the critical conditions of ventricular fibrillation or cardiac arrest. Then blood supply to the rest of the body, including the brain, ceases. The stress reaction that sets in at the onset of myocardial infarction can lead to very high plasma concentrations of NEFAs, and there is considerable evidence that these may themselves be a potent cause of ventricular dysrhythmias (Michael Oliver, a UK cardiologist, strongly argued for this view). It has been postulated that the stresses associated with modern living lead to sustained elevations of the plasma NEFA concentration that are now inappropriate, and may lead to, or potentiate, ventricular dysrhythmias. The UK biochemist Eric Newsholme suggested that in earlier times we would have oxidized the excess FAs in muscular activity; hence the hazards of stressful, sedentary living. There is an interesting recent development in this story. The American nutritionist Alexander Leaf has shown that, amongst FAs, the very long-chain n-3 PUFAs appear to have a unique ability to stabilize the heart rhythm. In several studies in experimental animals, he has shown that acute administration of very long chain n-3 PUFAs can prevent death from coronary artery obstruction. There is support now from human dietary trials (see Section 10.5.4.3).

10.5.3 Coagulation and lipids

Once atherosclerotic plaques have developed, further events must follow to lead to myocardial infarction or ischaemic stroke. These are aggregation of platelets, probably at the site of rupture of the atherosclerotic lesion, and then the formation of a thrombus. These processes may be related to aspects of lipid metabolism. Platelet aggregation, an early step, is propagated by the formation of the eicosanoid, thromboxane A2 (TXA_2), from arachidonic acid in platelets. Release of TXA_2 by activated platelets leads to a cascade effect on aggregation. Formation of TXA_2 is inhibited by aspirin (an inhibitor of the COX pathway – see Section 3.5.2), and this is the reason that low doses of aspirin can markedly reduce the risk of reinfarction in people who have already suffered one myocardial infarction. Platelet aggregation is also inhibited by very long-chain n-3 PUFAs. This is probably mediated via the generation of eicosanoids of the 3-series rather than the 2-series (see Section 3.5.2, Table 3.11, Fig. 3.43 & Fig. 10.2). The blood coagulation pathways are also related to lipid metabolism. High circulating concentrations of the activated form of coagulation factor VII, often known as VIIa (or, if measured by assay of its coagulant activity, as VIIc), are associated with increased risk of CHD. Generation of factor VIIa is related to TAG metabolism. There is a relationship between circulating concentrations of VIIa and of TAGs, and the elevation of serum TAG concentrations that occurs after a meal is associated with activation of factor VII.

Lp(a) is a lipoprotein fraction whose concentration is linked to CHD risk. As discussed in Section 7.2.3, its protein component, apolipoprotein(a), is related to plasminogen, precursor of the enzyme plasmin in the fibrinolytic system. It interacts with the coagulation system in a number of ways including inhibition of the formation of plasmin. (See Riches & Porter (2012) in **Further reading**, for more information.)

There is, of course, a further link between lipids and coagulation, in that vitamin K, a fat-soluble vitamin, is essential for coagulation (see Section 6.2.3.4). It is unlikely, however, that this is involved in relationships between plasma lipids and atherosclerosis.

10.5.4 Effects of diet on lipoprotein concentrations and risk of coronary heart disease

From the preceding description, it will be clear that the major lipid risk factors for CHD are: an elevated circulating LDL-cholesterol concentration; a low HDL-cholesterol concentration; a predominance of small, dense LDL particles; and elevated circulating TAG concentrations (particularly in the postprandial period). The total cholesterol concentration in serum is often listed as a risk factor, but this reflects the fact that LDL-cholesterol is the major component of total serum cholesterol. The ratio of total cholesterol/HDL-cholesterol captures some of this information simply, and is a strong predictor of CHD mortality (twice as strong as total serum cholesterol concentration, in the Prospective Studies Collaboration analysis of 2007: see Section 10.5.2.2 above and **Further reading**). These risk factors are modified to a great

extent by genetic and environmental factors, as described in the sections above, but they are also modulated to a considerable extent by dietary factors, which will be reviewed briefly here.

A word of caution is important here. Evidence for relationships between diet and lipoprotein concentrations, or between diet and cardiovascular disease, comes from two types of studies. One is *observational*, e.g. looking at differences between different populations, or between people within one population, in dietary intake and in outcome measures. There are clear issues here with confounding factors: just as one example, people who eat a diet with a high PUFA content may well also have a lifestyle that is 'healthier' in other ways. The other source of information is from *intervention* studies, where a change in diet is introduced deliberately. This is the 'gold standard' of proof of a relationship (especially when conducted as a randomized prospective study). But changes to diet are complicated. Suppose we want to change the total contribution of dietary fat to energy (e.g. reduce it from 40% to 30%). Something else must then increase its contribution to energy by 10%, and the answer to the study may depend as much on this other 'make-up' nutrient as on the change in dietary fat. So we should always look for what is changing in the diet, and it is rarely one factor alone.

Perhaps surprisingly, the effect of dietary cholesterol on serum cholesterol concentration is relatively weak over any reasonable range of intake. This can be understood if we remember the powerful systems that regulate cholesterol accumulation in cells (see Section 7.3.1): an increased intake may down-regulate endogenous biosynthesis. Also, we should remember that endogenous biosynthesis of cholesterol is of the order of 600 mg/day, comparable with typical dietary intake (maybe 1 g/day but only about 50% absorbed).

10.5.4.1 Dietary fat quantity and cardiovascular disease risk

The total quantity of dietary fat has an important influence on serum lipoprotein concentrations. It is a common observation that a change to a diet containing a lower proportion of energy from fat, and a correspondingly higher proportion from carbohydrate, is associated with a reduction in serum HDL-cholesterol concentration and an elevation of serum TAG concentration. There is usually also a reduction in total serum cholesterol concentration (but this will be at least partly due to the drop in HDL-cholesterol). The reduction in serum HDL-cholesterol concentration may reflect reduced flux of fat through the exogenous lipoprotein pathway (see Section 7.2.5), which involves the transfer of surface components from the chylomicrons, as they are lipolysed by LPL, to HDL particles. The elevation of serum TAG concentration is not clearly understood. It is thought to represent increased secretion of VLDL-TAGs from the liver, perhaps because of a change in the metabolic partitioning of FAs in the liver between oxidation (favoured when fat levels in the diet are high) and esterification (favoured when carbohydrate is plentiful, and insulin levels are high). These observations have led to questions about the safety of low-fat diets (i.e. high-carbohydrate diets, since carbohydrate usually replaces fat), since both these changes (depression of HDL-cholesterol, elevation of TAG concentrations) appear to be deleterious in terms of CHD risk. Such questioning is important, but many nutritionists feel that a shift towards lower fat diets may also play an important role in reducing the prevalence of obesity, which in the longer term will have beneficial effects on health. It may be that the changes in lipid concentrations are not maintained long term (there is some evidence that they are not, but there are few long-term studies); and in addition, the changes in lipid concentrations depend upon the nature of the dietary carbohydrate. In many short-term experimental studies, much of the dietary carbohydrate is in the form of simple sugars, which may have a particularly marked TAG-raising effect. In large epidemiological studies (e.g. the Nurses' Health Study, a long-running prospective study of >80,000 women in the US), total dietary fat intake (expressed as contribution to energy) did not predict CHD risk.

10.5.4.2 Dietary fat quality and cardiovascular disease risk

Much more marked effects, however, are seen when the nature (or 'quality') of the dietary fat is changed. This was first recognized by the American nutritionist, Ancel Keys. Keys had been travelling in the Mediterranean countries, investigating the apparently very low prevalence of chronic diseases including CHD. He questioned whether this might be related to aspects of the diet in those areas. This led to the initiation of the Seven Countries Study, an international study of CHD and associated factors in a number of countries (it soon expanded beyond seven) with widely differing prevalences of CHD. There were two major, early findings. Comparing average values in one country with another, there was a strong positive relationship between serum cholesterol concentration and prevalence of CHD, and a strong positive relationship between consumption of SFAs and serum cholesterol. Those observations led to the recognition that the nature of the FAs in the diet is an important influence on the serum cholesterol concentration: SFAs tend to raise it, PUFAs to lower it. If MUFAs replace SFAs in the diet, they reduce serum cholesterol concentration although not as strongly as do PUFAs. Amongst these FA classes, not all FAs have the same effect. Of the SFAs, for instance, those with chain length below 12C seem to have no effect; 12C, 14C and 16C lengths raise cholesterol concentration, whereas stearic acid, 18:0, may slightly reduce cholesterol concentration (perhaps because of rapid desaturation to oleic acid).

The relative effects of dietary FAs on cholesterol concentrations have been combined by many investigators to produce equations that predict the average change in serum cholesterol concentration in response to any dietary manipulation (summarized in Box 10.4). The mechanism for the effects of different dietary FAs on serum cholesterol concentration is not clearly understood. It appears to reflect a shift in the hepatic cholesterol pool between esterified and unesterified forms, with up-regulation of hepatic LDL-receptors (and a consequent reduction in serum LDL-cholesterol concentration) when PUFAs predominate. There are also suggestions of effects on VLDL secretion and removal of plasma TAGs. For more information, see Fernandez & West (2005) in **Further reading**.

It should be noted that these effects of changing dietary fat quality are based on mean changes in groups of subjects reported in the literature. Individual responses may differ: indeed, there will be some individuals who respond more than the mean, others less or not at all; and some whose serum cholesterol concentration will respond in the opposite way. There are relationships between responses to dietary change and the genotype of the individual. An important factor modifying dietary responses is the apolipoprotein E genotype. Apolipoprotein E was described in Section 7.2.2. It was mentioned earlier (see Section 10.5.2.1) that the gene can exist in different forms. There are three common alleles for the *APOE* gene, with products known as apoE$_2$, apoE$_3$ and apoE$_4$. These differ in two amino acid residues and affect remnant lipoprotein clearance. Each

Box 10.4 Effects of dietary fatty acids on serum cholesterol concentrations

In many experiments in healthy subjects, the FA composition of the diet has been manipulated to assess the effect on the serum cholesterol concentration. These studies have been summarized by a number of investigators to produce 'predictive equations'. Two examples are given here.

Hegsted *et al.* (1965) produced the equation:

$$\Delta serum\ cholesterol = 0.026 \times (2.16\Delta SFA - 1.65\Delta PUFA + 6.66\Delta Chol - 0.53)$$

where Δserum cholesterol represents the change in serum cholesterol concentration (mmol/l), ΔSFA and ΔPUFA are changes in the percentage of dietary energy derived from saturated and polyunsaturated fatty acids respectively, and ΔChol is the change in dietary cholesterol in 100 mg/day. Note that the term for ΔS is positive (an increase in SFA intake raises serum cholesterol) whereas that for ΔP is negative (an increase in PUFA intake reduces serum cholesterol). (The factor 0.026 converts from mg/dl to mmol/l.)

Yu *et al.* (1995) collated data from 18 studies in the literature that gave information on individual FAs in the diet. Their predictive equation was:

$$\Delta serum\ cholesterol = 0.0522\Delta\ (12:0\ to\ 16:0) - 0.0008\Delta 18:0 - 0.0124\Delta MUFA - 0.0248\Delta PUFA$$

where ΔMUFA is the change in the percentage of dietary energy derived from monounsaturated fatty acids (other terminology as above). Note that most SFAs are shown as raising serum cholesterol, but stearic acid (18:0) as slightly lowering it.

References

Hegsted DM, McGandy RB, Myers ML & Stare FJ. (1965) Quantitative effects of dietary fat on serum cholesterol in man. *Am J Clin Nutr* **17**:281–95.

Yu S, Derr J, Etherton TD & Kris-Etherton PM (1995) Plasma cholesterol-predictive equations demonstrate that stearic acid is neutral and monounsaturated fatty acids are hypocholesterolemic. *Am J Clin Nutr* **61**:1129–39.

individual carries two of these alleles – e.g. someone might have a phenotype apoE$_3$/apoE$_3$ (the most common). The presence of an E$_2$ or E$_4$ allele reduces the serum cholesterol response to change of dietary fat. There is also evidence for an effect of polymorphisms in the *APOA4* gene. See Schaefer (2002) in **Further reading** for more information.

Some dietary UFAs with unusual structures (*trans*-double bonds, or conjugated double bonds) may have particular effects. These are considered in Box 10.5.

10.5.4.3 *n*-3 Polyunsaturated fatty acids and cardiovascular disease risk

Dietary *n*-3 PUFAs have a special role. They are relatively neutral in terms of serum cholesterol, but they reduce serum TAG concentrations. They also reduce platelet aggregation (see Section 10.5.3), and may stabilize the myocardial rhythm (see Section 10.5.2.5). Their role in dietary protection from cardiovascular disease, or more specifically CHD, has been somewhat controversial. There is much epidemiological evidence linking high fish consumption with low rates of cardiovascular

disease. But supplementation trials, usually with purified very long-chain *n*-3 PUFAs, have not always found the expected beneficial effects.

Some large intervention trials have given credence to the beneficial effects of *n*-3 PUFA intake in protection against cardiovascular disease. In the Diet and Reinfarction (DART) study, conducted in the UK, survivors of myocardial infarction were randomized into two groups, one of which was given advice to eat oily fish (and if this advice was not followed, they received supplementation with fish-oil [very long-chain *n*-3 PUFA] capsules). Over the next two years, the 'fish-advice' group suffered the same rate of new myocardial infarctions as did the control group; but their death rate was significantly reduced, by 29%. This may reflect protection against the fatal dysrhythmias that often accompany myocardial infarction. The Lyon Diet-Heart Study also investigated the effects of dietary modification in people who had suffered from a myocardial infarction. In this case the 'intervention' group was asked to follow a 'Mediterranean-style' diet, rich in α-linolenic acid (long-chain PUFA: 18:3*n*-3). The trial

Box 10.5 Unusual isomers of dietary fatty acids may have particular health effects

The double bonds in dietary UFAs are mostly of the *cis* geometrical configuration. However, some foods contain isomeric fatty acids in which the double bonds are in the *trans* configuration (Section 2.1.3.1 & Box 11.4). These are produced naturally in ruminant animals and so enter the food chain in small amounts, but larger amounts enter the food chain from industrial hydrogenation of vegetable oils (discussed further in Box 11.4).

A number of epidemiological studies have shown a relationship between *trans*-FA intake and cardiovascular disease: in the Nurses' Health Study in the US, *trans*-FA intake was more strongly related to CHD risk than was SFA intake. Controlled feeding studies suggest that dietary *trans*-FAs raise serum cholesterol and reduce HDL-cholesterol concentrations to a similar extent to SFAs. Given the similarity of their molecular configurations, this is perhaps not surprising. (Vaccenic acid, a product of rumen fermentation, seems to be an exception, and some studies suggest that it may have beneficial effects on health.)

Trans-FAs can be replaced in food products with other fats, and moves are afoot worldwide to do so, discussed further in Box 11.4.

In most naturally occurring *cis*-PUFAs, the double bonds are separated by a methylene bridge; e.g. the most common form of linoleic acid in nature is c9,c12-18:2 (Section 2.1.3.2). However, a large number of isomers of linoleic acid is found. Some of these have the double bonds between consecutive pairs of carbon atoms (the most common is c9,t11-18:2, also known as rumenic acid). This arrangement of double bonds is known as conjugated. The group of isomers of linoleic acid with this configuration is known collectively as conjugated linoleic acid (CLA). CLA is formed in the rumen of ruminant animals, and is found in milk fat, cheese and beef.

CLA has come to prominence because of claims from animal studies that CLA can protect against some forms of cancer. Dietary CLA has also been shown to alter body composition in mice, with a loss of body fat, and in cultured adipocytes to reduce the activity of lipoprotein lipase. More recently, it has been claimed that dietary CLA may protect against atherosclerosis. However, the evidence in this respect is not clear-cut: there have also been demonstrations that high levels of CLA fed to rodents can predispose to the formation of fatty streaks in the aorta. Various mechanisms for the potential beneficial effects have been proposed, and may differ for the different isomers. For instance, some isomers are potent agonists of PPAR-α (Section 7.3.2). A quick internet search will show the availability many commercial preparations of CLA with claims such as 'helps promote fat loss' and 'increases energy' (!). As yet, however, there are no convincing studies showing beneficial effects of CLA on human health.

was stopped prematurely, after an average of 2.5 years on the diet, because the results were so clearly significant, with lower death rates in the 'Mediterranean-diet' group. A 4-year follow-up of the participants showed that even after the formal trial had finished, the difference in all-cause mortality between the groups was marked (about half in the test group compared with the control group). For 'cardiac deaths' the ratio was 3:1. The GISSI- (Gruppo Italiano per lo Studio della Sopravvivenza nell'Infarto)-Prevenzione study involved 11,000 patients studied for 3.5 yrs after a recent myocardial infarction. They were randomized to receive either 1 g/day of very long-chain *n*-3 PUFAs (EPA and DHA in a ratio of 1:2) in capsules or no supplement. The very long-chain *n*-3 PUFA group showed a 10–15% reduction in deaths and nonfatal cardiovascular events. Another arm of the study investigated vitamin E (300 mg/day) but did not find significant benefits: in fact there was increased risk of developing heart failure in those with evidence of cardiovascular disease at the beginning of the study.

Against these apparently very impressive results, however, was a follow-up of the DART study, called DART-2. In this study in 3000 men with angina (i.e. signs of CHD), advice to eat oily fish or take fish oil capsules did not affect mortality, but was associated with increased sudden cardiac death (especially in the subgroup taking capsules). Similarly, in the recent, large ORIGIN trial 12,000 patients who were at high risk of cardiovascular disease were randomized to capsules containing ethyl esters of *n*–3 PUFAs or placebo daily. There was no effect on incidence of cardiovascular events over 2 years. The lead investigator of the DART and DART2 studies, Michael Burr of Cardiff University, UK, concluded that 'nutritional supplements do not necessarily have the same effects as the foods from which they are derived'.

Summarizing all studies of fish or fish oil consumption and cardiovascular disease, there is some compelling evidence for a protective effect of fish consumption, less for supplementation with fish oil capsules, and some contradictory findings. See **Further reading** for more information. None of this negates the evidence for beneficial effects of *n*-3 PUFA intake on other conditions such as rheumatoid arthritis (see Section 10.3.2).

KEY POINTS

- Lipids play important roles in health and in disease. Lipids in the diet are important for health and may be useful in treatment of diseases. Many diseases are characterized by alterations in lipid metabolism, either of genetic or environmental origin. Many steps of lipid metabolism provide targets for drug development in the treatment of a wide range of diseases.

Inborn errors of lipid metabolism

- There is a wide range of disorders of lipid metabolism, often with serious consequences, that arise from mutations in single genes, and so are inherited in Mendelian fashion. Many of these result in excessive lipid accumulation. They affect sphingolipid catabolism, FA oxidation and acyl lipid biosynthesis amongst other pathways.

Lipids and cancer

- Cancer results from the uncontrolled proliferation of cells. Lipids are involved in many aspects of cancer development and treatment. Alterations in sphingolipid metabolism have been recognized for many years in cancer cells. If these affect cell-surface expression of lipids, they may provide opportunities for targeting treatments. More recently it has been recognized that tumour cells are characterized by high rates of FA biosynthesis and this may provide another opportunity for drug targeting.

Lipids and the immune system

- The immune system protects the body from infectious agents and other noxious insults. It has two general components, the innate and the acquired immune systems. Inflammation, which may be chronic and low-grade, represents activation of the innate immune system and may become a destructive process. Lipids are involved at many stages in immune processes.

- Dietary lipids play important roles in modulating the activity of the immune system, leading to the term 'immunonutrition'. Very long-chain PUFAs, especially those of the n-3 family, may suppress excess inflammation, through the generation of eicosanoids and related compounds.

- Very long-chain PUFAs may also affect the immune system through modulation of gene expression. Nuclear factor-κB (NF-κB) is a transcription factor central to immune responses and is regulated both positively and negatively by various PUFAs (positively by arachidonic acid (n-6) but negatively by the very long-chain n-3 PUFAs EPA and DHA). Very long-chain PUFAs, especially those of the n-3 family, are also ligands for the PPARs, which may have anti-inflammatory effects.

- Some specific lipids are involved in the immune system or more generally in health and disease. These include lipopolysaccharide (endotoxin) from the cell envelope of Gram-negative bacteria, which is responsible for toxic effects in the mammalian host, and PAF, an ether-linked phosphoglyceride that is produced by many cells involved in host defence and acts as a mediator of many aspects of immune function and inflammation. Pulmonary surfactant is a PtdCho-enriched lipoprotein mixture that lowers the surface tension at the alveolar air–liquid interface and reduces the work involved in breathing. Lack of pulmonary surfactant in premature infants is a serious condition, respiratory distress syndrome of the new-born. Synthetic or animal-derived surfactant preparations may be used to counteract it.

Obesity and lipodystrophy

- Lipids, especially TAGs, are stored as an energy reserve in adipose tissue (see also Chapter 9). Excessive adipose tissue lipid storage is the condition of obesity. Obesity has a strong genetic component, the genes involved being mainly those of the pathway of appetite regulation.

- Obesity is a risk factor for many diseases, especially type 2 diabetes and cardiovascular disease. These conditions are associated with elevated concentrations of insulin, in the state known as insulin resistance. This in turn may arise from 'overspill' of FAs from overfull adipose depots, and deposition of TAGs in other tissues ('ectopic fat deposition').

- The risk of these metabolic complications is dependent upon where the excess fat is stored. Abdominal fat deposition ('apple shape') is associated with much higher risk of metabolic disorders than is fat accumulation on buttocks and thighs ('pear shape'). This probably reflects different metabolic properties of adipocytes in the different fat depots. In addition, abdominal fat deposition includes a component within the abdomen, so-called visceral fat, which appears to have particularly adverse metabolic effects.

- Some people cannot form adipocytes normally and are deficient in adipose tissue: the condition of lipodystrophy. This may be genetic or acquired (the latter especially associated with HIV infection and its treatment). People with this condition often suffer metabolic problems, similar to those seen in obesity, especially development of type 2 diabetes.

Disorders of circulating lipoprotein concentrations

- Cardiovascular disease involves accumulation of lipids and other components in the process of atherosclerosis, development of lesions in the walls of arteries. This may lead to narrowing of arteries and risk of complete blockage through thrombus (blood clot) formation. If this occurs in a coronary artery, it results in loss of blood supply to a portion of the heart muscle and MI (heart attack). If it occurs in an artery supplying the brain, the result is a stroke.

- Atherosclerosis often arises from high concentrations of lipids in the bloodstream. Of particular importance is a high concentration of cholesterol in the LDL fraction. However, increased risk of cardiovascular disease is also associated with low serum concentrations of HDL-cholesterol, or complex changes often involving low HDL-cholesterol concentrations and elevated TAG concentrations in blood.

- There are some single-gene mutations that raise serum lipoprotein concentrations. The most common is a mutation in the *LDLR* gene encoding the LDL receptor. This causes the condition of familial hypercholesterolaemia, which confers high risk of early myocardial infarction. The common, heterozygous form of familial hypercholesterolaemia is readily treated with statin drugs (inhibitors of HMG-CoA reductase). Some mutations give rise to lower-than-normal lipoprotein concentrations.

- Much more common are complex, polygenic and environmental influences giving rise to a variety of changes in lipoprotein pattern that are usually classified according to a scheme developed by Fredrickson. Lipids (especially the NEFAs) may also affect the electrical stability of the heart and coagulation.

- Lipoprotein concentrations may be altered by drugs (e.g. the statins, referred to above) and by diet. In particular, the quality of dietary fat (the FA composition) can affect lipoprotein concentrations. In general, dietary SFAs tend to raise serum cholesterol concentration, whereas dietary PUFAs tend to reduce it. Genetic background may modify the lipoprotein responses to dietary changes.

Further reading

Inborn errors of lipid metabolism

Eckhardt M (2010) Pathology and current treatment of neuro-degenerative sphingolipidoses, *Neuromolecular Med* **12**:362–82.

Houten SM & Wanders RJ (2010) A general introduction to the biochemistry of mitochondrial fatty acid β-oxidation, *J Inherit Metab Dis* **33**:469–77.

Lamari F, Mochel F, Sedel F & Saudubray JM (2013) Disorders of phospholipids, sphingolipids and fatty acids biosynthesis: toward a new category of inherited metabolic diseases, *J Inherit Metab Dis* **36**:411–25.

Özkara HA (2004) Recent advances in the biochemistry and genetics of sphingolipidoses, *Brain Dev* **26**:497–505.

Platt FM (2014) Sphingolipid lysosomal storage disorders, *Nature* **510**:68–75.

Rinaldo P, Matern D & Bennett MJ (2002) Fatty acid oxidation disorders, *Annu Rev Physiol* **64**:477–502.

Schweiger M, Lass A, Zimmermann R, Eichmann TO & Zechner R (2009) Neutral lipid storage disease: genetic disorders caused by mutations in adipose triglyceride lipase/PNPLA2 or CGI-58/ABHD5, *Am J Physiol Endocr Metab* **297**:E289–E296.

Scriver's The Online Metabolic & Molecular Bases of Inherited Disease: http://www.ommbid.com/.

Spiekerkoetter U (2010) Mitochondrial fatty acid oxidation disorders: clinical presentation of long-chain fatty acid oxidation defects before and after newborn screening, *J Inherit Metab Dis* **33**:527–32.

Diet and cancer

Currie E, Schulze A, Zechner R, Walther TC & Farese RV Jr (2013) Cellular fatty acid metabolism and cancer, *Cell Metab* **18**:153–61.

Hakomori, S (2008) Structure and function of glycosphingolipids and sphingolipids: recollections and future trends, *Biochim Biophys Acta* **1780**:325–46.

Flavin R, Peluso S, Nguyen PL & Loda M (2010) Fatty acid synthase as a potential therapeutic target in cancer, *Future Oncol* **6**:551–62.

Leyssens C, Verlinden L & Verstuyf A (2013) Antineoplastic effects of 1,25(OH)$_2$D$_3$ and its analogs in breast, prostate and colorectal cancer, *Endocr Relat Cancer* **20**:R31–R47.

Morad SA & Cabot MC (2013) Ceramide-orchestrated signalling in cancer cells, *Nat Rev Cancer* **13**:51–65.

van Blitterswijk WJ & Verheij M (2013) Anticancer mechanisms and clinical application of alkylphospholipids, *Biochim Biophys Acta* **1831**:663–74.

World Cancer Research Fund/American Institute for Cancer Research 2nd report, 2007: http://www.dietandcancerreport.org/expert_report/report_contents/index.php.

Lipids and immunity

Calder PC (2012) Mechanisms of action of (*n*-3) fatty acids, *J Nutr* **142**:592S–599S.

Krasity BC, Troll JV, Weiss JP & McFall-Ngai MJ (2011) LBP/BPI proteins and their relatives: Conservation over evolution and roles in mutualism, *Biochem Soc Trans* **39**:1039–44.

Maceyka M & Spiegel S (2014) Sphingolipid metabolites in inflammatory disease, *Nature* **510**:58–67.

Moreira APB, Texeira TFS, Ferreira AB, Peluzio MDCG & Alfenas RCG (2012) Influence of a high-fat diet on gut microbiota, intestinal permeability and metabolic endotoxaemia, *Br J Nutr* **108**:801–9.

Peri F, Piazza M, Calabrese V, Damore G & Cighetti R (2010) Exploring the LPS/TLR4 signal pathway with small molecules, *Biochem Soc Trans* **38**:1390–5.

Pond CM (2005) Adipose tissue and the immune system, *Prostag Leukotr Ess* **73**:17–30.

Sampath H & Ntambi JM (2005) Polyunsaturated fatty acid regulation of genes of lipid metabolism, *Annu Rev Nutr* **25**:317–40.

Wang X & Quinn PJ (2010) Lipopolysaccharide: Biosynthetic pathway and structure modification, *Prog Lipid Res* **49**:97–107.

Weylandt KH, Chiu CY, Gomolka B, Waechter SF & Wiedenmann B (2012) Omega-3 fatty acids and their lipid mediators: Towards an understanding of resolvin and protectin formation, *Prostag Oth Lipid M* **97**:73–82.

Yaqoob P & Calder PC (2011) The immune and inflammatory systems. In SA Lanham-New et al., eds, *Nutrition and Metabolism*, 2nd edn. Wiley-Blackwell, Oxford, pp. 312–38.

PAF

Stafforini DM, McIntyre TM, Zimmerman GA & Prescott SM (2003) Platelet-activating factor, a pleiotrophic mediator of physiological and pathological processes, *Crit Rev Clin Lab Sci* **40**:643–72.

Yamashita Y, Hayashi Y, Nemoto-Sasaki Y, *et al.* (2014) Acyltransferases and transacylases that determine the fatty acid composition of glycerolipids amd the metabolism of bioactive mediators in mammalian cells and model organisms, *Prog Lipid Res* **53**:18–81.

Dietary n-3/n-6 PUFA ratio

Cunnane SC (2003) Problems with essential fatty acids: time for a new paradigm?, *Prog Lipid Res* **42**:544–68.

Griffin BA (2008) How relevant is the ratio of dietary n-6 to *n*-3 polyunsaturated fatty acids to cardiovascular disease risk? Evidence from the OPTILIP study, *Curr Opin Lipidol* **19**:57–62.

Kuipers RS, Luxwolda MF, Dijck-Brouwer DA, *et al.* (2010) Estimated macronutrient and fatty acid intakes from an East African Paleolithic diet, *Br J Nutr* **104**:1666–87.

Stanley JC, Elsom RL, Calder PC, *et al.* (2007) UK Food Standards Agency Workshop Report: the effects of the dietary *n*-6:*n*-3 fatty acid ratio on cardiovascular health, *Br J Nutr* **98**:1305–10.

Adipose tissue, health and disease

Dubé JJ, Amati F, Stefanovic-Racic M, *et al.* (2008) Exercise-induced alterations in intramyocellular lipids and insulin resistance: the athlete's paradox revisited, *Am J Physiol-Endoc M* **294**:E882–E888.

Frayn KN (2002) Adipose tissue as a buffer for daily lipid flux, *Diabetologia* **45**:1201–10.

Manolopoulos KN, Karpe F & Frayn KN (2010) Gluteofemoral body fat as a determinant of metabolic health, *Int J Obes* **34**:949–59.

O'Rahilly S & Farooqi IS (2008) Human obesity as a heritable disorder of the central control of energy balance, *Int J Obes* **32**Suppl 7:S55–61.

Perry RJ, Samuel VT, Petersen KF & Shulman GI (2014) The role of hepatic lipids in hepatic insulin resistance and type 2 diabetes, *Nature,* **510**:84–91.

Spalding KL, Arner E, Westermark PO, *et al.* (2008) Dynamics of fat cell turnover in humans, *Nature* **453**:783–7.

Virtue S & Vidal-Puig A (2010) Adipose tissue expandability, lipotoxicity and the metabolic syndrome – an allostatic perspective, *Biochim Biophys Acta* **1801**:338–49.

Lipodystrophy

Garg A (2011) Lipodystrophies: genetic and acquired body fat disorders, *J Clin Endocr Metab* **96**:3313–25.

Lipoproteins and atherosclerosis

Fernandez ML & West KL (2005) Mechanisms by which dietary fatty acids modulate plasma lipids, *J Nutr* **135**:2075–8.

Gould AL, Davies GM, Alemao E, Yin DD & Cook JR (2007) Cholesterol reduction yields clinical benefits: meta-analysis including recent trials, *Clin Ther* **29**:778–94.

Holmes MV, Asselbergs FW, Palmer TM, *et al.* (2015) Mendelian randomization of blood lipids for coronary heart disease. *Eur Heart J* **36**:539–50.

Hu FB, Stampfer MJ, Manson JE, *et al.* (1997) Dietary fat intake and the risk of coronary heart disease in women, *N Engl J Med* **337**:1491–9.

Kwak SM, Myung SK, Lee YJ & Seo HG; Korean Meta-analysis Study Group (2012) Efficacy of omega-3 fatty acid supplements (eicosapentaenoic acid and docosahexaenoic acid) in the secondary prevention of cardiovascular disease: a meta-analysis of randomized, double-blind, placebo-controlled trials, *Arch Intern Med* **172**:686–94.

Prospective Studies Collaboration (2007) Blood cholesterol and vascular mortality by age, sex, and blood pressure: a meta-analysis of individual data from 61 prospective studies with 55,000 vascular deaths, *Lancet* **370**:1829–39.

Riches K & Porter KE (2012) Lipoprotein(a): cellular effects and molecular mechanisms, *Cholesterol* doi: 10.1155/2012/923289: 1–10.

Schaefer EJ (2002) Lipoproteins, nutrition, and heart disease, *Am J Clin Nutr* **75**:191–212.

Willett WC (2012) Dietary fats and coronary heart disease, *J Intern Med* **272**:13–24.

CHAPTER 11
Lipid technology and biotechnology

11.1 Introduction

Biologically derived lipids are important raw materials for a huge range of products and materials that are widely used in everyday life both in the home and in industry. Many different types of technology and biotechnology (see Box 11.1 for definitions of these terms) are used to make finished products from lipid-based feedstocks. Examples of edible lipid-based products in widespread use include margarine, dairy products, food emulsions such as mayonnaise and sauces, and animal feed additives such as astaxanthin. Nonedible products include soaps, biodiesel fuels, lipid-based paints and varnishes, and a wide range of oleochemicals.

As discussed in Chapter 6, lipids make up the second largest component in a typical human diet and ideally should constitute 20–30% of total energy intake. However, lipids are rarely consumed in foods as raw products. Instead, edible lipids are normally first extracted from their biological tissue of origin before being processed and mixed with other products to manufacture a particular type of food. Dietary lipids are derived from animals, plants, fungi, and even bacteria with the type of lipid obtained from these different sources varying considerably in its physicochemical properties and nutritional quality. Lipids are also crucial ingredients in the food industry where they enable manufacturers to produce the optimal textures, tastes and aromas in a wide range of foods that includes sauces, spreads, confectionary, soups, and hundreds of 'ready meal' products.

In addition to their roles as edible ingredients, lipids are also raw materials for a host of nonfood products such as fuels, polymers, paints, lubricants, solvents and many composite materials from vehicle interiors to furniture items. The anaerobic decay of billions of tonnes of plant biomass, mostly during the Carboniferous Era around 300 million years ago, led to the formation of the fossil-derived hydrocarbons that make up all our existing petroleum, natural gas and coal reserves. Fossil-derived hydrocarbons are still our main source of fuel for transport and electricity generation. These hydrocarbons are also raw materials in the manufacture of a wide range of petrochemicals, including plastics and lubricants.

Unfortunately, however, we are using up these nonrenewable hydrocarbons at such a high rate that they are likely to run out in the next few centuries or maybe sooner. This means that we will eventually need to obtain hydrocarbons from alternative sources, with plant or algal lipids as the most likely candidates. Unlike fossil hydrocarbons, plant or algal lipids are renewable because they are products of recent photosynthesis. As long as there is sunlight, algae and land plants can be grown and harvested to produce these lipids on an indefinite basis.

In this chapter, we will discuss some of the ways in which lipids are being exploited via various technologies and biotechnologies for a range of practical purposes.

11.2 Lipid technologies: from surfactants to biofuels

Naturally occurring lipids have been manipulated via various empirical (i.e. 'trial and error') processes to generate a wide range of food and nonfood products for thousands of years (see Box 11.2). During the 19th and 20th centuries, scientific advances enabled us to understand many of the chemical and biochemical mechanisms involved in these processes. At the same time, technological advances meant that many production processes could be dramatically scaled up. For example, the manufacture of lipid-based

---

Lipids: Biochemistry, Biotechnology and Health, Sixth Edition. Michael I. Gurr, John L. Harwood, Keith N. Frayn, Denis J. Murphy and Robert H. Michell.

Box 11.1 Definitions of technology and biotechnology

Technology in its modern sense can be defined as *the application of scientific knowledge for practical purposes, especially in industry*. For example, knowledge of physics underpins inventions such as powered flight and computers while chemistry is the basis of developing materials such as plastics and solvents. Many types of technology are used to modify and manipulate lipids in order to manufacture useful products. One of the most diverse examples is the oleochemicals industry that uses plant and animal fats and oils as raw materials to create a host of products ranging from toothpaste to varnishes.

The term 'biotechnology' was invented in 1919 by Hungarian agricultural engineer, Karl Ereky, to describe large-scale microbial-based manufacturing processes. At the same time, progress in structural engineering enabled many microbial fermentation processes to be carried out more efficiently and reliably in sealed and controlled systems, such as very large (>50,000 litre) steel vats or glass vessels. Biotechnological processes based on FA acid interesterification using isolated enzymes are now used to make everyday products such as ice cream, while microbial fermentation is used in the manufacture of higher value lipid-based products such as vitamin supplements.

Since the 1980s, the term biotechnology has also been used to describe the genetic engineering of biological organisms. This normally involves the transfer into a recipient organism of synthetic DNA encoding one or more genes that was copied from a donor organism. The recipient organism, which can be an animal, plant or microbe, is then termed 'transgenic'. Transgenic organisms are often popularly known as 'genetically modified' or GM. It should be noted, however, that biotechnology can involve the use of a broad range of tools and is by no means restricted to genetic engineering. A commonly used definition of modern biotechnology developed by the United Nations Convention on Biological Diversity is as follows: *Any technological application that uses biological systems, living organisms, or derivatives thereof, to make or modify products or processes for specific use.*

items such as soaps and margarines changed from being traditional, locally based cottage industries into globalized large-scale manufacturing processes based firmly on scientific knowledge. The manufacture of both soap and margarine is now performed using highly sophisticated industrial processes. These processes require millions of tonnes of raw materials such as tropical vegetable oils and tallow (beef fat) to be shipped across the world before being converted into finished products in large factories that are normally situated close to seaports.

11.2.1 Surfactants, detergents, soaps and greases
11.2.1.1 Surfactants

A surfactant is a compound that lowers the surface tension between two immiscible liquids or between a liquid and a solid. This property makes surfactants ideal for use as cleaning agents, detergents and dispersants, as well as key functional components in emulsifiers and foaming agents. Most surfactants are carbon-based amphipathic molecules, which means that many are

Box 11.2 Ancient and modern types of lipid-related technology

People have used lipids in the form of plant-or animal-derived fats and oils (mainly comprising TAGs) for many purposes for tens of thousands of years. Well before the advent of agriculture, people used these oils and greases to lubricate simple machines or as skin applications for cosmetic or medical purposes. For many millennia before the introduction of electricity in the 20th century, relatively slow burning vegetable fats and oils, such as beeswax or tallow, were often the primary forms of household illumination.

The first major commercial technological innovation for lipids came in 1869 when a French chemist, Hippolyte Mège-Mouriès, produced a substitute for butter that he called 'margarine'. The raw material was a solid FA fraction called margaric acid. This name comes from the lustrous pearly-white drops of the crystalline form of margaric acid that are reminiscent of oyster pearls, which are called margarites in Greek (this is also the derivation of the name, Margaret).

Two technical advances helped margarine to become an effective competitor with butter. Firstly, improved refining methods allowed for the purification of a greater variety of liquid oils and solid vegetable fats that could be blended to make good spreadable margarines. Secondly, the process of hydrogenation, which was invented in 1901 by English chemist William Normann, allowed the large scale conversion of relatively cheap plant oils into solid fats. Not only did the hydrogenation process produce a good, inexpensive butter substitute, it also significantly reduced the amount of oxidation-prone PUFAs in the solid margarine, which greatly extended its shelf life and therefore its utility for shoppers. An undesirable side effect of incomplete FA hydrogenation is the accumulation of *trans*-fats as discussed in Box 11.4.

derived from naturally occurring acyl lipids. Examples of widely used surfactants include solid and liquid soaps plus a huge range of synthetic chemicals such as alkyl-sulphonates and acyl ethoxyates.

11.2.1.2 Detergents

A detergent is a type of surfactant used in dilute solution for its cleaning properties as well as for many scientific applications, especially in biochemistry. Many widely used commercial detergents are based on medium-chain fatty acids (FAs) while other milder detergents commonly have phenolic or steroidal backbones (see Fig. 11.1). One of the most widely used groups of ionic detergents is based on laurate (12:0), a FA that is generally derived from palm kernel or coconut oils. The most common formulation of laurate in household detergents is in the form of its sodium ether sulphate derivative – often referred to on product labels as 'sodium laureth sulphate'. This detergent is the principal active ingredient in many domestic cleaning products including washing up liquids, shampoos, liquid soaps, shower gels and toothpastes. Two of the most important characteristics of such products are their effectiveness as foaming agents and their ability to emulsify and disperse fats and oils.

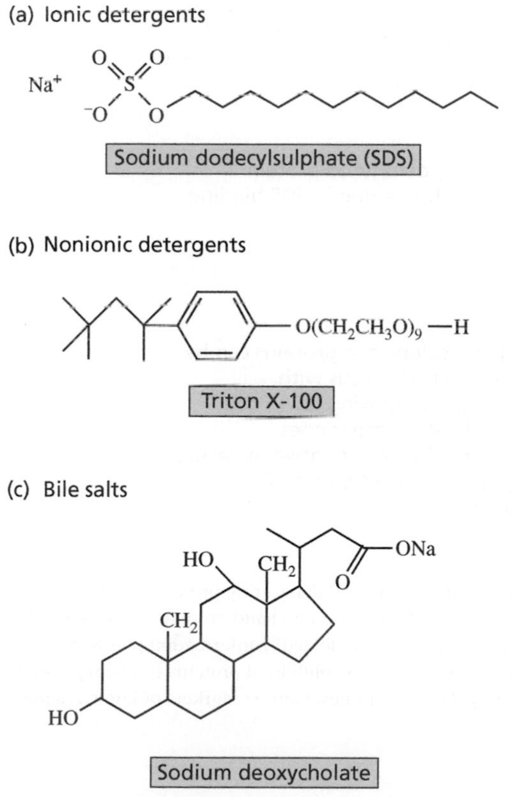

In scientific research applications, sodium laureth sulphate is better known as the ionic detergent, sodium dodecyl sulphate (SDS). This versatile compound is widely used as an analytical detergent especially for the solubilization of biological membranes and the fractionation of complex mixtures of large biomolecules such as proteins or nucleic acids. The solubilization of bio-membranes involves the disruption of lipid bilayer organization by detergent molecules. A bilayer is able to accommodate only a small amount of detergents, such as FAs or lysolipids, before these compounds form a separate micellar phase into which membrane components are partitioned (see Chapter 5). The eventual result is the disintegration of the membrane and the dispersal of its lipid and protein components into many small micelles as depicted in Fig. 11.2(a).

SDS is also used for protein separation via electrophoresis. As shown in Fig. 11.2(b), native proteins tend to be folded into a tightly packed tertiary structure that makes it difficult to separate the different types of proteins on the basis of their size. When SDS is added, each polypeptide unwinds into a linear structure, which carries a negative charge that is proportional to the molecular mass of the polypeptide. If a mixture of these SDS-solubilized polypeptides is then loaded onto an acrylamide gel and subjected to an electric current, the polypeptides will move down the gel at a velocity proportional to their molecular mass. As shown in Fig. 11.2(c), the end result is a series of separated polypeptide bands where those of lowest mass run fastest and are found near the bottom of the gel while the slower higher mass components are nearer the top of the gel.

SDS is an example of an ionic detergent that carries a single negative charge on each molecule. Ionic detergents are relatively harsh and tend to denature most proteins irreversibly, which limits their usefulness in some biological applications. Milder types of synthetic detergent include nonionic compounds such as octyl-glucoside (*n*-octyl-ß-D-gluco-pyranoside) and zwitterionic compounds such as CHAPS (3-[(3-cholamido-propyl)dimethylammonio]-1-propanesulphonate). These relatively mild detergents will solubilize membranes without denaturing the constituent proteins, which can then

(a) Ionic detergents

Sodium dodecylsulphate (SDS)

(b) Nonionic detergents

Triton X-100

(c) Bile salts

Sodium deoxycholate

Fig. 11.1 Major classes of lipid-based detergents: (a) ionic detergents; (b) nonionic detergents; (c) bile salts.

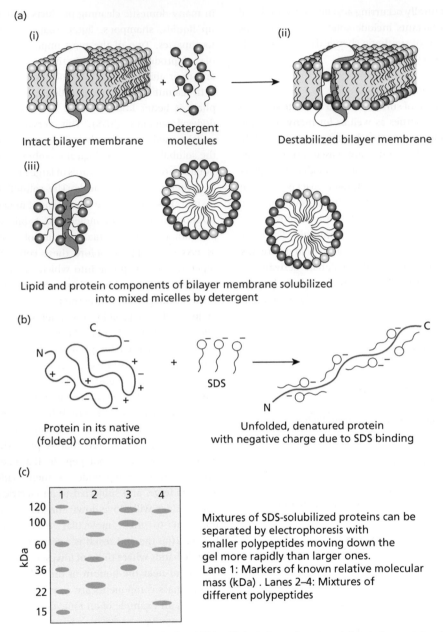

Fig. 11.2 Principal mechanisms of detergent action in destabilizing biological membranes and denaturing proteins. (a) (i) Intact bilayer mechanism plus detergent molecules results in (ii) destabilized bilayer membrane. (iii) Lipid and protein components of bilayer membrane are solubilized into mixed micelles by detergent. (b) Protein in its native (folded) conformation plus SDS results in unfolded, denatured protein with negative charge due to SDS binding. (c) Mixtures of SDS-solubilized proteins can be separated by electrophoresis with smaller polypeptides moving down the gel more rapidly than larger ones. Lane 1: Markers of know relative molecular mass (kDa). Lanes 2–4: Mixtures of different polypeptides.

be extracted and studied while often retaining their biological function. A final class of detergent is another group of ionic compounds called bile salts. Unlike the fatty acyl backbone of SDS, bile salts contain a rigid steroidal group with a polar and a nonpolar face. This structure means that bile salts form small kidney-shaped aggregates instead of the spherical micelles formed by linear-chain ionic detergents such as SDS. As their name implies, bile salts such as sodium cholate are components of the bile that is released from the pancreas into the intestine in order to help solubilize and digest fatty foods (see Section 7.1.1, Box 7.1 & Fig. 7.2).

11.2.1.3 Soaps

Soaps are metal salts of FAs and their main uses are as detergents or surfactants for washing, bathing and cleaning. The widespread availability of cheap soaps in the 19th century is credited with reducing the incidence of some infectious diseases in populations that often lived and worked in overcrowded and dirty conditions. The industrial scale manufacture of soaps involves the treatment of triacylglycerols (TAGs) with strong alkali at high temperature to produce FA salts plus unesterified glycerol (see Fig. 11.3). The most common alkalis used are hydroxides of sodium, calcium, aluminium and lithium and such soaps are the basis of many domestic cleaning products that are still in widespread use.

The physical form of a soap is largely determined by its FA composition. Hence long-chain saturated fatty acids (SFA) such as palmitate (16:0) or stearate (18:0) produce hard solid soaps while monounsaturated fatty acids (MUFAs), such as oleate (18:1) or shorter-chain SFA, such as laurate (12:0), produce semisolid or liquid soaps. By mixing several FAs it is possible to produce blended soaps with intermediate properties. One of the best-known examples of such a blend is Palmolive, which

was the best-selling cleaning product in the world in the early 20th century. This soap was originally made from a mixture of palm and olive oils with the major fatty acids being palmitate and oleate. The resulting blend was a solid bar that was more easily transported than a liquid soap while the presence of unsaturated fatty acids (UFA) meant that it dissolved in warm water much more readily than alternative soaps based mainly on saturated fats. Modern versions of Palmolive tend to be made from a mixture of palm oil and palm kernel oil that contains an even broader range of fatty acids from 8C to 18C. This mixture produces a soap that lathers well and has a creamy skin feel. Some other modern soaps are made of a mixture of tallow and coconut oil, which has a similarly broad range of FAs from 12C to 18C.

Glycerol, often also called glycerine, is a by-product of soap manufacture. In large-scale soap making, glycerol is normally separated from the FA salts and is then used as an ingredient in many personal care products, pharmaceuticals, chemical intermediates, alkyd resins and processed foods. However, it is also possible to leave some of the glycerol in the final soap to produce relatively soft and translucent soap bars that are often sold as speciality products. As mentioned below, glycerol is also a by-product of the production of oleochemicals and most forms of commercial biodiesel fuel. With the increasing nonfood use of TAG-based fats and oils, supplies of glycerol are likely to increase well beyond current requirements and technologists are now trying to find more uses for this trihydric alcohol.

11.2.1.4 Greases

Greases are traditionally semisolid emulsions of soaps and mineral or vegetable oils that have widespread uses as lubricants. Greases have the property of a having a high initial viscosity that is then reduced when subjected to shear stress, for example by moving surfaces such as

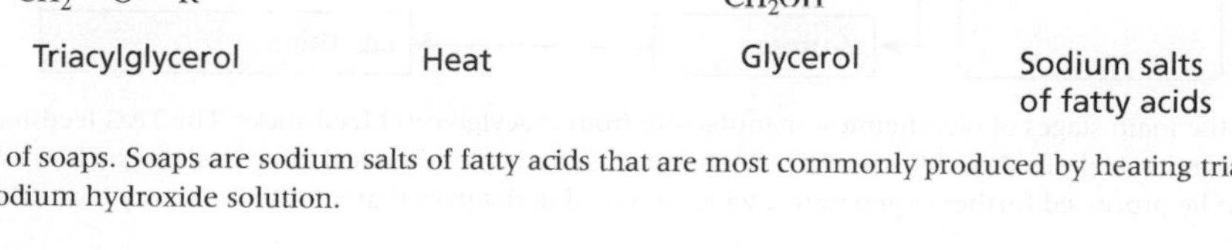

Fig. 11.3 Synthesis of soaps. Soaps are sodium salts of fatty acids that are most commonly produced by heating triacylglycerol oils in the presence of a sodium hydroxide solution.

metallic cogwheels. In this way the grease enables such hard surfaces to move smoothly without undue wear or damage. Lubricating greases are commonly used in most devices with moving parts such as industrial machinery and vehicles. Although non-oil-based greases are widely used in specialized applications, traditional oil-based greases remain the most common class of semisolid lubricants.

11.2.2 Oleochemicals

Oleochemicals are industrial intermediates or end products that are synthesized from animal, plant or microbial lipids. In contrast, petrochemicals are synthesized from fossil-derived raw materials such as petroleum or coal. In both cases the most useful property of the raw materials is their enrichment in medium- and long-chain hydrocarbons. The most common raw materials for oleochemical production are TAG-rich vegetable oils or animal fats, although in a few cases long chain waxes may be used instead. In the first stage of oleochemical production, TAGs are normally hydrolysed to their FA and glycerol components followed by fractionation of the different FAs based mainly on chain length and degree of unsaturation.

Plant-derived FAs can be converted into methyl esters, alcohols, and amines from which intermediate chemicals, such as alcohol sulphates, alcohol ethoxylates, alcohol ether sulphates, quaternary ammonium salts, monoacylglycerols (MAGs), diacylglycerols (DAGs), tailored TAGs, and glycoesters may be produced (see Fig. 11.4). Detergent alcohols are used mainly to produce surface-active compounds for personal care and for cleaning processes. Other major oleochemical intermediates include fatty esters, lubricant esters, and solvent replacers. These chemicals are used in the manufacture of paints, inks, lubricants, and production of polyols to make polyurethanes. Oleochemicals are also used to produce many fine chemicals, cosmetics, pharmaceuticals, coatings, textiles, polymer, paper, additives for fuel and rubber, biodegradable polymers, and offshore drilling muds.

Many, but by no means all, of the end products synthesized from oleochemicals can also be produced from fossil-derived petrochemicals. For much of the 20th century it was far cheaper to make products such as oil-based paints, polymers, textiles and lubricating oils from petrochemicals. However, it is now recognized that the use of petrochemicals contributes to net greenhouse gas emissions and will be limited in the long term because they are not renewable. This has stimulated an increased interest in the development of alternative and more efficient biological sources of oleochemicals and improved processes for their conversion into a wider

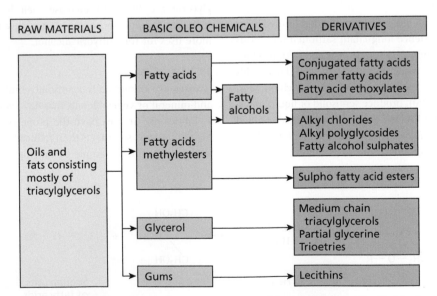

Fig. 11.4 Outline of the main stages of oleochemical manufacture from triacylglycerol feedstocks. The TAG feedstocks can be derived from animal, plant or microbial oils. Breakdown of the TAGs produces basic oleochemical intermediates such as fatty acids, glycerols and gums. These can be processed further to generate a wide variety of derivatives that are used in many industrial processes.

variety of useful end products. Some of these plant-based products, such as bioplastics, are described in more detail in Section 11.5.

11.2.3 Biofuels

Biofuels are renewable fuels derived from living organisms as distinct from fossil fuels, which are derived from nonrenewable hydrocarbon deposits such as coal, natural gas or petroleum. Biofuels are the products of recent photosynthesis by plants or algae that can be regrown each year. This means that, unlike fossil fuels, the combustion of biofuels does not result in any net CO_2 emission, i.e. they are carbon-neutral. Because they are both renewable and carbon-neutral, biofuels were initially regarded as a good alternative to fossil fuels and, since 2000, many governments have set up schemes to promote their use. More recently the suitability of these so-called first generation biofuels has been questioned. Some of the practical and ethical issues relating to biofuels are discussed in Box 11.3.

The most common types of biofuel are 'bioethanol' and 'biodiesel'. Bioethanol is obtained from sugar- or starch-rich crops such as sugar cane or maize. The starch or sugar from these crops is fermented on a large scale to produce ethanol, which can then be added to gasoline to power vehicle engines. Typically 10% w/w bioethanol is added to gasoline to produce an E10 biofuel that can be used directly in all gasoline engines. The vast majority of bioethanol is produced either in the USA (from maize) or Brazil (from sugar cane).

Box 11.3 Ethical dilemmas of lipid (bio)technology: genetically modified crops and biofuels

Traditional forms of biotechnology that are based on microbial fermentation have not been particularly controversial. In contrast, some of the more recent methods such as genetic engineering and animal cloning have attracted much more public attention and in some cases have been criticized on ethical, safety and/or environmental grounds. In general most of this criticism has been levelled at the use of GM crops for human consumption but concerns have also been expressed about the use of such crops for animal feed and, to a much lesser extent, for industrial purposes. None of the GM crops commercially released on a large scale since 1996 contains modified lipids but, as discussed in Section 11.6, there are now several potential future GM crops that have modified lipid profiles.

All new technologies involve risks as well as benefits; for example there are reports of possible effects on the brain of prolonged use of mobile phones but for most people these uncertain risks are more than offset by the obvious benefits from such devices. The current versions of GM crops also carry uncertain risks but for most people these are not offset by clear benefits. If GM crops are to be more acceptable to the public in the future they will need to provide clear benefits. GM crops modified to produce industrial lipid-derived products, such as renewable and biodegradable plastics like PHAs, are likely to be among the most promising. They will not enter the human food chain and will have clear environmental benefits compared with similar products derived from fossil fuels. GM crops modified to produce edible lipids might be more controversial, especially in Europe where GM-foods are currently banned by most supermarket chains. However, if future GM crops provide more nutritious foods, such as carotenoid-rich golden rice or spreads enriched in n-3 fats, they may become more acceptable to most shoppers.

Biofuels are renewable, carbon-neutral sources of fuel that can partially replace nonrenewable fossil fuels and potentially mitigate some of the factors contributing to climate change. This means that, at first sight, biofuels should be ethically and environmentally superior to other types of fuel. This was a major justification for many governments to set targets to increase biofuel production in the early 2000s. Unfortunately, almost all the so-called 'first generation' biofuels produced so far have been obtained from crops that were previously grown for food. The result has been increased prices of some foods and concerns that biofuels might exacerbate food shortages and environmental degradation in some developing countries. For example, the production of biodiesel from oil crops increased from 4 to 25 million tonnes during the decade from 2003 to 2013. The two major sources of biodiesel are rapeseed and oil palm, both of which are also important food crops.

In the case of oil palm, the diversion of millions of tonnes of oil from food use to biodiesel resulted in large increases in the prices of vegetable oil and the food products containing such oils. These food price increases have occurred mostly in relatively poor Asian countries, whereas most of the biodiesel is exported to enable people in relatively rich countries to use biofuel in their vehicles. Clearly this creates an ethical dilemma and most environmental groups that previously supported the introduction of biofuels are now among the strongest critics of the policy. A second unintended consequence of the increased demand for biofuels is the stimulus that it provides for the conversion of unused land into oil palm plantations. Much of this land is ecologically sensitive tropical rainforest or peatland that harbours high levels of plant and animal biodiversity including iconic species such as orang-utan, elephant and teak and ironwood trees. Since 2000, millions of hectares of these irreplaceable habitats have been lost to new plantations. While biodiesel demand is only one factor in the loss of these tropical landscapes, it has played a role and is another reason for demands that policies that encourage biofuel use should be reassessed.

Biodiesel is obtained from oil-rich crops such as rape-seed, sunflower, soybean and oil palm. The oil in such crops is largely made up of long-chain TAGs. Although unmodified vegetable oil can be used directly in diesel engines, it is far more efficient to convert the TAGs into their methyl ester derivatives. This is achieved via the process of methanolysis whereby the vegetable oil is heated with methanol at about 60 °C in the presence of an alkaline catalyst such as sodium hydroxide or alkoxide. The result is the formation of methyl esters and glycerol. Methyl esters can be used alone as a diesel fuel normally labelled as B100. More commonly, how-ever, methyl esters are mixed with petroleum diesel to produce either 5% (B5) or 10% (B10) biodiesel blends. The leading producers of biodiesel are the European Union (mostly from rapeseed oil), USA and Brazil (both mostly from soybean oil).

11.2.4 Interesterification and transesterification

Interesterification and transesterification are widely used processes for lipid modification in the food industry. In both cases the esterification reactions can be achieved either via chemical or biological (enzyme) catalysts. In general, chemical catalysis is cheaper but less specific while the use of enzymes enables esterification reactions to be based on regioselectivity (discriminating according to the bond to be cleaved), enantioselectivity (optical activity), chemoselectivity (based on functional group), and FA chain length.

11.2.4.1 Interesterification

Interesterification involves the exchange of fatty acyl side chains between two acyl ester molecules. In the form of interesterification considered here, two different TAG molecules are able to exchange acyl residues. As with transesterification (see below), this reaction can be car-ried out in a relatively nonspecific manner by using chemical catalysts or more specifically using lipases. Interesterification is commonly used in the food industry in the manufacture of a wide range of products including spreads, creams, pastes, as well as many of the fat substitutes discussed in Section 11.4.1. More recently, interesterification has become more widely used as a way of reducing or eliminating *trans*-fats in foods as discussed in Section 11.4.2.

An example of the usefulness of interesterification is in margarine manufacture. If a relatively unsaturated liquid vegetable oil is simply blended in the right proportions with a relatively saturated fat, the resulting mixture is a soft solid that is suitable for use as a spread. However, if the spread is stored at low temperature (e.g. in a refrig-erator) the different fats will begin to settle out into two phases comprising an unappetizing gritty solid (contain-ing the more saturated fat) and a cloudy liquid (contain-ing the more unsaturated oil). This unpleasant mixture can no longer be used as a spread and therefore limits the useful shelf life of the product. Such phase separations can be avoided by interesterifying the initial mixture of fats and oils so that most TAG molecules now contain both saturated and unsaturated acyl groups.

Interesterification can also be used to produce a syn-thetic TAG with the same acyl composition as a naturally occurring one. An example is cocoa butter, which is the solid fat from cocoa beans used in the manufacture of chocolate. Cocoa butter is made up of a mixture of TAG species but the most common form and the one mainly responsible for its desirable properties is, 1-palmitoyl, 2-oleoyl, 3-stearoyl glycerol, or 'POS'. Cocoa butter is relatively expensive compared to many other vegetable oils that also contain similar FAs, but not in the POS form. Therefore it is advantageous to use interesterification with a mixture of such oils to produce POS TAG species. For example, palm oil enriched in palmitate and oleate can be interesterified enzymically or chemically with the stearate-rich fat from the tropical shea tree, *Butyrosper-mum parkii* to give a new TAG mixture that is very similar in composition and physical properties to cocoa butter. This synthetic fat can then be mixed with genuine cocoa butter to give a chocolate product that is much cheaper to manufacture than one made from pure cocoa butter, although it would not have the same rich flavour as a chocolate made mainly from cocoa butter.

11.2.4.2 Transesterification

Transesterification in its broadest sense means the exchange of the organic R^1 group of an ester with the organic R^4 group of an alcohol as shown below. The methanolysis process by which FA methyl esters are produced for biodiesel (see above) is an example of transesterification. If a long-chain fatty alcohol is used instead of methanol, it is possible to replace one of the acyl groups of a TAG with the acyl group from the fatty alcohol. Such exchanges are often used in the food industry where it might be desirable to make a particular fat or oil either less or more saturated for a specific downstream application. For example, more saturation might be desirable to create a more solid fat in margarine production, while less saturation might be desirable to create a softer fat as part of a creamy filling in chocolates.

Transesterification can be achieved by using chemical catalysts as shown below. Alternatively, a more precise biotechnological form of the reaction can be done using lipases purified from various fungi or bacteria.

11.3 Lipids in foods

Lipids are present in the vast majority of edible products and their presence often enhances flavour and palatability and increases the desirability of our foods. Indeed, there is a well-known correlation between *per capita* income and the consumption of fats and oils in the human diet. The reason is that dietary fats are particularly attractive to people who have historically subsisted mainly on starch- and vegetable-based diets made up of relatively dull and tasteless foodstuffs, such as boiled rice, manioc or potatoes. For example, it was found that, following rising income levels across much of the developing world in the 1990s, vegetable oil consumption in such countries increased much faster than general food intake. Hence, during the 1990s, *per capita* vegetable oil consumption rose by 64% in China, 65% in Indonesia, and 94% in India.

As a rule when people become more affluent, they tend to switch to a more satisfying diet containing a higher proportion of fats and oils. Interestingly, the reverse is also true and when times are hard people tend to cut back on relatively expensive 'luxuries' like fats and oils. Such an effect was seen during the economic collapse that followed the fall of the Soviet Union when, between 1990 and 1994, consumption of food lipids (mostly in the form of vegetable oil and products thereof) in Russia fell by 35%. Edible lipid consumption in Russia then rose once again during the 2000s as the economy recovered and average incomes increased. The downside to this craving for dietary lipid is seen in the dramatic rise in obesity that is now occurring around the world (see Chapter 10). The rise in dietary fat consumption has been exacerbated by the addition of lipid components to many processed foods, which then become so-called 'invisible fats' that are not obviously present in the food. One example is a form of packaged seaweed that appears at first sight to be a 'healthy', low-fat product. However, closer examination reveals that two thirds of the weight of the product comes from added vegetable oil and that lipid-derived foods account for over 80% of the total energy content. In contrast with more obvious fatty foods such as butter, margarine, or cooking oil, these 'invisible fats' can be difficult to monitor in the diet and can be readily consumed without realizing it.

11.3.1 Lipids as functional agents in foods

Lipids are important functional agents in many food products. Their most obvious role is to increase the energy content of foods due to their high energy value. Dietary lipids have an average energy content of 37 kJ/g whereas carbohydrates and proteins yield only about 17 kJ/g. Before they can be absorbed by the gut, dietary lipids must be solubilized into a fine emulsion (see Section 7.1.1). This means that most lipids take a lot longer to digest than carbohydrates or proteins. Because lipids are present in the gut longer than other foods they produce a more extended period of dietary satiety before we begin to feel hungry again. During the relatively lengthy period of digestion of fatty foods, more blood is diverted from voluntary muscles, especially in the limbs, to the gut. This means that highly active endurance athletes, such as marathon runners or cyclists, tend to avoid consuming very fatty foods during intense exercise – often focusing instead on much more easily digestible simple carbohydrate foods such as high calorie products known as 'energy gel bars'.

11.3.1.1 Vitamin carriers

One of the key roles of dietary lipids is to act as carriers for fat-soluble vitamins, the most important of which are vitamin groups A, D, E and K (see Section 6.2.3). These vitamins are insoluble in water and can be obtained in the diet only by consuming foods that contain lipids. In many cases, fat-soluble vitamins are removed during the processing of lipid-based food products. For example, most commercial vegetable oils are manufactured by means of a high-temperature fractional distillation process using a hexane solvent. This results in a clear and pure liquid TAG with no vitamins or pigments. Therefore, the micronutrient content of most commercial vegetable oils is quite low. Ironically many people wishing to replace these missing vitamins choose to take vitamin supplement capsules that might be made up of vitamins recovered from unprocessed plant oils. Biotechnological attempts to increase the content of fat-soluble vitamins in food are discussed below in Section 11.4.5.

So-called 'virgin' vegetable oils are extracted by crushing the seeds or fruits at room temperature (known as cold pressing). The resultant oil is normally a cloudy and coloured mixture that still contains most of the vitamins and pigments present in the original plant tissue. Because cold pressing is a much less efficient method for oil extraction than fractional distillation, cold-pressed oils are not only more expensive but tend to have shorter shelf lives than their processed counterparts. Another problem with some virgin oils is their colour, which may be unappealing for some shoppers and can also result in the colouring of foods prepared using such oils. For example, virgin olive oil is a cloudy dark green, while virgin palm oil is bright red and not everybody would welcome green or red crisps or chips. Therefore, although virgin oils are nutritionally superior to processed oils, they have several practical drawbacks that mean they are rarely bought by the average supermarket shopper.

11.3.1.2 Taste, odour and texture

The use of lipids in cooking greatly enhances the taste and odour of foods. This is because heated fats and oils can produce a complex bouquet of attractive flavour compounds, many of them volatile. Fats also solubilize, and thereby enhance, otherwise cryptic flavours that may be present in nonfatty foodstuffs, such as some herbs and spices. Fats tend to coat the sensory receptors on the tongue and stick to them far longer than aqueous or solid foods. This allows flavours to linger in the mouth for longer and can produce a more pleasant and satisfying sensory experience compared with many nonfatty foods. Another common use of lipids to modify food texture is their use as shortening agents in baked goods. If flour is kneaded with water, long gluten strands are formed and the resulting dough is elastic, which produces bread with a chewy texture. But when fat is added to the dough mixture, gluten strand formation is reduced, which results in flaky and tender products, such as pie crusts, pastries and biscuits.

Lipids play key roles in the appearance and texture of many foods and in the way that they are cooked. Edible fats have much higher boiling points than water and hot oil is able to transfer heat quickly to the surface of food without overheating the interior portions. This can facilitate crust formation and browning such as in French fries. The refractive properties of lipids can give food an appealing glossy appearance and are responsible for the opaque appearance of fresh milk. Many food products consist of either water-in-oil (see next section) or oil-in-water emulsions. Among some of the many examples of oil-in-water emulsions are salad dressings, mayonnaise, sauces, gravies, ice cream and some yoghurts. In most cases the physical properties of these lipidic emulsions are responsible for the creamy texture that is essential for their appealing taste and flavour qualities.

11.3.2 Butter, margarine and other spreads

Butter, margarine and other spreads are examples of water-in-oil emulsions. The major phase is a liquid or semisolid fat or oil, in which many small droplets of an aqueous phase are suspended. By varying the composition of the lipid phase and the proportion of aqueous phase, different forms of semisolid emulsion can be produced. Hence commercial spreads can range from highly solid slabs of lard, consisting mainly of long-chain saturated TAGs and very little water, to very soft plant-based spreads enriched in unsaturated TAGs and containing as much as 50% water.

11.3.2.1 Butter

Butter is a semisolid water-in-oil emulsion made by churning cream obtained from animal (most commonly cow's) milk. Churning forces the TAG droplets in cream to coalesce to produce a semisolid lipid phase in which some water droplets are trapped. Commercial butter typically contains about 85% fat and 15% water. Butter is solid at room temperature because most mammalian milks contain high proportions of SFAs with overall melting points above 25 °C. Refrigerated butter at 4–10 °C is very solid but becomes softer and 'spreadable' at a normal room temperature above 15 °C. When eaten, butter melts in the human mouth, which is at 32–34 °C, to produce a satisfyingly creamy 'mouth feel' that also improves our perception of relatively dry foods like bread once they are buttered. Unlike unprocessed milk, which goes sour in a few days, butter can be stored for many weeks, especially if it is tightly packaged to exclude air and kept at low temperature.

Modern butter comes in a bewildering variety of forms that include mixtures with nondairy fats and various low-fat versions. Because traditional butter is relatively hard, especially if kept refrigerated, so-called 'soft' butters are marketed as products that are more 'spread-able', for example onto bread. These 'soft' butters are either hybrid products made up of a butter/vegetable oil mixture or are produced by interesterification of butter with unsaturated fats. Some extra-soft types of butter are produced with much higher water contents than the 15% normally found in traditional butter. In some cases, low-fat butter products have been manufactured with half the fat of normal butter and 50% or more of the solid emulsion is water. Due to their high water contents, early versions of low-fat butter and other nondairy spreads performed poorly at high temperatures but modern versions contain stabilizers, such as alginates, pectin and carrageenans that stabilize the emulsion over a wide temperature range.

11.3.2.2 Cheese

Cheese is similar to butter in being a semisolid water-in-oil emulsion that is mainly made up of milk fat. The major difference is that bacterial and/or fungal cultures are added to the milk fat, which is then aged for a relatively long period that can extend to several months for some speciality cheeses. Cheeses vary considerably in their fat content. Hard cheeses, like Cheddar, typically contain over 60% fat, while softer, more runny, spread-able cheeses, like Brie, might contain only 30% fat. Depending on the type of milk and microbial cultures used in the maturing process, each cheese has its own characteristic flavour and colour. Modern commercial cheeses tend to have a uniform appearance because they are made from strictly controlled ingredients under sterile conditions. However, some of the more traditional homemade cheeses can vary considerably in taste and appearance from batch to batch, depending on the type of microbial culture added and the particular conditions under which they were made.

11.3.2.3 Margarine

Like butter and cheese, margarine is a semisolid water-in-oil emulsion but it is normally made from vegetable oils rather than from milk. Margarine was invented in the late 19th century as a nondairy spread that could be made from virtually any fat or oil of animal and/or vegetable origin. This removed the need to use milk fat, which was more expensive than most other fats, although small amounts of fat-free skimmed milk are added to margarine in order to create a butterlike taste and odour. One drawback about the use of most

vegetable oils in spreads is their high content of UFAs, which makes them more likely to be liquid rather than solid at room temperature. Hence, unmodified vegetable oils are commonly used as liquid ingredients in foods such as salad oils or cooking oils. However, a few tropical vegetable oils have relatively high saturated fat contents and are solid at room temperature. A well-known example is the lipid component of cocoa seeds, commonly called cocoa butter, which is a major ingredient of chocolate. However, cocoa fat is an exception and most vegetable oils are unsuitable to be used in solid spreads in their unmodified unsaturated state.

The problem of how to produce a solid spread from vegetable oils was resolved early in the 20th century when catalytic hydrogenation was invented (see also Section 6.1.2.3). This process involves the chemical conversion of double bonds in a FA into single bonds as shown below. The most commonly used catalyst is reduced nickel, which is immobilized on an inert supporting material such as silica. The liquid oil is then passed across the catalyst together with hydrogen gas for 40–60 minutes in a pressurized vessel at 150–200 °C. By changing the reaction conditions, different degrees of hydrogenation can be achieved, resulting in TAG products with differing degrees of solidity. For most margarines the result is a solid fat that can be blended with small amounts of water to produce something that closely resembles butter in its physical properties.

According to commonly displayed health claims, margarines have several nutritional advantages over butter. Firstly, many types of margarine contain much higher amounts of polyunsaturated fats than butter. Polyunsaturated fatty acids (PUFAs) can be included in margarines as long as they are balanced by fully hydrogenated fats in order to generate a final product that is solid. A second advantage is that butter contains animal-derived cholesterol whereas most margarines are plant products that are free of cholesterol. As discussed in Chapter 10 (Box 10.4), dietary intake of foods that are low in cholesterol and enriched in PUFAs is associated with a putative reduction in risk of cardiovascular diseases. These features of margarine have led to it being widely promoted as a 'healthy alternative' to butter. However, in the 1990s the nutritional status of margarine was challenged when it was realized that its high *trans*-FA acid content might have adverse consequences (see Box 11.4 & Fig. 11.5, as well as Section 10.5.4.2 & Box 10.5).

Box 11.4 *Trans*-fats – man-made risk factors in disease?

Most FA double bonds in plant and animal lipids are *cis* isomers in which the acyl chain is bent in a way that increases its volume and flexibility resulting in a more fluid lipid with a lower melting point (Fig. 2.1). In contrast, double bonds that are *trans* isomers do not cause such chain bending and the resulting lipid behaves more like its saturated counterpart (Table 2.2). There are some examples of naturally occurring *trans*-FAs, such as vaccenic acid, *t*11-18:1. Vaccenic is the most abundant *trans* MUFA in ruminant lipids and results from biohydrogenation of linoleic and α-linolenic acids. However, such naturally occurring *trans*-FAs only make up a tiny proportion of dietary fat in humans.

Most of our dietary *trans* fats are created by the food industry as a side effect of partial catalytic hydrogenation of unsaturated vegetable oils with *cis* carbon-carbon double bonds as shown in Fig. 11.5. These partially hydrogenated fats have displaced naturally occurring solid fats and liquid oils in many products in the fast food, snack food, baked good and fried food sectors.

This situation changed early in the 20th century when the partial hydrogenation processes used for margarine manufacture resulted in a high proportion of *cis* double bonds being converted into *trans* isomers. In some foods derived from hydrogenated fat, the amount of *trans*-FAs can be as much as 40% of the total lipid content. Because they had similar physical properties to saturated fats these *trans*-fats were not initially regarded as posing a nutritional problem. This perception gradually changed due to evidence from animal, and later human, studies demonstrating that high amounts of dietary *trans*-fats were associated with elevated plasma concentrations of cholesterol and LDL and reduced amounts of HDL.

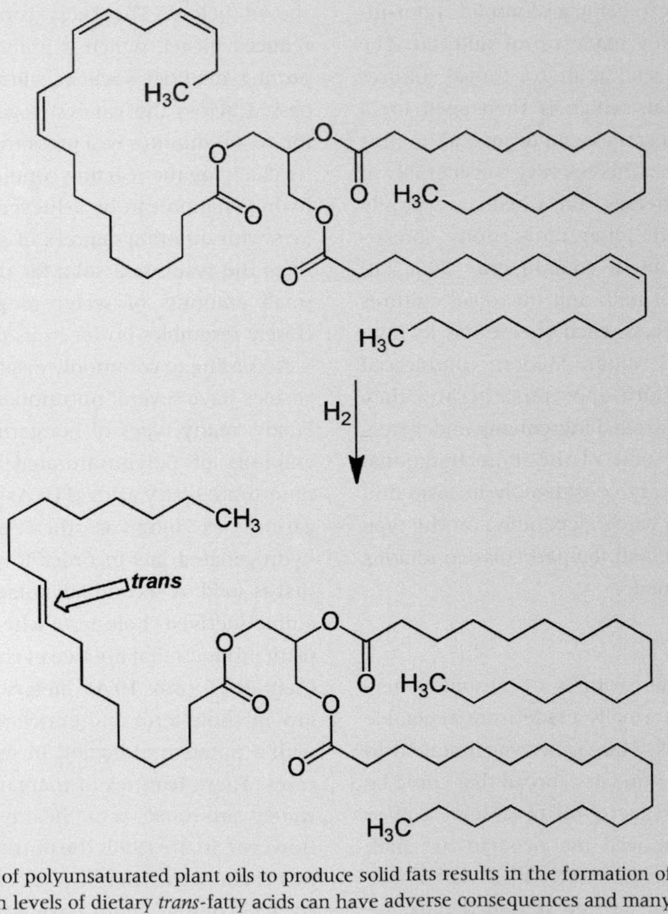

Fig. 11.5 Hydrogenation of polyunsaturated plant oils to produce solid fats results in the formation of *trans*-fatty acids. It is now recognized that high levels of dietary *trans*-fatty acids can have adverse consequences and many margarines are now produced by alternative methods that enable them to be marketed as *trans*-free products.

By the early 21st century it was realized that an increasing number of food products contained appreciable amounts of hydrogenated fat. In addition to obvious examples such as margarines, this included many processed and convenience foods. The possible adverse health effects of *trans*-FAs have led to the imposition of food labelling requirements in some countries. For example, in 2003 Denmark banned all but trace amounts from the food supply. From 2006 all foods in the USA had to be labelled as containing *trans*-FAs if they contain in excess of a given threshold of these FAs. Typical threshold levels of *trans*-FAs that would trigger compulsory labelling are in the region of 0.5 to 1.0% w/w. In 2013, the US government announced a consultation on a proposed outright ban on *trans*-fats in foods. The United Nations FAO/WHO reviewed the issue in 2007 and concluded that there was sufficient evidence 'to recommend the need to significantly reduce or to virtually eliminate industrially produced *trans*-FAs from the food supply'. As discussed in Section 11.4.2, the perceived health risks of *trans*-FAs are currently driving efforts to replace high-PUFA oils with high-oleate oils because the latter require little or no hydrogenation during food manufacture.

11.4 Modifiying lipids in foods

Current patterns of dietary lipid intake have led to increasing concerns about the quantity and type (or quality) of edible lipids being consumed (see Chapters 6 & 10). Some of the major lipid-related public health concerns include the following:

- too much overall dietary lipid intake;
- too much saturated fat and *trans*-fat;
- too little ω-3 (*n*-3) and too much ω-6 (*n*-6) fat;
- an excess of cholesterol;
- deficiencies in fatty vitamins such as A, D and E.

These concerns have led to the use of modern technological and biotechnological approaches to manipulate the lipid content of several food sources, including livestock, crop plants, fungi, and microalgae.

11.4.1 Fat substitutes in foods

By far the best way to reduce overall dietary lipid intake is to cut down on the consumption of fatty foods. However, many people crave the abundant, cheap, and readily available fat-rich foods that surround them in their daily lives. In many cases, it is only the presence of the lipid components that makes such foods so appealing and satisfying. One way the food industry has responded to the overconsumption of fat is by using fat substitutes in many foods. As a result, such foods can be labelled as 'low fat' or even 'zero fat' but still – the manufacturers hope – have the satisfying taste and appearance of the original 'high fat' versions.

Some of the most important roles of lipids in food technology involve the creation of products such as creams, pastes, emulsions and spreads, as discussed previously in Section 11.3. In some of these products, the functional role of the lipid component can be fulfilled to a greater or lesser extent by a nonlipid compound. A good example is yoghurt, where a semisolid creamy texture is integral to the appeal of the product. Traditionally such a texture was obtained by using dairy cream but such products have high amounts of total fat, saturated fat, and high energy contents, all of which now need to be declared on the label. In contrast, 'low fat' yoghurts use nonlipidic gelling agents, such as modified starches and gelatin, as well as a high sugar content to provide a semisolid appearance and a semblance of a creamy mouth feel. These products can be labelled as having very low amounts of total and saturated fat (although they may be rich in simple sugars). Also many of the gelling agents have lower energy values than cream so such yoghurts might also be labelled as 'low calorie' or 'light' or 'diet' products, which can appeal to some people.

There are three main categories of dietary fat substitutes based respectively on carbohydrates, proteins, and nondigested fats. Carbohydrate-based substitutes include cellulose, dextrins, modified starches, gums, pectin, gelatin, maltodextrin, carrageenan and agar. These agents are primarily thickeners that generate a creamy-looking product but they rarely fully reproduce the exact mouth feel of lipid-based 'creamers'. Moreover, although they might replace much or all of the fat in a particular food, some of these compounds also have high energy contents and if they are used in conjunction with large amounts of simple sugars like sucrose or fructose, as is frequently the case, they can simply replace one nutritional problem with another.

There are several protein-based fat substitutes based on whey from milk and/or egg white (albumin). The raw protein is mechanically dispersed to form a microcolloidal suspension of tiny particulate proteinaceous aggregates in aqueous solution. The particles are so small that the suspension is perceived in the mouth as having a smooth creamy texture that is very similar to a fatty emulsion. However, because the protein in such products will denature to a gel when heated, it can be used only in foods that do not require cooking, such as salad dressings, ice cream, mayonnaise and creamy dips. Protein-based products can substitute for as much as 80% of the fat in a food and can contain as little as 15% of the energy of the full fat version.

Fat replacements based on nondigested or poorly digested fats have an advantage over carbohydrate and protein substitutes in that they tend to be much better at mimicking the mouth feel of normal dietary fats. However, such products have had a mixed reception in the food industry and have yet to be widely adopted. One of the most highly promoted of these compounds is the group of sucrose polyesters (see also Section 6.1.2.4). The polyester molecule consists of a core sucrose moiety esterified to between 5 and 8 fatty acyl groups that are normally saturated. The resulting fatty compound behaves in many respects like a TAG-based oil and can be used in high temperature cooking such as frying. Its mouth feel is also very close to that of TAG oils.

These synthetic compounds are not digested in the gut and pass through without contributing either energy or nutrition. This has made them attractive as components of otherwise high fat, high energy foods such as potato chips or French fries that could instead be marketed as 'low fat' or 'low calorie' products. On the negative side there has been concern that sucrose esters can dissolve the fat-soluble vitamins in foods and reduce their availability to the body, which could contribute to vitamin deficiency. A further drawback was the reported incidence of abdominal cramps and diarrhoea as the undigested sucrose esters were passed out of the gut. These compounds can still be found in a few types of 'light' potato chips sold in the USA but have not met with widespread acceptance by shoppers.

11.4.2 Polyunsaturated, monounsaturated, saturated, and *trans* fatty acids

The ongoing debate about the nutritional status of dietary SFAs versus UFAs is discussed in Chapters 6 and 10. This is a complex topic and new research findings can sometimes contradict previous assumptions. For example, in the late 20th century evidence that dietary PUFAs helped reduce serum cholesterol concentrations (Chapter 10) led to their promotion as 'healthy fats'. However, later findings in animal models showed that high intake of PUFAs might be associated with the release of free radicals derived from double bond oxidation and an increased risk of cancer spread (metastasis). While such risks could be reduced substantially by the addition of lipophilic antioxidants such as ubiquinone (coenzyme Q_{10}), they highlight the dangers in presenting an overly simplistic message about a particular type of dietary lipid being either 'good' or 'bad'. In general, unsaturated fats have a more positive nutritional image than saturated fats. However, due to its higher oxidative stability, the 18C MUFA, oleate, is sometimes more highly recommended than the 18C PUFAs, linoleate and α-linolenate.

The concern about overly simplistic messages discussed above also applies to the widespread perception that all saturated fats are 'bad'. This has led to the mandatory labelling of the SFA content of most packaged foods (see Box 11.5). In reality, research has shown that not all SFAs or UFAs are equivalent. However, all SFAs tend to be lumped together and most food labels do not distinguish between them (see Box 11.5). The adverse image of SFAs and their prominence on food labels is one of the drivers towards finding fat substitutes as discussed above in Section 11.4.1.

As discussed in Box 11.4, health concerns about *trans*-FAs have led to the imposition of labelling requirements in the USA. Such labels reveal whether a product contains over a given threshold of these FAs, typically in the region of 0.5 to 1.0%. In contrast, some existing foods can contain as much as 40% by mass of their FA complement as *trans*-FAs. This is leading the food industry to develop new sources of high-oleic vegetable oil either via conventional breeding or by using genetic engineering. Other changes to FA composition occur during industrial or domestic processing and are summarized in Table 11.1.

Trans double bonds also occur in some natural fats, but are much less abundant than *cis* bonds. Some seed oils have a significant content of FAs with *trans* unsaturation, and all green plant tissues contain small quantities of *t*3-hexadecenoic acid, which is a component of the photosynthetic membranes of chloroplasts (see Table 11.2). The most abundant naturally occurring isomeric FAs in animal-based foods are found in fats derived from ruminants. In principle, the chemical outcome of rumen

Box 11.5 Lipids and food labelling

In many modern food labels there tends to be more information about the amount and type of the lipid ingredients than any other food component. For example most labels will list the total fat and SFA in grams as well as the energy value of the total fat and SFA in calories and/or joules. In addition some labels might list the amount of monounsaturates and/or polyunsaturates as well as the presence of components such as *trans*-FAs, omega-3 and omega-6 FAs plus nonacyl lipids such as phytosterols.

These labelling systems are driven mainly by concerns about the nutritional consequences of consuming the various types of lipid that are named. In some cases, such as the 'traffic light' system operated by certain retailers, the presence of a particularly high amount of total fat and saturated fat means that the food will carry a cautionary red label. Such labels act as a clear warning to shoppers and might result in them choosing not to buy such a product. As discussed in Box 11.4, concerns about *trans*-FAs are increasingly resulting in labels such as '*trans*-fat free' or 'low in *trans*-fats' on certain products including some snack foods and margarines.

In other cases, labels like 'fat-free' or 'low-fat' are included to indicate that the fat content of a food product has been reduced or eliminated. In such cases the fat is normally replaced by a substitute such as a sugar derivative (see Section 11.4.1) where the energy content might be as high or higher than the original fat. Foods containing vegetable lipids might be labelled 'cholesterol free' to distinguish them from those containing animal (especially dairy) lipids that contain cholesterol. In most cases such labels are used as more as marketing tools than as genuine health warnings.

There are some labels that relate to perceived positive nutritional aspects of dietary lipids. For example, MUFA such as oleate (found abundantly in olive and rapeseed oils) are associated with the so called 'Mediterranean diet' and are preferred by some people to the more saturated animal/dairy fats. Other labels might include 'high in polyunsaturates' because of evidence that dietary PUFAs are associated with lower concentrations of plasma cholesterol and HDL. Another type of FA with an increasingly positive image that is found on some food labels is omega-3 (*n*-3) group that includes DHA and EPA. Finally, evidence that phytosterols and stanols can reduce plasma cholesterol and HDL has resulted in the inclusion of these plant-derived lipids in some brands of margarine. In such cases, label such as 'contains phytosterols' or 'contains stanols' might be supplemented by information about the putative cholesterol-lowering effects of these added lipids.

The increasing use of such labels on such a wide range of food products is a highly effective marketing tool that directly influences customer behaviour. In turn this is driving the food industry to modify the lipid content of many food products via a range of technologies and biotechnologies from chemical modification to genetic engineering as discussed in the main text.

biohydrogenation (see Section 3.1.9) is similar to that of industrial hydrogenation. However, whereas in ruminant fats there is a preponderance of the MUFA with the *trans* bond at position 11 (vaccenic acid, *t*11-18:1), in industrially hydrogenated fats, MUFA with double bonds at positions 8, 9, 10, 11 and 12 make similar contributions.

11.4.3 *n*-3 (ω-3) and *n*-6 (ω-6) polyunsaturated fatty acids

Oils rich in *n*-3 (ω-3) FAs include the so-called 'fish oils' (or more correctly 'marine oils'), which are characterized by relatively high levels of very long-chain PUFAs such as eicosapentaenoic acid (20:5*n*-3, EPA) and docosahexaenoic acid (22:6*n*-3, DHA). These compounds are part of the group of *n*-3 FAs that are essential components of mammalian cell membranes, as well as being precursors of biologically active signalling molecules such as eicosanoids, docosanoids and the prostaglandin group of autocoids. (NB: As indicated in Chapter 2, the approved *biochemical* nomenclature for these PUFA 'families' is *n*-3, *n*-6 etc. However, the terms 'ω-3' and 'ω-6' have been in

alternative use, especially by nutritionists, so that these terms have been used in food labelling and should be understood by students.) There have been numerous reports concerning the importance of dietary supplementation with these FAs for human health and well-being. For example, they have been shown to confer protection against common chronic diseases such as cardiovascular disease, metabolic syndrome and inflammatory disorders, as well as enhancing the performance of the eyes, brain and nervous system. As noted in Chapter 10, it is important to be cautious in assessing the real life implications of such studies in the light of additional factors such as the effects of other dietary components, and of the genetic background and lifestyles of different individuals.

Allowing for the above caveats, consumption of oily fish is currently recommended in most Western countries as part of a balanced diet, with much of the nutritional benefit coming from the very long-chain polyunsaturated oils. These FAs can be synthesized only partly by the fish themselves from precursor EFAs and

Table 11.1 Industrial and domestic processes that modify dietary lipid composition.

Process	Brief description	Changes produced
Heating	Makes food more palatable and kills pathogens	Depends on presence of O_2: see below.
Heating without O_2	As in deep-fat frying	Accumulation of cyclic monomers of triacylglycerols followed by polymeric products. As long as oils are not reused excessively, such polymeric material does not significantly reduce the functional properties of the oil, nor is it digested and absorbed.
Heating in the presence of O_2 and trace metal catalysts (e.g. iron, copper)		Oxidation of the double bonds of unsaturated fatty acids leads to formation of lipid peroxides (Section 3.3). More highly unsaturated fatty acids are more susceptible to oxidation. Lipid peroxidation can reduce concentrations of essential fatty acids, vitamin A and vitamin E and may cause damage to proteins (including enzymes) and DNA. Some products may be carcinogenic. Lipid peroxidation may be reduced by the presence of lipid-soluble antioxidants, either natural (e.g. vitamin E, carotenoids) or synthetic (butylated hydroxytoluene, BHT, or butylated hydroxyanisole, BHA).
Irradiation	Used to kill pathogens	May generate lipid radicals and thus cause peroxidative damage. Vitamins E and K are particularly susceptible but, surprisingly, not carotenes. In practice, however, very little direct nutritional damage occurs as a result of food irradiation because of its limited application.
Inter-esterification	Involves treatment of a fat or fat blend with a catalyst (e.g. sodium methoxide) at high temperature so as to randomize the fatty acids among the triacylglycerol molecules	Changes physical properties of lipids (crystal structure, hence melting point). Used as an alternative to catalytic hydrogenation because of concerns about potential adverse health effects of *trans*-unsaturated fatty acids (Section 10.5.5.2 and Box 11.5).
Fractionation	Separation with various physical methods of individual lipid species from the complex mixtures of triacylglycerols with different fatty acid distributions found in natural fats and oils.	Fractions enriched in more highly saturated or unsaturated molecular species can be obtained, e.g. to improve physical, textural and metabolic properties. For example, fractions of triacylglycerols enriched in saturated fatty acids of chain lengths 8C and 10C can be obtained from coconut oil (Table 2.7). These find clinical use for patients with fat malabsorption and disorders of long-chain fatty oxidation (Section 10.1.2). Spreadable butters may be produced by selective removal of higher-melting triacylglycerols.

must otherwise be derived directly from microorganisms, especially photosynthetic microalgae that are ingested as part of their diet. As an alternative to fish consumption, therefore, it is possible to purchase dietary supplements that are derived from cultured microalgae or fungi. However, low oil yields and high costs of oil extraction have limited the scope for this production method, and ever-dwindling fish stocks are also threatening supplies of the main source of marine oils. This situation has led to renewed interest in the possibility of breeding oilseed crops that are capable of producing significant quantities of very long-chain PUFAs in their storage oils as discussed in Section 11.6.1.

11.4.4 Phytosterols and stanols

Phytosterols and stanols are found in a range of plant sources such as vegetables, vegetable oils, nuts, grains, seeds and legumes. There is some evidence that they can help lower plasma low density lipoprotein (LDL) concentrations and possibly reduce the risk of heart disease if they are consumed them as part of a healthy diet. The structures of phytosterols and stanols are partially similar to cholesterol and they are thought to decrease the absorption of dietary cholesterol in the intestine (see Box 10.3). The reduction in cholesterol absorption results in increased uptake of LDL cholesterol by the liver leading to lower plasma LDL. Phytosterols and stanols seem to have no effect on plasma concentrations of the so-called 'good' high density lipoprotein (HDL) or TAGs.

An average 'healthy' diet that includes foods naturally containing phytosterols and stanols will deliver less than 160–400 mg/day, which falls short of the estimated 2 g per day needed to lower plasma LDL. As a result phytosterols and stanols have been added to some foods, such as yoghurts, yoghurt drinks, spreads and soft cheeses to help meet the 2 g/day recommendation. Margarines enriched in phytosterols extracted from wood pulp or vegetable oils have recently been marketed. They have enjoyed modest commercial success despite being more expensive than conventional margarines.

Such products could be made more cheaply if higher levels of phytosterols were synthesized in the same seeds as the oil from which the margarine is derived. Efforts are now underway to up-regulate phytosterol biosynthetic pathways in genetically modified (GM) plants (see Section 11.6.1). It has been claimed that the widespread availability and consumption of low-cost, phytosterol-enriched margarines could eventually lead to a significant reduction in rates of cardiovascular disease although this has yet to be demonstrated in practice.

11.4.5 Fat-soluble vitamins (A, D, E, K) in animals and plants

Vitamins A, D, E, K and their dietary precursors plus useful nutrients such as coenzyme Q_{10} are all water-insoluble lipid molecules that are often most efficiently ingested in fatty foods (see Section 6.2.3 for more detailed discussion of the metabolism and dietary roles of these vitamins). Deficiencies in some of these vitamins in some modern diets have led to various strategies for their supplementation that include the addition of naturally occurring or chemically synthesized versions to foods or the biotechnological modification of edible plants or animals in order to express them at higher levels. The former approach is commonly used in many brands of margarine where vitamins D and E and sometimes vitamin A are often added as a legal requirement. The latter biotechnological approach is discussed in more detail in Section 11.6.1.

Vitamin A is retinol, which is required for eyesight and a variety of neurological functions. The major vitamin A precursors are carotenoids, which are yellow/orange pigments found in a wide range of coloured plant and animal foods. Some dietary carotenoids can also be deficient in farmed fish. Due to overfishing and the collapse of several marine fisheries, an increasing proportion of dietary fish now comes from farms where the fish are fed rations instead of their normal food. Farmed salmonid fish, such as trout and salmon, are frequently deficient in a class of dietary carotenoids, known as xanthins. The absence of xanthins can leave their flesh with a pallid and unattractive appearance. Because they are unable to forage for their normal xanthin-rich food, farmed fish are unable to accumulate xanthins, which

are also important for their growth and development. Xanthins and other carotenoids are useful nutrients in the human diet where they act as antioxidants and serve as additional sources of vitamin A. To ensure the adequate development of farmed salmonids, dietary supplementation with astaxanthin is required. Astaxanthin can be produced by fermentation using the microalga *Haematococcus pluvialis* or the pink yeast *Xanthophyllomyces dendrorhous*, or it can be synthesized chemically. If sufficient astaxanthin is provided to the farmed fish, their flesh will have a healthy pink colour that can deliver useful amounts of dietary vitamin A precursors.

Vitamin D group compounds include ergocalciferol and cholecalciferol, which can be synthesized from cholesterol in the skin in the presence of light. In the 21st century, a rise in vitamin D deficiency symptoms, including childhood rickets, in some developed countries has caused concern that some diets are deficient in vitamin D precursors. In some countries, staple foods such as milk and some breakfast cereals have been supplemented with vitamin D for many years and there are pressures for this practice to be adopted more widely in order to ensure that young people get adequate dietary vitamin D.

Vitamin E comprises a group of compounds that includes four tocopherols, and four tocotrienols, all of which have significant antioxidant properties. These lipophilic vitamins can be found in most nonprocessed or 'virgin' seed oils but are often lacking in foods made from processed oils. For example, virgin palm oil contains significant amounts of vitamin E group compounds and several varieties of oil palm have been identified that produce oil that is highly enriched in tocols to levels in excess of 1500 ppm, which would be of great interest as potential health food products. Most people prefer to consume clear processed vegetable oils rather than cloudy coloured virgin oils so food technologists tend to remove lipophilic vitamins (mainly A and E) from oils as they are processed and then repackage these compounds as oily capsules that are then sold separately as vitamin supplements.

11.5 Modifying lipids in nonedible products

The vast majority of biologically derived lipids are obtained from a limited range of agricultural crops and

Table 11.2 Percentage fatty acid composition of the 'big four' commodity oil crops. Note the predominance of palmitic, oleic and linoleic acids.

Fatty acid[a]	Oil palm[b]	Soybean	Rapeseed/canola	Sunflower
Palmitic 16:0	45	11	5	6
Stearic 18:0	5	4	1	5
Oleic 18:1	38	22	61	20
Linoleic 18:2	11	53	22	69
Linolenic 18:3	0.2	8	10	0.1

[a] Fatty acids are denoted by their carbon chain length followed by the number of double bonds;
[b] mesocarp oil (palm oil).

livestock. As a result, the chemical composition of such lipids is not very diverse (see Table 11.2). As discussed in Section 11.2, there is much interest in using biologically derived lipids as industrial feedstocks for the manufacture of renewable hydrocarbon-based oleochemical products that could partially replace the current dependence on fossil-derived petrochemicals. Two of the most promising examples of such renewable resources are bioplastics and biofuels.

11.5.1 Biodegradable plastics from bacteria

The vast majority of plastic products are polymers of nonrenewable petroleum-derived feedstocks such as adipic acid and vinyl chloride. Manufacture of such plastics is energy intensive and frequently produces undesirable byproducts, which are costly and difficult to dispose of. In addition, many petroleum-based plastic products are notoriously difficult to recycle and can sometimes take decades or longer before they break down in landfill sites. An attractive alternative to conventional plastics is to use naturally occurring biopolymers that are renewable, biodegradable, nonpolluting and less energy-intensive to manufacture. Certain soil bacteria such as *Ralstonia eutrophus* can accumulate up to 80% of their mass in the form of nontoxic biodegradable polymers called polyhydroxyalkanoates (PHAs). These biopolymers are made up of β-hydroxyalkanoate

(a)

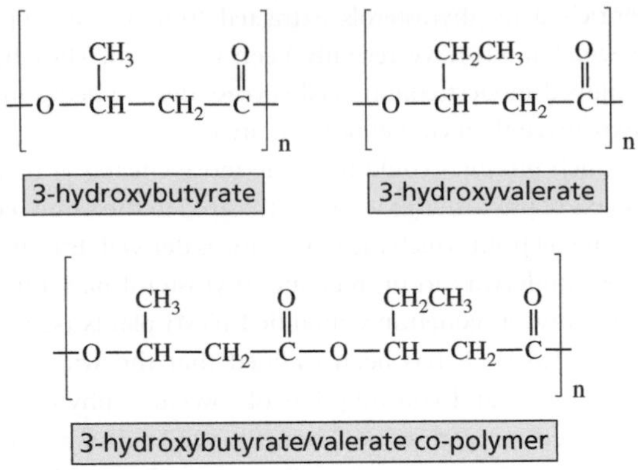

Polyhydroxy alkanoates can be built up from single monomers such 4-carbon hydroxybutrate or 5-carbon hydroxyvalerate. Alternatively, a co-polymer can be made from a mixture of these two monomers.

(b)

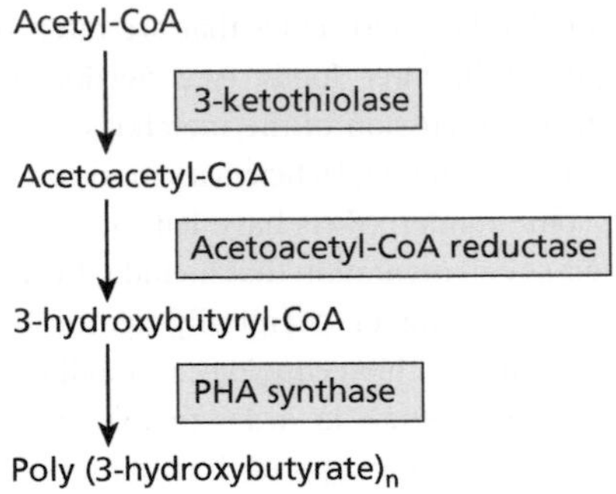

Biosynthesis of the simplest polyhydroxyalkanoate, poly (3-hydroxybutyrate).

Fig. 11.6 (a) Structures of monomeric units of the simple bioplastics 3-hydroxybutyrate (PH3B), 3- hydroxyvalerate (PHV) and of the more complex co-polymer, 3-hydroxybutyrate-co-3-hydroxyvalerate (PHBV). (b) Biosynthetic pathway for the conversion of acetyl-CoA to poly (3-hydroxybutyrate).

subunits that are synthesized from acetyl-CoA via a short pathway involving as few as three enzymes as shown in Fig. 11.6.

The PHA polymers accumulate as dozens of tiny solid granules within the cytoplasm of bacterial cells. A typical 500 nM granule of PHA contains about 40,000 polymer molecules, each of which is made up of about 30,000 PHA monomers. These form a semisolid lipid core that is surrounded by a phospholipid monolayer into which several specific granule-binding proteins are embedded.

PHAs can be made industrially in bioreactors where bacterial cultures are incubated with carbon sources such as sugars, plant oils or crop byproducts. Once the bacterial cultures reach stationary phase and accumulate the biopolymers, the cells are broken up and the granules extracted and purified.

Native PHA-synthesizing bacteria have relatively specific carbon source requirements in bioreactor production, which limits their utility, but this drawback has been overcome by transferring the PHA-synthetic genes into established bioreactor species such as *E. coli*. The assembly of polyhydroxybutryate from acetyl-CoA involves three enzymes respectively encoded by the *phaA*, *phaB* and *phaC* genes, located in a single operon. The most important enzyme in the pathway is PHA synthase which assembles monomers such as hydroxy butyrate or hydroxyvalerate into either homopolymers, such as polyhydroxybutryate, or copolymers, such as poly(3-hydroxybutyrate-co-3-hydroxyvalerate).

11.5.2 Using micro-algae and bacteria for biodiesel production

Algae range from simple unicellular organisms to comparatively complex multicellular species such as seaweeds. As with many other unicellular organisms, some algal species accumulate large numbers of TAG-rich lipid droplets (LDs) as storage reserves in response to certain forms of nutrient limitation or abiotic stress (see Section 5.5.2). In some cases these lipids can make up as much as 86% of cell dry weight. Oleogenic marine microalgae are of considerable biotechnological interest both for their ability to synthesize large amounts of high-value lipids and for their possible use as feedstocks for the production of renewable biofuels. Other commercially useful lipids accumulated from algae include very long-chain PUFAs such as DHA or pigments such as astaxanthin. Other novel lipids include very long-chain polyunsaturated alkenones, alkenoates, and alkenes.

Microalgae can be grown in liquid culture and, unlike higher plants, they do not need to synthesize large amounts of structural compounds such as cellulose for leaves, stems and roots. This means that their theoretical maximum yield of useful oil is much higher than current sources of biodiesel such as rapeseed or palm oil. It has been calculated that if some of these species of microalgae can be optimized for TAG production, they might be capable of yields in the region of 10–20 tonnes per

hectare. If such yields could be scaled up, it would be possible to produce the entire 1100 million tonnes global annual production of diesel fuel by growing microalgae on about 57 million hectares or 570,000 square kilometres. In addition to their high oil yield potential, microalgae can be grown in relatively inhospitable areas, such as semideserts, that are unsuitable for food crops (although they do require plenty of water). Some microalgae can also be grown on waste materials such as animal plant or human waste and can use waste CO_2 emissions from industrial processes.

In addition to algae, some bacteria can accumulate large amounts of TAGs. For example, the Gram-positive, nonspore-forming actinomycete, *Rhodococcus opacus* PD630, grown in media with a low carbon:nitrogen ratio, can accumulate TAGs to more than 75% of cellular dry weight, with a potential daily production from organic wastes of almost 60 mg.l^{-1}. Several TAG-accumulating cyanobacteria, including *Dunaliella salina* and *Synechocystis* spp, are currently being assessed for their potential to act as solar-powered sources of renewable hydrocarbons, especially in the context of so-called 'third generation' biofuels. Once the oil is extracted from bacteria or micro algae, the organic residue can be used as animal feedstock or a soil fertilizer.

Because of their great potential as renewable sources of biofuels that do not compete with existing food crops, there is a great deal of interest in developing large-scale lipid production systems based on both bacteria and microalgae. Some of the major challenges that need to be resolved include scaling up from several litre systems to huge outdoor ponds extending to several square kilometres in area, the control of contamination by other microbes that might compete with the lipid-producing strains, and development of efficient methods for harvesting, transporting and processing the huge biomass into diesel fuel.

11.6 Lipids and genetically modified organisms

As discussed previously in this chapter, there is much interest in manipulating the lipid composition of crops and livestock to produce more nutritious foods or more useful industrial applications. The first transgenic (genetically modified or GM) plants were produced in the early 1980s and one of the earliest commercial GM crops was a type of rapeseed, released in 1995, that was modified to produce a novel medium-chain FA in its seed oil. Since then, there have been many attempts to use GM methods to alter the lipid compositions of several mainstream crops as well as to seek potential new sources of lipids such as microalgae. The production of GM livestock is also gradually moving closer to reality and several breeds of animals with modified lipid compositions are currently under development.

11.6.1 Genetically modified crops with novel lipid profiles

11.6.1.1 High-lauric oils

The vast majority of commercial vegetable oils are highly enriched in 16C and 18C fatty acyl groups but for some applications it may be advantageous if shorter or longer-chain FAs could be accumulated instead. For example, laurate (12:0), has a wide range of edible and nonedible (see Section 11.2.1) uses. Plants such as rapeseed do not accumulate laurate because they have a fatty acid synthase (FAS) that efficiently elongates acyl-ACP esters from 2C all the way to 16C with very little accumulation of acyl species with intermediate chain lengths. However, a few plants such as the Californian bay, *Umbelluria californica*, can accumulate as much as 60–80% laurate in their seed oils. This is possible because such plants contain a 12C thioesterase that releases laurate from lauroyl-ACP, preventing its further elongation. The presence of 12C-specific acyltransferases ensures that the laurate is efficiently esterified to glycerol to form a laurate-rich TAG seed oil.

In the early 1990s, the small biotechnology start-up company, Calgene, transferred copies of genes encoding its 12C thioesterase and acyltransferase from Californian bay into rape plants. These plants had previously produced oil containing >90% 18C FAs but the new type of GM rapeseed instead produced oil with as much as 60% laurate. Although this was an outstanding scientific achievement, the high-laurate rapeseed was a commercial failure. The main reason was that there are far cheaper and more productive sources of laurate oils than rapeseed. In particular, the tropical oil-rich tree crops, palm kernel and coconut, are much more cost-effective sources of laurate oils. After a few years of production in the USA, the cultivation of high laurate rapeseed was discontinued.

11.6.1.2 Very long-chain polyunsaturated oils

As outlined in Section 11.4.3, very long-chain polyunsaturated oils enriched in *n*-3 FAs are widely regarded as having nutritional benefits (see also Section 10.5.4.3). The major source of such oils is fish, in particular species such as trout and salmon. As with many marine species, stocks of these fish are becoming depleted. They are increasingly being grown in fish farms but are still relatively expensive and are not suitable for all types of diet. One alternative is to produce very long-chain PUFAs in crop plants instead of fish. However, plants do not normally accumulate such FAs and numerous additional enzymes are needed for the conversion of a typical plant 18C PUFA to very long-chain PUFAs such as DHA or EPA. In one experiment, nine genes from various fungi, algae, and higher plants were inserted into the oilseed, *Brassica juncea*, with the resultant accumulation of as much as 25% arachidonic acid and 15% EPA. While this was a notable achievement, much higher levels of these lipids will need to be produced before GM plants can be viable alternative sources of marine oils.

11.6.1.3 Other novel oils

The relatively restricted lipid profiles of the major oilseed crops are shown in Table 11.2. These crops are highly productive commercial sources of vegetable oils but are almost exclusively made up of 16C and 18C acyl species with between 0 and 3 double bonds. This narrow range of FAs is not always optimal for various downstream uses, which is why there is great interest in extending the diversity of FAs in commercially available sources of lipids. In contrast to the commonly used crop and livestock sources of lipids, there are many other species that accumulate more exotic FAs that could be used in many types of industrial applications. As shown in Table 11.3, such novel FAs can range from 8C to 24C chains with a host of interesting functional groups including conjugated double bonds, triple bonds, hydroxy and epoxy groups. Most of these plants are unsuitable to be grown as crops themselves but they could be sources of genes encoding the enzymes responsible for their ability to accumulate the novel acyl groups. If such genes could be successfully transferred to some of the major oil crop species then it might be possible to produce commercially viable quantities of the novel acyl lipids.

Table 11.3 Diversity of fatty acyl composition in selected plant storage oils.

Fatty acid		% FA in oil	Plant species	Uses
Chain length/functionality	Common name			
8:0[a]	Caprylic	94%	Cuphea avigera	Fuel, food
10:0	Capric	95%	Cuphea koehneana	Detergents, food
12:0	Lauric	94%	Betel nut laurel Litsea stocksii	Detergents, food
14:0	Myristic	92%	Knema globularia	Soaps, cosmetics
16:0	Palmitic	75%	Chinese Tallow Triadica sebifera	Food, soaps
18:0	Stearic	65%	Kokum Garcinia cornea	Food, confectionery
16:1	Palmitoleic	40%	Sea Buckthorn Hippophae rhamnoides	Cosmetics
18:1 18:1 9c	Oleic	78%	Olive Olea europea	Food, lubricants, inks
18:1 18:1 6c	Petroselinic	76%	Coriander Coriandrum sativum	Nylons, detergents
cyclopropene-18:1 18:1 9,10-me, 9c	Sterculic	50%	Sterculia foetida	Insecticides, herbicides
18:2 18:2 9c,12c,	Linoleic	75%	Sunflower Helianthus annuus	Food, coatings
α-18:3 18:3 9c,12c,15c	α-linolenic	60%	Linseed Linum usitatissimum	Paints, varnishes
γ-18:3 18:3 6c, 9c,12c	γ-linolenic	25%	Borage Borago officianalis	Therapeutic products
OH-18:1 18:1 9c,12OH	Ricinoleic	90%	Castor Ricinus communis	Plasticizers, lubricants
epoxy-18:1 18:1 9c,12epx	Vernolic	70%	Ironweed Vernonia galamensis	Resins, coatings
triple-18:2 18:2 9c,12trp	Crepenynic	70%	Crepis alpina	Coatings, lubricants
oxo-18.3 4-oxo-9c11t13t	Licanic	78%	Oiticica Licania rigida	Paints, inks
18:3 8t,10t,12c	Calendic	60%	Calendula Calendula officinalis	Paints, coatings
18:3 9c,11t,13t	α-Eleostearic	70%	Tung Vernicia fordii	Enamels, varnishes, resins, coatings
18:3 9c,11t,13c	Punicic	70%	Pomegranate Punica granatum	Varnishes, resins, coatings
20:0	Arachidic	35%	Rambutan Nephelium lappaceum	Lubricants
20:1	Eicosenoic	67%	Meadowfoam Limnanthes alba	Polymers, cosmetics
22:0	Behenic	48%	Asian mustard Brassica tournefortii	Lubricants
22:1	Erucic	56%	Crambe Crambe abyssinica	Polymers, inks
wax 20:1/22:1	Jojoba wax	95%	Jojoba Simmondsia chinensis	Cosmetics, lubricants
24:1	Nervonic	24%	Honesty Lunaria biennis	Pharmaceuticals

[a] c, cis double bond; t, trans double bond; epx, epoxy group; trp, triple bond; me, methylene

Despite over 20 years of trying, it has proved much harder to achieve this aim of 'designer GM oil crops' than first imagined. In many cases small amounts of the novel FAs are produced but are not efficiently transferred to form TAGs, or accumulate instead on membrane lipids with undesirable side effects. In other cases, the novel FAs are broken down by oxidation before they can accumulate in storage oils. It also appears that different plant species often have different mechanisms for processing FAs for storage, so that different strategies for the identification and transfer of suitable genes must be used in each case. While many challenges remain, the goal of producing commercially viable designer GM oil crops is still an enticing prospect that is gradually moving closer to being realized.

While GM approaches to oil modification have mainly focused on the introduction of exotic FAs, they have also been used to downregulate existing genes in order to reduce levels of unwanted FAs. For example, linoleate desaturase genes have been suppressed in order to reduce levels of α-linolenate in seed oils. This approach has been used by several companies to complement conventional breeding programs aimed at developing high-oleate, low-PUFA oils. The following high-oleate transgenic lines have been developed but not necessarily commercialized to date: rapeseed with 89% oleate; soybean with 90% oleate; and cottonseed with 78% oleate.

11.6.1.4 Golden rice

Normal rice grains are white and are almost entirely deficient in the lipophilic pigment, β-carotene (provitamin A). In some parts of Asia where poorer people often mainly subsist on rice, with very low intakes of coloured vegetables, the incidence of vitamin A deficiency (leading to night blindness) is very high and is estimated to affect some 124 million children. This led a group of Swiss and German biotechnologists to develop so-called 'golden rice'. The grains of this GM rice variety are yellow because they have been engineered to accumulate high levels of β-carotene.

As shown in Fig. 11.6, the transgenic rice contains two inserted genes encoding the two enzymes responsible for conversion of geranyl geranyl diphosphate to β-carotene that are missing in normal white rice. In the initial variety of golden rice produced in the late 1990s, the two transferred genes were phytoene synthase from the daffodil, *Narcissus pseudonarcissus*, and carotene desaturase from the soil bacterium *Erwinia uredovora*. These two genes were inserted into rice under the control of an endosperm-specific gene promoter to ensure that they were only expressed in developing grains and not in other tissues. This early variety of golden rice accumulated 1.6 μg/g of carotenoids in its grains and was light yellow in colour, but an adult would need to eat several kilograms of this rice each day in order to obtain their recommended dietary intake of provitamin A.

The relatively low β-carotene yields in the first version of golden rice led to the substitution of a phytoene synthase gene from maize, which was much more effective than the original daffodil gene. This new version of golden rice contained 37 μg/g of carotenoids in its grains and was a darker orange in colour. It would require only 144 g of this improved rice to provide the recommended daily requirement for provitamin A. Although the first versions of golden rice were available by 1999, the extensive food safety and environmental checks, plus the need to backcross into local rice varieties have meant that it has so far taken well over a decade for this GM crop to be developed for public release. Advanced field trials have been carried out at the International Rice Research Institute in the Philippines but as of early 2016 the release date for GM golden rice was still uncertain.

11.6.1.5 Biopolymers from genetically modified plants

As discussed above in Section 11.5.1, biopolymer-producing bacteria can be grown in large scale fermentation vessels in order to generate biodegradable plastics based on PHA. An alternative approach would be to transfer the three bacterial genes responsible for PHA into crops, which would then be expected to accumulate PHAs on an agricultural scale. Crop-based PHA production would eliminate the need for a costly, energy-requiring indoor bioreactor process. Moreover, the carbon feedstocks would be free, since they would be derived directly from photosynthesis rather than from existing organic feedstocks, such as crop-based sugars, used to sustain bacterial growth. Hence, plant-derived PHAs have the prospect of being much cheaper and capable of production on a much larger scale, which might make them more competitive with petroleum-derived plastics. Several crops, including the major oil-producing plant rape, have been modified by the insertion of bacterial PHA genes and modest amounts of PHA production have been reported.

A major unresolved technical hurdle is how to extract biopolymer beads from plant tissues efficiently and

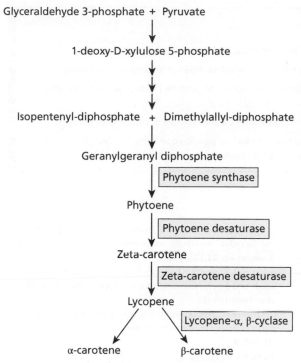

Glyceraldehyde 3-phosphate + Pyruvate

1-deoxy-D-xylulose 5-phosphate

Isopentenyl-diphosphate + Dimethylallyl-diphosphate

Geranylgeranyl diphosphate

Phytoene synthase

Phytoene

Phytoene desaturase

Zeta-carotene

Zeta-carotene desaturase

Lycopene

Lycopene-α, β-cyclase

α-carotene   β-carotene

Fig. 11.7 Biosynthetic pathway for the formation of the major carotenes that serve as precursors of vitamin A. Rice grains lack some of these enzyme activities but if the missing desaturase genes are transferred from other species, this results in an accumulation of carotenes to produce a so-called 'golden rice' that has elevated levels of pro-vitamin A.

cost-effectively. Unlike bacteria, plant tissues often contain complex crosslinked polysaccharides, polyphenols and highly lignified structures that can severely disrupt the normal extraction procedures used to obtain biopolymers such as PHAs. Several PHA extraction methods from plants are being trialled. These include nonhalogenated solvent extraction at high-temperature and differential solvent extraction, but commercial-scale extraction of PHA or other biopolymers from plants has yet to be achieved.

11.6.2 Genetically modified livestock with novel lipid profiles

Livestock are important sources of SFAs and cholesterol in the human diet – lipids that have been associated with conditions such as cardiovascular disease and type-2 diabetes (see Section 10.5). This is one of the factors behind research into the production of GM livestock with

novel lipid profiles in meat, milk, eggs or fatty tissue. The first example of the transfer of a gene to produce a 'new GM animal' was reported in 1982, when a mouse with enhanced growth rates was created. Since then there has been some progress in producing GM livestock but this has yet to be translated into commercial reality, particularly in the case of lipid-related phenotypes. However, experimental results over the past few years suggest that such genetically engineered animals might be fast becoming a practical proposition.

For example, it appears that lipid composition of most bodily tissues, including meat and fat, can be modified by the insertion of one or a few transgenes. Hence, in the case of a mouse model, the transfer of a roundworm (*Caenorhabditis elegans*) gene encoding an n-3 FA desaturase can result in appreciable decreases in the ratios of n-6/n-3 FAs in all tissues and organs that were examined. If such work could be extended to cattle, it might enable the future production of a whole range of animal products enriched in 'desirable' n-3 FAs.

In other cases, the addition of a single stearoyl-CoA desaturase transgene could lead to useful modifications to milk lipid composition. For example, the expression of a rat stearoyl-CoA desaturase gene under the control of a mammary gland-specific bovine β–lactoglobulin promoter led to the accumulation of elevated levels of MUFAs in the milk of GM goats and mice. Another approach has been to overexpress genes encoding milk proteins, such as β-casein and κ-casein, to reduce the ratio of fat/protein in the final milk. These studies encourage the view that it is likely to be feasible to use genetic engineering to alter the amount and quality of lipid in milk. Similar studies have been reported on the insertion of 18C desaturase genes in pigs. For example a humanized version of the *fat-1* gene was transferred into pigs by nuclear transfer of somatic cells and the resulting GM pigs not only have a 3-fold increase in n-3 PUFAs but also a 25% reduction in their n-6 PUFAs. As a result, the n-6/n-3 ratio in such animals was reduced 5-fold (from 8.52 to 1.69) compared to controls. For further details of basic UFA biosynthesis, see Section 3.1.8.

As with GM plants, attempts to produce GM animals with commercially useful alterations in their lipid compositions are still at a relatively early stage. Moreover, even if such plant or animal products are created, they are not necessarily assured of acceptance by the general public where concerns about GM technologies are still present.

KEY POINTS

- Lipids serve as raw materials for a host of useful products from foods and fuels to detergents, paints and polymers.

- The manufacture of lipid-based spreads such as margarine was one of the earliest examples of modern food technology.

- Lipids can be manipulated by a variety of industrially based technological processes to generate products such as emulsions, soaps and coatings.

- Lipid modifications are major processes within the food industry, especially for the manufacture of convenience foods.

- More recently lipids have been manipulated by biotechnological processes such as genetic engineering to produce biodegradable plastics and novel food products, although these are not yet available on a commercial scale.

Further reading

Beyer P (2010) Golden Rice and 'Golden' crops for human nutrition, *New Biotechnol* **27**:478–81.

Chapman KD, Dyer JM & Mullen RT (2013) Commentary: Why don't plant leaves get fat? *Plant Science* **207**:128–34.

Knothe G (2011) Will biodiesel derived from algal oils live up to its promise? A fuel property assessment, *Lipid Technol* **23**:247–9.

Mensink RP & Katan MB (1990) Effect of dietary *trans* fatty acids on high-density and low-density lipoprotein cholesterol levels in healthy subjects, *New England J Med* **323**:439–45.

Paine JA, Shipton CA, Chaggar S, *et al.* (2005) Improving the nutritional value of Golden Rice through increased pro-vitamin A content, *Nature Biotechnol* **23**:482–7.

Prather RS (2008) Genetically modified pigs for medicine and agriculture, *Biotechnol Genet Eng* **25**:245–66.

Noemi Ruiz-Lopez N, Haslam RP, Venegas-Calerón M, *et al.* (2012) Enhancing the accumulation of omega-3 long chain polyunsaturated fatty acids in transgenic Arabidopsis thaliana via iterative metabolic engineering and genetic crossing, *Transgen Res* **21**:1233–43.

Seddon AM, Curnow P & Booth PJ (2004) Membrane proteins, lipids and detergents: not just a soap opera, *Biochim Biophysica Acta* **1666**:105–17.

Siri-Tarino PW, Sun Q, Hu FB & Krauss RM (2010) Saturated fat, carbohydrate, and cardiovascular disease, *Am J Clin Nutr* **91**:502–9.

Tang G, Hu Y, Yin S *et al.* (2012) ß-carotene in golden rice is as good as ß-carotene in oil at providing vitamin A to children, *Am J Clin Nutr* **96**:658–64.

Volek JS, Volk BM & Phinney SD (2012) The twisted tale of saturated fat, *Lipid Technol* **24**:106–7.

Wu G, Truksa M, Datla N, *et al.* (2005) Stepwise engineering to produce high yields of very long-chain polyunsaturated fatty acids in plants, *Nature Biotechnol* **23**:1013–17.

Index and list of abbreviations

Main Index entries and the sub-entries under each main index entry are listed alphabetically, as are Abbreviations (which are in bold). Abbreviations are in the form: **ACP:** Acyl Carrier Protein (with letters that form part of an acronym capitalized in the definition). Page numbers that are followed by 'F' 'B' or 'T' refer, respectively, to a Figure, Box or Table on that page. When an entry term is prefaced by letter(s) or number(s) that qualify some element of a molecule's structure (as in N-acetylglucosamine, S-adenosylmethionine or sn-glycerol-3-phosphate) the entry is alphabetized according to the name of the core chemical entity

---

Lipids: Biochemistry, Biotechnology and Health, Sixth Edition. Michael I. Gurr, John L. Harwood, Keith N. Frayn, Denis J. Murphy and Robert H. Michell.
© 2016 John Wiley & Sons, Ltd. Published 2016 by John Wiley & Sons, Ltd.

H

Haematoside (Cer-glc-gal-NANA) 170

Hajra AK 127–128, 154

Hakomori S 325

Halophile (*see also* Archaea) 40, 41, 41F

Hamburg M 105, 108

Hanahan DJ 156

Harwood JL 164–165

HDL: High **D**ensity **L**ipoprotein – *see* Lipoprotein, high density

Heart disease – *see* Cardiovascular disease

Heinz E 166

Hepatic steatosis, *see* Fatty liver

Herbicide 127, 338

HETE: Hydroperoxy**E**icosa**T**etra**E**noic acid

Hevesy G 146

Hexadecatrienoate 17T, 30T, 31T, 74, 163, 164T

Hexadecenoic acid/hexadecenoate, -*trans*-3- 14, 16T, 70, 380

Hexenal 98F, 99

Hexosaminidase 170

Hildebrand 99

Hilditch T 22

Histamine 105, 106

HIV: Human **I**mmunodeficiency **V**irus

HMG-: Hydroxy**M**ethyl**G**lutaryl-

Hokin LE & Hokin MR 290

Holloway PW 68

Holman R 119

Holo-ACP synthase 54

Hopanoid 39, 40F, 177, 187

Hormone-sensitive lipase – *see* Lipase, hormone-sensitive

HPETE: Hydro**P**eroxy**E**icosa**T**etra**E**noic acid

HPLC: High **P**erformance **L**iquid **C**hromatography

HSL: Hormone-**S**ensitive **L**ipase

Huang A 135

Hübscher G 132

Human Immunodeficiency Virus (HIV) 347, 353

Hydratase, in β-oxidation 87, 87F, 88T

Hydrocarbons
  biosynthesis 146
  chemical synthesis from CO and H 194
  long-chain 25, 146
  as non-polar molecules 188
  'renewable, from microorganisms 385
  in surface lipids 24–25, 146, 224

Hydrogen bonds in lipid assemblies 188

Hydrogen ^{2}H (deuterium), as metabolic tracer 47, 68, 146

Hydrogen peroxide 92–94

Hydrogenation
  in margarine manufacture 377
  of PUFA in rumen 16–17, 75–76, 77F, 124

Hydroperoxide 96, 99

Hydroperoxy

alkenals 233

eicosapentaenoate (HPETE) 108, 109F, 110

eicosatetraenoate (HETE) 108, 109F, 110

  -endoperoxide 102, 103F

  -palmitate 94

  -pentadiene 96

3-Hydroxyacyl-CoA dehydrogenase 87F, 88T, 90, 90F, 91, 92F, 320

Hydroxyapatite, in bone 247

3-Hydroxybutyrate (β-hydroxy)
  in FA β-oxidation 89
  polymers 384F, 385
  PHA precursor 215, 384F, 385

25-Hydroxycholecalciferol (calcidiol) 241, 242F

Hydroxy fatty acid – see Fatty acid, hydroxy

Hydroxy intermediates in desaturation 68

3-Hydroxy-3-methylglutaryl-CoA (HMG-CoA)
  biosynthesis 89, 174, 175, 175F, 176, 177F
  in cholesterol/sterol biosynthesis 174–177, 175F, 177F
  linking FA & cholesterol metabolism 89–90
  lyase 174
  in β-Oxidation of branched-chain FA 89
  reductase 175–176, 175F, 176T, 276–277, 278T, 357
  synthase 174–175, 175F, 276, 278T

Hydroxy-myristate 333

Hydroxy-valerate 385

Hydroxylation/hydroxylase 67–68, 94, 112, 241, 242F

Hypercalcaemia 327

Hypercholesterolaemia, familial 254T, 255

Hyperlipidaemia – *see* Hyperlipoproteinaemia

Hyperlipoproteinaemia 352–357
  and cardiovascular disease 350, 352–357, 352F, 353T, 354T, 355B
  drug targets 353T, 355, 355B
  familial combined (FCHL) 354T, 355–356
  primary (genetic) 352–356, 353T, 354T
  secondary (environmental) 352
  type III 353T, 356

Hyperthermophile 199

Hypertriacylglycerolaemia, and cardiovascular disease 353

Hypertriglyceridaemia – *see* Hypertriacylglycerolaemia

Hypobetalipoproteinaemia 356

Hypoketotic hypoglycaemia 320

Hypolipidaemic drugs (*see also* clofibrate) 93, 127

Hypothalamus, in metabolic regulation 80

Hypothyroidism, in obesity 339

I

Ibuprofen 103, 127

I-cells, of gastrointestinal tract 254

Icthyosis 322

Printed and bound by CPI Group (UK) Ltd, Croydon, CR0 4YY

27/10/2024

14580165-0005